Correlative Microscopy in Biology

Instrumentation and Methods

Correlative Microscopy in Biology

Instrumentation and Methods

Edited by

M. A. HAYAT

Department of Biology
Kean College of New Jersey
Union, New Jersey

ACADEMIC PRESS, INC.
Harcourt Brace Jovanovich, Publishers
Orlando San Diego New York Austin
Boston London Sydney Tokyo Toronto

ACADEMIC PRESS, INC.
Orlando, Florida 32887

United Kingdom Edition published by
ACADEMIC PRESS INC. (LONDON) LTD.
24–28 Oval Road, London NW1 7DX

Library of Congress Cataloging in Publication Data

Correlative microscopy in biology.

Includes index.
1. Microscope and microscopy—Technique. I. Hayat, M. A. Date
QH207.C64 1987 578 86-32039
ISBN 0–12–333922–7 (alk. paper)

PRINTED IN THE UNITED STATES OF AMERICA

87 88 89 90 9 8 7 6 5 4 3 2 1

Contents

3 Confocal Microscopy in Comparison with Electron and Conventional Light Microscopy

H. T. M. van der Voort, J. A. C. Valkenburg, E. A. van Spronsen, C. L. Woldringh, and G. J. Brakenhoff

4 Combined Light and Electron Microscopy of Central Nervous System Neurons and Their Afferent and Efferent Synaptic Connections

F. G. Wouterlood

5 Intracellular Staining of Neurons in the Central Nervous System: Combined Physiological, Morphological, and Cytological Analysis of Identified Neurons

Georgia A. Bishop and James S. King

6 Correlative Microscopy of Catecholamines and Neuropeptides in Adrenal Medulla

Markku Pelto-Huikko

7 Sequential Immunocytochemical and Silver Staining of Neuroendocrine Cells in the Same Section

Erik Wilander and Monalill Lundqvist

8 Correlative Light and Electron Microscopic Method for the Visualization of the Same *in Vitro* Cell Using Radioautography and Serial Sectioning

Jacqueline Vuillet

9 Correlative Light and Transmission Electron Microscopy of Mitotic Cells in the Nervous System

R. R. Sturrock

Preface

The primary objective of this volume is to present detailed methodology concerning biological correlative microscopy, a technology that allows the acquisition of multiple data from a single tissue block, cell, or section. Each type of microscope has both advantages and limitations. The limitations of one type of microscopy can be circumvented by using an additional microscopical system. The resolution of certain problems may necessitate the application of three or four forms of microscopy. Correlation methods are useful in resolving doubts in the correct interpretation of a microscopic image. Thus, the advantage of corroborative data is apparent. It can also be used with advantage in carrying out electron microscopy of small experimental regions in large pieces of tissue that can be serially sectioned at about 1 μm and from which suitable sections can be selected for thin sectioning. In certain cases, more than one type of information is needed from a single specimen. This objective can be achieved by employing different systems of microscopy.

Although low resolving power is a serious limitation in light microscopy, it allows the visualization of large, whole structures, cells, or tissues. An additional advantage is the use of chromatic staining, which can be combined with scanning electron microscopy. A cell having a particular location in the tissue or a material present in a specific region of the cell can be easily observed in the light microscope (LM) by selective staining. The same cell or material can be identified in a thin section cut serially to the semithin section. Alternatively, the same semithin section can be reembedded for thin sectioning. To save time and effort in locating desired areas for thin sectioning, this approach should be routinely used. It is also possible to observe the ultrastructure of a cell that has been studied in the living state in the LM.

The availability of the combined light and electron microscope (LEM-2000) allows rapid observation of the same specimen in both the light and transmission electron modes. At intermediate magnifications (about 1500×) the same structure can be compared in the LM and transmission electron microscope (TEM). The large grid size (7-mm diameter) and the built-in microprocessor for recording areas of interest make this instrument very useful in routine surgical pathology. The large grid allows the examination of large tissue sizes for histopathological diagnosis.

Scanning electron microscopy (secondary or backscattered imaging) serves as a useful intermediate between the LM and the TEM. Scanning electron microscopy of thick resin sections provides excellent orientation for the TEM. The scanning electron microscope (SEM) can be used to obtain a clear profile of a particular cell observed chromatically in the LM. Preparatory methods are available that allow examination of the same specimen with the LM, the SEM, and the TEM. Resin sections on glass slides can be selectively stained and then observed in the LM and the SEM. Specimens that have been studied with the SEM can be processed for light and transmission electron microscopy.

In the case of transmission electron microscopy, accurate sampling of the specimen is an inherent difficulty. This microscopy permits the observation of only a very small area of the specimen. This limitation becomes obvious when electron microscopic analysis is needed of infrequent objects such as pathological lesions in a biopsy sample. This point is well illustrated by the following examples. The detection of amyloid in the biopsy sample with the TEM is difficult. However, amyloid can be easily demonstrated with the TEM, provided its deposition is first localized in semithin resin sections by using congo red staining and polarizing microscopy. The thin section is cut serially to the semithin section.

Correlative microscopy allows preselection of specific tissue areas of interest for chemical microanalysis. In some studies, correlation between morphological and elemental information is necessary. For example, in the diagnosis of human breast lesions, elementary composition of calcifications must be related to the appearance of the surrounding tissue so that an association of microcalcification with the malignant part of the tissue can be made. A possible solution is to examine the same section by the LM and/or the SEM, followed by X-ray microanalysis. Similar approaches are applicable to solving divergent problems in diagnostic pathology.

Cells on a grid can be examined first by fluorescence microscopy, and after staining, the same grid can be observed in the TEM. Correlation between the distribution of immunofluorescently labeled antigens and the ultrastructure in the same section has been accomplished. Resin-embedded specimens, sectioned for conventional or high-voltage electron microscopy, can be visualized within the same section by phase-contrast or dark-field light microscopy. After thin and semithin sections for content are screened with the LM, only those grids containing sections of interest need to be stained and examined in the TEM. The same paraffin section that has been examined in the LM can be studied with the SEM combined with X-ray microanalysis. Correlative methods also include scanning transmission electron microscopy and ion microscopy. These and other methods (simple as well as complex) and their applications are explained in detail in this volume.

All chapters contain detailed and complete instructions on the preparatory procedures. The step-by-step approach of presenting the methods enables the reader to carry out processing rapidly and without outside help. For additional details of methodology, the reader is referred to "Fixation for Electron Microscopy" and "Basic Techniques for Transmission Electron Microscopy," published by Academic Press.

This book has developed through the efforts of 40 scientists representing 13 countries. All of the contributors are eminent authorities in their respective fields of specialty. I am grateful for their participation and promptness in completing the chapters.

M. A. Hayat

1

Biological Specimen Preparation for Correlative Light and Electron Microscopy

DAVID T. MORAN AND J. CARTER ROWLEY

Department of Cellular and Structural Biology
University of Colorado School of Medicine
Denver, Colorado

CORRELATIVE MICROSCOPY IN BIOLOGY:
INSTRUMENTATION AND METHODS

INTRODUCTION

The objective of this chapter is to provide the reader with a detailed set of instructions on how to obtain a matched set of light and electron micrographs of serial thick and thin sections of a single epoxy-embedded biological specimen. Once the serial sections are obtained, the photography is simple. Consequently, the bulk of this chapter is devoted to obtaining serial thick and thin sections of a single tissue sample.

Why would one want to obtain serial "thick" and "thin" sections of the same tissue sample? The answer lies in the fact that, quite frequently, a given block of tissue will contain a particular cell or type of cell that is of special interest to the investigator. If the cell type under investigation is quite common, one can cut a thin section and obtain an electron micrograph of the field of interest. If the cell type of interest is uncommon, or of uneven distribution within the tissue, however, a very real problem arises: It is likely that that the "target" cell will *not* appear within the confines of a given thin section. It is difficult and tedious to explore a block of tissue by cutting serial thin sections and viewing them sequentially with the electron microscope. The simplest way to proceed is to cut thick sections, observe them with the light microscope, and, when the desired cell is found, immediately cut a serial thin section and observe it at high resolution in the electron microscope.

The aforementioned procedure is theoretically feasible, but in practice it can be technically difficult. If, for example, there is only one cell of interest in the section, it is likely that, once in the electron microscope, it may be hidden beneath a grid bar or spoiled by an artifactual precipitate. In setting up to take thin sections, one may, before proper sections are achieved, have passed entirely through the cell of interest, thereby losing it. For this reason, it would be most desirable to have at hand a method by which one could (1) cut thick sections and find the cell of interest with the light microscope, (2) immediately cut a serial thin section that contains the very same cell for viewing by electron microscopy, and (3) ensure that the serial thin section is flat, totally free from contamination, and mounted on a Formvar-covered "slot" grid that leaves the entire field free for photography. We have developed such a method which is described in detail in this chapter.

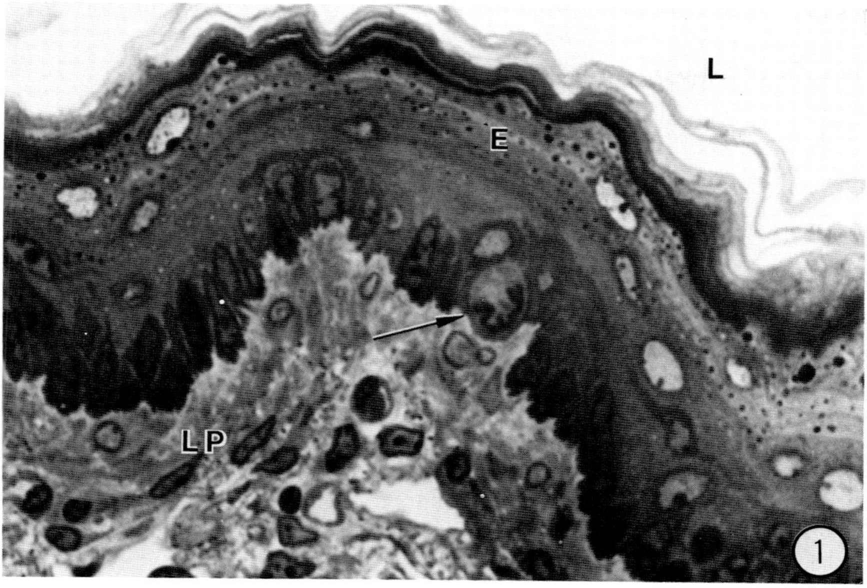

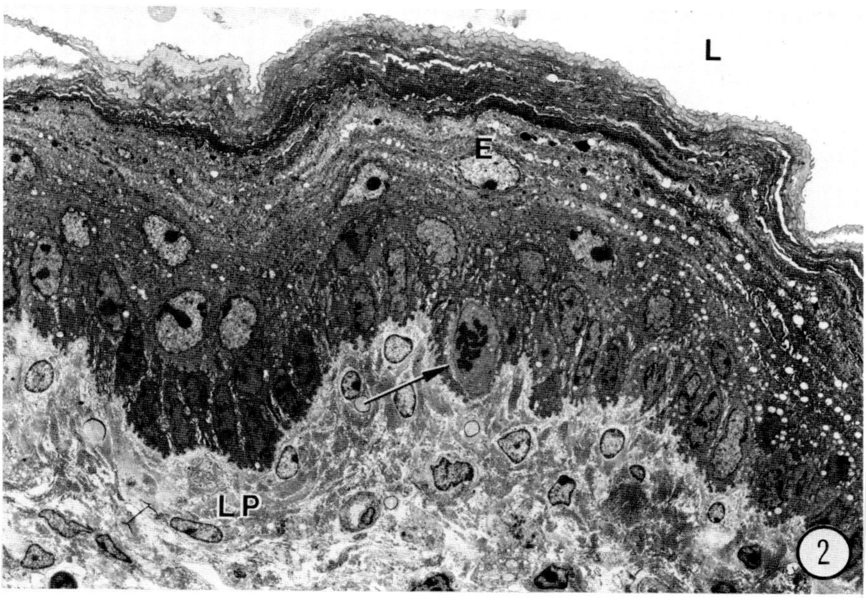

Figs. 1 and 2. Matched pair of light (Fig. 1) and electron (Fig. 2) micrographs of serial thick and thin sections taken through the esophagus of the mouse. E, epithelium; L, lumen; LP, lamina propria; arrow, cell in mitosis; ×1400.

An example of this type of correlative light and electron microscopy is shown in Fig. 1. In this experiment, the objective was to investigate the ultrastructure of cells undergoing mitosis in the basal layer of the epithelium of the esophagus. Since cell division is quite rapid—mitosis takes half an hour—the chances of finding a mitotic figure in any given thin section through the esophagus are very small. Consequently, we utilized a large block of tissue (to increase the odds of finding a mitotic figure), and cut large sections (1 × 2 mm) for examination by light microscopy. In this way, the better part of the mouse esophagus would be caught within a single cross section. One-micrometer-thick sections were cut and observed under the light microscope until a mitotic figure was found (see arrow, Fig. 1). Then, using the method described below, a large, serial, thin (50 nm) section, also measuring 1 × 2 mm, was cut and mounted on a Formvar-coated slot grid and photographed by electron microscopy. The very same mitotic figure shown by light microscopy in Fig. 1 is evident in the electron micrograph in Fig. 2 (see arrow, Fig. 2).

The method described in this chapter has been refined to the point where it takes very little time and is extremely reliable.

METHODS

Fixation for Light and Electron Microscopy

Since the methodology described below is designed to facilitate the production of serial thick and thin sections of the same specimen, the procedures for fixation, dehydration, and infiltration/embedding apply to both light and electron microscopy.

There are myriad fixatives currently in use in biological electron microscopy; these have been covered in detail by Glauert (1974), Hayat (1981a,b, 1986), and others. Unfortunately for the investigator, the "ideal" fixative seems to vary from animal to animal—and from tissue to tissue within the same animal. We have found the mixture developed by Karnovsky (1965) to be an excellent general-purpose fixative. It is made up as follows:

1. Make up 0.2 *M* Na-cacodylate buffer and adjust the pH to 7.2. This solution will be used for making Karnovsky's fixative and will serve as the buffer used to rinse the specimen following fixation.
2. Dissolve 4 g of paraformaldehyde powder in 40 ml of distilled water.

Heat slowly and carefully in a fume hood to 70°C. Stir occasionally; do not boil.

3. Remove from hot plate and add 2–6 drops of 1 *N* NaOH and stir until solution clears.
4. Cool and add 20 ml of 25% glutaraldehyde.
5. Add 0.2 *M* Na-cacodylate buffer to make a final fixative volume of 100 ml.
6. Add 50 mg of anhydrous calcium chloride.
7. Adjust the final pH of the fixative to 7.2.

Specimens are *immediately* immersed in Karnovsky's fixative either at room temperature or at 4°C. Should full-strength Karnovsky's fixative bring about shrinkage and tissue distortion due to its extremely high osmolarity (2010 mosmol), it may be diluted to one-half or one-quarter strength to achieve better fixation. Fixation times can vary from 1 hr to overnight, depending on the size and nature of the specimen. Fixation can be improved if the specimens are agitated slowly while immersed in fixative. Karnovsky's fixative works best if prepared immediately before use. If Karnovsky's fixative is to be stored and used at a later time, it is essential to reheat it to 70°C for 15 min prior to use. Failure to do this may permit polymerization of the paraformaldehyde, which can ruin the fixation.

Following fixation, specimens are rinsed in 0.2 *M* cacodylate buffer three times for 10 min per rinse. Specimens are postfixed in 2% osmium tetroxide in 0.2 *M* cacodylate buffer for 1 hr at room temperature in a fume hood. Following three more 10-min buffer rinses (in a fume hood), specimens are ready for dehydration.

Dehydration

Specimens are dehydrated in an acetone series according to the following schedule:

30% Acetone	10 min
50% Acetone	10 min
70% Acetone	10 min
95% Acetone	10 min
100% Acetone	Three rinses, 15 min per rinse

All dehydration steps should be carried out while the specimen is being agitated slowly on a rotator. For samples larger than 1 mm, or those containing dense material, a fourth rinse in 100% acetone for 1 hr should be added.

Infiltration and Embedment

Tissues are embedded in Spurr's (1969) low-viscosity epoxy resin (medium hardness). Gloves are worn during all stages of embedment, which is carried out in a fume hood. To ensure even distribution of the catalyst in the resin throughout all stages of infiltration and embedding, we mix up batches of resin and catalyst and store them in polyethylene bottles in the freezer. Resin with catalyst is thawed and dispensed as needed. Specimens are infiltrated and embedded according to the following schedule:

1 Part resin	1 Part 100% acetone, 30 min, on rotator
2 Parts resin	1 Part 100% acetone, 30 min, on rotator
100% resin	Under vacuum in vacuum oven at room temperature for 30 min

Flat embed tissues in 100% Spurr's resin in an oven set at 68°C; leave overnight.

Tissues can best be oriented for sectioning if they are flat-embedded and later mounted in a vise-type chuck. Tissues can be manipulated and oriented quite easily if they are flat-embedded in a Petri dish cover. To do this, the Petri dish cover *must first be coated with a mold release compound;* otherwise the resin will fuse to the glass. Spray a light coating of Hysol AC4-4368 Mold Release (Hysol Division, Dexter Corporation, Olean, NY) into the Petri dish cover, spread evenly with a piece of lens paper, and add enough resin to cover the tissues. Tissues may be placed wherever the investigator wishes. When hard, the tissues are embedded in a disk of resin in which they may be easily stored and from which they may be readily cut with a jeweler's saw. The polymerized disk of resin will usually fall out of the Petri dish cover. If it does not, the glass may be carefully broken away.

Sectioning

Assuming the reader knows how to cut ordinary thin sections with an ultramicrotome, the following procedure will permit the retrieval of clean, flat, serial thick and thin sections for light and electron microscopy.

1. Remove the disk of resin (containing the flat-embedded pieces of tissue) from the Petri dish cover.

2. Cut out the desired piece of tissue with a jeweler's saw; cut the piece in such a way that it can be mounted directly into a vise-type chunk.

Should the tissue require reorientation, it may be cut out and glued onto a blank of resin.

3. Insert the flat-embedded piece of tissue into the chuck, carry it to the fume hood, and trim away the excess resin. This may be done rapidly and effectively with a Dremel Moto-Tool (Model 280, series 66-2) fitted with an abrasive stone cylinder (#8175). Caution should be exercised lest tissue be ground away with the resin.

4. Insert the tissue block in the ultramicrotome and face the block with a glass knife.

5. At this point, cut several crude 1-μm-thick sections, and stain them with Millipore-filtered toluidine blue. To do this, pick up the thick sections floating on the water in the boat with a clean wire loop. Touch the wire loop, with its contained droplet of water and sections, to a microscope slide that has been sitting on a hot plate set at 95°C. When the sections dry down and attach to the slide, raise the wire loop. This will draw off any remaining water. At this point, place several large drops of stain (0.5% toluidine blue in 1% borax) on top of the sections. Wait until the stain starts to dry (about 15 sec). Remove the slide from the hot plate, place over a beaker, and flood the slide with distilled water delivered by a squeeze bottle. Aim the stream of water above the sections to avoid washing them away; let the water run down over them. Replace the slide on the hot plate to dry the sections. At this point, the stain may be differentiated with ethanol (although this is usually not necessary). View the sections with the light microscope to ensure that the tissue is adequate for further study. If necessary, continue taking thick sections until the particular area of interest is exposed. Then, having selected the area of intended investigation, the block is now ready for final trimming in preparation for taking serial thick and thin sections.

6. Mechanically trim the edges of the block with the glass knife. Most microtomes permit this by allowing the stage to be rotated, thus causing the knife to meet the block at an angle selected by the operator. About 25° or less works quite well; thus set, the knife trims the block at a shallow angle, conserving potentially valuable tissues. Since the objective is to cut serial thick and thin sections, the size of the block face that can be sectioned will be determined by the size of the aperture in the "slot" grid upon which the thin section will be mounted. To establish the maximum section dimensions, hold the grid up to the block face and trim to size.

7. Insert a diamond knife into the ultramicrotome. At this point, extreme care must be taken not to introduce any possible contaminants into the boat of the diamond knife. Henceforth, any water or solution introduced into the boat must be passed through a Millipore filter with a pore size of 0.22 μm (available from the Millipore Corp., Bedford, MA). This is

easily achieved by delivering fluids with 10-ml glass syringes fitted with disposable Millex GS filter units.

8. Place two to three drops of soapy water (i.e., 1% Triton X100 in Millipore-filtered, double-distilled water), delivered by a filter-fitted syringe as described above, into the boat of the diamond knife. This prevents thin sections from skating away from the grid during section pickup.

9. Fill the boat to the desired height with double-distilled, Millipored water delivered by a filter-fitted syringe. Keep the meniscus as low as possible to prevent accidental wetting of the block face during sectioning.

10. Having approached the block face carefully with the diamond knife, cut 0.5- to 1-μm-thick sections, stain them with toluidine blue as described above, and examine them with the light microscope. When a specific cell targeted for ultrastructural investigation is located, cut a thin section. When the section is floating freely atop the water in the boat, pick it up on an uncoated slot grid and place it on a Formvar film mounted on a Domino Rack (Rowley and Moran, 1975). The procedure for preparing Formvar films and mounting them on Domino Racks is described in detail in the following section. We will complete the description of "Sectioning" later.

Preparation of Formvar Films

Materials

In order to consistently obtain serial thick and thin sections, it is essential to prepare clean, hole-free, thin Formvar support films and mount them on a suitable substrate, the Domino Rack. To do this, the following materials, many of which are illustrated in Fig. 3, are necessary.

1. Domino Rack: (A, Fig. 3; Fig. 4). The investigator will need a supply of Domino Racks to support the Formvar films upon which slot grids, bearing sections, are placed. Domino Racks are available from Sundance Technology, 150 South Cherry St., Denver, CO 80222.
2. Water surface apparatus: (B, Fig. 3). This device helps to provide clean water for casting Formvar films by capturing dust and oils in a water reservoir (E.F. Fullam, Latham, NY).
3. Glass microscope slides: plain, nonfrosted ends, precleaned (Fisher Scientific, Pittsburgh, PA).
4. Slide tray and staining dish: (C and D, Fig. 3).
5. Coplin jar, with cap: (E, Fig. 3).
6. 500-ml Beaker and Petri dish cover: (F, Fig. 3).
7. Glass funnel: (G, Fig. 3).
8. High-intensity lamp: (H, Fig. 3).

Fig. 3. Items of equipment used to cast and mount Formvar films. A, Domino Rack; B, water surface apparatus; C, slide tray; D, staining dish; E, covered coplin jar; F, 500-ml beaker with Petri dish cover; G, glass funnel; H, high-intensity light; I, reverse-action self-locking forceps; J, diamond scribe.

9. Reverse-action self-locking forceps: (I, Fig. 3).
10. Diamond scribe: (J, Fig. 3).

All of these materials must be kept scrupulously clean and reserved for Formvar film-making only. Set them aside and store them in a dust-free environment.

Preparation of Formvar Solution

Formvar can be bought either in powder form or as a premixed solution. We find that the premixed solutions, available in concentrations of 0.5% and 0.25% (E.F. Fullam), produce cleaner, more uniform films than can be achieved by dissolving the powder in solvent. Final film thickness can be controlled by careful volumetric mixing of the 0.5% and 0.25% solutions. We use a final concentration of 0.35%, obtained by mixing the two solutions in approximately equal parts. To prevent unwanted contamination by water vapor (which produces holes in films) and airborne particles, the final solution must be stored in a closed container in a dessicator. With careful handling, the Formvar solution has a shelf life of up to 9 months.

Microscope Slide Cleaning

Clean hands carefully before handling slides; fingerprints are a major source of contamination. Clean slides individually with a mild soap and water solution. Place soapy slides in a slide tray (C, Fig. 3), place the slide tray with slides into the staining dish (D, Fig. 3), and allow distilled water to run over them to rinse them for 15 min. After rinsing, allow slides to dry in a dust-free atmosphere.

Coating Slides with Formvar

1. Once again, clean hands thoroughly.
2. With slides in the slide rack, mark one end of each slide with a laboratory marker. Henceforth, handle each slide by the marked end only: this prevents getting fingerprints on the end that will be coated with Formvar.
3. Pour final Formvar solution into clean coplin jar and cover.
4. Holding slide by marked end, immerse lower two-thirds of slide into the Formvar solution. Ensure that Formvar solution does not contact fingers.
5. Remove slide, hold it with its short axis vertical and its long axis set horizontally, and rapidly place the lower (long) edge of the slide on a filter paper under the high-intensity lamp. This allows the Formvar to flow downward across the short axis of the slide, producing a film of uniform thickness.

6. Place the Formvar-coated slide on end in a beaker (whose bottom is lined with filter paper) and cover with a Petri dish cover for final clean drying.

Casting Formvar Films on a Water Surface

1. Fill the reservoir of the water surface apparatus with double-distilled, Millipore-filtered water.
2. Pick up the dry, Formvar-covered slide by the marked end and place the opposite end on a piece of filter paper. Using the diamond (or hardened steel) scribe, very lightly trace around the outer margins of the surface of the slide covered with Formvar. Be careful to cut through the Formvar but not cut the glass; undue pressure will create tiny glass chips that will contaminate the film.
3. Go back to the water surface apparatus, pump water into the funnel, create a positive meniscus, and run a clean glass rod across the water surface to remove any remaining particulates.
4. Take a deep breath and exhale across the slide surface. This not only aids in release of the film upon the water surface, but also renders the film slightly opaque, which enhances its visibility.
5. Lower the slide slowly into the water at 45°. At this point, the Formvar film should release from the slide and peel off onto the water surface. Slide release characteristics may vary from one laboratory to the next. Should films fail to release, release may be facilitated by wiping the clean slide with Ross lens tissue prior to coating the slide with liquid Formvar. For preparing films directly on water, see Hayat (1986).

Mounting Formvar Film on the Domino Rack

The Domino Rack, illustrated in Fig. 4, is a machined, perforated rectangle of aircraft alloy sheetstock coated with a thermally bonded space-age copolymer. It provides an ideal surface that grips the Formvar film and stretches it across a series of smooth-edged, 5-mm holes that can receive several dozen slot grids. The Domino Rack allows the sections within the slot grids to dry down flat and wrinkle-free, and reduces film and section contamination to negligible levels.

1. Clean the Domino Rack with soap and water; wipe it lightly with ethanol to remove any residual oils.
2. Pick up the Formvar film from the water surface as shown in Fig. 5. Note: the rack should be grasped by one leg and the upper surface placed upon the floating film. Touch one end of the rack to one end of the film; with a sweeping motion, glide the rack across the water surface so the film adheres to it uniformly.

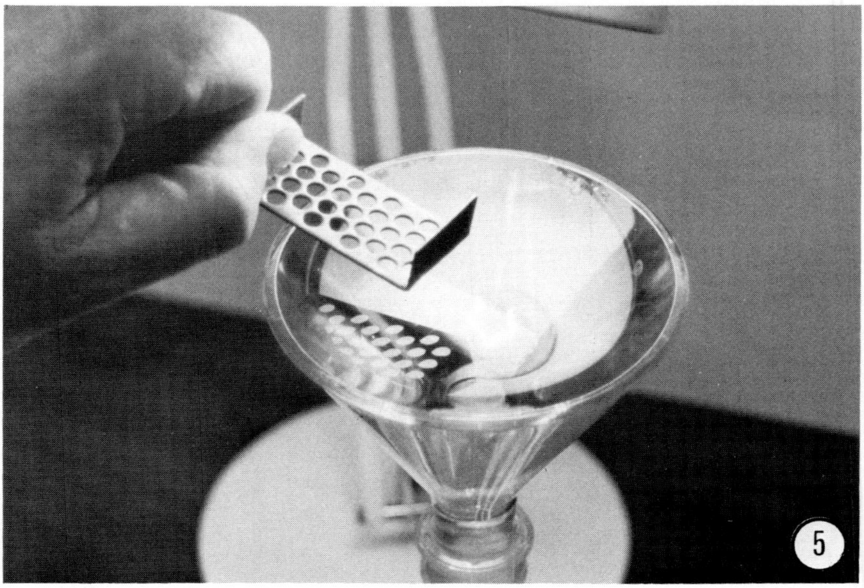

Fig. 4. The Domino Rack upon which Formvar films are suspended.
Fig. 5. Picking up a floating Formvar film with the Domino Rack.

3. Remove excess water from the film-covered Domino Rack. This may be done by gently tapping the rack—or by capillary withdrawal with a filter paper.

4. Place the film-covered Domino Rack in a covered Petri dish for final drying and storage prior to use.

Mounting Slot Grids with Sections on the Film-Covered Domino Rack

Now that one has prepared a film-covered rack, return to the description of "Sectioning."

1. With the film-covered rack at the ready in a Petri dish, pick up a slot grid in a pair of reverse-action self-locking forceps. (Veco slot grids, available from E.F. Fullam, Latham, NY, are ideal; they are thick, and permit manipulation without untimely bending. Self-locking forceps, such as Dumont Reverse-Action Tweezers, are available from Fullam as well.) As shown in Fig. 6, use the slot grid as you would a wire loop for section pickup. Lower the slot grid over the section; the section will be drawn into the slot. Prior to use, slot grids are dipped in a Formvar solution to promote adhesion of the grid to the Formvar film on the rack in the next step. Use a *separate* stock of Formvar for this purpose to avoid contamination of the solution in which slides are dipped.

2. Take the slot grid with its droplet of water and the section and place it gently upon the Formvar film that spans a hole in the rack (Fig. 7).

3. When sectioning is complete, and all slot grids are in place on the rack, cover the Petri dish in which the rack sits and let the sections air-dry onto the film. When dry, they are ready to stain.

Staining

Thin sections are double-stained for electron microscopy with alcoholic uranyl acetate (2% uranyl acetate in equal parts of 100% methanol and 70% ethanol) and lead citrate (Reynolds, 1963). Since the Formvar-covered slot grids are very fragile, much care is required during staining to prevent breakage and contamination or stain precipitation. To this end, we recommend the following procedure.

First, remove the slot grids from the Domino Rack. To do this, use the point on the self-locking forceps and perforate the film around the perimeter of the grid with six to seven small holes. Grasp the grid with the forceps; with a gentle motion, push the grid *down* through the Formvar film. This will effectively detach the grid from the surrounding film.

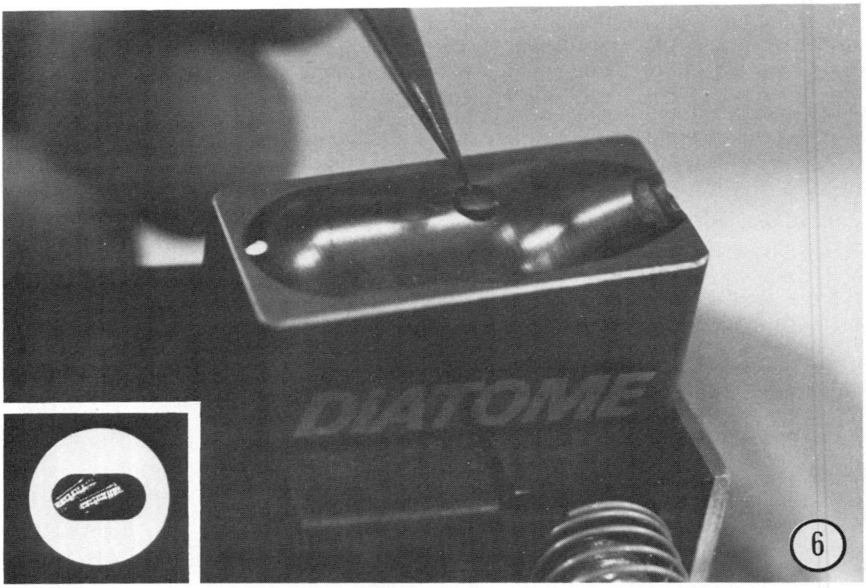

Fig. 6. Picking up a section from the boat of the diamond knife with an empty, uncoated slot grid. Inset shows slot grid afloat on surface of boat water with sections in slot.

Fig. 7. Slot grids, with sections, in place on the Formvar film spread across the surface of the Domino Rack.

Second, stain the grids using the Hiraoka Staining Kit (SPI Supplies, West Chester, PA). This kit is ideal, for it centers its function around a flexible support plate that holds many slot grids at one time. After insertion into the support plate, grids are never handled during staining. The support plate itself, with its contained grids, is transferred to the stain-containing dishes. Place the grids in the support plate, immerse the plate in a staining dish filled with alcoholic uranyl acetate for 10 min, and then rinse the grids with a (gentle) stream of Millipore-filtered, double-distilled water. Immerse the grid-bearing support plate in Millipore-filtered Reynolds (1963) lead citrate for 5 min and then rinse the grids in distilled water as above. Grids are now ready to be viewed in the electron microscope. Sections can be stabilized with a light coating of carbon if desired; this reduces specimen movement in the beam, to which sections mounted on Formvar films are prone.

Correlative Light and Electron Microscopy of the Same Section

Cutting Thin Sections from a Single Epoxy-Embedded Thick Section

Di Sant'Agnese and De Mesy-Jensen (1984) have described an interesting method by which single "thick" sections (1–2.5 μm), mounted on microscope slides, can be reembedded and "popped-off" for thin sectioning. Following their procedure, conventional "thick" sections are cut, mounted on microscope slides, and stained with basic fuchsin and methylene blue. After the sections have been viewed and/or photographed by light microscopy, a size 3 BEEM capsule, filled with partially polymerized Spurr's (1969) resin, is inverted and placed over the section. The slide, with the inverted resin-containing BEEM capsule in place over the section, is polymerized in an oven for 3 hr at 90°C. After removal from the oven, the slide is placed on a hot plate at high heat until the resin on the glass surface begins to vaporize. At this point, the capsule is gently rocked back and forth until the polymerized block, which now has the reembedded section on its flat surface, "pops" off. The BEEM capsule is then placed in the ultramicrotome, and thin sections are cut from the reembedded "thick" section.

Photography of the Same 1-μm-Thick Section by Light and Electron Microscopy

Wells and La Cour (1971) have devised a method by which a single 0.5- to 1.0-μm thick section can be viewed by both light and electron microscopy. In their procedure, a section is cut, mounted on a slot grid, and stained with toluidine blue. The slot grid bearing the stained thick section is then placed upon a microscope slide, and a coverslip is mounted over the grid and its contained section using hexachlorabuta-1,3-diene as a mounting medium. The section may then be viewed and photographed by conventional light microscopy. After light microscopy is completed, the coverslip is removed and the highly volatile mounting medium is allowed to evaporate. The slot grid and its thick section may now be stained with heavy metals and placed directly into the electron microscope for observation and photography. It is interesting to note that the toluidine blue does not interfere with the electron image; very little contamination from the dye is evident.

The same end may be accomplished by cutting a 0.5- to 1.0-μm thick section, mounting it on a Teflon-coated slide, and staining for light microscopy with toluidine blue. Following light microscopy, the section may be removed from the Teflon-coated slide, mounted upon a slot grid, and viewed by electron microscopy. The limitation of both of the techniques cited above lies in the relatively low penetrating power of the 100 kV electron beam available in most transmission electron microscopes. Use of the high voltage electron microscope (HVEM), equipped with an accelerating voltage of 1 million MeV (1000 kV), permits high-resolution images of thick sections. Unfortunately, such microscopes are rare, and special arrangements must be made to use them.

Recently, however, conventional transmission electron microscopes have been put on the market in which advances in electron optics have been incorporated into the lens systems that permit the generation of high-contrast, high-resolution images of thick sections photographed at 80–100 kV. Notable in this respect is the Zeiss EM 902, which incorporates a spectrometer into the column. The spectrometer allows the orderly separation of elastically and inelastically scattered electrons, which permits the generation of high-contrast, high-resolution images of thick sections. In addition, Philips, in the 401 LS and CM-10 models, incorporated a long focal-length objective lens that, taken together with other lenses in the system, is able to correct much of the aberration induced by thick sections. Surprisingly good and clear images of thick sections are obtainable with the modern Philips instruments under normal operating conditions.

In addition to the methods described above, it is possible to achieve correlative phase-contrast, immunofluorescensce, and electron microscopy on the same thick section. These techniques are described in detail by Reider and Bowser, in Chapter 11, of this volume.

APPLICATIONS OF CORRELATIVE LIGHT AND ELECTRON MICROSCOPY

Applications in Biological Research

During the course of investigating the fine structure of biological systems, it is often desirable to be able to locate single cells, cut thin sections through them, and examine their ultrastructure in the transmission electron microscope. When many cells of interest are present within a single tissue block, this is a relatively simple matter. When few of the cells of interest are present, however, it is necessary to locate the "target" cell by light microscopy, and then cut a thin section through it for ultrastructural analysis.

An example of this type of correlative light and electron microscopy is illustrated in Figs. 8 and 9. These micrographs were taken in the course of the investigation of the morphological changes that insect mechanoreceptors undergo during the moulting cycle. Here, the receptor in question was the campaniform sensillum, a proprioceptor associated with the insect's exoskeleton. Each campaniform sensillum is innervated by a single bipolar neuron, from whose dendrite tip emerges a modified cilium called the sensory process. The tip of the sensory process attaches to a thin cap of cuticle in the exoskeleton (Moran *et al.*, 1971). When the animal walks, the cap in the cuticle acts as a strain gauge (Zill and Moran, 1981); deformation of the cuticular cap squeezes the tip of the sensory process, which stimulates the bipolar neuron. The geometry of the campaniform sensillum raised an important question: What happens to the sensory process of the receptor when, prior to shedding its skin during the moulting cycle, the animal has two exoskeletons—a "new," inner exoskeleton, and an "old," outer exoskeleton? To answer this question, it was essential to locate a campaniform sensillum oriented in longitudinal section with the light microscope (which takes time, because only a few are present). Sections of 0.5 μm were cut until the tiny sensory process was located. This was followed by a serial thin section cut right through the center of the long axis of the sensory process for examination by electron microscopy. The tip of the sensory process is less than 0.2 μm in diameter; analysis of its fine structure provided a true test of the "Domino Rack" method of correlative microscopy described in this chapter.

The correlative light and electron microscopy that led to the elucidation of the ultrastructural changes undergone by insect mechanoreceptors during the moulting cycle (Moran *et al.*, 1976) are shown in Figs. 8 and 9. Figure 8 is a light micrograph of the integument of a cockroach just prior

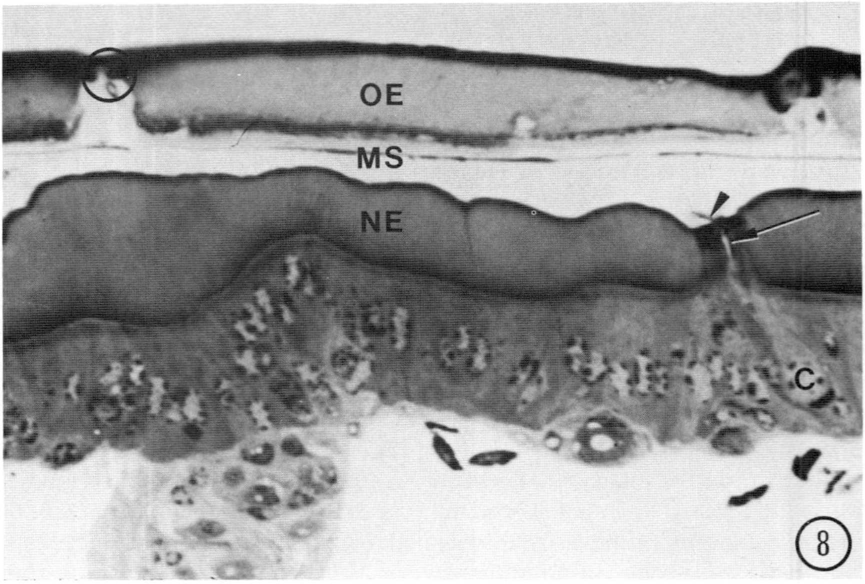

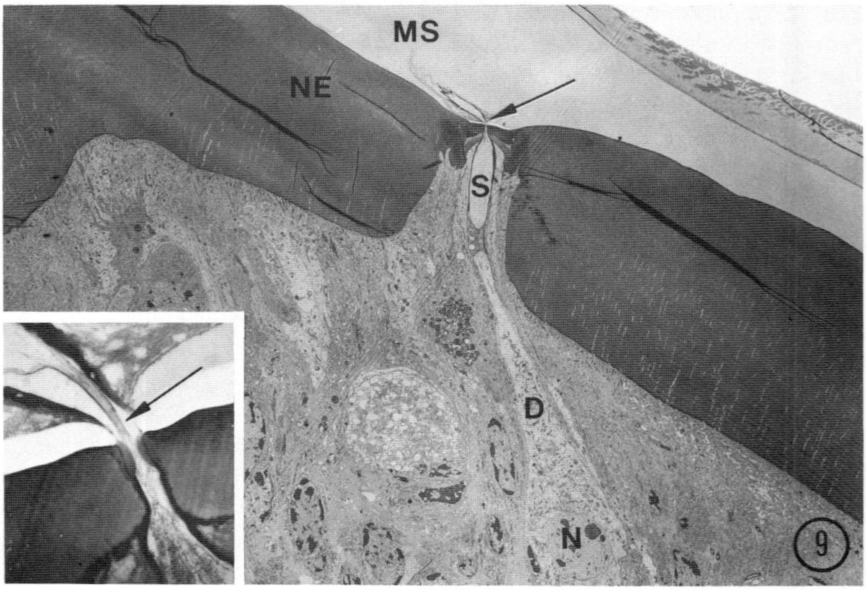

Fig. 8. Light micrograph of the exoskeleton of a cockroach showing a longitudinally sectioned sensory receptor, the campaniform sensillum. C, campaniform sensillum; MS, moulting space; NE, new exoskeleton; OE, old exoskeleton; arrow, sensory process; arrowhead, sensory process passing through cap in old exoskeleton; circle, tip of campaniform sensillum in old exoskeleton to which extension of sensory process (arrowhead) attaches; ×500.

to moulting. The outer, "old" exoskeleton (OE) is at the top of the micrograph; the inner, "new" exoskeleton (NE) lies beneath it. The two exoskeletons are separated by a moulting space (MS). Here, a single campaniform sensillum (C) has been caught in longitudinal section. The sensory process (arrow), which inserts into the cap of cuticle (arrowhead), appears to pass through the cap of cuticle, traverses the moulting space (MS) between the two exoskeletons, and inserts into a cap in the old exoskeleton (circle). Details of the geometry of the sensory process are revealed by electron microscopy of a single serial thin section prepared by the method described here (Fig. 8). The sensory process (S) is evident as it passes through a tiny hole in the cap (arrow), enters the moulting space, and travels on to make a functional connection with the cap in the outer exoskeleton. In this way, insects maintain functional continuity of their cuticular mechanoreceptors during the moulting cycle. Details of the tip of the sensory process and its passage through the cap in the inner exoskeleton are shown in the inset in Fig. 9.

The very same procedures used to solve the problem cited above can be used in a wide variety of situations in which single cells must be located and their ultrastructure investigated.

Applications in Biomedical Education

The teaching of the microscopic structure of cells and tissues, usually called histology or microanatomy, can be quite difficult. Histology is an intensely visual discipline in which the instructor is charged with generating a three-dimensional model of the cells and tissues that make up the body in the student's "mind's eye." One of the factors that makes histology difficult to teach, be it at the undergraduate or graduate level, lies in the limitations imposed by the resolving power of the very tool the student uses—the light microscope.

The light microscope cannot resolve structures smaller than 0.2 μm in diameter. Unfortunately, the cell membrane that forms the outer boundary of the cell is only 8 nm in diameter. Consequently, the boundary of the cell itself lies far beneath the resolving power of the light microscope. This puts the teacher of histology in somewhat of a predicament, since implicit in his/her task is the description of where one cell ends and another begins.

Fig. 9. Electron micrograph of serial thin section through same specimen shown in Fig. 8. D, dendrite of campaniform sensillum; MS, moulting space; N, nucleus of campaniform sensillum; NE, new exoskeleton; S, sensory process; arrow, point where sensory process passes through cap in new exoskeleton en route to cap in old exoskeleton (see inset for details); $\times$900; inset, $\times$12,700.

Most biological researchers experienced in electron microscopy have no problem in determining cellular boundaries with the light microscope, since they routinely take thick sections through a tissue block, examine them with the light microscope, and, when an area of interest is found, take thin sections and observe them with the electron microscope. By looking at the same block of tissue by both light and electron microscopy, investigators develop a kind of "X-ray vision" that allows them to visually interpolate the presence of cell boundaries in the light microscope despite their physical invisibility.

Students of histology usually do not have the opportunity to develop the kind of "X-ray vision" described above, since most of their experience is limited to light microscopy alone. Students can, however, develop the capacity for visual interpolation, heretofore the province of the experienced researcher, by judicious use of correlative microscopy. Experience with first-year medical students in the microanatomy course at the University of Colorado School of Medicine indicates that students can make very rapid progress in their study of histology by looking at matched pairs of light and electron micrographs of serial thick and thin sections photographed by light and electron microscopy at the same magnification.

An example of such an application of correlative microscopy is shown in Figs. 10 and 11. Serial thick and thin sections taken through the monkey retina have been photographed by light (Fig. 10) and electron (Fig. 11) microscopy. By aligning the illustrations side-by-side, one achieves a point-to-point correspondence of structures in the photographs. The fuzzy images presented by the light microscope are made clear by the electron microscope—even at this relatively low (600×) magnification. The pigment cell layer (P) that abuts against the choroid (C), for example, looks somewhat blurry in the light micrograph (Fig. 10). A glance at the same structure with the electron microscope (Fig. 11) shows at once that it consists of a layer of cells filled with small, distinct pigment granules that extend into cytoplasmic extensions of the pigment cells which, in turn, embrace the outer segments (OS) of the rods and cones of the retinal photoreceptors. The other classical "layers" of the retina, again somewhat blurry in the light micrograph, are very clearly depicted in the matched electron micrograph. This greatly facilitates the learning—and hence teaching—of the microanatomy of the retina. Needless to say, a series of electron micrographs of the same thin section, taken at progressively higher magnifications, can be used to clarify further the ultrastructure of specific retinal components.

The single example cited above underlines the effectiveness of correlative microscopy in the teaching of histology. The technique described in this chapter for the preparation of biological tissue for correlative light

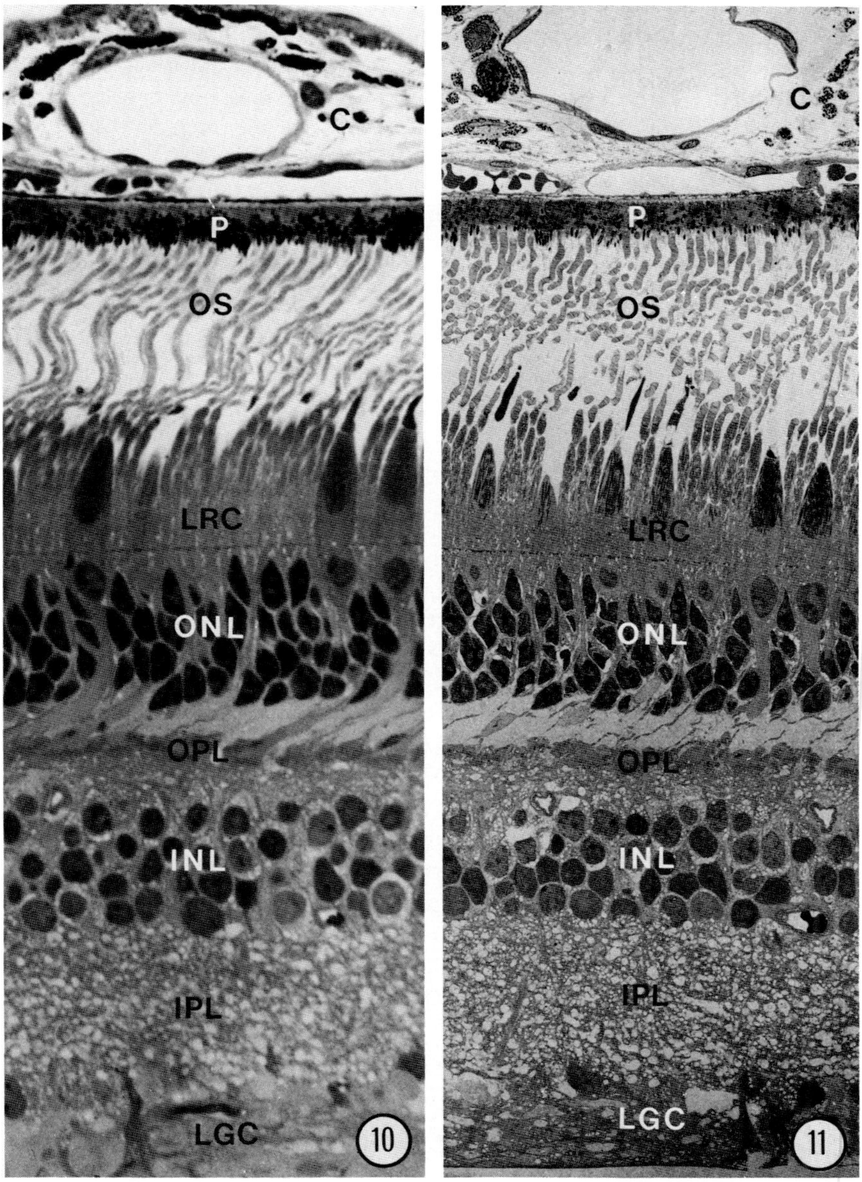

Figs. 10 and 11. Matched set of light (Fig. 10) and electron (Fig. 11) micrographs of serial thick and thin sections taken through the retina of the monkey. From top to bottom, abbreviations are C, choroid; P, pigment cell layer; OS, outer segments of photoreceptors; LRC, layer of rods and cones; ONL, outer nuclear layer; OPL, outer plexiform layer; INL, inner nuclear layer; IPL, inner plexiform layer; LGC, layer of ganglion cells; ×600.

and electron microscopy has proven so effective that an entire histology atlas, based on matched pairs of light and electron micrographs of serial thick and thin sections, has recently become available (Moran and Rowley, 1987).

The authors gratefully acknowledge the generous support provided by the Tilloo Cay Foundation for Biological Research. In addition, they wish to thank Dr. George Dickel for his spirited discussion and encouragement throughout the development of the methodologies described in this chapter. Also, the authors wish to thank their publishers, Lea & Febiger, Philadelphia, for granting permission to use four figures (Figs. 1, 2, 10, and 11 in this chapter) from: "Visual Histology."

REFERENCES

di Sant'Agnese, P. A., and De Mesy-Jensen, K. L. (1984). Diagnostic electron microscopy on reembedded ("Popped off") areas of large Spurr epoxy sections. *Ultrastruct. Pathol.* **6,** 247.

Glauert, A. M. (1974). "Practical Methods in Electron Microscopy," Vol. 1. Am. Elsevier, New York.

Hayat, M. A. (1981a). "Principles and Techniques of Electron Microscopy: Biological Applications," Vol. 1. Aspen Publishers, Inc., Rockville, Maryland.

Hayat, M. A. (1981b). "Fixation for Electron Microscopy." Academic Press, New York.

Hayat, M. A. (1986). "Basic Techniques for Transmission Electron Microscopy." Academic Press, Orlando, Florida.

Karnovsky, M. J. (1965). A formaldehyde-glutaraldehyde fixative of high osmolality for use in electron microscopy. *J. Cell Biol.* **27,** 137a.

Moran, D. T., and Rowley, J. C., III (1987). "Visual Histology." Lea & Febiger, Philadelphia, Pennsylvania.

Moran, D. T., Chapman, K. M., and Ellis, R. A. (1971). The fine structure of cockroach campaniform sensilla. *J. Cell Biol.* **48,** 155.

Moran, D. T., Rowley, J. C., III, Zill, S. N., and Varela, F. G. (1976). The mechanism of sensory transduction in a mechanoreceptor: Functional stages in campaniform sensilla during the molting cycle. *J. Cell Biol.* **71,** 832.

Reynolds, E. S. (1963). The use of lead citrate at high pH as an electron-opaque stain in electron microscopy. *J. Cell Biol.* **17,** 208.

Rowley, J. C., III, and Moran, D. T. (1975). A simple procedure for mounting wrinkle free sections on Formvar-coated slot grids. *Ultramicroscopy* **1,** 151.

Spurr, A. R. (1969). A low-viscosity epoxy resin embedding medium for electron microscopy. *J. Ultrastruct. Res.* **26,** 31.

Wells, B., and La Cour, L. F. (1971). A technique for studying one and the same section of a cell in sequence with the light and electron microscope. *J. Microsc.* (*Oxford*) **93,** 43.

Zill, S. N., and Moran, D. T. (1981). The exoskeleton and insect proprioception. I. Responses of tibial campaniform sensilla to external and muscle-generated forces in the American cockroach *Periplaneta americana*. *J. Exp. Biol.* **91,** 1.

2

Light and Scanning Electron Microscopy in a Combined Instrument

CORNELIA H. WOUTERS AND J. S. PLOEM

Department of Histochemistry and Cytochemistry
University of Leiden
Leiden, The Netherlands

INTRODUCTION

With scanning electron microscopy (SEM)* it is possible to study the fine surface features of a specimen. The image is produced by focusing a

* The abbreviation SEM is used for scanning electron microscopy as well as for the scanning electron microscope. The abbreviation LM is used for light microscopy as well as for the light microscope.

beam of electrons (coming from an electron gun at one end of a vacuum column) to a small spot. This probe (10 nm or less) is moved in a regular pattern over the specimen. The primary electrons hit the specimen with a high velocity and impinge with the atoms of the specimen, giving rise to several types of radiation. One of these are secondary electrons, which are produced by inelastic collisions between the primary electrons and the electrons in the target area of the specimen. These electrons have a low energy and can only escape from the top 5- to 10-nm zone, revealing no information concerning the internal structures (e.g., nuclear chromatin patterns and cytoplasmic granules).

Wetzel *et al.* (1973) pointed out that the identification of individual cells with a scanning electron microscope alone is sometimes impossible. Light microscopy (LM), on the other hand, can provide information concerning the internal cell structure and the presence of special macromolecules after an appropriate cytochemical staining. Several approaches for correlating LM and SEM have been proposed.

This correlation was first performed on stained microtome sections, which were coated with a platinum–palladium alloy and investigated with the SEM. The preparations were mounted on glass slides and studied with the LM (McDonald *et al.,* 1967). The same correlation was later performed on sections which were first investigated with an LM and afterward with an SEM (Geissinger, 1971). A similar type of investigation was performed on isolated cells (McDonald and Hayes, 1969). Like many other methods where SEM and LM are combined, this implied transferring a specimen from one microscope to another, and hence required an accurate relocation of the same cells. This has been accomplished in many ways.

A frequently used method consists of repositioning the specimen based on identical coordinates of the two microscope stages. For example, Geissinger (1974) constructed a module for an SEM stage which could carry microscope slides of standard dimensions. Another relocation method consists of markers on the supporting glass. These markers varied from a gold shadow produced by gold coating of a grid, which was placed on a cover glass and removed afterward (Michaelis *et al.,* 1971), or India ink (Wetzel *et al.,* 1973) to etching (Bahr *et al.,* 1976) or printing a pattern on the supporting glass (Thornthwaite *et al.,* 1976). The bleaching effect of the electron beam on a Millipore filter was also used (Lytton *et al.,* 1978). Photographs sometimes served as maps (Becker *et al.,* 1981). Shelton and Orenstein (1975) marked the supporting Millipore filter.

All these methods are rather time-consuming, with limited possibilities for alternating the LM and the SEM for reinvestigations. When interesting cells had been individually marked (van Ewijk and Mulder, 1976), only a limited number of cells could be investigated with LM and SEM for reasons of efficiency. Furthermore, in many cases the removal of a cover glass (used for LM examination) can cause cells to be lost or damaged.

To circumvent all the problems caused by using two separate microscopes, a combined instrument has been developed (Hartmann *et al.*, 1978; Ploem, 1981; Ploem and Thaer, 1981). A microfluorometer unit (MPV-II, Leitz, W Germany) was attached (Wouters and Koerten, 1982), which offered the possibility of determining the DNA content of a cell after an appropriate staining (Wouters *et al.*, 1986a; Wouters and Ploem, 1985).

DESCRIPTION OF THE COMBINED INSTRUMENT

The instrument that combines SEM with LM, the "LM/SEM" (Fig. 1), consists of an SEM (Leitz AMR 1200 B) with a special specimen stage in the vacuum chamber. The SEM is connected by means of an optical bridge to an LM (Leitz, West Germany) without objective and condenser. The LM optics are build onto the specimen stage (Fig. 2) inside the vacuum chamber. The oil-immersion objective is exchangeable and the condenser consists of a specially modified medium power objective with a long free working distance. This condenser can be swung out of the light path of the optical axis to permit undisturbed electron-beam passage (Fig. 3).

The path of the electron beam and the two light paths respectively for transmitted illumination (absorption) and epi-illumination (fluorescence) are shown in Fig. 4. Specimens are prepared on cover glasses (described

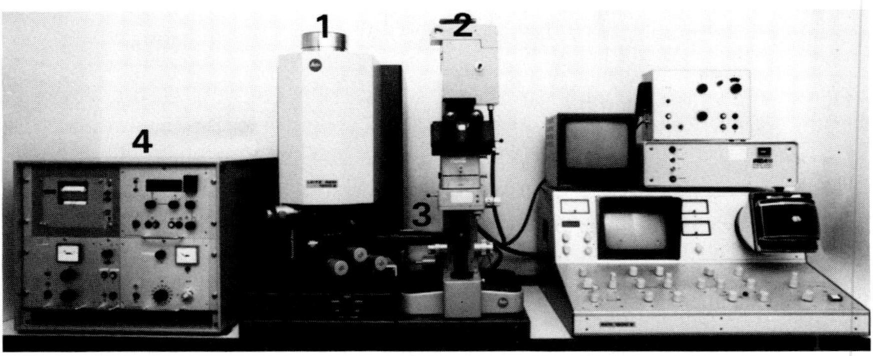

Fig. 1. The complete setup of the combined LM/SEM microscope, consisting of an SEM (1) and an LM (2) connected by an optical bridge (3). A photometer unit (4) is attached to the LM/SEM, allowing fluorescence measurements.

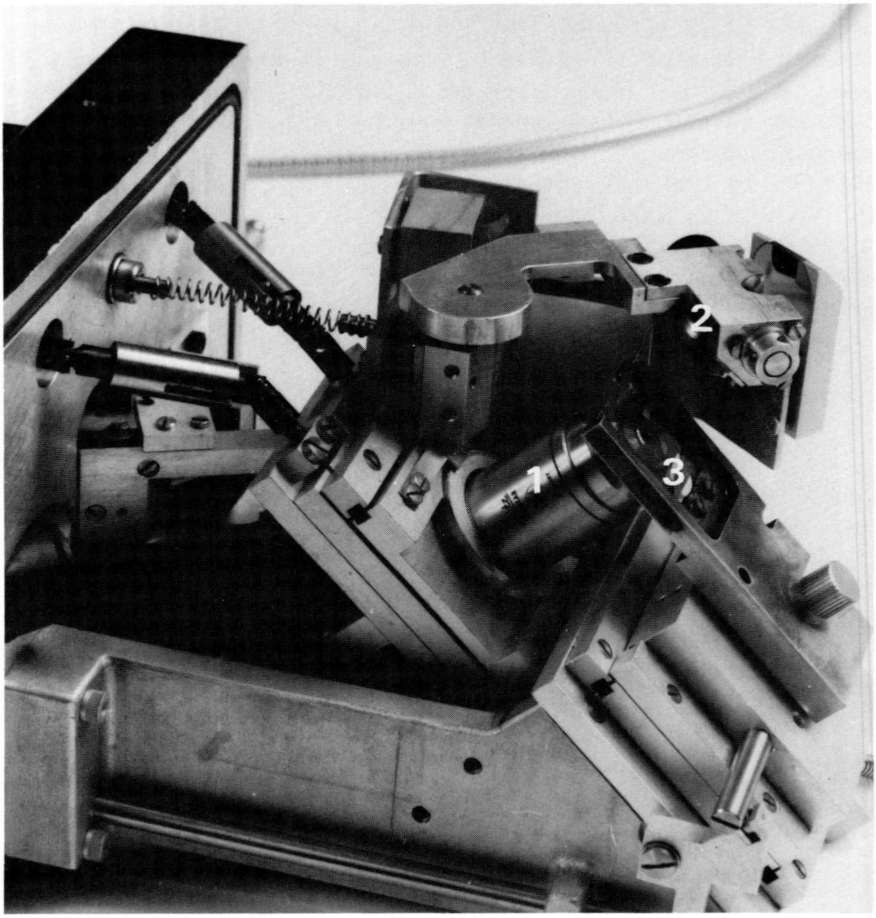

Fig. 2. The special specimen stage, showing the exchangeable oil-immersion objective (1), the condenser (2), and the specimen prepared on a cover glass in its aluminum holder (3).

below) and placed above the objective. This objective is orientated with its optical axis at an angle of 135° to the electron beam and 90° to the cover glass. A vacuum-resistant fluorescent-free immersion oil is applied between specimen and objective. Above the cover glass the condenser is positioned in the same optical axis as that of the first objective.

The specimen is inclined 45° with respect to the electron beam and can be moved in the X and Y directions (up to 6 mm). The objective can be moved in three directions: the Z direction for focusing, the X and Y directions for arranging the objective exactly under the electron beam.

Fig. 3. The front panel of the vacuum chamber, showing the devices for X and Y movement of the specimen table (1), the focusing device (2), and the centering device (3) for the objective and the condenser handle (4).

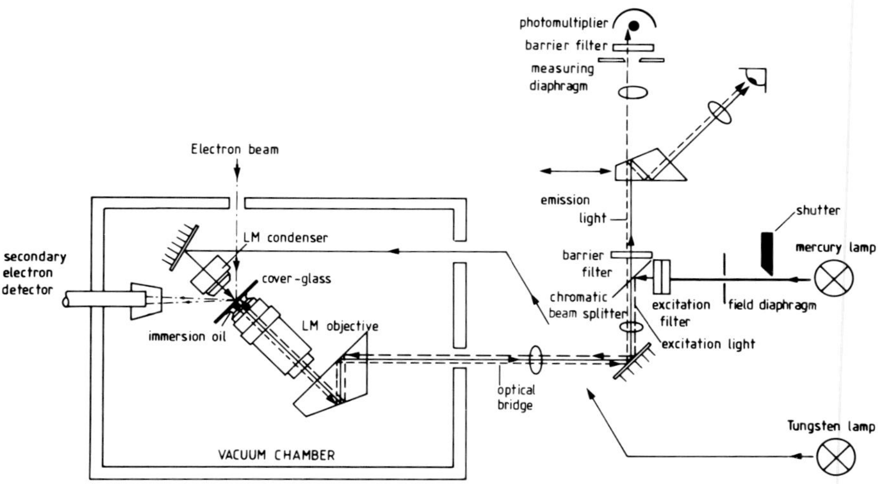

Fig. 4. The schematic arrangement in the vacuum chamber and the LM part of the instrument. The path of the electron beam is visible (primary and secondary electrons) and the two light paths for transmitted illumination (absorption) and epi-illumination (fluorescence) are shown. The optical bridge connects the SEM and LM part.

These movements are controllable from outside the vacuum chamber. Figure 3 shows the front panel of this vacuum chamber with the centering and focusing devices for the objective and the handle for swinging the condenser in and out of the light path.

Figure 4 shows the electron beam impinging on the specimen and producing, among others, secondary electrons which are detected by means of the secondary electron detector. A tungsten lamp (attached to the LM part of the LM/SEM) produces the light for the transmitted illumination. By means of a glass fiber cable it is guided into the vacuum chamber. Here the light is led by mirrors and a prism inside the chamber to the condenser. After traversing the condenser, the light illuminates the specimen and is collected by the oil-immersion objective.

The light leaves the vacuum chamber through the optical bridge and enters the LM part of the LM/SEM. Here it can be studied through the eyepieces or photomicrographs can be taken. For fluorescence microscopy, light is obtained from a mercury high-pressure lamp (again attached to the LM part of the LM/SEM). The light passes through the optical bridge into the vacuum chamber. Here the exciting light is concentrated by the objective onto the specimen (epi-illumination). The emitted fluorescence light returns to the LM part of the LM/SEM the same way. The attachment of a microfluorometer unit (Leitz MPV-II, West Germany) to the LM part of the instrument provided with a fluorescence illuminator (Ploem, 1969) also allows measurement of the intensity of the emitted light from a certain object (e.g., the quantitatively stained nucleus of a cell).

PREPARATION AND STAINING METHOD FOR LM/SEM SPECIMENS

Since two microscopical techniques are combined in one instrument, the preparative techniques to which the specimens are subjected must fulfill the general requirements for both SEM and LM. Generally, biological specimens (e.g., cells or tissue) which are going to be investigated in an SEM must be fixed and the water in the specimen has to be replaced by a polar organic solvent. Since the specimens are placed inside the vacuum chamber, they must be dried. Another requirement for SEM is the conductivity of the specimen, to avoid irregular charging of the specimen surface which can seriously and unpredictably affect the secondary electrons. To accomplish this, specimens are usually coated with gold or other conductive materials. Since transparency is essential for the LM part of the LM/SEM, preparations are not made on SEM stubs, but on cover glasses.

Some biological specimens possess the ability to emit electromagnetic radiation in the form of light as a result of (and only during) the absorption of light with a shorter wavelength from another source. An example of this autofluorescence (staining with a fluorochrome is unnecessary) can be seen in Fig. 5. Here a handmade section of *Vinca rosea* was critical-point dried, glued to the cover glass with fluoromount (Gurr, England), and coated with gold. The xylem, which transports the water in the plant, is visible by the autofluorescence of the cell walls. Other biological specimens have to be stained, for instance, for visualization of internal cell structures. The staining can be performed with an absorption stain or with a fluorochrome.

If cells from a suspension are to be examined with the LM/SEM, they have to be brought onto a cover glass. This can be achieved by several methods. For instance, unfixed cells which have the capacity of active attachment (e.g., macrophages) will attach to glass (Linthicum *et al.*, 1974; Sanders *et al.*, 1975). This has been described as the sedimentation method. However, fine details of the cell surface might be influenced by large variations in cell size due to the flattening of certain cell types placed onto the glass in an unfixed state. For this reason, cells are sometimes fixed prior to attachment; however, then their attachment will be less strong. The same is the case for cells that do not or hardly actively attach to glass, even in an unfixed state. Poly-L-lysine coating of the cover glass will promote cell adhesion, but many cells will still be lost (leading to low recovery) under conditions of unit gravity sedimentation. Moreover such methods are time-consuming (Sanders *et al.*, 1975).

Another preparation method in which fixed cells (or nonadherent cells) can be transferred to cover glasses for LM/SEM examination has been described (Wouters *et al.*, 1984). This method requires a few minutes and, even after cytochemical staining, the recovery is high. In short, prefixed cells are centrifuged on poly-L-lysine-coated cover glasses using a centrifugation bucket (van Driel-Kulker *et al.*, 1980). During the centrifugation the cells remain wet. The bucket (Fig. 6) consists of two parts: a steel holder that fits into the centrifuge and a plastic inner part that contains two cylindrical holes with a diameter of 1.5 cm each. Two cover glasses (1.8 × 1.8 cm) are placed on the lower side of the holes onto the wetted rubber rings to prevent leakage. The inner part of the bucket with the cover glasses is placed on a slide inside the steel holder.

A suspension of fixed cells is placed in each hole of the bucket. Centrifugation is performed, time and force depending on the cell type. For example, when peripheral blood mononuclear cells are brought onto the glass, a centrifugal force of 70 *g* during 10 min is sufficient for a high recovery (Wouters *et al.*, 1984). Cervical epithelial cells need a centrifugal force of 1200 *g* during 15 min (Wouters *et al.*, 1986b). To prevent prema-

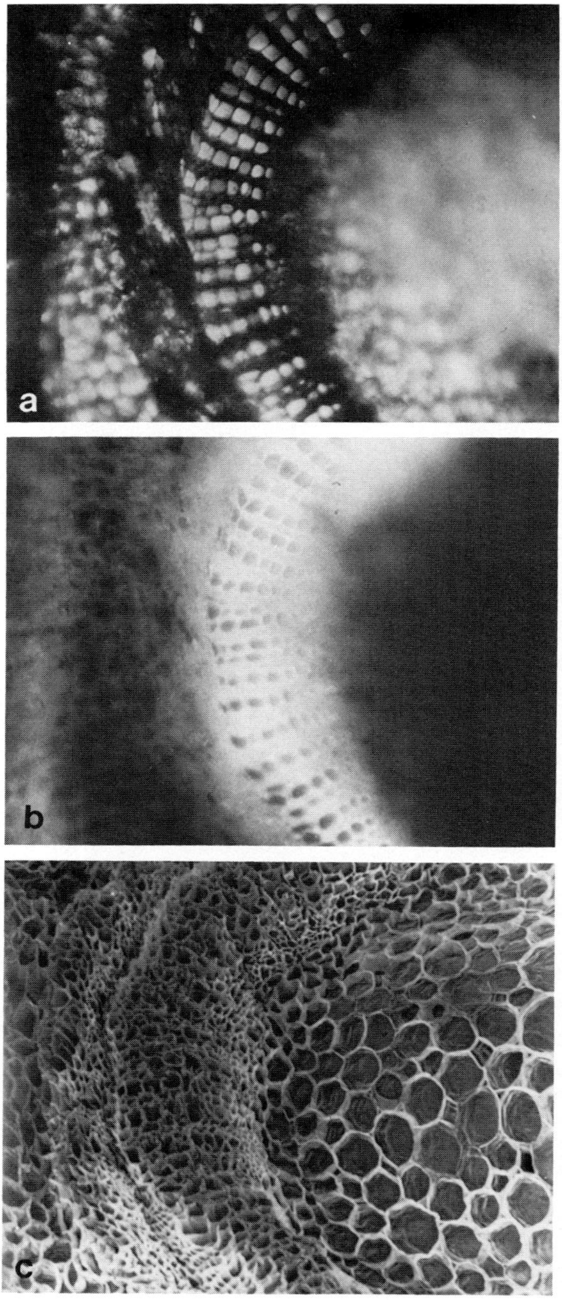

Fig. 5. Combined LM/SEM micrographs of a handmade section of *Vinca rosea* glued to a cover glass (fluoromount), critical-point dried, and coated with gold. a, LM absorption image. b, LM fluorescence image showing the autofluorescence of the xylem. c, SEM image.

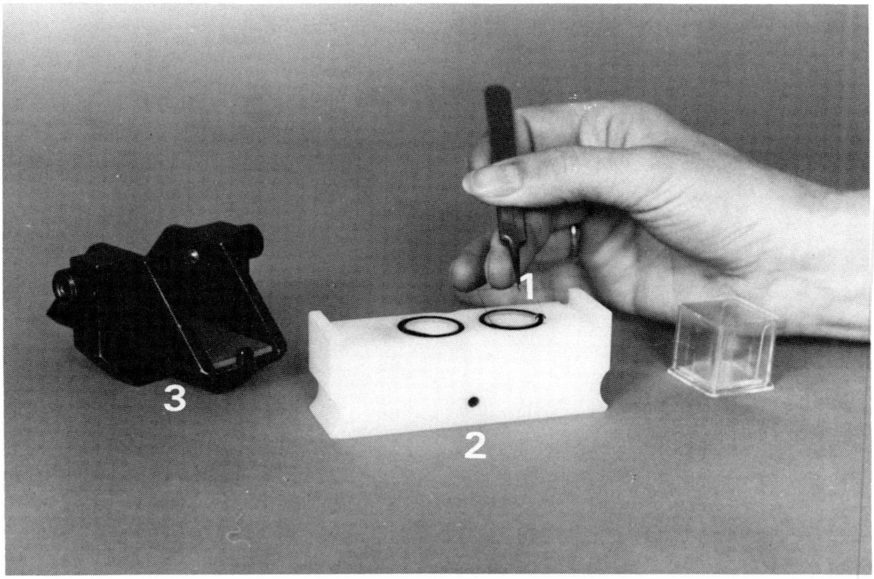

Fig. 6. The centrifugation bucket used for the sedimentation of fixed cells on cover glasses for the combined LM/SEM microscope. Poly-L-lysine-coated cover glasses (1) are placed onto the wet rubber rings, encircling two cylindrical holes (1.5 cm diameter) of a polyethylene inner part (2), which is placed in a steel holder (3).

ture drying the bucket is disassembled under liquid (Fig. 7) and the cover glasses containing the cells are placed quickly in small Petri dishes for staining and dehydration.

The fixation (either before or after bringing the cells onto the glass) also depends on the cell type and possible staining. For peripheral blood mononuclear cells and mouse bone marrow cells, a mixture of 0.1% glutaraldehyde and 1% paraformaldehyde in 0.1 *M* cacodylate buffer is used for 30 min. The cervical epithelial cells are prefixed in suspension with carbowax [2%, (w/w) polyethylene glycol, mol.wt. 1500, in 50% ethanol] and postfixed with 100% ethanol for 30 min.

The cells are quantitatively stained for DNA with a Feulgen acriflavine-SO_2 stain (Tanke and van Ingen, 1980) with the following modification (Fig. 8). The Feulgen hydrolysis is performed in 5.0 *N* HCl for 15 and 30 min for peripheral blood mononuclear cells (or mouse bone marrow cells) and cervical epithelial cells, respectively. The hydrolysis is stopped by rinsing for 3 min in demineralized water. Subsequently, the preparations are stained in a solution of 0.01 or 0.0085%, respectively, (for peripheral blood mononuclear cells and mouse bone marrow cells or cervical epithelial cells) (w/v) acriflavine-SO_2 (Chroma, West Germany) in 14 m*M* HCl

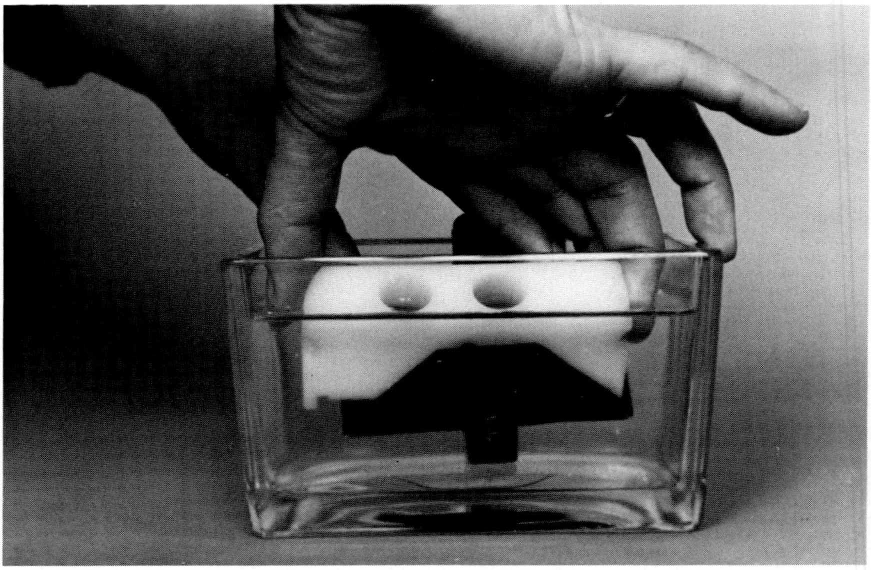

Fig. 7. The opening of the centrifugation bucket under demineralized water to prevent premature drying.

5 MIN DEMINERALIZED WATER

15 OR 30 MIN 5.0 *N* HYDROCHLORIC ACID

3 MIN DEMINERALIZED WATER

15 MIN 0.01% OR 0.0085% (w/v) ACRIFLAVINE-SO_2 SOLUTION IN 14 m*M* HYDROCHLORIC ACID, 6 m*M* POTASSIUM METABISULPHITE

6 MIN DEMINERALIZED WATER

10 MIN ACID ETHANOL (1 ml CONCENTRATED HYDROCHLORIC ACID in 99 ml 70% ETHANOL)

5 MIN DEMINERALIZED WATER

5 MIN PHOSPHATE BUFFER (pH 6.8)

5 MIN DEMINERALIZED WATER

5 OR 10 MIN ETHANOL 70%

5 OR 10 MIN ETHANOL 80%

5 OR 10 MIN ETHANOL 90%

15 MIN ETHANOL 100% (2 X)

30 MIN ETHANOL 100%

Fig. 8. Flow chart of staining and dehydration procedure for LM/SEM preparations of peripheral blood mononuclear cells, mouse bone marrow cells, or cervical epithelial cells.

and 6 m*M* potassium metabisulfite for 15 min, and rinsed for 6 min in demineralized water. To remove noncovalently bound dye, the Petri dishes are filled with acid ethanol [1% HCl (w/v) in 70% ethanol] for 10 min. Then the preparations are washed first with demineralized water, followed by 0.1 *M* phosphate buffer (pH 6.8), and finally again with demineralized water (5 min each). This is followed by a dehydration in 70% ethanol, 80% ethanol, and 90% ethanol for 5 min each for lymphocytes and bone marrow cells and 10 min each for cervical epithelial cells, 100% ethanol (twice) 15 min each and 100% ethanol for 30 min.

The preparations are critical-point dried with CO_2, glued with conductive carbon cement (Neubauer, West Germany) to special aluminum holders with a hole in the middle (diameter 20 mm) and coated with gold by means of a sputter coater. These preparations are investigated in the LM/SEM.

QUANTITATIVE MEASUREMENTS IN THE LM/SEM

Beside the visualization of internal cell structures (e.g., nucleus) after an appropriate staining, the attached photometer unit provides the LM/SEM with the possibility to obtain quantitative information on the presence of fluorescing substrates. An example of such a quantitative staining has been described earlier where the cell nucleus was stained using Feulgen acriflavine (DNA-specific fluorochrome).

To test the accuracy of quantitative DNA measurements in the LM/SEM, several experiments were performed on different cell types, i.e., chicken erythrocytes, human lymphocytes, and mouse liver cells (Wouters *et al.*, 1986a). Chicken erythrocytes have been widely used as internal standards in cytometry (Diamond and Braylan, 1980; Hoehn and Callis, 1978; Noguchi and Browne, 1978; Tannenbaum *et al.*, 1978). A comparison was made between Feulgen acriflavine-SO_2-stained cells embedded in fluoromount and measured with a conventional fluorescence microscope connected to a semiautomated microfluorometer and the same stained cells not embedded but processed for SEM and measured in the LM/SEM (again with a microfluorometer). The normally embedded chicken erythrocytes possessed a standard amount of DNA with a coefficient of variation (cv) of 5.7% (Fig. 9). This was comparable to the cv of completely dry and gold-covered cells measured in high vacuum (cv 7.1%). The cv values in the specimens are due to intraanimal and instrumental variations (Tanke and van Ingen, 1980).

It was found that both the gold layer and the vacuum condition consistently diminished the mean signal of the cells compared to cells not cov-

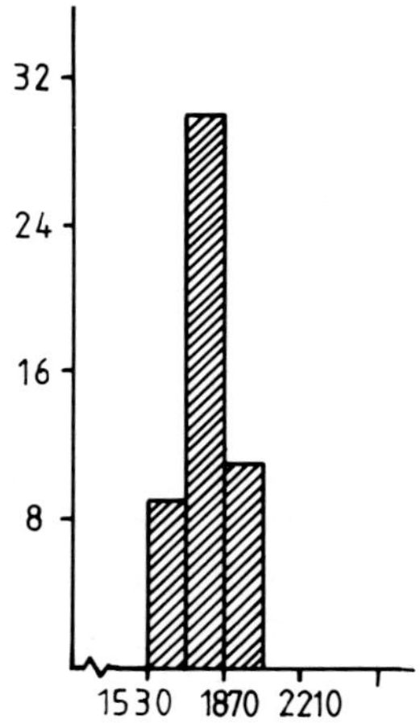

Fig. 9. Distribution of Feulgen–DNA values expressed in arbitrary units of normally embedded chicken erythrocytes measured with a microfluorometer connected to a conventional fluorescence microscope. Abscissa: Feulgen acriflavine-SO_2 fluorescence. Ordinate: number of cells.

ered with gold and measured in atmospheric conditions (Fig. 10). The reduction caused by measuring in high vacuum was reversible. It means that a cell measured without high vacuum showed a specific fluorescence signal; in high vacuum the signal of this cell was lower but when measuring this same cell again without high vacuum the first signal was found again. The reduction in the fluorescence signal caused by the gold layer might be explained as follows: the thin layer of gold will probably reflect the excitation light entering the cell in the spherical part not directly attached to the glass and, in addition, the returning emission light will again be partly reflected when leaving the cell (Fig. 11).

The same influence of the high vacuum condition and the gold layer was found for peripheral blood mononuclear cells (Fig. 10). The SEM morphology of the measured lymphocytes remained satisfactory after a quantitative DNA staining (Fig. 12). Here again the cv of normally embedded lymphocytes (7.1%) was comparable to the cv of completely dry, gold-coated and in high vacuum measured lymphocytes (6.1%). Although the remaining signal was sometimes only 16% of the original signal, it was still strong enough to be measured and the cv was acceptable.

Mouse liver cells were used to verify the maintenance of linearity of the measured fluorescence signal with widely varying DNA content. The LM/SEM specimens revealed the expected three peaks of cells with a diploid, tetraploid, and octaploid DNA content with acceptable cv's (9.4%, 8.4%, and 5.1%, respectively).

	Chicken erythrocytes	Lymphocytes
− gold − high vacuum	1.00 (8.3%)	1.00 (5.5%)
+ gold − high vacuum	0.81 (11.7%)	0.76 (8.9)
− gold + high vacuum	0.60 (7.9%)	0.25 (7.6)
+ gold + high vacuum	0.32 (7.1%)	0.16 (6.1)

Fig. 10. Distribution of mean Feulgen–DNA values expressed in arbitrary units (and coefficient of variation) of 50 critical-point dried chicken erythrocytes and lymphocytes measured with a microfluorometer connected to the LM/SEM, showing the influence of high vacuum and gold covering.

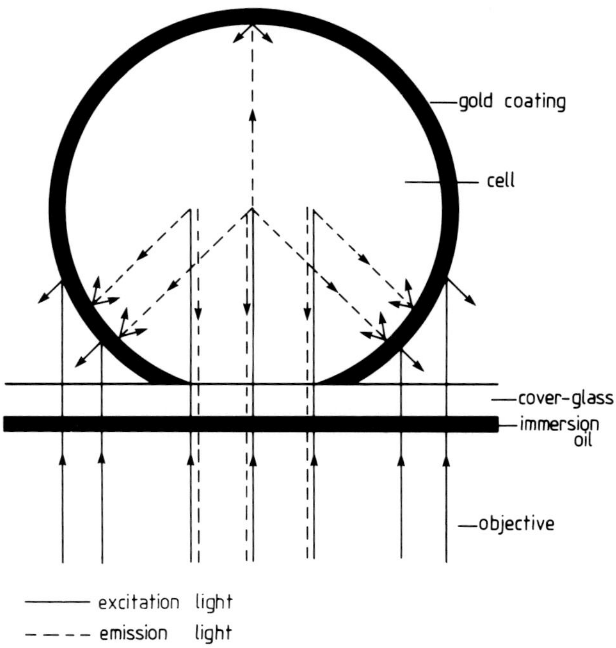

Fig. 11. The influence of the gold coating around a cell on the excitation and emission light.

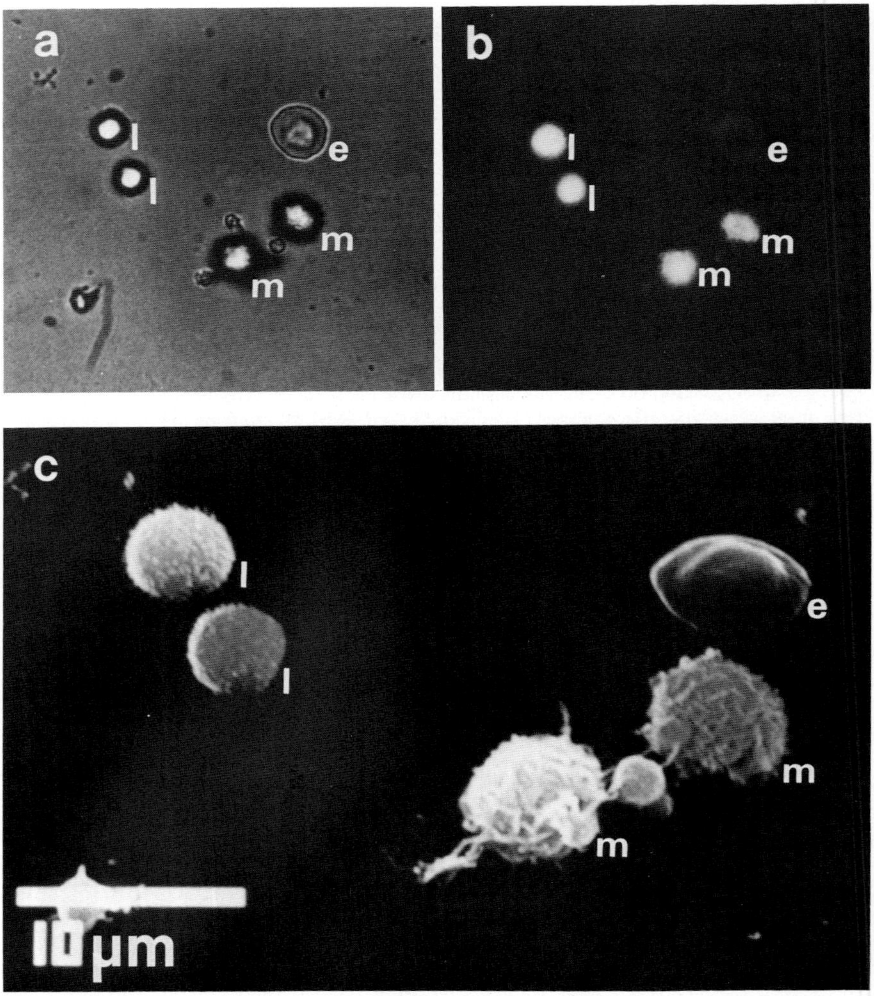

Fig. 12. Combined LM/SEM micrographs of human peripheral blood cells prepared on cover glass by means of centrifugation. Quantitatively stained with Feulgen acriflavine-SO_2, critical-point dried and covered with gold. a, LM absorption image showing lymphocytes (l), monocytes (m), and an erythrocyte (e). b, LM fluorescence image (blue excitation). c, SEM image.

CATHODOLUMINESCENCE IN THE LM/SEM

The detection of cathodoluminescence with the present optics is under investigation. Cathodoluminescence is the light (UV and visible) emitted by certain types of materials when they are bombarded by high-energy electrons. In a conventional SEM, a light collecting lens can be placed inside the vacuum chamber for focusing the emitted light onto the end of a light pipe. The light is then transferred to a photomultiplier. This signal is amplified and used to produce an image on the monitor. In the LM/SEM vacuum chamber the objective is present directly under the specimen and, without the intervention of a light pipe or photomultiplier, the cathodoluminescence can be seen in color through the eyepieces of the LM/SEM.

Glass produces a strong cathodoluminescence when bombarded by high-energy electrons. This cathodoluminescence is used to center the objective exactly under the beam. Several chemical compounds and stains have been tested in the LM/SEM. The cathodoluminescence of the glass in the background was so intense compared to the dye that although these compounds were known to demonstrate cathodoluminescence (Bröcker and Pfefferkorn, 1979; de Mets and Lagasse, 1971; Schmidt *et al.*, 1976; Soni *et al.*, 1975), it could not be seen in the LM/SEM. Polystyrene spheres containing various fluorochrome molecules used as immunological markers (Molday *et al.*, 1974) to cells also showed no cathodoluminescence in the LM/SEM.

A commonly used biomedical application of fluorescence microscopy, the immunofluorescence technique, was also tested in the LM/SEM. Several fluorochromes (e.g., FITC and TRITC) were tested as immunological markers in the LM/SEM. Although a nucleus could be quantitatively stained and measured in the LM/SEM (as described above), the immunofluorescence of the specific cell surface label could not be visualized in the LM/SEM after critical-point drying. When an enzyme was used as label, the reaction product could be seen in the absorption LM image of the LM/SEM. An example can be seen in Fig. 13 (van Blitterswijk *et al.*, 1986). Here rat middle-ear epithelium was cultivated on a cover glass and fixed with methanol. To investigate the presence of cytokeratin, the cells were stained with antikeratin monoclonal antibody, clone 80 (van Muijen *et al.*, 1984). Rabbit–anti-mouse Ig-conjugated horseradish peroxidase was used as second antibody and diaminobenzidine as substrate. The comparison with the SEM images showed that the immunoperoxidase labeling was more intense in cells lying partially or totally above the basal culture plane (van Blitterswijk *et al.*, 1986).

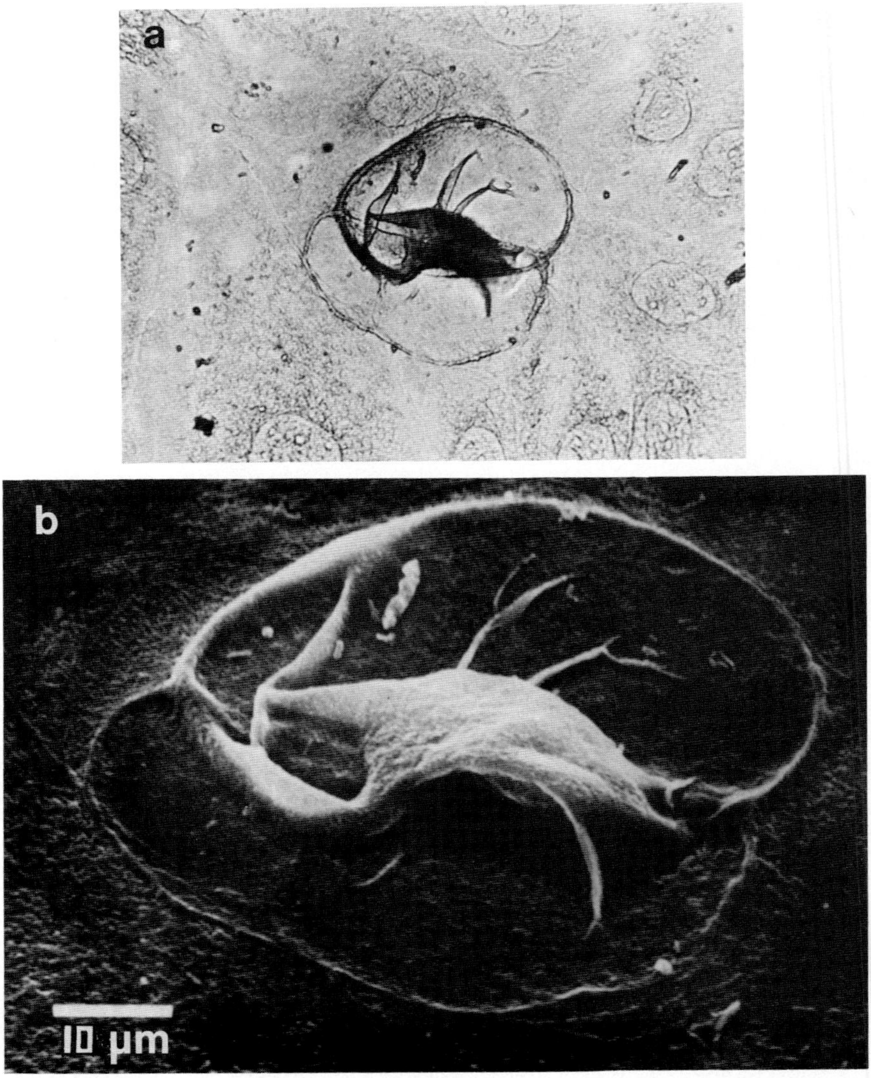

Fig. 13. Combined LM/SEM micrographs of cultivated rat middle-ear epithelium fixed with methanol and stained with an antikeratin monoclonal antibody, clone 80 (recognizing keratin, van Muijen *et al.*, 1984), as second antibody rabbit–anti-mouse Ig-conjugated with horseradish peroxidase was used. Diaminobenzidine was used as substrate. a, LM absorption image showing the cytokeratin in the epithelium cell. b, SEM image of the same cell.

In our laboratory the investigation of a new immunological marker is in progress (Tanke *et al.*, 1985), which has the quality of cathodoluminescence among others. This marker consists of phosphors in a crystal structure to which a macromolecule (e.g., antibody) can be bound. The luminescence is blue, green, or red depending on the phosphor used. The size varies from 6 to 0.01 μm. A major advantage of these markers compared to the conventional fluorochromes used as immunological markers is that they do not fade (reduction in fluorescence intensity during excitation). This fading can be the result of photochemical reactions which cause decomposition of fluorescing molecules (bleaching) or the presence of other fluorophores, oxidizing agents, or compounds which influence the electronic configuration of the fluorescing fluorophore (quenching). The cathodoluminescence of these phosphor markers is extremely strong and some preliminary results showed that labeling to lymphocytes and detection in the LM/SEM is possible.

BIOMEDICAL APPLICATIONS

In biomedical research several applications are possible with the LM/SEM. The general problem of specimen relocation and identification of cells can be solved. Cytochemical staining performed on a tissue section will reveal the position of nuclei and/or other cell organelles, which makes the location and identification of these cells (after LM examination) in the SEM part of the LM/SEM simple. The same applies for the examination of cell suspensions with the LM/SEM.

A possible application is the combination of the surface morphology with the DNA content of a cell. The DNA content is measured with the microfluorometer after a quantitative cytochemical staining (as described above) and the surface morphology is studied with the SEM system of the instrument. Several examples, concerning lymphocytes, cervical cells, and macrophages are given below.

The response of peripheral blood mononuclear cells (PBMC) to pokeweed mitogen (PWM) stimulation was investigated using the LM/SEM (Wouters *et al.*, 1986c). Human PBMC were isolated (Ficoll/Isopaque; Boyum, 1968), cultured in Teflon bags (de Vries *et al.*, 1980), and stimulated with PWM. Cell samples were taken at 24-hr intervals, fixed, and preparations were made as described above using centrifugation, Feulgen staining, and processing for SEM. The DNA content of the nucleus of a cell was compared directly with the diameter and morphology (by SEM). It was found that lymphocytes transformed into blast cells upon PWM stimulation without notable increase of DNA content. Cells with a diameter less than 5 μm having tetraploid DNA content (4C) were not seen (Fig. 14). In control specimens, the diameter of lymphocytes possessing a diploid DNA content was less than 5 μm. DNA fluorescence measurements were comparable with ^{3}H-thymidine uptake profiles, both having peak values at days 4 and 5 after the start of stimulation.

Lymphocyte-enriched populations, stimulated by PWM, showed a small but distinct population of cells with a diameter of approximately 4–5 μm having 4C DNA content at days 3, 4, and 5 after the initiation of stimulation. This indicates the potentiation of the cells to divide upon mitogen stimulation. By SEM, no particular cell surface morphology was found for blast cells with increasing DNA content compared to blast cells with a diploid DNA content. The blast cells had a diameter larger than 5 μm and were either covered with varying numbers of fingerlike or conical microvilli (Fig. 15); others had ridged profiles or uropods or had smooth surfaces.

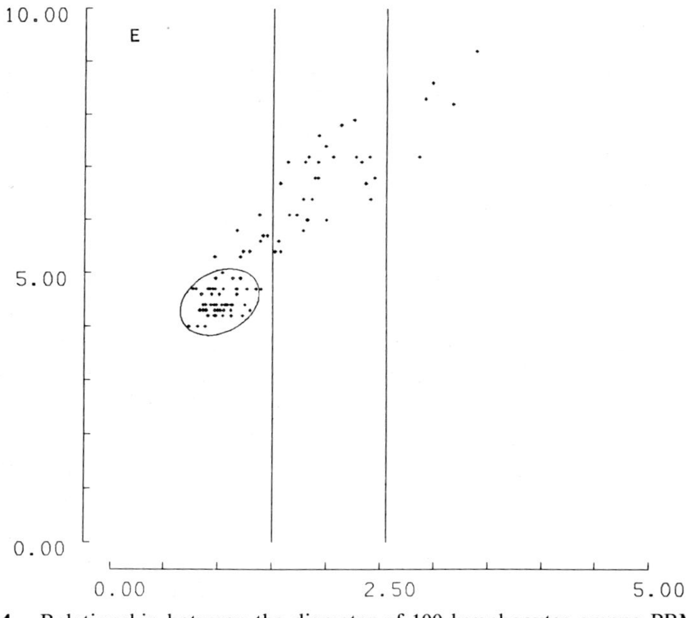

Fig. 14. Relationship between the diameter of 100 lymphocytes among PBMC stimulated 5 days with PWM and their relative fluorescence after Feulgen acriflavine staining. The data inside the ellipse (calculated according to Bartels, 1980) pertain to normally distributed lymphocytes (2C DNA); those between the vertical bars pertain to cells containing 4C DNA. Abscissa: Feulgen–DNA values (in arbitrary units); ordinate: cell diameter (in micrometers).

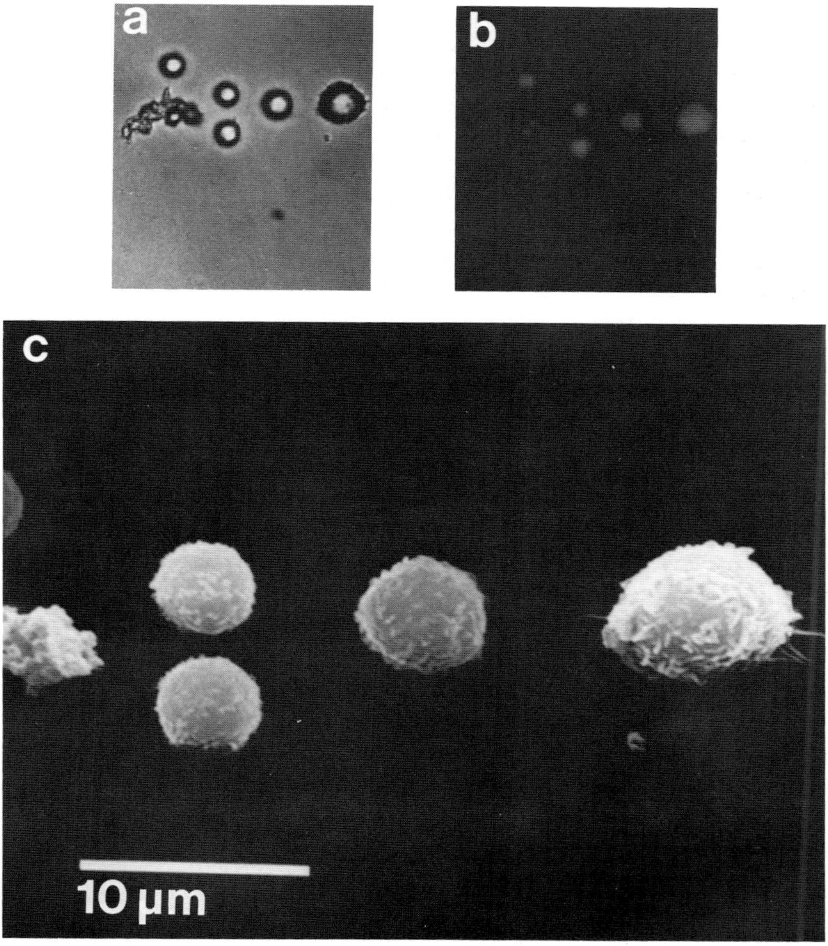

Fig. 15. Combined LM/SEM micrographs of PWM stimulated PBMC sampled 72 hr after the start of the stimulation. a, LM absorption image. b, LM fluorescence image (blue excitation). c, SEM image.

Another example is the study of the surface morphology of benign and malignant cervical cells with the LM/SEM (Wouters *et al.*, 1986b). Previously, other authors (Llanes *et al.*, 1973; Williams *et al.*, 1973; Murphy *et al.*, 1975; Kenemans *et al.*, 1981; Ito and Kudo, 1982; de Haan *et al.*, 1983; Domagala and Koss, 1983) used LM to determine the benignity or malignancy of the cervical cells with SEM. For this purpose morphological criteria such as size and shape of the nucleus, the chromatin pattern, the nucleus/cytoplasmic ratio, and the size and number of nucleoli were used. Only a limited number of cells could be investigated since preselection and relocation (both time-consuming) are necessary. The LM/SEM makes preselection and relocation redundant. Furthermore, the DNA content of each individual cell is used as an objective cellular parameter with a strong correlation with severe dysplasia, carcinoma *in situ,* and carcinoma (increased aneuploid DNA content, >5C; Ploem-Zaaijer *et al.*, 1979; Böcking *et al.*, 1984).

Cervical cells were sampled with an acryl cotton-tipped applicator from healthy volunteers or from patients with a cytological abnormality known from a previous smear. After fixation they were transferred to cover glasses together with fixed chicken erythrocytes as internal standard cells (Vindeløv *et al.*, 1983) by means of centrifugation (described above). Feulgen staining and SEM processing followed. The control specimens showed that most of the benign cells contained a diploid amount of DNA. The surface morphology of the intermediate or superficial cervical cells depends on the cell side. The luminal side was divided into various regions by crests (elevated borders) and covered with microridges, varying from sparse to numerous (Fig. 16). The basal side was divided into various regions by furrows (grooves) and covered with microvilli, again varying from sparse to numerous, but always a regular shape (Fig. 17).

The DNA measurements performed in the LM/SEM on abnormal specimens (from moderate to severe dysplasia, carcinoma *in situ* to carcinoma) showed cells with varying DNA contents. A histogram of DNA measurements from a carcinoma specimen is shown in Fig. 18. The highest DNA content measured was 10.0C. The surface morphology of the suspect cells (>5C) differed from the morphology of the benign intermediate cells (<5C). The suspect cells showed no crests, furrows, or microridges. Often the surface structures were flattened. Sometimes a roof-tile appearance and/or wrinkled cell surface could be seen (Fig. 19). Cells originating from a dysplastic or carcinoma *in situ* specimen with a DNA content <5C also showed a different morphology. No microridges, flattened surface structures, or a roof-tile appearance could be detected. Microvilli of different shapes and sizes (anisovillosis) were often present (Fig. 20). They sometimes had the tendency to grow together (anastomosis).

Fig. 16. Combined LM/SEM micrographs of a normal intermediate cervical cell brought on a cover glass by means of centrifugation, quantitatively stained with Feulgen acriflavine and processed for SEM. a, LM absorption image. b, LM fluorescence image. c, SEM image showing the luminal side. Crests can be seen as elevated borders and microridges are present. d, Detail of the surface structure.

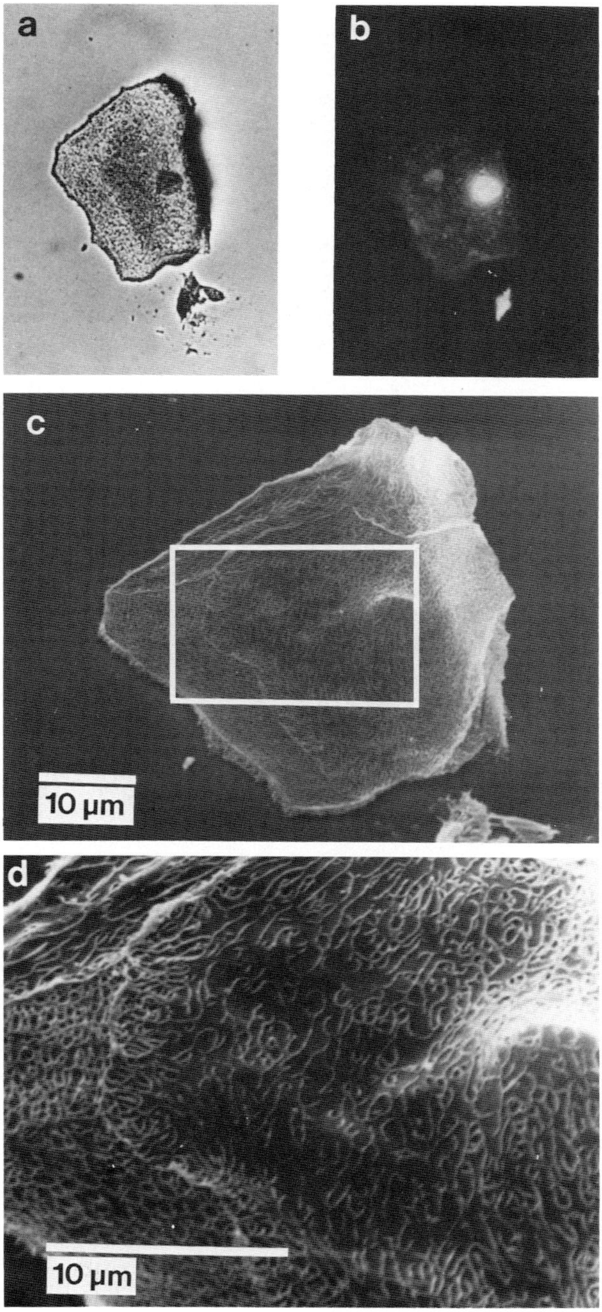
a
b
c
10 µm
d
10 µm

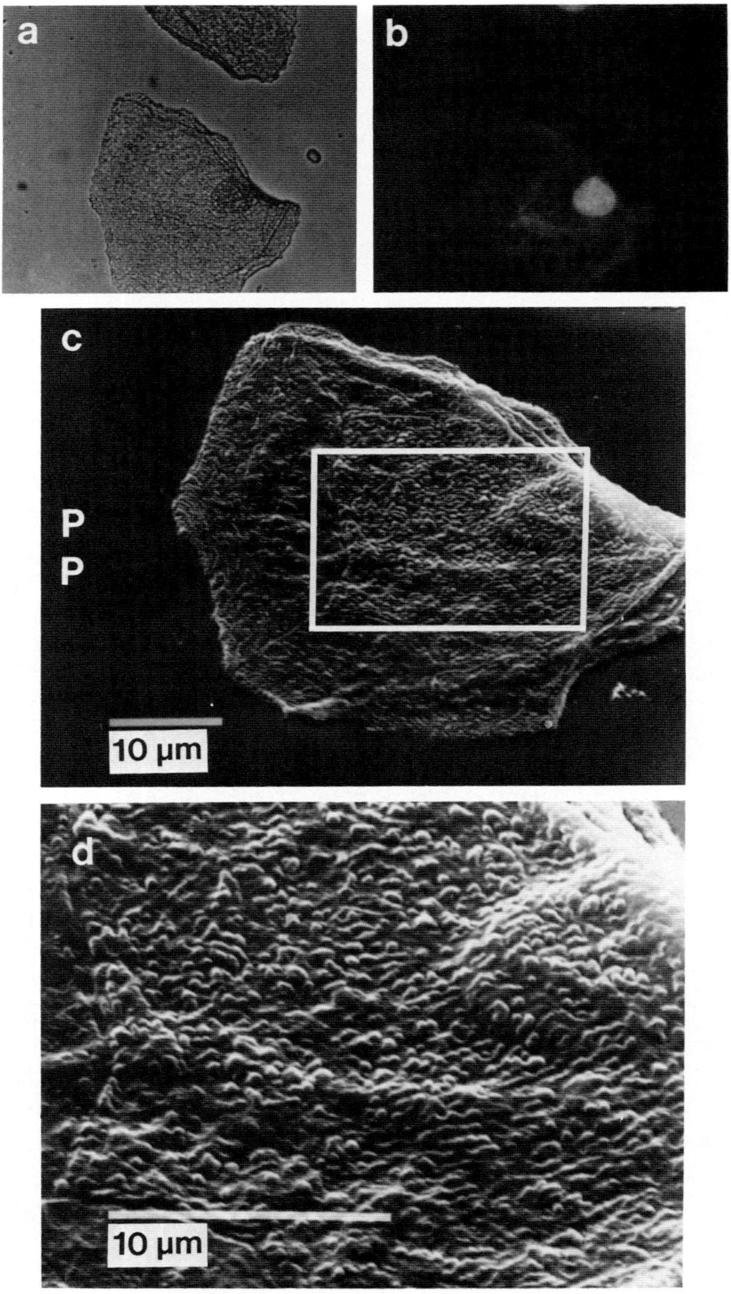

Fig. 17. Combined LM/SEM micrographs of a normal intermediate cervical cell prepared as in Fig. 15, showing the basal side. a, LM absorption image. b, LM fluorescence image. c, SEM image. Furrows can be seen as grooves and microvilli are present. d, Detail of the surface structure.

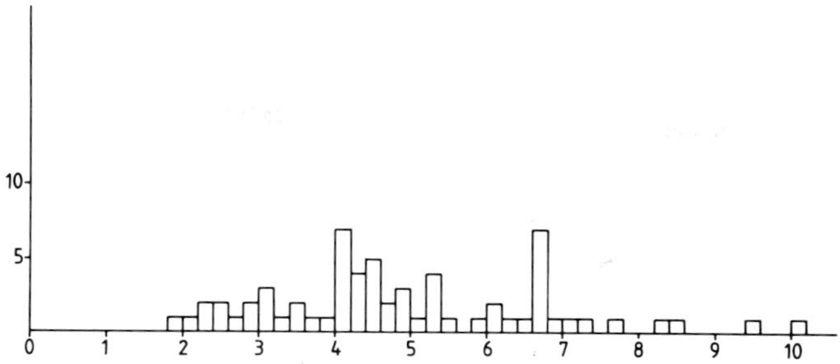

Fig. 18. DNA measurements of cervical epithelial cells from carcinoma lesion after the preparation method described in Fig. 15, performed in high vacuum of the LM/SEM. Abscissa: Feulgen acriflavine-SO_2 fluorescence expressed as C value. Ordinate: number of cells (Wouters *et al.*, 1986b).

The cell surface characteristics in correlation to the DNA content of macrophages in murine bone marrow cultures were also studied with the LM/SEM (Wouters *et al.*, 1987). Monocytes from the bone marrow stretched immediately on glass and transformed into macrophages and were therefore not recognized as monocytes under the LM. Part of these cells divided in liquid culture (van der Meer *et al.*, 1983), but did not show any particular morphology in the LM. The LM/SEM was used to find out whether these dividing macrophages represented a subpopulation of perhaps more immature macrophages recognizable with SEM. Bone marrow mononuclear phagocytes were isolated from the mouse femur (Goud *et al.*, 1975; Goud and van Furth, 1975) and cultured for 7 days on cover glasses. The cells were subsequently fixed, Feulgen-stained, and processed for SEM. The DNA content of each cell was measured and the surface was examined (Fig. 21). The cells were divided into five different categories according to their number of surface structures. The results showed that macrophages with diploid and more than diploid DNA content were equally distributed among the cell types with different cell surface structures. A slight correlation was present between the DNA content and the cell surface morphology. Cells with more surface structures had a tendency to contain more DNA.

The LM/SEM can also be used in geology or botany research. An example is the autofluorescence of pollen, which can be used in the palynological age determination. It is assumed that the vegetation of a sedimentation area is reflected in the pollen content, provided that the circumstances during the depositing time were favorable for fossilization. The vegetational changes will then be reflected in the changing pollen composition by which zones in a pollen diagram can be distinguished. Pollen and spores are highly resistant to temperature, pressure, and nonoxidative

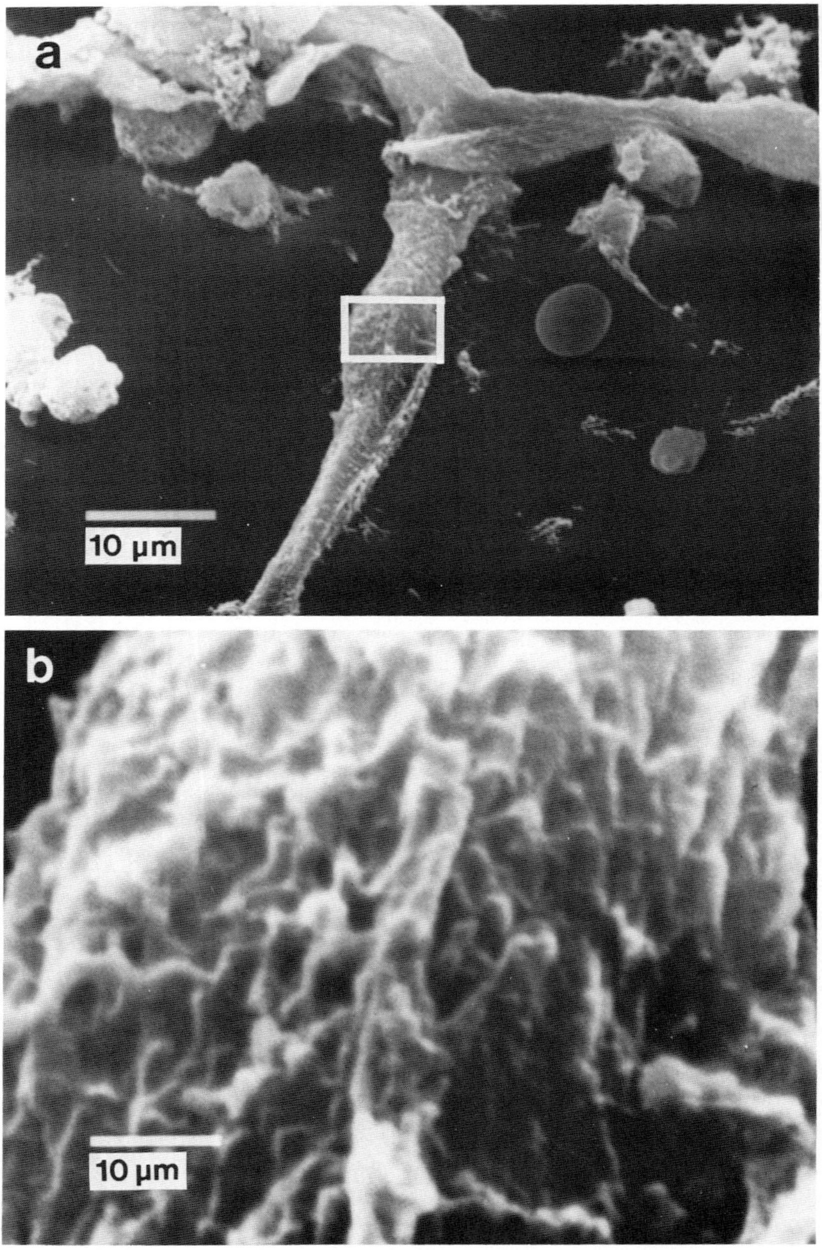

Fig. 19. a, Scanning electron micrograph of a suspect cervical cell (DNA content >5C) from a carcinoma lesion. Flattened surface structures and a roof-tile appearance is shown. b, Detail of the surface structure.

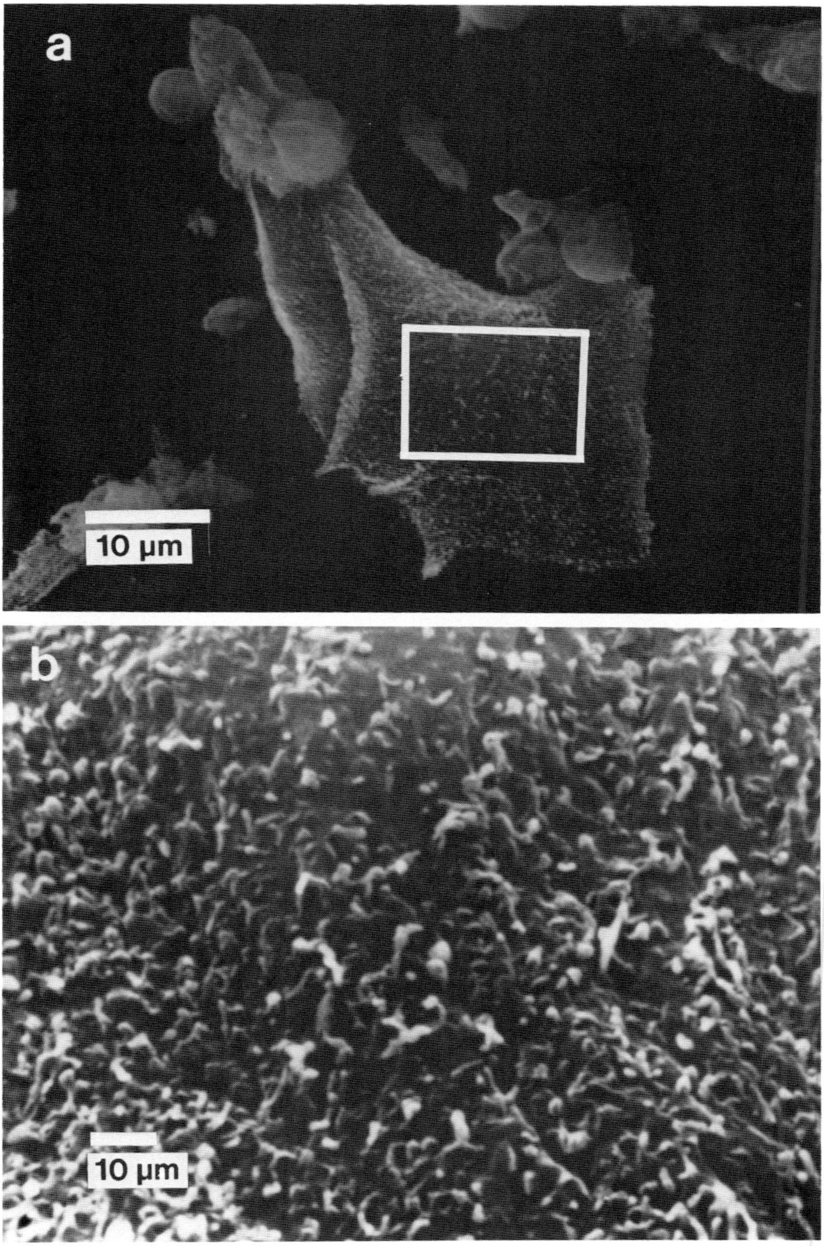

Fig. 20. a, Scanning electron micrograph of a cervical cell (DNA content <5C) from a dysplastic lesion. Microvilli show anisovillosis and anastomosis. b, Detail of the surface structure.

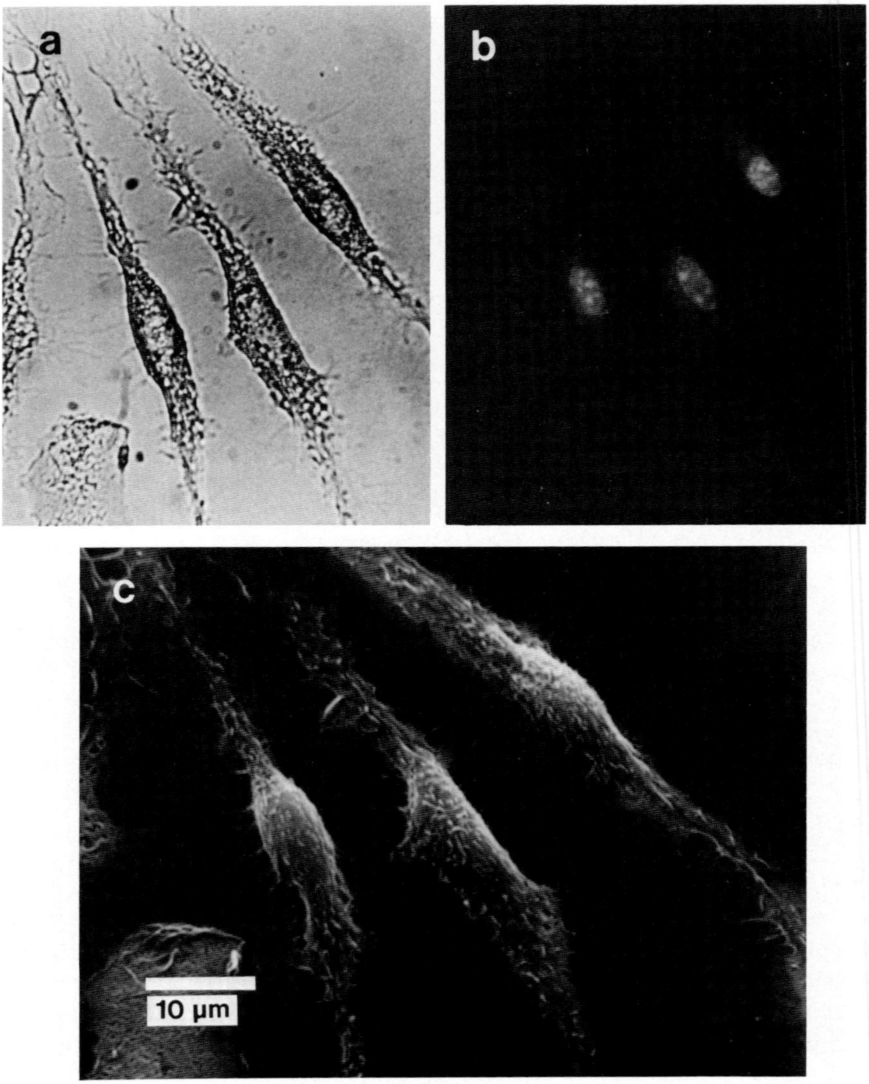

Fig. 21. Combined LM/SEM micrographs of mouse bone marrow macrophages cultured on cover glass for 7 days, quantitatively stained for DNA (Feulgen acriflavine) and processed for SEM. a, LM absorption image. b, LM fluorescence image. c, SEM image showing the cell surface structures.

chemical treatments (Sengupta and Muir, 1977) because they possess exceptionally tough organic outer cell walls (exines). In the SEM, pollen can be identified by their cell walls (Peat, 1981). However, sometimes the autochthonous vegetation can be contaminated with reworked pollen grains (secondary). Then the autofluorescence of the walls will help in the determination of pollen (van Gijzel, 1967). The protoplasm (intine) of fresh pollen contains a bright fluorescence, which is dominant over the exine (Fig. 22); however, during fossilization the intine disappears and only the exine remains fluorescent. With increasing geological age there is a change in the fluorescence color from blue or green to red or brown, due to the increase of pressure and/or temperature with increasing depth of burial. Thus, in addition to the morphology of the cell wall as can be seen with the SEM, the fluorescence of the cell wall helps to determine the fossil palynomorphs.

DEVELOPMENTS WITH ANOTHER TYPE OF SCANNING ELECTRON MICROSCOPE

A similar LM system has been developed in our laboratory for a Cambridge S180 SEM. The Cambridge instrument is more powerful in comparison to the Leitz instrument. Possibilities like backscatter electron detection and X-ray microanalysis are present, and the secondary electron image has a higher resolution than that of the AMR 1200.

A new specimen stage with integrated LM optics has been developed for the Cambridge S180 SEM. Compared to the Leitz AMR 1200, no LM stand is necessary.

The modified incident light illuminator for fluorescence (Zeiss IVF reflector 2 FL), with filter holder and binocular tubes (Zeiss), are attached to the front side of the specimen stage (Fig. 23). For photography a camera is attached to a phototube. The mercury lamp, producing the excitation light for the fluorescence, is permanently connected to the side plate of the SEM (Fig. 23). When changing a specimen the stage is moved while the mercury lamp remains on the same place.

For transmitted light two glass fibers are used. The first one guides the light from a halogen tungsten lamp to the vacuum chamber. To preserve the vacuum, a glass window separates the glass fiber from the chamber. The second glass fiber is inside the vacuum chamber extending from the glass window directly to the condenser (Fig. 24).

The optical tube length of the LM part is corrected by means of lenses and a prism (Fig. 25). Two correction lenses on the emission side are

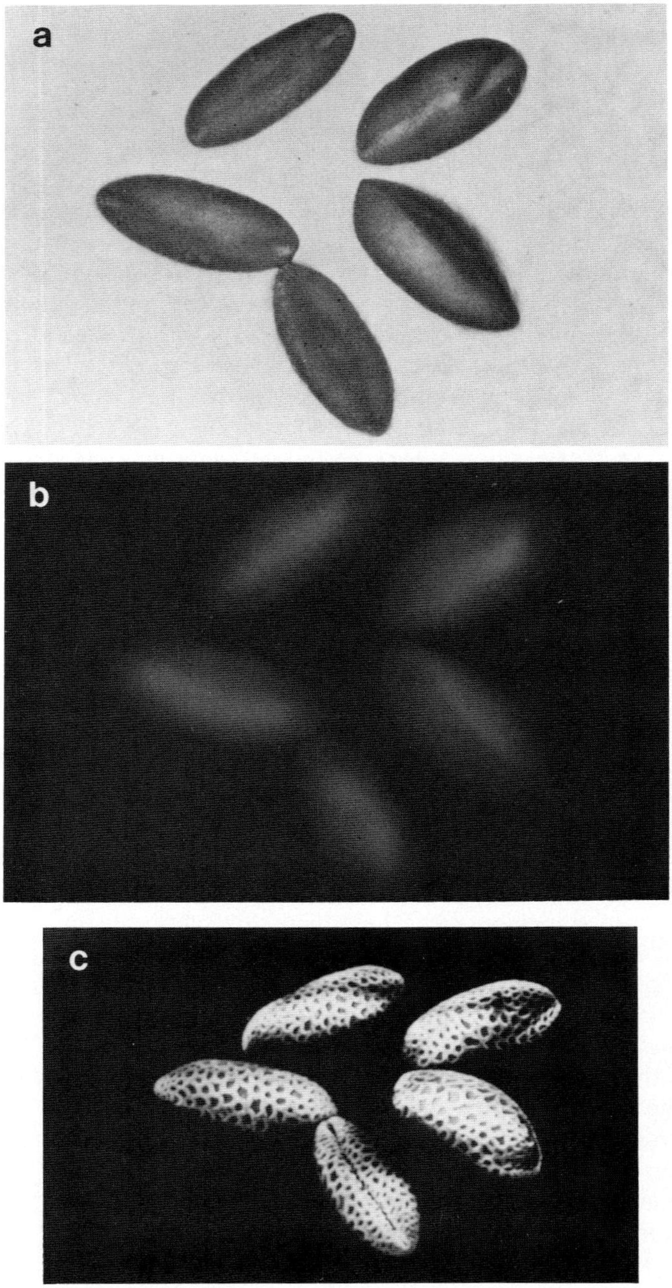

Fig. 22. Combined LM/SEM micrographs of fresh pollen from *Frittilaria meleagris* glued onto a cover glass with fluoromount. a, LM absorption image. b, LM fluorescence image showing the autofluorescence. c, SEM image of the same pollen showing the cell wall structure (Wouters and Ploem, 1985).

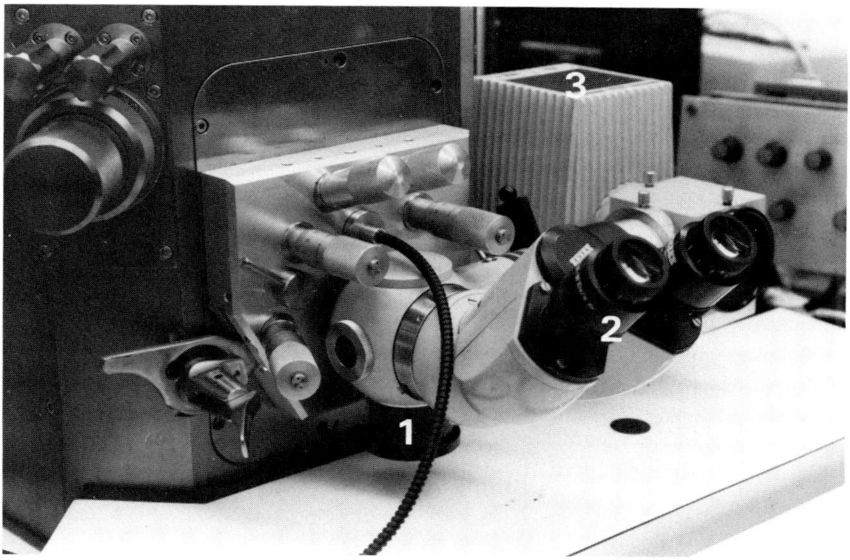

Fig. 23. The specimen stage designed for the Cambridge S180 SEM, allowing combined LM and SEM. The illuminator with filter holder (1), the binocular stage (2), and the mercury lamp (3) attached to the side plate of the Cambridge S180 SEM are shown.

Fig. 24. The specimen stage seen from the inside, showing the objective (1), condenser (2), specimen (3), and glass fiber (4).

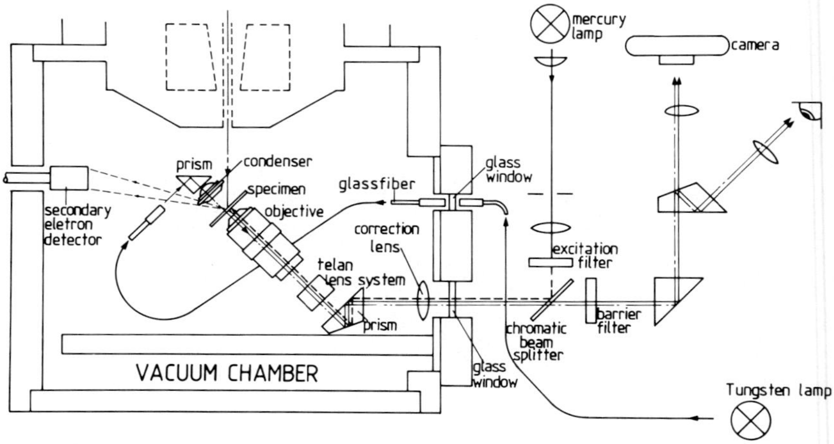

Fig. 25. A drawing of the specimen stage seen from the inside, showing the objective, condenser, specimen, two glass fibers, and prism. The path of the electrons and of the light necessary for absorption and fluorescence are visible.

removed from the Zeiss illuminator. In the vacuum chamber two correction lenses (taken from a Universal Zeiss microscope) are added. The first (telan, Zeiss) is placed directly behind the objective and converts the 16-cm optical tube length to infinitive. The second is mounted after the prism behind the objective and will form the primary image. The control system outside the vacuum chamber remains roughly the same. The objective can be focused and moved in X and Y directions up to 3 mm. The specimen table can be moved in X and Y directions up to 20 mm. The condenser can be taken out of the light path for a better secondary electron detection.

CONCLUSIONS

The development of a special stage incorporating LM optics in a vacuum chamber of an SEM, which is attached by means of an optical bridge to an LM, opened up new possibilities for SEM. An advantage is the identification without relocation of cells based on characteristics seen with LM optics after an appropriate cytochemical staining. The attachment of a photometer system offered the possibility of quantitative measurements (fluorescence), allowing the investigator to determine part of the stage of the cell cycle of each individual cell, combined to the cell surface morphology. Besides the known immunological markers in SEM such as latex spheres (LoBuglio *et al.*, 1972; Molday *et al.*, 1974), silica spheres (Peters *et al.*, 1976), tobacco mosaic virus (Hammerling *et al.*,

1975), bacteriophage T4 (Kumon *et al.*, 1976), and colloidal gold (Horisberger *et al.*, 1975), other markers (enzymes) can be used which can be detected with an LM. A special place might be occupied in the future by phosphor crystals as markers for antibodies, since they can be detected by using the electron beam for producing cathodoluminescence in different colors (visualized by the LM part).

The gold coating produces a slightly different LM absorption image than with a conventional LM. This image offers the advantage that unstained specimens such as chromosomes, which are not visible in a conventional LM, can be seen because of a phase phenomena in the LM/SEM.

Since transparency is essential for the combination LM/SEM, the applications will be restricted mostly to biomedical studies. On the other hand, it is possible to use this combination for inspection of specimens such as wafers where the cathodoluminescence (in color) can be measured and used beside the secondary electron image (Hund *et al.*, 1979). The design and development of a similar stage carrying all necessary equipments for LM (so no LM frame is needed) makes the combination LM and SEM in one instrument for owners of an SEM possible.

The authors wish to thank W. Th. Daems, H. K. Koerten, E. de Vries, and J. W. M. van der Meer for advice and S. C. Hesseling, J. P. van der Weij, J. S. van der Gevel, and B. K. Kunkeler for technical assistance. The authors are indebted to J. Bonnet and G. Vletter for design and fabrication of the specimen stage for the Cambridge S180 scanning electron microscope. The authors acknowledge the great help of J. J. Beentjes, L. D. C. Verschragen, Th. J. B. Wand, and N. Kotten for photography and F. Balke for drawing the figures. These investigations were supported by the Foundation for Medical Research FUNGO (grant No. 13-54-28).

REFERENCES

Bahr, G. F., Bibbo, M., Mikel, U., Engler, W., Rao, C., and Wied, G. L. (1976). Correlation of light and scanning electron microscopy, a new method for exfoliative cytology. *Acta Cytol.* **20,** 239.

Bartels, P. (1980). Numerical evaluation of cytologic data's V. Bivariate distributions and the Bayesian decision boundery. *Anal. Quant. Cytol.* **2,** 77.

Becker, S. N., Wong, J. Y., Marchiondo, A. A., and Davis, C. P. (1981). Scanning electron microscopy of alcohol-fixed cytopathology specimens. *Acta Cytol.* **25,** 578.

Böcking, A., Adler, C.-P., Common, H. H., Hilgarth, M., Granzen, B., and Auffermann, W. (1984). Algorithm for a DNA-cytophotometric diagnosis and grading of malignancy. *Anal. Quant. Cytol.* **6,** 1.

Boyum, A. (1968). Isolation of mononuclear cells and granulocytes from human blood. *Scand. J. Clin. Lab. Invest.* **21,** Suppl. 97, 77.

Bröcker, W., and Pfefferkorn, G. (1979). Applications of the cathodoluminescence method in biology and medicine. *Scanning Electron Microsc.* **2,** 125.

de Haan, R. W., van der Zanden, P. H. T., Vooys, G. P., and Kenemans, P. (1983). Scanning electron microscopy of exfoliative cervical cytology. *Scanning Electron Microsc.* **4,** 1899.

de Mets, M., and Lagasse, A. (1971). An investigation of some organic chemicals as cathodoluminescent dyes using the scanning electron microscope. *J. Microsc.* (*Oxford*) **94,** 151.

de Vries, E., Lafeber, G. J. M., van der Weij, J. P., van Buysen, A. C., Leijh, P. C. J., and Cats, A. (1980). Pokeweed-mitogen induced lymphocyte proliferation: The effect of stimulation on mononuclear cells. *Immunology* **40,** 177.

Diamond, L. W., and Braylan, R. C. (1980). Flow analysis of DNA content and cell size in non-Hodgkin's lymphoma. *Cancer Res.* **40,** 703.

Domagala, W., and Koss, L. G. (1983). Scanning electron microscopy in cytodiagnosis. *Scanning Electron Microsc.* **3,** 1307.

Geissinger, H. D. (1971). Correlated light optical and scanning electron microscopy of Gram smears of bacteria and paraffin sections of cardiac muscle. *J. Microsc.* (*Oxford*) **93,** 109.

Geissinger, H. D. (1974). A precise stage arrangement for correlative microscopy for specimens mounted on glass slides, stubs or EM grids. *J. Microsc.* (*Oxford*) **100,** 113.

Goud, Th. J. L. M., and van Furth, R. (1975). Proliferative characteristics of monoblasts grown in vitro. *J. Exp. Med.* **142,** 1200.

Goud, Th. J. L. M., Schotte, C., and van Furth, R. (1975). Identification and characterization of the monoblast in mononuclear phagocyte colonies grown in vitro. *J. Exp. Med.* **142,** 1180.

Hammerling, U., Polliack, A., Lampen, N., Sabety, M., and de Harven, E. (1975). Scanning electron microscopy of tobacco mosaic virus labeled lymphocyte surface antigens. *J. Exp. Med.* **141,** 518.

Hartmann, H., Hund, A., Moll, S. H., and Thaer, A. (1978). Attachment for combined scanning electron and light microscopical examinations. *Beitr. Elektronenmikrosk. Direktabb. Oberfl.* **11,** 381.

Hoehn, H., and Callis, J. (1978). Flow-cytometric DNA content differences as a function of chromosome constitution in resting human lymphocytes. *In* "Molecular Human Cytogenetics" (R. S. Sparkes, D. E. Comings, and C. F. Fox, eds.), p. 243.

Horisberger, M., Rosset, J., and Bauer, H. (1975). Colloidal gold granules as markers for cell surface receptors in the scanning electron microscope. *Experientia* **31,** 1147.

Hund, A., Thaer, A., and Noli, W. (1979). Untersuchung von opto-elektronischen Bauelementen mittels einer Lichtmikroskop-Rasterelektronenmikroskop-Anordnung. *GIT Fachz. Lab.* **23,** 15.

Ito, E., and Kudo, R. (1982). Scanning electron microscopy of normal cells, dyskaryotic cells and malignant cells exfoliated from the uterine cervix. *Acta Cytol.* **26,** 457.

Kenemans, P., Davina, J. H. M., de Haan, R. W., van der Zanden, P., Vooys, G. P., Stolk, J. G., and Stadhouders, A. M. (1981). Cell surface morphology in epithelial malignancy and its precursor lesions. *Scanning Electron Microsc.* **3,** 23.

Kumon, H., Uno, F., and Tawara, J. (1976). Morphological studies on viruses by SEM and an approach to labeling. *In* "Proceedings of the Ninth Annual SEM Symposium" (O. Johari, ed.), p. 85. IITRI, Chicago, Illinois.

Linthicum, D. S., Sell, S., Wagner, R. M., and Trefts, P. (1974). Scanning immunoelectron microscopy of mouse B and T lymphocytes. *Nature (London)* **252,** 173.

Llanes, A. T., Farre, C. B., Ferenczy, A., and Richart, R. M. (1973). Scanning electron microscopy of normal exfoliated squamous cervical cells. *Acta Cytol.* **17,** 507.

LoBuglio, A. F., Rhinehart, J. J., and Balcerzak, S. P. (1972). A new immunologic marker for scanning electron microscopy. *In* "Proceedings of the Fifth Annual SEM Symposium" (O. Johari, ed.), p. 313. IITRI, Chicago, Illinois.

Lytton, D. G., Yuen, E., and Rickard, K. A. (1978). Scanning electron and light microscope correlation of individual human bone marrow cells before and after culture in nutrient agar. *J. Microsc. (Oxford)* **115,** 35.

McDonald, L. W., and Hayes, T. L. (1969). Correlation of scanning electron microscope and light microscope images of individual cells in human blood and blood clots. *Exp. Mol. Pathol.* **10,** 186.

McDonald, L. W., Pease, R. F. W., and Hayes, T. L. (1967). Scanning electron microscopy of sectioned tissue. *Lab. Invest.* **16,** 532.

Michaelis, T. W., Larrimer, N. R., Metz, E. N., and Balcerzak, S. P. (1971). Surface morphology of human leukocytes. *Blood* **37,** 23.

Molday, R. S., Dreyer, W. J., Rembaum, A., and Yen, S. P. S. (1974). Latex spheres as markers for studies of cell surface receptors by scanning electron microscopy. *Nature (London)* **249,** 81.

Murphy, J. F., Allen, J. M., Jordan, J. A., and Williams, A. E. (1975). Scanning electron microscopy of normal and abnormal exfoliated cervical squamous cells. *Br. J. Obstet. Gynecol.* **82,** 44.

Noguchi, P. D., and Browne, W. C. (1978). The use of chicken erythrocyte nuclei as a biological standard for flow cytometry. *J. Histochem. Cytochem.* **26,** 761.

Peat, C. J. (1981). Comparative light microscopy, scanning electron microscopy and transmission electron microscopy of selected organic walled microfossils. *J. Microsc. (Oxford)* **122,** 287.

Peters, K. R., Gschwender, H. H., Haller, W., and Rutter, G. (1976). Utilization of a high resolution spherical marker for labeling of virus antigens at the cell membrane in conventional scanning electron microscopy. *In* "Proceedings of the Ninth Annual SEM Symposium" (O. Johari, ed.), p. 75. IITRI, Chicago, Illinois.

Ploem, J. S. (1969). The use of a vertical illuminator with interchangeable dichroic mirrors for fluorescence microscopy with incident light. *Z. Wiss. Mikrosk. Tech.* **68,** 129.

Ploem, J. S. (1981). A new instrument permitting simultaneous scanning electron microscopy and fluorescence microscopy of the same specimen by integrating a LM in a SEM vacuum chamber. *Cytometry* **2,** 121.

Ploem, J. S., and Thaer, A. (1981). A light microscope integrated in the vacuum chamber of a SEM. *Proc. R. Microsc. Soc.* **16,** 253.

Ploem-Zaaijer, J. J., Beyer-Boon, M. E., Leyte-Veldstra, L., and Ploem, J. S. (1979). Cytofluorometric and cytophotometric DNA measurements of cervical smears stained using a new bi-color method. *In* "Automation of Cancer Cytology and Cell Image Analysis. Tutorials of Cytology" (N. J. Pressman and G. L. Wied, eds.), p. 225. Chicago Univ. Press, Chicago, Illinois.

Sanders, S. K., Alexander, E. L., and Braylan, R. C. (1975). A high-yield technique for preparing cells fixed in suspension for scanning electron microscopy. *J. Cell Biol.* **67,** 476.

Schmidt, E. H., Bröcker, W., and Pfautsch, M. (1976). Cathodoluminescence properties of frequently used fluorescent stains in vaginal smears. *Scanning Electron Microsc.* **2,** 327.

Sengupta, S., and Muir, M. D. (1977). The use of light and electron microscopy in the study of experimentally altered spores and pollen grains. *J. Microsc.* (*Oxford*) **109,** 153.

Shelton, A., and Orenstein, J. M. (1975). A plating method for preparation of cells for culture and for observation by light or electron microscopy. *Exp. Mol. Pathol.* **23,** 220.

Soni, S. L., Kalnins, V. I., and Haggis, G. H. (1975). Localization of caps on mouse B lymphocytes by scanning electron microscopy. *Nature* (*London*) **255,** 717.

Tanke, H. J., and van Ingen, E. M. (1980). A reliable Feulgen acriflavine-SO2 staining procedure for quantitative DNA measurements. *J. Histochem. Cytochem.* **28,** 1007.

Tanke, H. J., Slats, J. C. M., and Ploem, J. S. (1985). A new immuno-phosphorescent marker for fluorescence microscopy, scanning electron microscopy and flow cytometry. *Abstr., Int. Conf. Anal. Cytol. 11th, 1985,* p. 602.

Tannenbaum, E., Cassidy, M., Alabaster, O., and Herman, C. (1978). Measurements of cellular DNA mass by flow microfluorometry with use of a biological internal standard. *J. Histochem. Cytochem.* **26,** 145.

Thornthwaite, J. T., Cayer, M. L., Cameron, B. F., Leif, S. B., and Leif, R. C. (1976). A technique for combined light and scanning electron microscopy of cells. *Scanning Electron Microsc.* **5,** 127.

van Blitterswijk, C. A., Ponec, M., van Muijen, G. N. P., Wijsman, M. C., Koerten, H. K., and Grote, J. J. (1986). Culture and characterization of rat middle-ear epithelium. *Acta Oto-Laryngol.* **101,** 453.

van der Meer, J. W. M., van de Gevel, J. S., and van Furth, R. (1983). Characteristics of long-term cultures of proliferating, mononuclear phagocytes from bone marrow. *RES: J. Reticuloendothel. Soc.* **34,** 203.

van Driel-Kulker, A. M. J., Ploem-Zaaijer, J. J., van der Zwan-van der Zwan, M., and Tanke, H. J. (1980). A preparative technique for exfoliated and aspirated cells showing different staining procedures. *Anal. Quant. Cytol.* **2,** 243.

van Ewijk, W., and Mulder, M. P. (1976). A new preparation method for scanning electron microscopic studies of single selected cells cultured on a plastic film. *Scanning Electron Microsc.* **2,** 131.

van Gijzel, P. (1967). Autofluorescence of fossil pollen and spores with special references to age determination and coalification. *Leidse Geol. Meded.* **40,** 263.

van Muijen, G. N., Ruiter, D. J., Ponec, M., Huiskens-van der Mey, C., and Warnaar, S. (1984). Monoclonal antibodies with different specifities against cytokeratins. *Am. J. Pathol.* **114,** 9.

Vindeløv, L. L., Christensen, I. J., and Nissen, N. I. (1983). Standardization of high-resolution flow cytometric DNA analysis by the stimulateous use of chicken and trout red blood cells as internal reference standards. *Cytometry* **3,** 328.

Wetzel, B., Erickson, B. W., Jr., and Levis, W. R. (1973). The need for positive identification of leukocytes examined by SEM. *Scanning Electron Microsc.* **3,** 535.

Williams, A. E., Jordan, J. A., Murphy, J. F., and Allen, J. M. (1973). The surface ultrastructure of normal and abnormal cervical epithelia. *Scanning Electron Microsc.* **3,** 597.

Wouters, C. H., and Koerten, H. K. (1982). Combined light microscope and scanning electron microscope, a new instrument for cell biology. *Cell Biol. Int. Rep.* **6,** 955.

Wouters, C. H., and Ploem, J. S. (1985). A new instrument combining simultaneous light and scanning electron microscopy. *In* "Advances in Microscopy" (R. R. Cowden and F. W. Harrison, eds.), p. 115. Alan R. Liss, Inc., New York.

Wouters, C. H., Hesseling, S., Daems, W. Th., and Ploem, J. S. (1984). A centrifugation method for standardized sedimentation of mononuclear human blood cells on glass for scanning electron microscopy. *J. Microsc.* (*Oxford*) **136,** 315.

Wouters, C. H., Koerten, H. K., Bonnet, J., Daems, W. Th., and Ploem, J. S. (1986a). Quantitative DNA measurements in an instrument combining scanning electron microscopy and light microscopy. *J. Microsc.* (*Oxford*) **141,** 41.

Wouters, C. H., Hesseling, S. C., Daems, W. Th., and Ploem, J. S. (1986b). DNA measurements of cervical epithelial cells in combination with scanning electron microscopy. *J. Histochemistry* **84,** 509.

Wouters, C. H., van der Weij, J. P., Koper, G. J. M., Daems, W. Th., de Vries, E., and Ploem, J. S. (1986c). Measurements of single-cell DNA synthesis by pokeweed mitogen stimulated mononuclear cells with combined light and scanning electron microscopy. *Immunology* **58,** 79.

Wouters, C. H., van de Gevel, J. S., van der Meer, J. N. M., Daems, W. Th., Furth, R., and Ploem, J. S. (1987). Cell surface characteristics and DNA contents of macrophages in murine bone marrow cultures: A study using simultaneous scanning electron microscopy and fluorescence microscopy. *Histochemistry* **86,** 433.

3

Confocal Microscopy in Comparison with Electron and Conventional Light Microscopy

H. T. M. VAN DER VOORT, J. A. C. VALKENBURG, E. A. VAN SPRONSEN, C. L. WOLDRINGH, AND G. J. BRAKENHOFF

Department of Electron Microscopy and Molecular Cytology
University of Amsterdam
Amsterdam, The Netherlands

CORRELATIVE MICROSCOPY IN BIOLOGY:
INSTRUMENTATION AND METHODS

INTRODUCTION

Confocal scanning laser microscopy (CSLM), as recently come into being, possesses a number of specific advantages in comparison with conventional microscopy. The most important of these are an extended dynamic range and a basically improved resolving power, both transverse as well as along the optical axis. The manner in which data are collected in this instrument is very suitable for storage, display, and subsequent data processing by digital computer. When applied to the study of biological structures, this technique will result in images which are different from those obtained by the usual methods. The aim of this chapter is to present an insight in the new possibilities offered by this technique and their physical background, as well as to correlate the information obtained by the confocal technique in the case of *Escherichia coli* bacteria to the information produced by conventional light microscopy and transmission electron microscopy. We will also demonstrate the extra possibilities presented by the use of computers for the realization of a series of measurements (cell shape and associated form criteria) on a population of hydrated fluorescent *E. coli* cells.

Further capabilities of the system, like the generation of three-dimensional (3-D) images of living cells with submicron resolution, are unique in so far that only by the confocal technique such spatial information can be obtained from hydrated material. These 3-D images are correlated, as far as possible, with data obtained by the serial sectioning technique of hydrated and fixed biological specimen. The time-consuming and tedious nature of the sectioning process constitutes, together with the preparation steps especially necessary for electron microscopy, the major drawback of the techniques used until now.

We think that confocal scanning microscopy with its various imaging modes will prove invaluable for applications to the field of biology both for its 3-D imaging capabilities and for the possibility it offers to observe live objects in their natural environment. In addition, the existence of sophisticated fluorescent probes, especially in combination with monoclonal antibodies, enables us to visualize the spatial organization of specific substances inside (possibly live) biological material.

THE CSLM INSTRUMENT

The image formation in a scanning microscope is realized by probing the specimen point for point with a finely focused laser beam. For each probed point a physical property of the radiation emanating from this point is measured. The results of this measurement can either be written directly on photographic film in analogue form, or, after a digitation step, be displayed by way of a video memory on a monitor. We have a transmission reflection or a fluorescence microscope depending on the type of detected radiation used for image formation, as illustrated in Fig. 1a and b.

In the instrument operating in our laboratory, the probing is done by scanning the specimen mechanically, while keeping the optics and the (diffraction limited) illumination spot by which the probing is realized stationary. Usage of a stationary optical system ensures optimal experimental freedom on the configuration of the detection and illumination systems. We have used this freedom to realize a confocal microscope with transmission, reflection, and fluorescent modes. In a confocal microscope, detection and illumination systems are optically highly identical, resulting in an improved (narrowed) point response of the overall system (Fig. 2).

Theoretical comparison in paraxial approximation of confocal and nonconfocal microscope systems was first carried out by Sheppard and Choudhury (1977). They showed that a narrowing down of the point response by a factor of 1.4 could be expected for a confocal system as compared to a conventional nonconfocal system.

In both systems the point response is proportional to the wavelength of the light source and inversely proportional to the numerical aperture of the objectives (Fig. 1a, 1b for the location of the objectives in a confocal system). It is therefore quite obvious that confocal microscopy only makes sense when microscope objectives with the highest possible numerical aperture are used. Moreover, the wavelength of the laser light should also be as short as the specimen permits.

Brakenhoff *et al.* (1979) showed that this theoretically expected improvement could indeed be realized with high numerical aperture objectives (NA = 1.3), resulting in an observed point response of 190 nm in a transmission confocal system operating at a wavelength of 633 nm. Later work with a HeCd laser, operating at 325 and 443 nm, resulted in a point response of 140–150 nm.

Important to notice here is that these optical characteristics are, in all modes, identical for all points of the image and independent of the se-

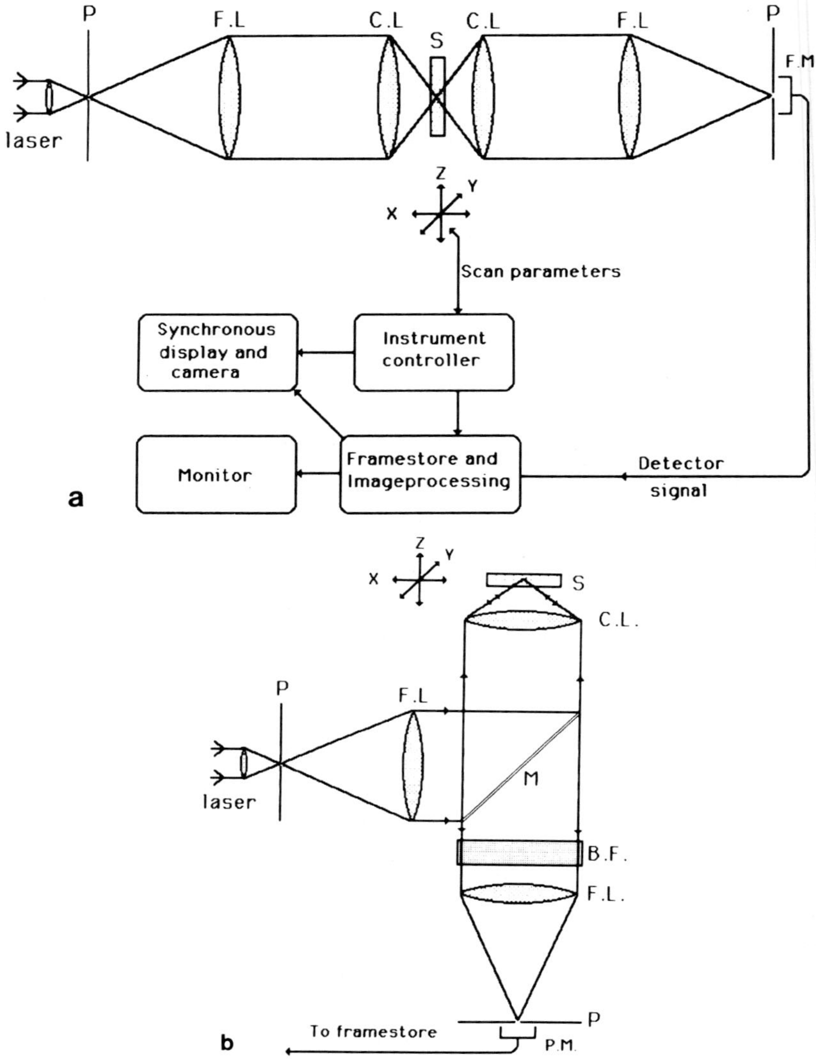

Fig. 1a. In a transmission type confocal microscope, light originating from a laser-illuminated pinhole is focused in a certain point of the specimen by the illuminating confocal lens. The same point is imaged on the detector pinhole by the collector confocal lens. The specimen is scanned mechanically through this common focal point, modulating the light intensity which reaches the detector. The image processing computer, acting as a framestore, stores this signal after a digitation step in a $256^2 \times 8$ video memory. In addition to this continuously displayed image a high definition analogue image is available on a high-resolution display running synchronously with the mechanical scan.

Fig. 1b. Light path in fluorescence mode. M is a dichroic mirror, B a low pass blocking filter. To operate in the reflection mode, the blocking filter is removed and the dichroic mirror is replaced by a beamsplitter.

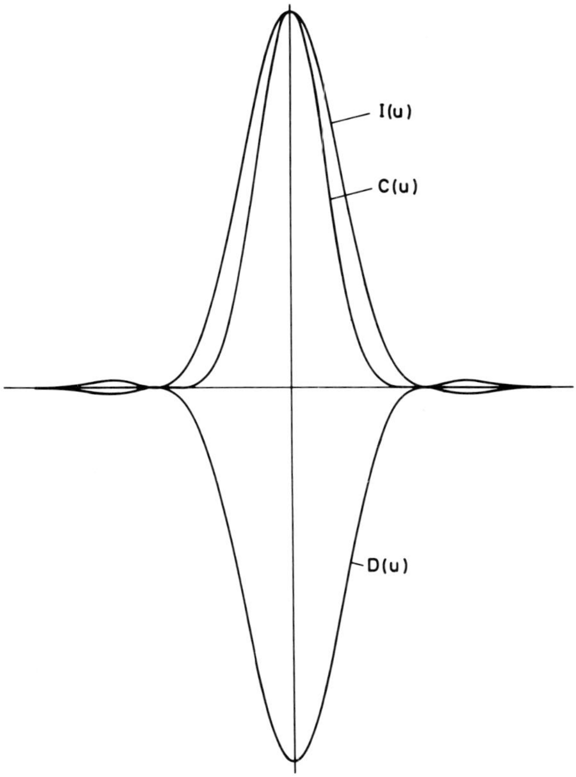

Fig. 2. Confocal point response in transmission can be represented by $(u) = [J_1(u)/u]^4$ with $u = (\mathrm{NA}\ 2\pi r)/\lambda$, J_1 the first-order bessel function, NA the numerical aperture of the diffraction limited confocal lenses, λ the wavelength of the light used, and r the distance from the optical axis. (u) is equal to the point response of a conventional microscope squared: the resulting narrowing down of the point response is by a factor of 1.4. The same factor applies to the reduction of the depth of field along the optical axis and therefore the volume sampled in the specimen by confocal microscopy is reduced by a factor of $(1.4)^3 \approx 3$. This result represents a fundamental improvement with respect to conventional microscopy.

lected magnification because of our use of stationary optics in combination with a mechanical scanner.

In the fluorescence mode a confocal system exhibits optical sectioning capabilities (Wijnaends van Resandt *et al.*, 1985; van der Voort *et al.*, 1985). The reason of this capability is explained in Fig. 3 by the use of geometric optics. This property, together with the importance of fluorescent labeling as pointed out in the introduction, makes the fluorescence mode well suited for imaging of 3-D structures.

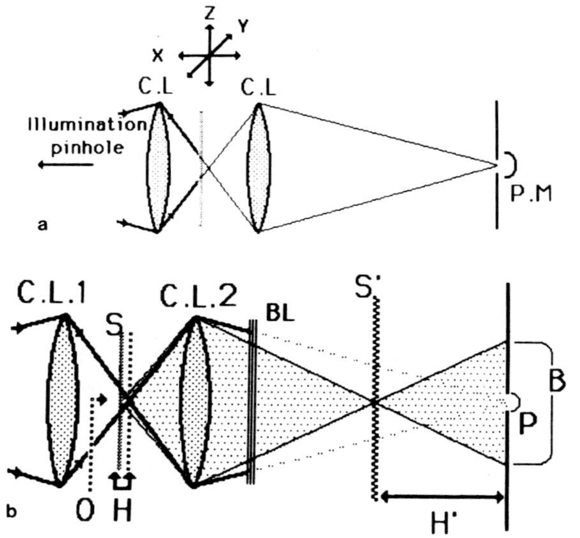

Fig. 3. Optical sectioning in a confocal fluorescence system. Suppose S is a homogeneous thin absorbent layer and no blocking filter (B) present (3a). No matter the position of this layer, in a transmission system always the same amount of radiation is absorbed, as the total energy flux through this surface is constant. Therefore, in a transmission system we cannot determine the position of this layer. When we are imaging a plane of a 3-D object, (blurred) off-focus layers will therefore all contribute equally to the in-focus image. A confocal fluorescence system behaves completely different (Fig. 3b). S is a fluorescent layer at distance H from the confocal plane. The amount of radiation passing the pinhole can be expressed as $I_p = \text{const.}\ (P/B)^2O^2I_h$, where P^2 is the area of the pinhole and B^2 the magnitude of the circular area which is formed by the intersection of the detector plane and the cone of fluorescent light. O is the region of S (with area O^2) which is within the illumination cone. The intensity of illumination light at O is I_h. Note that for all points in O there is an overlap of the cone of fluorescent light and the pinhole. Now, as $B \sim H'^2 \sim H^2$, $O^2 \sim H^2$ and $I_h \sim H^2$, we find that I_p varies inversely proportional to H^2 (Wijnaends van Resandt *et al.*, 1985), and thus the optical sectioning capability is explained.

Cox *et al.* (1982) showed that a further improvement in resolution may be expected due to the incoherence of the fluorescence radiation. However, signal to noise considerations and limitations of radiation damage (bleaching) dictate in most applications the use of a somewhat enlarged detector pinhole. Resolution in the x–y plane is in that case mainly determined by the dimensions of the excitation spot. This resolution is still superior to the resolution of a conventional fluorescence microscope in which the resolution is determined by the longer emission wavelength. Resolution at excitation wavelength $\lambda = 488$ nm and a 250-nm detector pinhole (geometrical image in confocal plane) is 220 nm in the x–y plane and 730 nm along the optical axis. Laser light of this wavelength can be generated by a relatively low cost argon–ion laser and is very suitable for exciting fluoresceineisothiocyanate (FITC)-labeled specimen.

Image Processing System and Instrument Controller

Without the use of a microcomputer system as an intermediate between the operator and the optical/mechanical part of the microscope, the CSLM would be a difficult instrument to work with. Furthermore, the advantages offered by a CSLM can only be fully exploited by the use of image processing techniques. Therefore, the complete electronic system consists of an instrument-controlling microcomputer coupled to an image processing computer. The first computer takes care of all scanning, calibrating, and optical alignment tasks and supplies the second one with the preprocessed detector signals and the scan parameters. The image processing computer uses these parameters to build the image and store the 3-D data of the scanned object, all in real time. Acquisition of a 3-D image takes place as a series of x–y scans; the z position of the specimen is increased after the completion of each x–y scan. The image acquisition program, which controls this process, has an averaging mode to increase the signal to noise ratio of the image. Each x–y plane is temporarily stored in a video memory so that the program can perform the following operation:

$$I_{\mathrm{i,j,n+1}} = (L_{\mathrm{i}} + (R - 1)I_{\mathrm{i,j,n}})\, R^{-1}$$

In this equation I (i, j: 0,1 . . . 255 and n is a positive integer) denotes the values of the i,j pixel in the nth frame, L_{i} denotes the samples along the scanned line. The constant R (1, 2, or 4) determines the relaxation time and can be adjusted interactively. The result is a continuously displayed average over a number of frames.

All image processing and data acquisition software runs under the Multitasking Image Processing Disk Operating System (MIDOS) operating system, which is a real time system for the MC 68000/10 microprocessor optimized for image processing. The main part of all software is written in Pascal. The system allows time-consuming Pascal procedures to be replaced by fast assembler-written procedures. Programs can consist of a number of independently compiled modules to facilitate program development by a team. We used this latter possibility to set up a software library of "standard" image processing routines such as spatial filters, global and local contrast manipulating algorithms, and several routines used in the construction of stereoscopic pairs. All routines in the library can be called from any program just by sending a mnemonic of the desired operation to a module which will then interpret the mnemonic. This module will then translate the mnemonic to more elaborate commands and parameters to control one or several assembler-written modules.

THE MORPHOLOGY OF THE NUCLEOID OF *ESCHERICHIA COLI*

The Fixation-Dependent Morphology of the Nucleoid

The bacterial DNA is contained within an area in the cell called nucleoid (for reviews on nucleoid morphology, see Nanninga and Woldringh, 1981; Woldringh and Nanninga, 1985). Although readily visualized by phase contrast microscopy (Mason and Powelson, 1956), its fine structure cannot be determined with this instrument because resolution is too low. In the electron microscope, nucleoid morphology is dependent on the preparation method used. In thin sections the nucleoid takes on a confined shape after OsO_4 fixation (Kellenberger *et al.*, 1985), whereas after glutaraldehyde fixation it becomes dispersed (Séchaud and Kellenberger, 1972; Woldringh, 1973). Freeze-fracturing cannot confirm the results obtained with either fixative because ice crystals prevent visualization and because freeze protecting agents probably cause rearrangement of nucleoplasmic and cytoplasmic components (Nanninga and Woldringh, 1981).

The CSLM has a higher point resolution than the phase contrast microscope. Furthermore, with the CSLM it is possible to study cells without the need to fix or dehydrate them as with electron microscopy. These properties make the CSLM the instrument of choice to study nucleoid morphology and the effects of fixation. In this section some of the results of our comparison of CSLM, phase contrast, and electron micrographs of the nucleoid of *Escherichia coli* are presented. It also gives an impression of the possibilities of the CSLM in the transmission mode.

Specimen Preparation

Escherichia coli B/r H266 cells were grown at 37°C in synthetic medium according to Helmstetter and Cooper (1968) supplemented with 1% tryptone and 0.5% yeast extract. Cell were prefixed by adding to the culture medium either OsO_4 or glutaraldehyde to a final concentration of 0.1 and 1%, respectively. Postfixation was performed according to the Ryter/Kellenberger technique (Kellenberger *et al.*, 1958) by suspending the cells in 1% OsO_4 in acetate–veronal buffer (pH 6.0) containing 0.12 *M* NaCl,

0.01% $MgCl_2$, and 0.1% tryptone, followed by further fixation in 1.0% uranyl acetate in the same buffer. For CSLM and phase contrast microscopy, cells were suspended in buffer. For electron microscopy cells were dehydrated in ethanol and embedded in araldite. Sections of the embedded material were cut on a LKB ultramicrotome (LKB instruments Inc., Stockholm). After being stained with uranyl acetate and lead citrate (Reynolds, 1963), the sections were examined in a Philips 300 electron microscope. Three-dimensional models of the nucleoid were constructed based on prints of serial-sectioned cells.

For imaging of the bacterial nucleoid in CSLM and phase contrast microscopy, several conditions had to be met. Due to the optical situation in the CSLM, the specimen had to be enclosed between two coverslips. Cells were immobilized by allowing them to attach to poly-L-lysine-coated coverslips according to a modification of Binnerts *et al.* (1982). A drop of 0.01% poly-L-lysine was placed on top of a coverslip and allowed to dry. A drop of bacterial suspension was then placed on the coverslip. After 1 min unattached cells were washed away with medium. Attachment probably took place through interaction of negative sites on the bacterial surface with positive lysine sites on the coverslip (Tsutsui *et al.*, 1976). A second coverslip was placed on top of the first. The refractive index of the culture medium used for washing and embedding was heightened by the addition of bovine serum albumin (Sigma Chemical Co., St. Louis, MO). In phase contrast microscopy this was necessary to accomplish visualization of the nucleoid (Mason and Powelson, 1956; for a discussion on this subject, see Valkenburg and Woldringh, 1984). In the CSLM the refractive index was adjusted to prevent the formation of dark halos along the rim of the cell during imaging (Brakenhoff *et al.*, 1980). For phase contrast microscopy a Wild (Wild, Heerbrugg, Switzerland) positive Zernike contrast microscope was used. The CSLM was operated with a HeCd laser at a wavelength of 325 nm in the transmission mode.

Comparison of Images

To compare the images obtained with the phase contrast, CSLM, and electron microscope, one must be aware of their differences in optical resolution and depth of field (Fig. 4). In the phase contrast microscope, the depth of field is larger than the diameter of the bacteria used in our experiments. Therefore, a layer above and below the nucleoid was imaged as well, obscuring details in nucleoid morphology. Phase contrast micrographs of living, OsO_4- (Fig. 5d), and glutaraldehyde-fixed cells showed the nucleoid as a cloudlike structure without much detail. In the CSLM only about half the cell but most of the nucleoid was imaged.

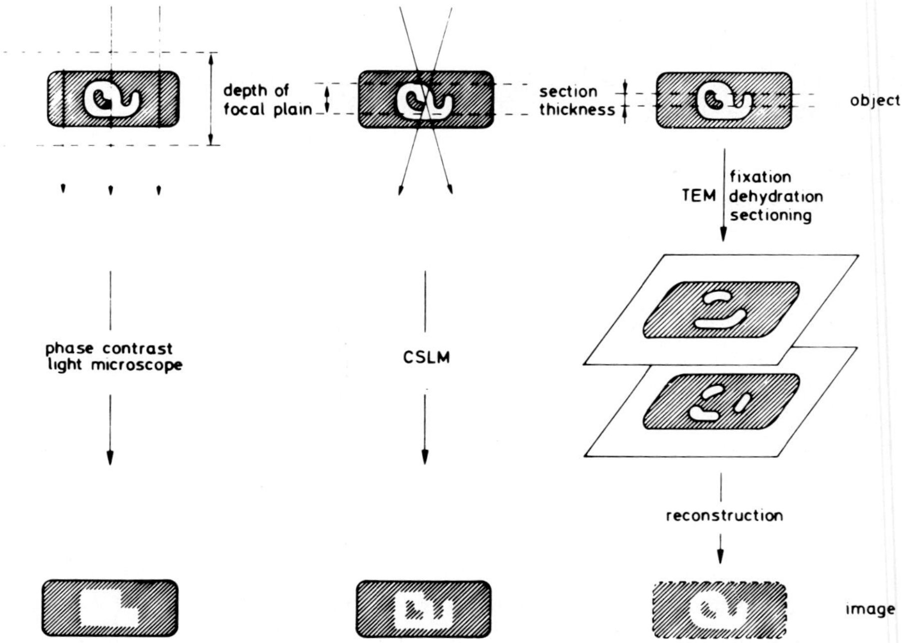

Fig. 4. Illustration of part of a bacterium with a diameter of 1 μm which has been visualized with the phase contrast, the CSLM, and after thin sectioning with the electron microscope.

Contribution of other layers of the preparation to the image was much less than in the phase contrast microscope. The combination of a higher point resolution and narrower depth of field resulted in more detailed images of the nucleoid in the CSLM as compared with the phase contrast microscope (Figs. 5c, 5d, 5g, and 5h, respectively). In the electron microscope, after thin sectioning, only a slice of about one-tenth (70 nm) of the cell diameter was imaged. No contribution of other layers of the preparation to the image took place. As, after thin sectioning only a fraction of the nucleoid was visualized, spatial models of nucleoid (Figs. 5b and 5f) were constructed from serial-sectioned cells (Figs. 5a and 5e). A good agreement could be observed between these spatial models and the CSLM micrographs (Figs. 5b and 5f, 5c, 5g, and 5h), showing the nucleoid to be a lobed structure with a central hole. Although differences in nucleoid appearance between OsO_4- and glutaraldehyde-fixed cells could be observed in sections (a confined versus a dispersed nucleoid, Figs. 5a and 5e), comparison of CSLM micrographs and spatial models showed the shape

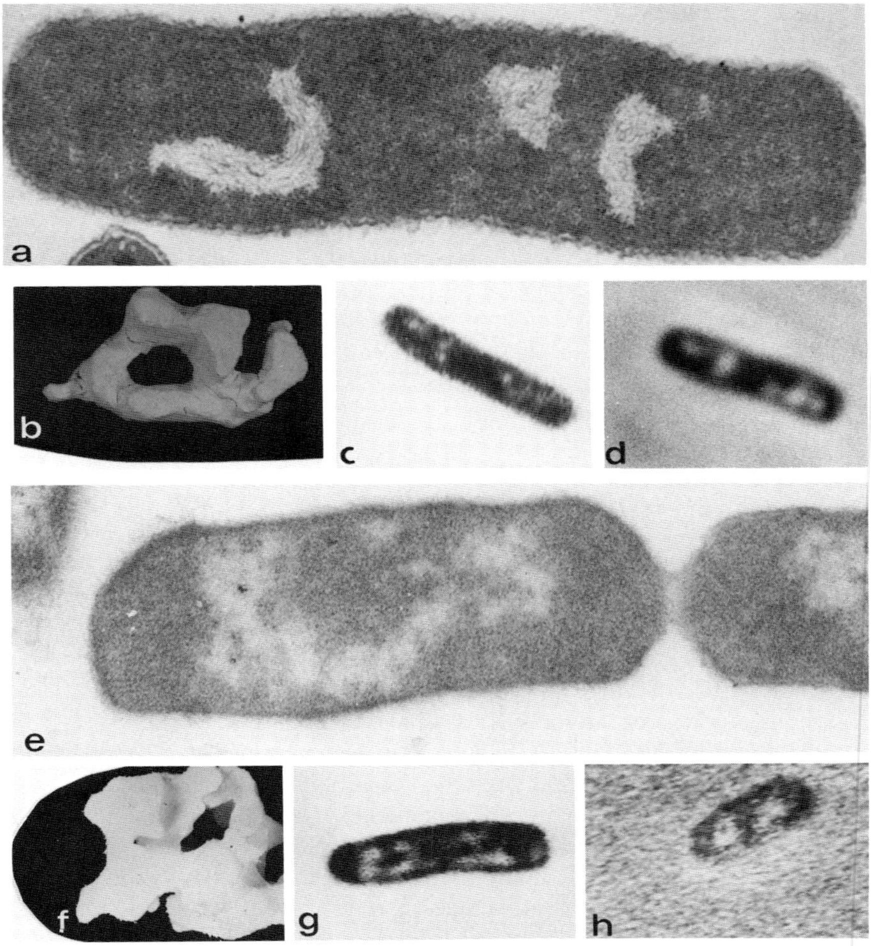

Fig. 5. Comparison of CSLM micrographs with electron microscope reconstructions of fast growing *E. coli* cells. c, d, g, CSLM micrographs of an OsO_4-fixed, a living, and a glutaraldehyde-fixed cell, respectively. b, f, Reconstructions of OsO_4- or glutaraldehyde-fixed cells based on serial sections (a, e). d, Phase contrast microscopy of OsO_4-fixed cell.

of the nucleoid as a whole to be the same in both living, OsO_4-, and glutaraldehyde-fixed cells.

Experiments like the one described above (see also Valkenburg *et al.*, 1985) demonstrate the high resolution and reliability of image formation in a CSLM in the transmission mode. Although resolution is higher in the electron microscope, a drawback is that only fixed, dehydrated, embedded material can be visualized. It is difficult to assess the artifacts introduced by these methods. The comparison of CSLM images of living with fixed cells (and with electron microscopic reconstructions) showed nucleoid shape to be unaffected by fixation. In electron microscopy such a comparison is impossible. Differences in nucleoid ultrastructure were shown to exist between sections of OsO_4- and glutaraldehyde-fixed cells.

The reliability of the results obtained with either fixative can in no way be affirmed by comparison with living cells. The advantage of the CSLM over the electron microscope is the possibility to study living hydrated cells—even more so as *E. coli* has been shown to grow and divide under the conditions present in the CSLM (Valkenburg, 1985) enabling the study of cells during the cell cycle as well.

CHARACTERIZATION OF CELL SHAPE

A great deal of effort has been expended to ascertain the mode of growth of individual cells in bacterial populations. Such knowledge is essential to answer fundamental questions on the regulation of growth and division of cells in general (for reviews, see Lloyd *et al.*, 1982; Nanninga and Woldringh, 1985).

In most studies bacteria like *E. coli* have been chosen because they can be easily cultured under steady-state conditions, which is a prerequisite for reproducible results. Although the collection of sufficient data is very time-consuming, electron microscopic techniques have been preferred over the use of electronic particle analyzers (Coulter counter, flowcytometer) because of the information which can be obtained on the state of division of individual cells. However, in spite of many detailed measurements and statistical analyses (Zaritsky *et al.*, 1982; Vicente, 1984; Trueba *et al.*, 1983; Koch, 1983; Kubitschek and Woldringh, 1983; Helmstetter *et al.*, 1979; Burdett and Kirkwood, 1983), no agreement could be obtained on the exact pattern of growth. This disagreement must be ascribed to unknown distortions in cell size and shape occurring during the dehydration of cells prepared for electron microscopy (Woldringh *et al.*, 1977).

The CSLM for the first time enables the study of hydrated cells with high-resolution digital imaging. The following section describes some techniques and makes a comparison with electron microscopic measurements.

Sample Preparation

E. coli B/rA (obtained from C. E. Helmstetter) were grown at 37°C in 1% tryptone, 0.5% yeast extract, and 0.5% NaCl. Cells were fixed by adding OsO_4 to the medium to a final concentration of 0.1%. For transmission electron microscopy, cells were prepared by the method of agar filtration as described previously (Woldringh *et al.*, 1977). For the CSLM, cells from the sample described above were washed and suspended in 1 M sodium carbonate buffer (pH 8.0). Cells were stained with the fluorochrome FITC. A fresh solution of FITC was prepared by solving 10 mg FITC in 1 ml dimethyl sulfoxide (DMSO). After centrifugation 10 μl of this solution was added to 1 ml cell suspension. After staining for 1 hr, the cells were washed with buffer. A preparation for the CSLM was made as described above. FITC reacts primarily with protein. About 50% of the dry weight of fast growing B/r cells consists of protein (Churchward and Bremer, 1982). Protein concentration in the DNA-containing area (nucleoid) is lower than in the cytoplasm (Valkenburg and Woldringh, 1984). However, nucleoid volume in fast growing cells is small compared to cytoplasmic volume and therefore cells are intensively and nearly homogeneously stained with FITC.

Data Acquisition

In order to carry out an automatic analysis on images successfully and fast, it is of prime importance to obtain a signal to noise ratio as high as possible. Also important are image fidelity and, as mentioned before, constant optical characteristics over the entire image. When one intends to analyze a cell population statistically, it is also wise to have as little as possible selection bias in the acquisition phase of the process. For this reason we used the CSLM in the fluorescence mode, in combination with a relatively large detector pinhole (backprojected in the confocal plane, 1 micron). Working in the fluorescence mode ensures that the chance for a given bacterium to be analyzed is independent of cell size and shape: The suppression of light from different focus levels causes the cells to be seen as bright objects against a dark background. The use of a pinhole of the size mentioned leads to the best possible signal to noise ratio, which is further increased by the averaging technique as described.

Sampling distance was chosen five to six times shorter (32 nm) than optical resolution distance to avoid the need for interpolating between the pixels. All images were digitized with 8-bit accuracy and stored permanently for later off-line processing.

Analysis of Confocal Images

After retrieval of the data from tape or disk the images were processed with a 5 × 5 Gaussian filter to remove some of the image noise, after which the largest measured intensities (pixel values) were normalized to 255.

Segmentation and contouring of the image were done at the same time. Starting at the bottom left, the image was scanned for a string of 10 pixels with value above the threshold. The start of such a string was considered a starting point for a contour follower, which encodes the contour in a Freeman chain code (Freeman, 1961), and an array of coordinates specifying each contour pixel. The area enclosed by this contour and the already searched-through parts of the image were then marked as forbidden to the contour follower as well as for searching for a starting point of a new contour. Contours with an area below a certain threshold (500 pixels) were considered dirt particles. The contour encoder was programmed in such a way that the difference between adjacent elements of the code was always smaller or equal to two, thus allowing the contour to take 90° turns at the most. From this chain code, contour length, enclosed surface, and bending energy (Bowie and Young, 1977) are easily calculated. Cell length was found by searching in the coordinate array for the pixel pair with the greatest mutual distance (Fig. 6a). In a similar way (see Fig. 6b) cell width, length, and width at a possible constriction site were found. From the quotient of the average width and constriction width, the constriction coefficient is defined.

We estimate that the relative accuracy of the present program is between 1.5 and 3% for cell length and width, corresponding with 30–60 nm. Further improvement will therefore be only necessary and meaningful when working with populations larger than 500 cells, and aiming at absolute measurements.

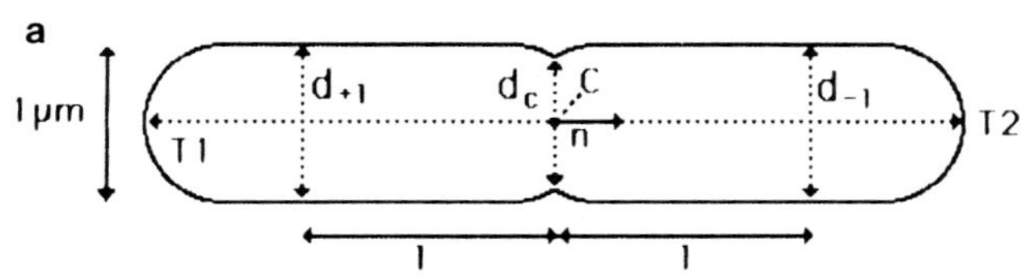

Fig. 6a. Geometric parameterization procedure. On entry of this procedure, the contour follower (see text) has already been applied to the image, and the resulting array of coordinates (A_k, K : O, K_{max}) is passed to the parameterization procedure. The first step is to determine the main axes of the bacterium by searching for the pixel pair on the contour with the greatest mutual distance ($|\mathbf{T}_1 - \mathbf{T}_2|$ with array indices k_{T1}, respectively, k_{T2}), then the center point $\boldsymbol{C}$ and a normalized orientation vector $\mathbf{n}$ are determined.

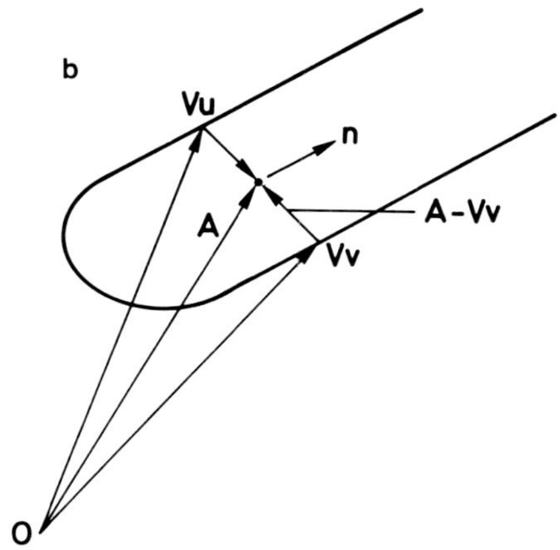

Fig. 6b. To assess the width of the bacterium, a given point [represented by the vector $A(1) = 1\mathbf{n}+C$, 1 an integer] on the line $\mathbf{T}_1\mathbf{T}_2$ at distance 1 from the center, the following procedure is executed: $\mathbf{T}_1$ and $\mathbf{T}_2$ divide the contour in two; each of these halves is searched for the pixel (represented by the vectors V_u and V_v) for which the absolute value of the inner product $\|[(A - V_u),\mathbf{n}]\|$, respectively, $\|[(A - V_v),\mathbf{n}]\|$ is minimal (u and v are integers values in the range $k_{T1} \ldots (k_{T1} + K_{max}/2 - 1) \bmod K_{max}$, respectively, $k_{T2} \ldots (k_{T2} + K_{max}/2 - 1) \bmod K_{max}$. We can now define cell width at $A(1)$ as the distance $d(1) = \min(|V_u - V_{v+n}|)$, $\mathbf{n}: -N \ldots N$, and N a small value corresponding to two times the optical resolution. Average bacterium width can be found as the average of $d(+1)$ and $d(-1)$ with $1 = 1/3\ (|\mathbf{T}_1 - \mathbf{T}_2| - P)$, P a number corresponding with the largest possible size of the polar region (1 μm). Cell width at a possible constriction site can be found as $d_c = \min\ (d_1)$ with $|1|1/10(|\mathbf{T}_1 - \mathbf{T}_2| - P)$.

Analysis of Electron Microscope Images

Electron micrographs of *E. coli* B/rA cells were projected on an electronic tablet digitizer (Summagraphics, Fairfield, CT), connected to a calculator (Hewlett Packard 9825). Length of the cells was measured at a magnification of 10,000–15,000 as reported elsewhere (Trueba and Woldringh, 1980). For comparison with CSLM length distributions, the length distributions were normalized by dividing all measurements by the sample mean.

Comparison of Results

Figure 7 plots a cumulative histogram of cell length distributions from electron micrographs and CSLM images. According to the Kolmogorov–Smirnov test, the difference between these distributions was not significant. A comparison was also made between the constriction coefficient and a subjective constriction assessment (Table I). The result clearly

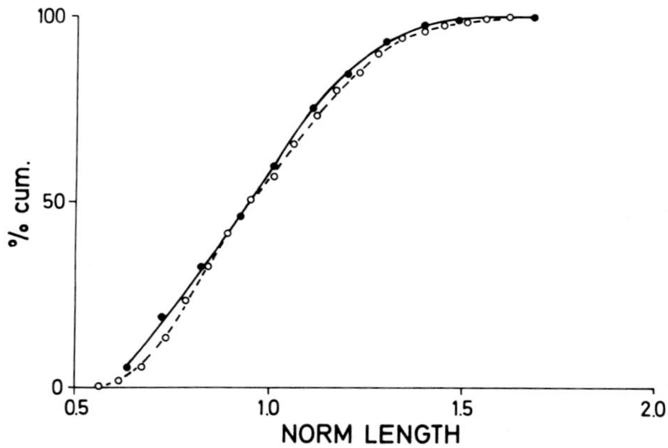

Fig. 7. Comparison of cumulative length distributions of the same culture of *E. coli* B/r obtained with the electron microscope (dotted line) and the CSLM (continuous line). The histograms were normalized by dividing all values by the sample mean. At the 5% confidence limit, no significant difference was observed between the distributions with the Kolmogorov-Smirnov test.

TABLE I

Constriction Coefficient (CC) versus Subjective Constriction Classes[a]

CC	0	1	2	3
91–95	2			
96–100	61			
101–105	2	6		
102–110		7		
115		3		
120		1	1	
125		2	5	
130			2	1
135			1	2
140			1	1
145			2	
146–200				2

[a] Samples were divided into four classes by three trained observers. Class 0, 1, 2, 3: no, slight, moderate, and deep constriction. The CC for unconstricted cells (class 0) was in the range 0.95 to 1.04 for all cells, which is equivalent to ± 1 pixel distance (32 nm). Also all constricted cells showed a CC of 100.

shows the reliability of the method. We consider the constriction coefficient a very useful instrument for the classification of individual cells or, in combination with other simple parameters like length and surface, of bacterial strains.

3-D IMAGING TECHNIQUES WITH CSLM

This section presents a comparison of conventional fluorescence microscopic images and images constructed by the confocal microscope. We have chosen two different cases: one at high and one at low magnification. In the former case, the high resolution was used to reveal the mitochondrial structure of the yeast *Saccharomyces cerevisiae;* in the latter case the chloroplast of the freshwater alga *Netrium digitus* was visualized. The use of a high numerical aperture objective (NA = 1.32, 100×) at low magnification (as is possible in this scanning microscope) allows a much better z resolution than can be realized by any standard low NA, low-magnification microscope objective. As is already explained, this z resolution is further increased by the confocal approach. The use of a high NA objective at low magnifications implies that it is possible to generate, for instance, a 2000^2 pixel image without any oversampling.

When operating the CSLM in fluorescence, the resolution along the optical axis (z axis) is three to four times the resolution in a direction transverse to the optical axis (x–y plane), as stated in the first section. The dimensions of the sampled volume represented by each pixel in the 3-D array are adapted to this resolution difference in x–y and z directions. To generate stereoscopic views we employ a very simple algorithm. We place the observer at infinity so we can use a parallel projection on the space represented by the 3-D array (Fig. 8). In practice the algorithm as mentioned in Fig. 8 is carried out by first calculating the projections of the x–y planes separately with a linear "mapper" which resides in the image processing library. This routine maps each pixel of the source buffer to a target buffer using a 2 × 2 matrix and a translation vector. If necessary, linear interpolation between neighboring pixels is used. Memory demands are reduced by the property of the algorithm so that it is possible to compare all projections pairwise. Moreover, comparing only two pixels in order to find the largest or smallest is a simple task for any processor and therefore results in fast execution times.

Visualization of Mitochondria

Saccharomyces cerevisiae cells were grown in a yeast pepton medium for at least ten generations. Steady-state conditions were checked by

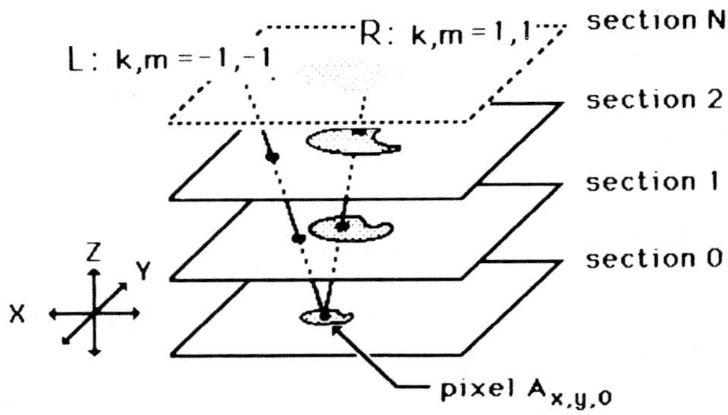

Fig. 8. The algorithm for the generation of stereo pairs assigns the maximum value of the pixels encountered along the "viewing" lines R and L to the corresponding pixels in the stereoscopic pair: a pixel $L_{x,y}$ (x,y: 1,0 . . . 255) will be the maximum of the pixels $A_{x+kz,y+mz,z}$ (z: 0 . . . $N-1$, N is the number of optical sections) where $A_{x,y,z}$ denotes a pixel of optical solution z with coordinates x, y. The variables k, m determine the viewing angle and need to be integers; see text.

monitoring the optical density of the culture. In order to stain them, cells were harvested and resuspended in 200 m*M* Tris-HCl buffer (pH = 8.0). The fluorescent dye 2-[4-(dimethylamino)styryl]-1-methylpyridinium iodide (DASPMI), which is known to be a vital stain of mitochondria (Bereiter-Hahn, 1976; Bereiter-Hahn *et al.,* 1983), was added in a final concentration of 10 μM. Exitation of the dye was performed with a krypton ion laser, tuned to a wavelength of 476 nm. Fast growing yeast cells appeared to have one or a few branched mitochondrial networks, rather than a large number of distinct ones (Fig. 9a and 9b). This observation is in accordance with results obtained through analysis of serial sections with the electron microscope (Hoffmann and Avers, 1973; Stevens, 1981).

Visualization of Chloroplasts

Green algae were collected from water samples from a fen. Because of the strong fluorescence of chlorophyll no dyes were added. The distinctive 3-D shape of *N. digitus*' chloroplast is poorly recognized in ordinary micrographs (Fig. 10a) of this organism due to the fact that it is much thicker than the depth of field of the microscope objective. The shallow depth of field of the CSLM, combined with the characteristics of the algorithm as explained in Fig. 8, results in an image which, paradoxically, presents a very large depth of field.

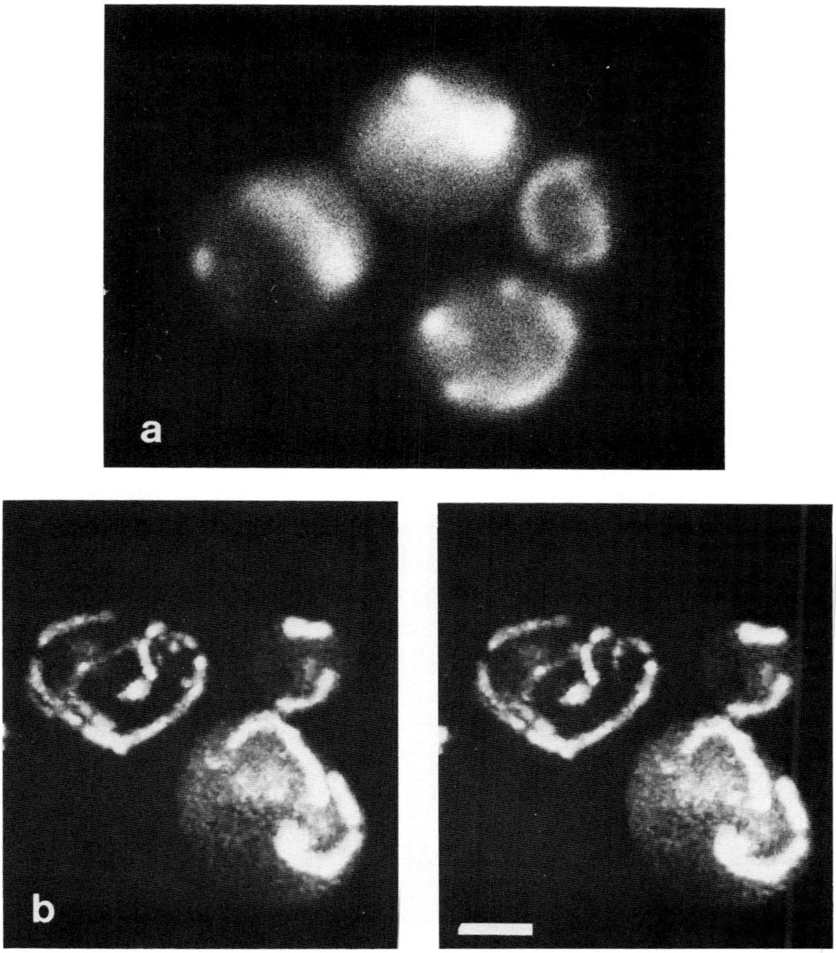

Fig. 9a. Cells of *S. cerevisiae* stained with the fluorescent dye DASPMI in order to reveal the structure of the mitochondrial network. Conventional fluorescence microscope with 100×/1.32 oil immersion objective.

Fig. 9b. Stereo picture of the same kind of cells as in Fig. 9a, visualized by confocal scanning laser microscopy (bar = 2 μm).

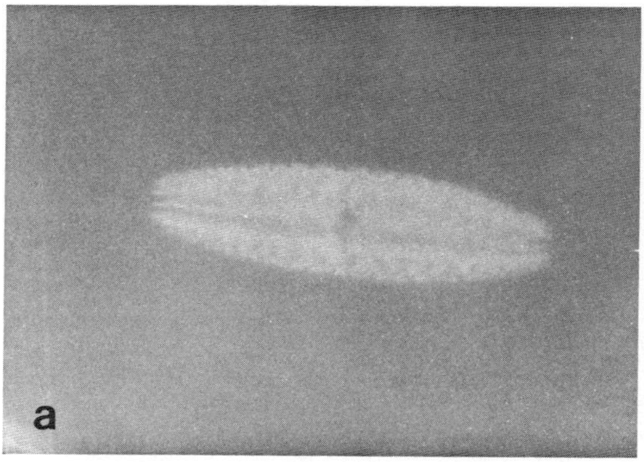

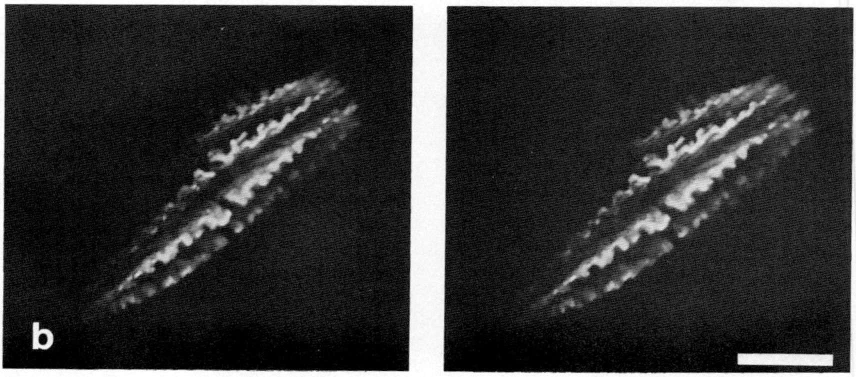

Fig. 10a. Low magnification fluorescent image of the chloroplast of *Netrium* as seen through a conventional fluorescence microscope with 16×/0.45 objective.

Fig. 10b. The same organism as in Fig. 10a visualized by the CSLM (bar = 50 μm).

CONCLUSION AND OUTLOOK

In this chapter we have shown some areas to which the confocal technique can be applied successfully. Without the application of any image processing techniques, the improved resolution can already reveal new information about biological structures. Through the integration of an image processing and analyzing system into the CSLM, some properties are enhanced while others are added. Large dynamic range images can be compressed in order to enhance details on both light and dark areas; 3-D image reconstruction is hardly imaginable without image processing.

The area of biological shape analysis with digital computers is quite old and is dominated by (2-D) video systems mounted on conventional microscopes. With the resolution and reliability of confocal imaging, especially in the fluorescence mode, smaller organisms can now be analyzed. In the future we will apply shape and texture analysis also on 3-D images.

The possibility to visualize fluorescently labeled 3-D structures is perhaps the most valuable property of the CSLM. In view of the potentialities for further development of this technique, both on the instrumentation side (solids modeling, time sequences) and specimen preparation side (double labeling), we expect confocal techniques to become very useful in a wide field of applications.

We are indebted to F. H. A. Aalst, O. Peters, J. A. W. Kalwij, and W. Takkenberg for their assistance in the development of the CSLM. We would like to thank J. Sikking and J. H. D. Leutscher for producing part of the figures, E. Pas for measuring the electron micrographs, and A. C. Ellis for providing the algae. This development was supported in part by the Stichting voor Technische Wetenschappen (STW) and by the Foundation for Fundamental Biological research (BION), which are subsidized by the Netherlands Organization for the Advancement of Pure Research (ZWO).

REFERENCES

Bereiter-Hahn, J. (1976). Dimethylaminostyrylmethylpiridiniumiodide (DASPMI) as a fluorescent probe for mitochondria *in situ*. *Biochim. Biophys. Acta* **423,** 1.

Bereiter-Hahn, J., Seipel, K. H., Vöth, M., and Ploem, J. S. (1983). Fluorimetry of mitochondria in cells vitally stained with DASPMI or rhodamine 6 GO. *Cell Biochem. Funct.* **1,** 147.

Binnerts, J. S., Woldringh, C. L., and Brakenhoff, G. J. (1982). Visualisation of the nucleoid in living bacteria on poly-lysine coated surfaces by the immersion technique. *J. Microsc.* (*Oxford*) **125,** 359.

Bowie, J. E., and Young, I. T. (1977). An analysis technique for biological shape. II. *Acta Cytol.* **3,** 455.

Brakenhoff, G. J., Blom, P., and Barends, P. J. (1979). Confocal scanning light microscopy with high aperture lenses. *J. Microsc.* (*Oxford*) **117,** 219.

Brakenhoff, G. J., Binnerts, J., and Woldringh, C. L. (1980). Developments in high resolution confocal scanning light microscopy (CSLM). *In* "Scanned Image Microscopy" (E. A. Ash, ed.), p. 183. Academic Press, London.

Burdett, I. D. J., and Kirkwood, T. B. L. (1983). How does a bacterium grow during the cell cycle. *J. Theor. Biol.* **103,** 11.

Churchward, G., and Bremer, H. (1982). Macromolecular composition of bacteria. *J. Theor. Biol.* **94,** 651.

Cox, I. J., Sheppard, C. J. R., and Wilson, T. (1982). Super resolution by confocal fluorescent microscopy. *Optik* **60,** 391.

Freeman, H. (1961). On the encoding of arbitrary geometric configurations. *IRE Trans. Electron. Comput.* **10,** 260.

Helmstetter, C. E., and Cooper, S. (1968). DNA synthesis during the division cycle of rapidly growing *Escherichia coli* B/r. *J. Mol. Biol.* **31,** 507.

Helmstetter, C. E., Pierucci, O., Weinberger, M., Holmes, H., and Tang, C. J. (1979). Control of cell division in *Escherichia coli. In* "The Bacteria" (J. R. Sokatch and L. N. Ornston, eds.), Vol. 7, p. 517. Academic Press, New York.

Hoffmann, H. P., and Avers, C. J. (1973). Mitochondrion of yeast: Ultrastructural evidence for one giant, branched organelle per cell. *Science* **181,** 749.

Kellenberger, E., Ryter, A., and Séchaud, J. (1958). Electron microscope study of DNA containing plasms. II. Vegetative and mature phage DNA as compared with normal bacterial nucleoid in different physiological states. *J. Biophys. Biochem. Cytol.* **4,** 671.

Koch, A. L. (1983). The surface stress theory of microbial moprhogenesis. *Adv. Microb. Physiol.* **24,** 301.

Kubitschek, H. E., and Woldringh, C. L. (1983). Cell elongation and division probability during the *Escherichia coli* growth cycle. *J. Bacteriol.* **153,** 1379.

Lloyd, D., Poole, R. K., and Edwards, S. W. (1982). "The Cell Division Cycle." Academic Press, London.

Mason, D. J., and Powelson, D. M. (1956). Nuclear division as observed in live bacteria by a new technique. *J. Bacteriol.* **71,** 474.

Nanninga, N., and Woldringh, C. L. (1981). The interpretation of chemically fixed and freeze-fractured bacterial nucleoplasm. *Acta Histochem., Suppl.* **23,** 39–53.

Nanninga, N., and Woldringh, C. L. (1985). Cell growth, genome duplication and cell division *In* "Molecular Cytology of Escherichia Coli" (N. Nanninga, ed.), p. 260. Academic Press, London.

Reynolds, E. S. (1963). The use of lead citrate at high pH as an electron opaque stain in electron microscopy. *J. Cell Biol.* **17,** 208.

Séchaud, J., and Kellenberger, E. (1972). Electron microscopy of DNA containing plasms. IV. Glutaraldehyde-uranyl fixation of virus infected bacteria for thin sectioning. *J. Ultrastruct. Res.* **39,** 598.

Sheppard, C. J. R., and Choudhury, A. (1977). Image formation in the scanning microscope. *Opt. Acta* **24,** 1051.

Stevens, B. (1981). Mitochondrial structure. *In* "The Molecular Biology of the Yeast *Saccharomyces cerevisiae:* Life Cycle and Inheritance" (J. N. Strathern, E. W. Jones, and J. R. Broach, eds.), pp. 471–504. Cold Spring Harbor Lab., Cold Spring Harbor, New York.

Trueba, F. J., and Woldringh, C. L. (1980). Changes in cell diameter during the division cycle of *Escherichia coli. J. Bacteriol.* **142,** 869.

Trueba, F. J., Neijssel, O. M., and Woldringh, C. L. (1983). Generality of growth kinetics of the average individual cell in different bacterial populations. *J. Bacteriol.* **150,** 1048.

Tsutsui, N., Kumon, H., Ichihawa, H., and Tawara, J. (1976). Preparative method for suspended biological materials for S.E.M. by using a polycationic substance layer. *J. Electron Microsc.* **25,** 163.

Valkenburg, J. A. C. (1985). The nucleoid of *Escherichia coli* as visualised by confocal scanning light microscopy, phase contrast microscopy and electron microscopy. Ph.D. Thesis, University of Amsterdam.

Valkenburg, J. A. C., and Woldringh, C. L. (1984). Phase separation between nucleoid and cytoplasm in *Escherichia coli* as defined by immersive refractometry. *J. Bacteriol.* **160,** 1151.

Valkenburg, J. A. C., Woldringh, C. L., Brakenhoff, G. J., and Nanninga, N. (1985). Confocal scanning light microscopy of the *Escherichia coli* nucleoid: Comparison with phase contrast and electron microscope images. *J. Bacteriol.* **161,** 478.

van der Voort, H. T. M., Brakenhoff, G. J., Valkenburg, J. A. C., and Nanninga, N. (1985). Design and use of a computer-controlled confocal microscope for biological applications. *Scanning* **7,** 66.

Vicente, M. (1984). The control of cell division in bacteria. *In* "The Microbial Cell Cycle" (E. Streiblova and P. Nurse, eds.), p. 29. CRC Press, Boca Raton, Florida.

Wijnaends van Resandt, R. W., Marsman, H. J. B., Kaplan, R., Davoust, J., Stelzes, E. H. K., and Sticker, R. (1985). Optical fluorescence microscopy in 3 dimensions. *J. Microsc. (Oxford)* **138,** 29.

Woldringh, C. L. (1973). Effect of cations on the organization of the nucleoplasm in *Escherichia coli* prefixed with osmium tetroxide or glutaraldehyde. *Cytobiologie* **8,** 97.

Woldringh, C. L., and Nanninga, N. (1985). Structure of nucleoid and cytoplasm in the intact cell. *In* "Molecular Cytology of Escherichia Coli" (N. Nanninga, ed.), p. 161. Academic Press, London.

Woldringh, C. L., de Jong, M. A., van den Berg, W., and Koppes, L. J. H. (1977). Morphological analysis of the division cycle of two *Escherichia coli* substrains during slow growth. *J. Bacteriol.* **131,** 270.

Zaritsky, A., Woldringh, C. L., Grover, N. B., Naaman, J., and Rosenberger, R. F. (1982). Growth and form in bacteria. *Comments Mol. Cell. Biophys.* **1,** 237.

4

Combined Light and Electron Microscopy of Central Nervous System Neurons and Their Afferent and Efferent Synaptic Connections

F. G. WOUTERLOOD

Department of Anatomy
Vrije Universiteit
Amsterdam, The Netherlands

INTRODUCTION

Correlative light and electron microscopy is a tool in neuroanatomy for the tracing of patterns of synaptic connectivity and for the detailed immunocytochemical visualization of neurons or parts of neurons which contain particular antigenic substances like structural proteins, neuroactive substances, or their synthesizing enzymes. The electron microscope can be used to investigate the morphological details of the neurons and their axon terminals, and to examine whether the processes of two identified neurons, which light microscopically appear closely apposed, indeed form synaptic contacts. In this chapter special attention will be given to combinations of light microscopic (LM) neuroanatomical tracing and immunocytochemistry, followed by electron microscopy (EM) of the same preparations.

Observations in Golgi-impregnated material have led to the distinction of many morphological classes of neurons (Ramón y Cajal, 1972). It is generally thought that the function of a neuron, in one way or another, is correlated with its morphology, and, conversely, that neurons of a similar morphology perform a common function. With most current neuroanatomical tracing techniques, label is delivered to and presumably taken up by all the cells located in and around the site of deposition. Since no region of the brain contains cells of only one neuronal class, it is reasonable to expect that always neurons of several types take up label. Studies by means of LM neuroanatomical techniques, therefore, are always carried out at the level of *populations* of neurons of mixed morphological classes. Thus, with the exception of particular double-labeling techniques and transmitter-specific tracing, the existing LM neuroanatomical techniques fail to provide answers to questions like, do all the neurons in area A that receive synaptic input from area B belong to one class? Are these target neurons the same neurons that project to area C? Does the synaptic input from area B arriving in area A converge on neurons which receive also synaptic input from area D? Which are the interneurons in area A and how is input channeled by these interneurons to the projection neurons? Which neurotransmitters are used by the various members of these microcircuits? Questions like these can only be answered using combinations of anatomical, immunohistochemical, and intracellular neurophysiological techniques, and EM.

The electron microscope is indispensable to visualize the sites of synaptic contact. However, it provides a two-dimensional image in which only small portions of neurons called "profiles" are observed. For an unequivocal identification of a synaptic connection between processes of two, previously light microscopically identified neurons, it is necessary to mark both the presynaptic and postsynaptic neurons with light and electron-opaque labels. The present chapter describes techniques aimed at this goal. For the sake of clarity, the morphology (structure)-oriented techniques will be treated separately from the immunology (function)-oriented techniques.

Apart from these techniques, an entirely new arsenal of techniques is rapidly emerging in which the morphological tracing technique based on uptake, orthograde transport, and visualization of the plant lectin *Phaseolus vulgaris*-leucoagglutinin (PHA-L; Gerfen and Sawchenko, 1984) is combined with other neuroanatomical tracing techniques and with EM. There are two major arguments to deal with the correlative PHA-L–LM-EM tracing in a separate section of this chapter. First, after PHA-L injection the perikarya at the injection site, their dendrites and efferent fibers, including their *en passant* and *terminal* varicosities are labeled in a Golgi-like fashion. This offers the opportunity to study in the light microscope the projections of neurons of which the morphology is already known. Second, since PHA-L tracing is based on an immunohistochemical detection system it can be combined with transmitter immunocytochemistry of these neurons, their projections (Gerfen and Sawchenko, 1985), and their target neurons (Wouterlood *et al.*, 1986a, 1987a). Since the PHA-L tracing technique can also be carried out at the EM level (Wouterlood and Groenewegen, 1985), it must be regarded as a method which is particularly suited to solve the questions concerning the synaptic connectivity of identified neurons as outlined above. Inevitably, the limited space of this chapter does not allow to review in detail all the possible combinations of morphological and immunological techniques. I have tried, wherever possible, to avoid duplication of protocols by cross-referencing within this chapter. The aspects of LM and EM visualization of intracellularly injected neurons, following intracellular neurophysiology, is dealt with in this volume in Chapter 5.

MORPHOLOGICAL TECHNIQUES: CORRELATIVE LIGHT AND ELECTRON MICROSCOPY OF GOLGI-IMPREGNATED NEURONS

Light microscopic neuroanatomical techniques are employed at the level of *populations* of neurons of mixed classes. This results in a description of anatomical connections which, with the exception of neurotransmitter-specific tracing and combined tracing–immunocytochemistry (see section on Immunocytochemical Techniques), offers little information concerning the morphology and the functional aspects of the individual neurons involved in these connections. From Golgi studies it is well known that most areas of the central nervous system (CNS) are composed of various classes of projection neurons and local circuit neurons (Ramón y Cajal, 1972). With immunocytochemical techniques it has been demonstrated that these brain areas are by no means homogeneous as regards neurotransmitter contents (see Björklund and Hökfelt, 1983-1985). Furthermore, it is becoming increasingly clear that anatomical connections between brain areas comprise fibers from neurons belonging to different chemical types. For instance, GABAergic as well as cholinergic projections have been observed in the fornix of the rat, transmitting information from neurons in the septal area to the hippocampus and entorhinal cortex (Alonso and Köhler, 1984; Köhler *et al.*, 1984).

Electron microscopic neuroanatomy operates at the level of the single cell, and the use of the electron microscope is primarily justified by its power to visualize synaptic junctions. The purpose of correlative LM–EM of Golgi-impregnated neurons, combined with neuroanatomical tracing techniques, is to classify presumed target neurons by their LM Golgi image. Subsequently, the electron-dense Golgi precipitate serves as a label to identify positively in the electron microscope the profiles belonging to the light microscopically identified neurons. The tracing technique serves to label entire presynaptic neurons or the axon terminals of the afferent fibers with an electron-dense label (which must be distinguishable from the Golgi precipitate). By subsequent EM it is possible to resolve the synaptic junctions. In this way, information is gained about the afferent synaptic connectivity of individual, light microscopically identified neurons. By means of retrograde tracing combined with Golgi EM, the synaptology can be examined of light microscopically identified neurons of which also the projection areas of their axons are known.

To successfully achieve correlative Golgi LM–EM in one preparation, the following conditions have to be met: (1) An electron-dense label must be attached to a selected presynaptic neuron, (2) the target neuron must be provided with a light and electron-opaque label, introduced by Golgi impregnation, (3) if information is desired about the connectivity of the target neuron, a retrogradely transported label must be introduced in the area to which the target neuron is supposed to project, and (4) the techniques by which the labels are applied must be compatible with the fixation and embedding procedures used for EM. At first sight, condition 4 may seem trivial, but it usually poses the most challenging technical problems. It is noteworthy that more than 20 years ago Blackstad (1965) published a pioneer study combining anterograde degeneration, Golgi impregnation, and EM. Blackstad fulfilled for the first time three of the aforementioned conditions in one preparation. Somogyi *et al.* (1979) were the first who succeeded in satisfying all four conditions in one experiment.

Combination with the Anterograde Degeneration Technique

The technique of anterograde degeneration is the most widely used method in combination with Golgi EM. This technique has firm roots in the experimental neuroanatomical tradition. Following an experimentally induced lesion of a nerve fiber tract or the cell bodies from which the tract arises, the fibers distal to the lesion undergo a chemical and physical breakdown into fragments. Lesions can be made either electrolytically, chemically, or surgically (for a survey, see Moore, 1981). The degenerating axons and terminals are engulfed by astroglial processes, and are then ingested and metabolically destroyed (Wiśniewski *et al.*, 1972; Wouterlood *et al.*, 1982).

In the electron microscope, the axon terminals belonging to the interrupted fibers can be distinguished from normal axon terminals by the ultrastructural changes accompanying the degeneration process. Three types of terminal degeneration have been described: neurofilamentous hypertrophy (Gray and Hamlyn, 1962), electron-dense degeneration (Colonnier, 1964), and electron-lucent degeneration (Westrum, 1973). Some authors suggest that the neurofilamentous hypertrophy is only a transient stage in the degeneration process, preceding the electron-dense stage (Mugnaini and Friedrich, 1981). During the process of degeneration, axon terminals can be recognized only during a limited time, as *terminals* forming intact synaptic junctions with postsynaptic neurons. It may occur that only the degenerating presynaptic boutons are ingested by reactive glia, leaving vacant postsynaptic membrane specializations (Westrum, 1969). Often, reactive glia appears to ingest not only the degenerating axon terminals but also the dendritic spines to which they are attached (Pinching, 1969; Wiśniewski *et al.*, 1972; Gentschev and Sotelo, 1973; Wouterlood *et al.*, 1982). In addition, trans-synaptic degeneration may occur, for instance in the olfactory tubercle of the rat after lesions of the olfactory bulb (Carlsen *et al.*, 1982).

The axons distal to the lesion, including the synaptic terminals, degenerate according to species- and system-dependent patterns and time courses. In poikilothermous vertebrates, the degeneration of the axons and terminals follows a time course markedly different from that in mammals; for example, in snakes, turtles, and lizards, survival periods after which degenerating synaptic terminals are observed, range between 3 and 93 days (Repérant and Rio, 1976; Smith *et al.*, 1980; Wouterlood *et al.*, 1982). In crocodiles, degenerating optic nerve fibers may still be present 20 months following enucleation (Kruger and Maxwell, 1969). These examples serve to emphasize the need, before any combination of anterograde degeneration and Golgi EM is attempted, to carry out a series of pilot experiments in order to determine the postsurgical survival period optimal for that particular system.

The pronouncedly increased osmiophilia occurring in electron-dense degeneration, first of the matrix of the synaptic terminals, and subsequently of the synaptic vesicles, provides a label which is excellent compared to the rather subtle changes accompanying the neurofilamentous hypertrophy and the electron-lucent degeneration. In Golgi EM preparations, suboptimal preservation of the ultrastructure is to be expected, caused by the deteriorating effect of the silver nitrate. In this and other combined procedures, the preservation of the neurofilaments and the protein matrix of the synaptic terminals is often inadequate to unequivocally distinguish normal terminals from axon terminals undergoing neurofilamentous hypertrophy or electron-lucent degeneration. This is the reason why only electron-dense degeneration has been used to mark specific presynaptic axon terminals in experiments combined with Golgi impregnation or immunocytochemistry (see the section Immunocytochemical Techniques). One particular disadvantage of combined anterograde degeneration–Golgi EM is that degenerating axon terminals, although visible in 0.5-μm-thick semithin sections (Holländer and Vaaland, 1968), cannot be observed in LM Golgi sections. Another disadvantage is that the Golgi-impregnation technique fails to label myelinated axons (Peters and Proskauer, 1980; Wouterlood and Mugnaini, 1984). This is a serious handicap in the study of chains of neurons.

Procedure

1. Surgery. A population of cells of origin of a fiber projection is destroyed, or a fiber bundle is interrupted. Lesions can be made by various means such as by suction, knife, cautery, ablation, or by the injection of chemical substances.
2. Postsurgical recovery and survival: species- and system-dependent.
3. Fixation. Standard EM fixation. It is important to use only perfectly fixed brain material for further processing, since the Golgi impregnation usually leads to loss of quality of ultrastructural details. For rats, the perfusion fixation procedure according to Friedrich and Mugnaini (1981) is recommended.
4. Golgi impregnation, either on blocks or whole brains. Commonly, a Golgi-rapid technique (Blackstad, 1965; Somogyi, 1978) or a Golgi–Kopsch technique (Wouterlood and Mugnaini, 1984; Wouterlood *et al.*, 1986b) is used.
5. Slicing of the blocks or brains into sections of approximately 100 μm thickness. The slices are treated chemically (Fairén *et al.*, 1977); Wouterlood *et al.*, 1983, 1987c) to produce a light and electron-dense deposit in the impregnated neurons, with better ultrasectioning properties than the original Golgi deposit.
6. Postfixation with osmium tetroxide and embedding in epoxy resin. After infiltrating the slices with epoxy resin they are placed with a drop of fresh resin between two strips of polyethylene plastic (cut from an ordinary trashbag). This sandwich is placed between two glass slides. On the upper slide a small weight is placed to obtain a perfectly flat polymerized section. After curing, the polyethylene strips are removed, the section is fastened with two small pieces of adhesive tape to a fresh slide, and stored in a slide box.
7. Correlative LM–EM: Sections prepared in this way can be studied in the light microscope in the same way as conventionally prepared sections (Fig. 1a). Impregnated neurons of interest are drawn under glycerin immersion using a drawing attachment, photographed in a through-focus

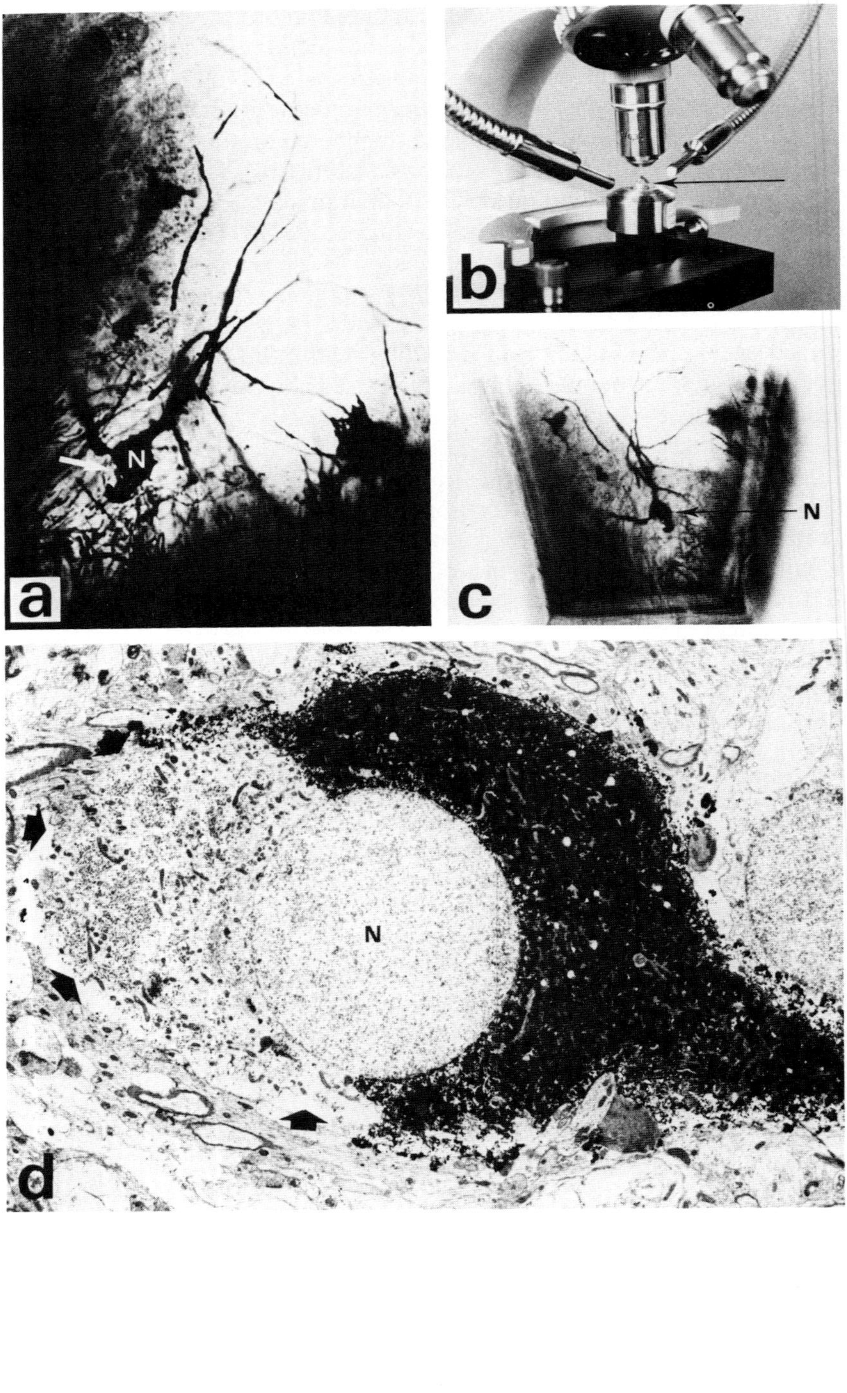
a
N
b
c
N
d
N

series, cut out of the section with a fine scalpel, and placed on a small drop of unpolymerized epoxy resin on top of a precured epon block. After 24 hr of curing at 60°, the block is mounted on the chuck of an ultramicrotome and serially sectioned. The first sections are used to inspect the quality of the ultrastructure and the presence of terminal degeneration. To accurately monitor and document the progress of sectioning through a Golgi-impregnated neuron, the chuck is taken out of the ultramicrotome at regular intervals and examined in a modified light microscope (Figs. 1b,c; see also Wouterlood, 1979). The ribbons of ultrathin sections are mounted on formvar-coated slot grids and contrasted with uranyl acetate and lead citrate.

8. Screening of the ultrathin sections in the EM (Fig. 1d). For positive identification of a degenerating axon terminal synapsing with a Golgi-impregnated profile, synaptic membrane specializations between the degenerating bouton and the impregnated profile must be clearly visible in several consecutive sections.

Applications

Anterograde degeneration combined with Golgi EM has been successfully used, among others, in cat (Somogyi, 1978; Frotscher *et al.*, 1981), mouse (White, 1978; White *et al.*, 1984), and rat (Peters *et al.*, 1979); Mihailoff and Watt, 1981; Blackstad, 1965, 1975; Wouterlood and Nederlof, 1983; Frotscher and Zimmer, 1983a,b; Seress and Ribak, 1984).

Combination with Retrograde Transport of Horseradish Peroxidase

Various techniques have been designed, based on the *in vivo* retrograde axonal transport of particular substances from the axon terminals to the perikarya of the parent nerve cells. Most widely used in LM–neuroana-

---

Fig. 1. Sequence of micrographs illustrating the procedure followed to obtain serial thin sections of individual, light microscopically visualized neurons (in this case a Golgi-impregnated hypothalamic neuron in a rat). a, Micrograph taken from a section following flat-embedding in epoxy resin. After photography, a drawing is made of the neuron (N) with a microscope equipped with a drawing tube. The neuron is then cut out of the section and glued on top of a prepolymerized block. Notice the partial impregnation of the perikaryon (arrow). b, During thin sectioning the position of the sectioned profiles is monitored by periodically inspecting the surface of the block (arrow) in a modified light microscope. c, At inspection, a photograph is taken of the block and a drawing is made of the processes still present in the block. d, After serial thin sectioning, the perikaryon (shown) and the processes of the neuron can be screened. Notice the unimpregnated portion of the soma (arrows); ×8000. (Figs. a and d with permission from Wiley and Sons, Chichester, England.)

tomical tracing are the methods based on the retrograde transport of horseradish peroxidase (HRP; Kristensson *et al.,* 1971; LaVail and LaVail., 1972; for reviews of the technique, see Warr *et al.,* 1981; Mesulam, 1982). More recently, several other retrogradely transported substances have been introduced such as antibodies (Fillenz *et al.,* 1976), ^{3}H-labeled neurotransmitters (Hunt and Künzle, 1976; Cuénod *et al.,* 1982), tetenus and cholera toxins (Stoeckel *et al.,* 1977), fluorescent tracers (Kuypers *et al.,* 1977, 1979; Schmued and Fallow, 1986), HRP conjugated with wheat germ agglutinin (WGA-HRP, Schwab *et al.,* 1978; Gonatas *et al.,* 1979), fluorescent latex microspheres (Beckstead, 1985), Herpes simplex virus (Bak *et al.,* 1978), and iron–dextran (Cesaro *et al.,* 1982). This section will deal only with HRP, which is the only retrogradely transported label used to date in correlative LM–EM studies, either in combination with the anterograde degeneration technique, Golgi EM, or with immunoelectron microscopy. Retrogradely transported HRP is visualized by the enzymatic oxidation of a substrate into a colored reaction product. Several substrates are presently in use for HRP histochemistry (Mesulam, 1982). The reaction products of five of these substrates can be seen as an electron-dense precipitate after treatment with osmium tetroxide: diaminobenzidine (DAB; Graham and Karnovsky, 1966), tetramethylbenzidine (TMB; Schönitzer and Holländer, 1981; Stürmer *et al.,* 1981; Carson and Mesulam, 1982; Rye *et al.,* 1984), *o*-tolidine (Somogyi *et al.,* 1979), paraphenylenediamine-pyrocatechol (PPC-PD; Spreafico *et al.,* 1982), and benzidine dihydrochloride (BDHC; Carson and Mesulam, 1982; Lakos and Basbaum, 1986). Detailed descriptions of the histochemistry involved in the visualization in the light and electron microscope of these five substrates are provided by Mesulam (1982) and Carson and Mesulam (1982).

Retrogradely transported HRP visualized with DAB as substrate is predominantly seen as electron-dense material located in lysosome-like bodies in the cytoplasm. Histochemistry with the other substrates often results in fine, granular, crystalline, or fuzzy electron-dense precipitate in the perikaryon. The light and electron-dense DAB-reaction product found trapped in lysosomes following retrograde transport of HRP and subsequent histochemistry, can easily be distinguished from the electron-dense, cytoplasmically located label introduced by Golgi impregnation or by immunocytochemistry. However, histochemistry of HRP can be combined with Golgi EM only, if the DAB histochemistry is performed before the Golgi-impregnation procedure (Somogyi *et al.,* 1979; Somogyi and Smith., 1979). If the reverse procedure is carried out, no HRP-based product is formed. Presumably, the enzymatic activity is lost by denaturation or by poisoning with heavy metal ions. Further, due to the limited penetration of the DAB, HRP histochemistry has to be performed on sections maximally 100–200 μm thick. Golgi-impregnation techniques on single 100- to 200-μm sections and on 100- to 200-μm Vibratome sections reassembled into blocks have recently been developed (Harris *et al.*, 1980; Landas and Phillips, 1982; Freund and Somogyi, 1983; Gabbott and Somogyi, 1984).

Procedure

An excellent survey of the protocols involving the retrograde transport–Golgi EM method is given by Somogyi *et al.* (1979).

Applications

The feasibility of the combined retrograde transport–Golgi EM technique has been elegantly demonstrated by Somogyi and Smith (1979) and Bolam *et al.* (1981). These authors studied the ultrastructural details of two types of striatonigral neurons. An even more complex combination of correlative LM–EM techniques—anterograde degeneration, retrograde transport of HRP, and Golgi EM—has been used by Somogyi *et al.* (1981a,b) to study the afferent synaptic relationships (labeled through lesions in the cerebral cortex and ventral striatum, respectively) of neurons in the dorsal striatum (labeled by Golgi impregnation) which project to the substantia nigra (demonstrated by retrograde transport of HRP, visualized by DAB histochemistry). Finally, with a similar technique, Totterdell *et al.* (1984) investigated the synaptic relationships of striatal fibers with identified palladal neurons which, in turn, project to the substantia nigra.

Other Combinations

HRP histochemistry combined with the anterograde degeneration technique, and followed by EM, has been used by White *et al.* (1980), White and Hersch (1982), and Hendry and Jones (1983) to study the synaptic relationships between thalamocortical and corticothalamic neurons. Further, Williams and Faull (1985) used this combination in the rat to investigate the synaptic connections of fibers originating in the striatum with substantia nigra neurons projecting to the superior colliculus. Autoradiography following uptake of ^{3}H GABA, combined with Golgi impregnation and EM, has been used by Somogyi *et al.* (1981c), Bolam *et al.* (1983), and Somogyi *et al.* (1984). Transmitter-specific retrograde labeling of neurons has been developed by Hunt and Künzle (1976). In combination with EM radioautography (Cuénod *et al.*, 1982), the ultrastructural characteristics can be examined of neurons of which the neurotransmitter is known by the retrograde labeling.

IMMUNOCYTOCHEMICAL TECHNIQUES: CORRELATIVE LIGHT AND ELECTRON MICROSCOPY OF NEURONS LABELED IMMUNOCYTOCHEMICALLY FOR THE PRESENCE OF NEUROACTIVE SUBSTANCES OR THEIR SYNTHESIZING ENZYMES

Electron Microscopic Immunocytochemistry: Potentials and Pitfalls

By the Golgi-impregnation technique, the morphology of individual neurons is revealed, including all the fine details of their perikarya, dendrites, and axons. However, this technique does not provide any information with respect to the functional properties of the labeled neurons, because the Golgi precipitate serves only as a purely morphological marker. The application to brain sections of antibodies directed against neuroactive substances, their synthesizing enzymes, receptor proteins or other neuron-specific structural proteins, and the subsequent immunocytochemical visualization, leads to the deposition of a light and/or electron-dense label in those neurons which contain the specific antigen. In effect, the immunocytochemically introduced labels serve as morphological markers and simultaneously provide information with regard to the function of the immunoreactive neurons.

Immunocytochemistry as a tool to study the ultrastructural distribution of neuroactive substances and their synthesizing enzymes has been introduced by Pickel and colleagues (1975). Two types of technical approaches have been developed to visualize by means of EM the presence in neurons of antigenic substances, i.e., preembedding procedures (Nakane and Pierce, 1967; Pickel *et al.,* 1975; Sternberger, 1979; for review, see Pickel, 1981; Priestley, 1984) and postembedding procedures (for review, see Newman and Jasani, 1984). According to the preembedding techniques, mildly fixed tissue slices are incubated with antisera before postfixation in osmium tetroxide and embedding in epoxy resin. In the postembedding techniques, the material is first postfixed in osmium tetroxide, embedded in epoxy resin, and ultrasectioned before it is incubated with the antisera. Recently, for detecting two or more antigens in one structure an amalgam technique has been developed utilizing pre- and postembedding on the same sections (Van den Pol., 1985; Van den Pol *et al.,* 1985; see also section on double-labeling techniques). Perhaps, because LM preembedding procedures with the indirect, unlabeled antibody method of Sternberger (1979) can be applied without modifications to brain sections prepared according to procedures and with equipment available in the standard neuroanatomical laboratory, they have become very popular for the LM mapping of CNS neurons containing a particular antigen. Other major advantages of preembedding procedures are: (1) the antibodies are allowed to react with tissue antigens before these are radically changed by dehydration and mounting procedures; (2) the sensitivity is generally high; (3) brain sections stained according to preembedding immunocytochemical procedures can directly be compared with standard neuroanatomical sections; no sampling procedures are necessary because the sections are cut and, except for the immunocytochemical steps, are processed and counterstained in the same way as standard brain sections; (4) preembedding procedures have been shown to be fully compatible with LM–neuroanatomical tracing techniques (Ljungdahl *et al.*, 1975; Priestley *et al.*, 1981; Skirboll *et al.*, 1984; Wouterlood *et al.*, 1986a, 1987a).

Once the presence of an antigen has been visualized in the light microscope, it is tempting to study its ultrastructural localization. For correlative LM–EM studies combined with neuroanatomical tracing, preembedding immunocytochemistry provides a light and electron-dense label for pre- or postsynaptic neurons. In the light microscope, neurons of interest can be sampled for further EM examination. Simultaneously, the presence of a transported tracer substance provides information about the connectivity of the immunocytochemically characterized neurons.

Three major disadvantages are inherent to preembedding procedures:

1. Only thin, superficial areas in Vibratome slices contain sufficient immunolabeling. This phenomenon is thought to be caused by a slow and uneven penetration of the large antibody molecules into the slices (Piekut and Casey, 1983). Unfortunately, the ultrastructural details in these superficial areas are less well preserved than in the deeper areas.
2. Abundant nonspecific staining is most often found at the very surfaces of the Vibratome sections. It may not be surprising that, if no special technical measures are taken, only small areas underneath the surfaces of the Vibratome sections show both a balanced ultrastructure and sufficient, specific immunostaining. It is extremely difficult to obtain Golgi-impregnation-like images of neurons with simultaneous retention of good ultrastructural preservation (see also the following section).
3. The final resolution of the preembedding techniques is much lower than that of the postembedding techniques. The latter techniques are particularly suited for determining the exact localization of the antigens studied, for instance in cell organelles or on membranes (for a review, see Varndell and Polak, 1984).

A matter of concern of all immunocytochemical methods is the specificity of the label and the labeling technique. This means that the specificity of the primary antibody must be carefully tested and that an accurate estimation must be made as to what extent the labeling procedure contributes to false-positive staining. Thus, parallel incubations with absorbed primary antiserum or without primary antiserum are always necessary (see also Pool *et al.*, 1983).

Because the majority of correlative LM–EM immunocytochemical studies employ preembedding techniques, only these techniques will be considered. The general procedure of preembedding immunoelectron microscopy can be summarized as follows:

1. Fixation. Commonly the animals are perfused with saline followed by a fixative, in most cases formaldehyde to which a small amount of glutaraldehyde is added. The purpose of this fixative is to leave the antigenicity intact (Puchtler and Meloan, 1985), while simultaneously the tissue is stabilized by mildly cross-linking the protein components.
2. Sectioning on a Vibratome.
3. Optional treatments to enhance the penetration of the antibodies, in order to restore antigenicity or to suppress nonspecific staining. (see the section on The Compromise between Penetration of Antisera and Preservation of Ultrastructure).
4. Incubation of free-floating sections in the antisera according to the procedure of Sternberger (1979).
5. Optional silver or gold intensification (Gallyas *et al.*, 1982; Liposits *et al.*, 1985).
6. Postfixation in osmium tetroxide.
7. Optional *en bloc* staining with uranyl acetate.
8. Dehydration and embedding in epoxy resin.

The parameters (composition of the fixative, fixation procedure, buffers, temperatures, dilutions, incubation times, and optional treatments) have to be optimized for each type of antigen that is to be visualized, and for the particular immunocytochemical technique that is applied. In this chapter examples are given for the detection of glutamic acid decarboxylase (GAD) and the lectin tracer substance PHA-L. Protocols for a number of neuroactive substances, structural proteins, or their synthesizing enzymes also follow this general scheme: adrenocorticotropic hormone (ACTH; Léránth *et al.*, 1980; Palkovits *et al.*, 1982), cholecystokinin (CCK; Takagi *et al.*, 1984), choline acetyltransferase (Hattori *et al.*, 1976; Wainer *et al.*, 1984; Houser *et al.*, 1983, 1985; Phelps *et al.*, 1985; Phelps and Vaughn, 1986; Frotscher *et al.*, 1986), corticotropine-releasing factor (Liposits *et al.*, 1983), dopamine (Onteniente *et al.*, 1984), dopamine-β-hydroxylase (Olschowska *et al.*, 1981), γ-amino butyric acid

(Seguela *et al.*, 1985); Somogyi and Hodgson, 1985; Somogyi *et al.*, 1985), glutamic acid decarboxylase (GAD; Ribak, 1978; Ribak *et al.*, 1978, 1981; Wu *et al.*, 1982; Oertel *et al.*, 1982a; Mugnaini, 1985), histidine decarboxylase (Hayashi *et al.*, 1984; Wouterlood *et al.*, 1986a), Leu-enkephalin (DiFiglia *et al.*, 1982; Voorn and Buijs, 1983), neuron-specific enolase (Vinores *et al.*, 1984), neurophysin (Piekut and Casey, 1983), phenyletholamine N-methyltransferase (PNMT; Pickel *et al.*, 1986b), oxytocin (Voorn and Buijs, 1983), serotonin (Pickel *et al.*, 1976; Liposits *et al.*, 1985), somatostatin (DiFiglia and Aronin, 1982), substance-P, enkephalin, and tyrosine hydroxylase (Pickel *et al.*, 1977, 1979; Bouyer *et al.*, 1984); tyrosine hydroxylase (Van den Pol *et al.*, 1984), vasoactive intestinal polypeptide (VIP; Connor and Peters, 1984; Léránth *et al.*, 1984), and vasopressin (Buijs and Swaab, 1979; Voorn and Buijs, 1983).

Anterograde Degeneration Combined with Immunoelectron Microscopy

Labeling of synaptic terminals by anterograde degeneration following the interruption of their connections with the parent cell bodies is a well-established technique (see section on Combination with the Anterograde Degeneration Technique). It has been shown for the first time by Hattori *et al.* (1976) that the technique of anterograde degeneration is fully compatible with EM immunocytochemistry. The synaptic relationships of immunocytochemically stained neurons can systematically be investigated by placing lesions in areas suspected to house neurons which communicate synaptically with the antigen-containing neurons. When degenerating axon terminals are observed in synaptic contact with the immunocytochemically labeled cells, the evidence is obtained for the synaptic involvement of these cells with the afferent fibers. Following the pioneering effort of Hattori *et al.* (1976), several studies have been published, combining anterograde degeneration and immunoelectron microscopy (Barber *et al.*, 1978; Palkovits *et al.*, 1982; Léránth and Frotscher, 1983; Liposits *et al.*, 1983; Frotscher *et al.*, 1984; Záborszky *et al.*, 1984; Wouterlood *et al.*, 1985). The procedure of correlative LM–EM immunocytochemistry, combined with anterograde degeneration, parallels that of anterograde degeneration–Golgi EM (see the section on Combination with the Anterograde Degeneration Technique). A procedure is described here used by Wouterlood *et al.* (1985) to study the synaptic terminals of olfactory bulb fibers on GAD-immunoreactive neurons in the entorhinal cortex of rats. The steps of fixation, preincubation, and immunocytochemistry were modified after Oertel *et al.* (1982a).

Procedure

1. Surgery. Unilaterally, the olfactory bulb of rats was radically removed by suction. Directly afterward, 20 μg colchicine dissolved in 2 μl saline was injected in the ipsilateral entorhinal area. The purpose of the latter treatment is to block axoplasmic transport, leading to accumulation in the perikaryon of the antigen to detectable levels (Ribak *et al.*, 1978).

2. Postsurgical recovery and survival. Most authors recommend colchicine injection 48 hr prior to sacrifice. If this cannot be combined with an optimal survival time for degeneration, the animal has to be operated on a second time.

3. Fixation. For the immunoelectron microscopical detection of GAD, the perfusion fixation designed by Oertel *et al.* (1982a) is optimal.

4. Slicing of the entorhinal area with the Vibratome (Oxford Instruments), into sections 35 μm thick.

5. Immunocytochemistry. After three rinses in cold 50 m*M* Tris buffer, pH 7.6, the sections were preincubated with nonimmune rabbit whole serum in Tris-buffered saline (TBS), pH 7.6, with 100 m*M* DL-lysine monohydrochloride to suppress nonspecific staining. They were subsequently incubated according to the peroxidase-antiperoxidase (PAP procedure (Sternberger, 1979); (a) Incubation for 60 hr at 4°C with sheep anti-GAD (S3; fourth bleed; Oertel *et al.*, 1982a), diluted 1 : 1500 with TBS to which 2% nonimmune (normal) rabbit serum had been added; (b) incubation for 60 min in rabbit anti-sheep antiserum 1 : 50 diluted with TBS+2% normal rabbit serum, room temperature; (c) incubation for 60 min at room temperature in goat PAP complex diluted 1 : 100 in TBS+2% normal rabbit serum. After rinsing in several changes of TBS, the sections were then incubated for 5–15 min in freshly prepared 0.05% 3-3′-diaminobenzidine tetrahydrochloride (DAB) and 0.015% H_2O_2 in 50 m*M* Tris-HCl, pH 7.6. The reaction is stopped by rapid washing in several changes of ice-cold TBS. Sections for EM were rinsed three times in 100 m*M* cacodylate buffer, pH 7.3, and postfixed in ice-cold 2% OsO_4 in 100 m*M* cacodylate buffer, pH 7.3.

6. The steps of dehydration and embedding are identical to the procedure described previously on p. 89–91.

7. Correlative LM–EM. The only difference here with the procedure described on p. 89–91 is that relatively thick ultrathin sections are cut (yellow interference color; 100–120 nm thickness), and that contrasting is done only with lead citrate.

8. Screening of the ultrathin sections in the electron microscope.

In the light microscope, the GAD-immunoreactive neurons and terminals are labeled by the brown DAB-reaction product. By the procedure outlined on pp. 89–91, it is possible to select individual GAD-positive perikarya and to ultrasection serially the whole perikaryon, including those parts into which the antibodies have not penetrated. In the electron microscope the reaction product appears in perikarya and dendrites as a dispersed, fuzzy, cytoplasmic label, and in GAD-positive terminals as a fuzzy, electron-dense precipitate in the axoplasmic matrix. This precipitate can easily be distinguished from the profound osmiophilia of the axoplasmic matrix associated with electron-dense degeneration.

EM Autoradiography following Accumulation or Anterograde Transport of Radioactive Labels, Combined with Immunocytochemistry

EM autoradiography (EMA) is used in neuroanatomy as a marking technique to detect the presence of a certain neurotransmitter or enzyme system in a particular type of neuron (Sotelo, 1971; Descarries *et al.*, 1982), or to visualize the transport route in the CNS of radioactive tracers, in most cases ^{3}H-labeled amino acids, from the injection site to the synaptic terminals (for EMA techniques, see Hendrickson, 1969 and Behan, 1981). Briefly, EMA techniques involve two steps: (1) The introduction of a radioactive label which is taken up by neurons, incorporated in proteins and transported, either to the perikaryon or to the axon terminals; and (2) following fixation and embedding, ultrathin sections are covered with a photographic emulsion, which is subsequently hit ("exposure") by the β particles emitted from decaying radioactive isotopes in the thin section. By photographic development, the resulting latent image in the emulsion is converted into an image of silver grains overlying profiles in which radioactive substance is present. The major advantage of the EMA technique is that the application and visualization of the label does not interfere with the preservation of the ultrastructure. Hence, the rigid, standard fixation EM techniques can be used, without the need to compromise between fixation and subsequent treatments needed for immunoelectron microscopy. Also, the normal ultrastructure of axon terminals is retained. This is in contrast to the anterograde degeneration technique in which the marking is based on pathological changes in the synaptic terminals. Its major advantage in combination with immunoelectron microscopy is that EMA can be used to provide a second label in a fully independent way in addition to an immunocytochemically introduced label. The major disadvantage of EMA is the extremely long exposure time necessary to produce sufficient latent image in the photographic emulsion.

With a few adaptations with regard to the demands imposed on fixation by the immunoelectron microscopy, it has been shown that EMA in combination with immunocytochemistry can be very useful as a double-labeling technique. Such a combination has been used by Pickel *et al.* (1984; 1986a) to study the synaptic relationships between dopaminergic terminals (visualized by tyrosine hydroxylase immunocytochemistry) and serotoninergic neurons (visualized by EMA following uptake of ^{3}H-serotonin. Kiss and Halász (1985) used this technique to study the synaptic connections of LHRH-immunoreactive neurons and axon terminals labeled by uptake of 5-^{3}H-hydroxytryptamine. The combination of immunocytochemistry and visualization of synaptic terminals by EMA following anterograde tracing by injection of ^{3}H-amino acids has not been reported to date.

Combinations with Retrograde Tracing Techniques and Golgi Impregnation

By means of retrograde transport techniques combined with immunocytochemistry, the efferent connections of chemically identified cells can be examined. The great potential of these combinations for LM, specifically of the retrograde transport of fluorescent dyes combined with immunofluorescence, has been recognized by Ljungdahl *et al.* (1975), and has been extensively exploited (see review by Skirboll *et al.,* 1984). Since retrogradely transported HRP, visualized with DAB histochemistry, can be seen as a light and electron-dense deposit (see section on Combination with Retrograde Transport of HRP), it is no surprise that studies in which retrograde transport of HRP and immunoelectron microscopy were combined soon followed Ljungdahl *et al.*'s (1975) paper (Priestly *et al.,* 1981; Ingham *et al.,* 1985; Milner and Pickel, 1986; Lakos and Basbaum, 1986; Záborszky *et al.,* 1986).

An even more sophisticated method has been developed by Freund and Somogyi (1983), combining in one experiment the techniques of retrograde transport of HRP, Golgi impregnation, immunocytochemistry, and EM. It is thus possible to study with the light microscope the Golgi image of a neuron, and simultaneously to gain information about its neurotransmitter and its efferent connections (Somogyi *et al.,* 1982; Bolam *et al.,* 1984a,b; with acetylcholinesterase histochemistry instead of choline acetyltransferase immunocytochemistry, Somogyi and Hodgson, 1985; Frotscher and Léránth, 1986). Since this combination of retrograde transport, immunocytochemistry (or histochemistry), Golgi impregnation, and EM can also be used together with the anterograde degeneration technique with no radical technical modifications, future studies can be expected in which (degenerating) axon terminals are studied (by EM) which synapse with immunocytochemically characterized, Golgi-impregnated, HRP-containing (by retrograde tracing) neurons.

Double-Labeling Techniques at the Light and Electron Microscopic Levels

For LM purposes, several double-labeling immunocytochemical techniques have been developed. The combination immunofluorescence–retrograde tracing of fluorescent dyes has been mentioned in the section on Other Combinations, but techniques have also been introduced to demonstrate the colocalization of two antigens in one neuron. For the latter purpose, parallel immunocytochemical incubations can be performed on adjacent sections (Pickel *et al.,* 1979; Takeda *et al.,* 1984). The evidence of colocalization, however, is more convincing when on single sections two-color immunocytochemistry (Oertel *et al.,* 1982b) or double-labeling immunofluorescence (Takeda *et al.,* 1984) is applied.

For demonstrating colocalization of two or more antigens in single nerve terminals, an elegant technique has been developed by Van den Pol (1985), including preembedding immunocytochemistry to visualize a first antigen and postembedding immunoelectron microscopy on the same material to visualize a second or third, colocalized antigen. A double-preembedding immunoelectron microscopic procedure, including a detection system based on PAP detection of a first antigen and simultaneous avidin-biotin-ferritin detection of a second antigen, has been developed by Léránth *et al.* (1985). With this technique, different electron-dense labels are introduced in two-neuron systems of which each neuron utilizes a different neurotransmitter, for instance adrenergic terminals on corticotropin-releasing factor-synthesizing neurons (Liposits *et al.,* 1986). Another way to study the interaction between two chemically different neuron systems is by preembedding immunocytochemistry combined with EM autoradiography (Pickel *et al.,* 1984, 1986a).

CORRELATIVE LIGHT AND ELECTRON MICROSCOPY OF NEURONS LABELED WITH *PHASEOLUS VULGARIS* LEUCOAGGLUTININ (PHA-L)

Light and Electron Microscopy of PHA-L

With the recently developed LM anterograde tracing technique based on the injection, uptake, transport, and visualization of the plant lectin PHA-L, neuronal connections can be traced with a high degree of accuracy (Gerfen and Sawchenko, 1984; Groenewegen and Van Dijk., 1984). This technique offers several advantages over the LM–anterograde tracing methods utilizing radioactive amino acids (Ter Horst *et al.,* 1984), HRP, or wheat germ agglutinin conjugated to HRP (WGA-HRP). The injection sites are very small and there appears to be virtually no uptake of the tracer by fibers of passage or in areas adjacent to the injection site. The complete labeling of neurons at the injection site includes their somata, dendrites, and dendritic appendages and strongly resembles Golgi impregnation (Fig. 3a). Further, the axons, their *en passant* varicosities, and terminal specializations appear to be excellently marked. An advantage inherent to the immunocytochemical detection system is the possibility to combine PHA-L tracing with transmitter immunocytochemistry (Gerfen and Sawchenko, 1985; Wouterlood *et al.,* 1986a, 1987a).

We have recently developed a combined LM–EM approach, visualizing the cell bodies, dendritic processes, axons, and axonal varicosities of PHA-L-stained neurons (Groenewegen and Wouterlood, 1984; Wouterlood and Groenewegen, 1985). Three major adaptations of the original procedure of Gerfen and Sawchenko (1984) were necessary to obtain preparations suitable for combined LM–EM: (1) Modification of the fixation in order to obtain well-preserved ultrastructural details; (2) the use of Vibratome sections instead of frozen sections; (3) modification of the type and amount of detergent necessary to enhance the penetration of the antisera into the sections.

Procedure

1. Iontophoretical application of PHA-L. Under anesthesia of 1 ml/kg body weight (rat) of a mixture of four parts Ketaset (Bristol Laboratories, Syracuse, NY) and three parts Rompun (Bayer, Leverkusen, Germany), PHA-L (Vector Labs, Burlingame, CA), dissolved at a concentration of 25 μg/μl in TBS (50 m*M* Tris-HCl buffer to which 0.9% sodium chloride is

Fig. 2. LM and EM of PHA-L-labeled neurons. a, Light micrograph of a PHA-L labeled neuron in the septal area of a rat, taken from a section which was flat-embedded in epoxy resin. PHA-L immunocytochemistry; addition of 0.1% Photo-Flo 200 to the primary antiserum. B = blood vessel. b, Electron micrograph of the soma of the same neuron; ×6100. c, Dendrite (asterisk) of a PHA-L-labeled neuron in the nucleus accumbens of a rat showing labeled spines (arrows) on which axon terminals form synaptic contacts (AT); ×33,000. d, Following PHA-L injection in the nucleus accumbens of a rat, labeled axon terminals (AT) can be seen in the ventral pallidum, forming symmetric synaptic junctions (solid arrow) with dendrites (D) and cell bodies (not shown). The open arrows indicate the damage to myelinated axons caused by the detergent used in the PHA-L immunocytochemistry; ×43,700. In all cases shown here, 0.1% of the surfactant Photo-Flo has been added to the primary antiserum (see Procedure).

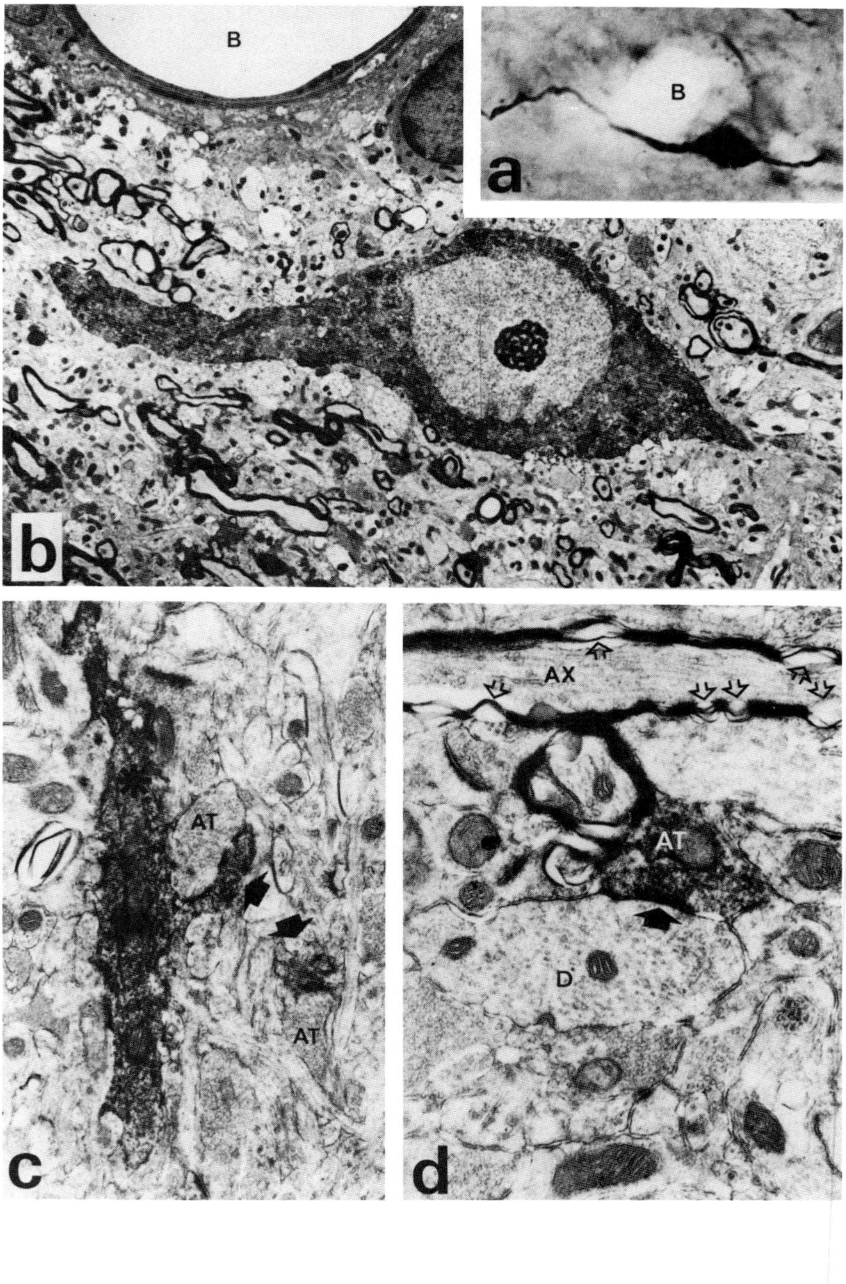
B
a
B
b
AT
AT
c
AX
AT
D
d

added), pH 7.4, is injected under stereotactic guidance, through glass micropipettes (tip diameter 10–20 μm), using a positive pulsed 5 μA DC current (7 sec on/ 7 sec off) for 15–30 min.

2. Survival time. Postsurgical survival varies between 2 and 6 days, depending on the purpose of the study, and based on the assumption of a transport of 4 mm per day in rat.

3. Fixation. Rats are reanesthetized with sodium pentobarbital (Nembutal; Ceva, Paris, France; 1 ml/kg body weight) and perfused transcardially at a constant hydrostatic pressure (Jonkers *et al.*, 1984) with 100 ml of physiological saline solution, pH 6.9, of 38°C, followed immediately by 1500 ml of a mixture of 2% freshly depolymerized paraformaldehyde and 0.1% glutaraldehyde, 0.1% Photo-Flo 200 (Eastman Kodak, Rochester, NY), and 0.2% picric acid in 125 m*M* phosphate buffer, pH 7.3 (room temperature).

4. Slicing. Two hours after perfusion, the brains are dissected and sectioned at a thickness of 35–40 μm on a Vibratome (Oxford Instruments), using chilled TBS, pH 8.6, as bath fluid, and rinsed three times.

5. Immunocytochemistry. The sections are pretreated (optional) for 30 min in 0.1% Triton X-100 (Sigma, St. Louis) in TBS, and incubated in the primary antibody (goat anti-*Phaseolus;* Vector Labs; dilution of 1 : 6000) in TBS, pH 8.6, to which 0.1% of the surfactant Photo-Flo 200 is added (TBS-D). Incubation is performed under continuous, gentle agitation in a refrigerator (60 hr) or at room temperature (14 hr). The sections are then rinsed 3 × 15 min in TBS, pH 8.6, and incubated with rabbit anti-goat whole serum (Sigma) in TBS-D, pH 8.6 (1 : 40) for 5 hr at room temperature. After a thorough rinse in TBS, pH 8.6, the sections are incubated overnight (room temperature) with goat PAP (Nordic Immmunology, Tilburg, The Netherlands), 1 : 400 in TBS, pH 8.6. After rinsing in several changes of TBS, pH 8.6 and TBS pH 7.6, the sections are subsequently incubated for 40–60 min with a solution of 0.04% 3,3′-diaminobenzidine tetrahydrochloride (DAB; BDH Chemicals, Poole, England) and 0.015% H_2O_2 in 50 m*M* Tris-HCl buffer (pH 7.6). After incubation, the sections are washed three times with Tris-HCl buffer (pH 7.6), and three times with 100 m*M* cacodylate buffer (pH 7.4). Sections selected for EM are postfixed for 1 hr in 1% OsO_4 in 100 m*M* cacodylate buffer, pH 7.4, and stored in 100 m*M* cacodylate buffer, pH 7.4, in a refrigerator; the other sections are mounted on slides, dried, counterstained, and coverslipped.

6. Embedding. After block staining in maleate-buffered uranyl acetate, the sections are dehydrated and embedded in Epon-Araldite on slides between polyethene foils (same procedure as step No. 6 listed on p. 89).

7. Correlative LM–EM (Figs. 2a,b) is performed in a similar manner as on Golgi-EM sections (same procedure as step No. 7 on p. 89). The thin sections are contrasted with lead citrate only.

Applications

With this modified method we studied the distribution of the PHA-L immunoprecipitate in labeled perikarya and dendrites of neurons in the nucleus accumbens of rats (Fig. 2c), and the ultrastructure of PHA-labeled terminals in the ventral pallidum of rats of fibers originating in the nucleus accumbens (Fig. 2d). The present and earlier (Wouterlood and Groenewegen, 1985) results show that PHA-L is useful as a label for correlative LM–EM neuroanatomical tracing studies. Recently, this has been confirmed by other authors (Zahm and Heimer, 1985). PHA-L labeling may also be used as a Golgi impregnation-like marker for LM and subsequent EM (see section on Combinations of PHA-L Visualization). Correlative LM–EM of PHA-L combined with anterograde degeneration and retrograde transport of HRP is reported here for the first time.

The Compromise between Penetration of Antisera and Preservation of Ultrastructure

It was shown previously that the majority of the correlative LM–EM immunocytochemical studies, exploiting the indirect, preembedding, unlabeled antibody method (PAP; Sternberger, 1979), use DAB as the substrate of choice for the visualization of the antigen. The final step of the PAP detection consists of the enzymatic oxidation of DAB into a light and electron-dense reaction product (Wu *et al.,* 1982; Oertel *et al.,* 1982a; Priestley, 1984). It must be noted that for EM immunocytochemistry, a compromise has to be reached between two radically opposite demands: one for strong fixation in order to preserve the ultrastructural details, and one for as little fixation as possible in order to retain full accessability and expression of the antigenic substances (being represented by the deposition of immunoreaction product *deep within* a Vibratome section as well as in the more superficial areas). Often, a choice is made in favor of the quality of the ultrastructural preservation, which, as a consequence, results only in staining of antigenic structures located at the surfaces of the Vibratome sections (Pickel, 1981; Buijs, 1982; Piekut and Casey, 1983; Priestley, 1984; Wouterlood *et al.,* 1985; Wouterlood *et al.,* 1987b). Conversely, when detection of antigenic structures is needed throughout the entire depth of the sections, and this applies especially to PHA-L tracing and to those immunocytochemical studies for which it is necessary to visualize the dendrites and axons of neurons, then the immunocytochemical procedures commonly include the use of detergent Triton X100 (Pickel, 1981; Houser *et al.,* 1985; Wouterlood and Groenewegen, 1985). The effect of the addition to the antisera of even low concentrations of Triton X100 on the preservation of the ultrastructural details is disastrous (Olschowska *et al.*, 1981; Voorn and Buijs, 1983; Priestley, 1984; Wouterlood *et al.*, 1987b).

In an attempt to compromise the incompromisable, i.e., to obtain sufficient ultrastructural preservation while simultaneously obtaining Golgi-impregnation-like visualization of neurons, we investigated the effects of various concentrations of Triton X100 and of various other detergents added to the incubation media. Incubation of Vibratome sections with antisera in the presence of the wetting agent Photo-Flo 200 (Eastman Kodak) results in improved penetration as compared to nondetergent-treated, immunostained sections, while simultaneously the ultrastructural details are sufficiently preserved (Wouterlood *et al.*, 1987b). For correlative LM–EM detection of PHA-L, addition of 0.1% Photo-Flo (Eastman Kodak) to the primary antiserum, in combination with prolonged incubation, results in the best compromise between a full LM visualization and retention of the ultrastructural preservation (Figs. 2b–d). With this procedure, the ultrastructural details are much better preserved than in material incubated in the presence of even the lowest concentration of Triton X100.

Procedure

Essentially the same procedure is followed as described in the section on LM and EM of PHA-L. The best compromise between penetration of the antisera and preservation of ultrastructure is achieved by incubating Vibratome sections under continuous agitation for 60 min at 4°C in the primary antibody: goat anti-PHA-L (Vector Labs) diluted 1 : 6000 in TBS with 0.1% Photo-Flo 200 (Eastman Kodak). The prolonged incubation with the second antibody and with the PAP complex, as mentioned in the protocol on p. 104 is necessary to obtain sufficient light microscopic visualization of the PHA-L labeled structures. The secondary and tertiary antibodies are dissolved in TBS, without adding a detergent, or with 0.1% Photo-Flo 200 added.

Control Sections

In parallel incubations, Photo-Flo 200 (Eastman Kodak) was either omitted or replaced by various amounts of Triton X100 (Sigma, St. Louis, MO), Saponin (Merck, Darmstadt, Germany), or Nonidet P-40 (Sigma, St. Louis, MO). To examine the preservation of the ultrastructural details as it appears without immuno-incubation, immediately after sectioning a few slices were stored in 100 m*M* cacodylate buffer, pH 7.4, in a refrigerator, while the other sections were treated immunocytochemically and further processed for EM together with the immunostained sections. The thin sections were contrasted with lead citrate and examined in a Philips

EM 301 electron microscope. When no detergent is added to the antisera, immunolabeling is seen only in thin zones immediately subjacent to the surfaces of the Vibratome slices. Such a low penetration has been reported also by others (Piekut and Casey, 1983; Priestley, 1984). Only the PHA-L-containing cell bodies lying at the surfaces of the slices as well as fragments of PHA-L-containing processes are stained. The PHA-L-immunoreactive neurons share a common feature, i.e., they are cut open by the Vibratome knife. At the very surface of the slices the ultrastructural preservation is poor. Much, apparently nonspecific, labeling is present in this area. Sections incubated with antisera to which Saponin (Merck; up to 2%) was added to show the same immunoreactivity and ultrastructural preservation as the nondetergent control sections. Addition of Triton X100 to the antisera at concentrations of 0.075% and higher markedly improves the penetration; however, this causes poor ultrastructural preservation. Nonidet P-40 (Sigma; 0.5%) added to the antisera results in poor contrast, high background, and poor ultrastructure.

The penetration of the antisera into the Vibratome sections may be enhanced also by other techniques, for example by a two-cycle incubation (Ordronneau *et al.*, 1981), rapid freezing-thawing of cryoprotected blocks or Vibratome sections (Somogyi and Takagi, 1982; Priestley and Cuello, 1983; Priestley, 1984), dehydration–rehydration before immunostaining (Eldred *et al.*, 1983), either or not in combination with specially designed fixatives (Mclean and Nakane, 1974; Berod *et al.*, 1981; Somogyi and Takagi, 1982; Eldred *et al.*, 1983; Willingham, 1983). Recently, freeze-thawing and sodium borohydride treatment have been successfully applied in EM to enhance penetration of anti-PHA-L antibodies (Zahm and Heimer, 1985).

Combinations of PHA-L Visualization with Anterograde and Retrograde Tracing Techniques

We have conducted a series of experiments to establish a protocol for correlative LM–EM of PHA-labeling of neurons, combined with anterograde degeneration and retrograde transport of HRP. In rats, PHA-L was injected at various loci in the septal area according to the procedure presented in the section on LM and EM of PHA-L. In several animals, 2 μl of 30% HRP (Boehringer) in saline was mechanically deposited in the mammillary body using a microsyringe attached to the stereotaxic apparatus, whereas in other animals the fornix was exposed by suction of the overlying cortex, and transected. In both experiments, the animals were allowed to survive for 3 days before fixation, PHA-L immunocytochemistry, and processing for combined LM–EM. By incubation with

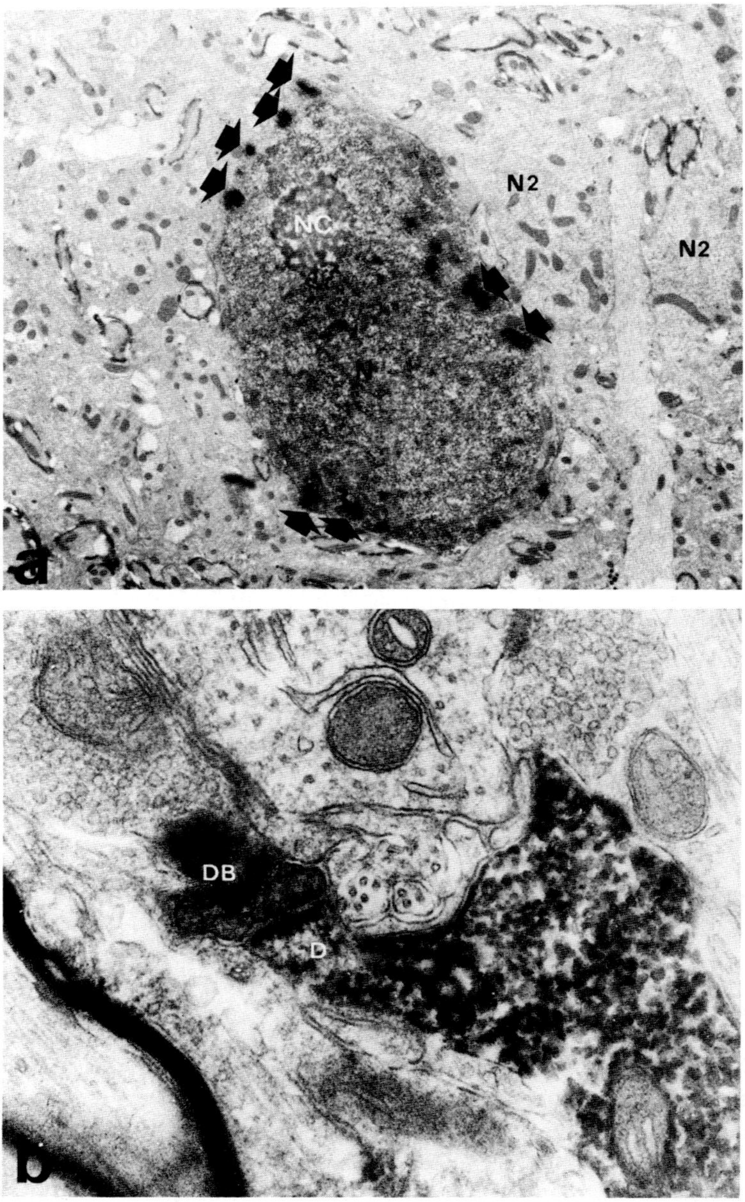
N2
NC
N2
a
DB
D
b

DAB, in one step both the immunocytochemically bound PAP complex and the retrogradely transported HRP are visualized. The immunocytochemical label appears in the electron microscope finely dispersed in the cytoplasm while the retrogradely transported HRP can be found in lysosome-like bodies scattered in the cytoplasm (Fig. 3a). Electron-dense degenerating axon terminals in synaptic contact with PHA-L-immunoreactive cell bodies and dendrites can be distinguished with certainty (Fig. 3b). In our opinion, the analysis of circuits of neurons by combinations of anterograde degeneration, PHA-L labeling of the target neurons, labeling of these same neurons with retrogradely transported HRP, and EM, has great potential value.

I would like to thank Enrico Mugnaini for teaching me the tricks of immunoelectron microscopy, and Anthony Lohman for carefully reviewing the manuscript. Expert technical assistance was provided by Janny Nederlof, Annaatje Pattiselanno, and Anne-Lise Dahl. Photographic help was kindly provided by Dirk de Jong (light microscopy) and Shimon Paniry (electron microscopy).

REFERENCES

Alonso, A., and Köhler, C. (1984). A study of the reciprocal connections between the septum and the entorhinal area using anterograde and retrograde axonal transport methods in the rat brain. *J. Comp. Neurol.* **225,** 327.

Bak, I. J., Markham, C. H., Cook, M. L., and Stevens, J. G. (1978). Ultrastructural and immunoperoxidase study of striatonigral neurons by means of retrograde axonal transport of Herpes simplex virus. *Brain Res.* **143,** 361.

Barber, R. P., Vaughn, J. E., Saito, K., McLaughlin, B. J., and Roberts, E. (1978). GABAergic terminals are presynaptic to primary afferent terminals in the substantia gelatinosa of the rat spinal cord. *Brain Res.* **141,** 35.

Beckstead, R. M. (1985). Striatal axons to globus pallidus, entopeduncular nucleus and substantia nigra come from separate cells: Retrograde double-labeling with HRP and fluorescent latex microspheres. *Abstr. Soc. Neurosci.* **11,** 208.

Fig. 3. Combination of PHA-L tracing, other neuroanatomical tracing techniques, and EM. a, PHA-L-labeled neuron (N) in the septal area of a rat, retrogradely labeled with HRP (arrows) following injection in the mammillary body. The PHA-L label is seen as the fuzzy precipitate in the cytoplasm (compare with "normal" cytoplasm of neighbouring neuron N2). NC = nucleolus. Incubation with primary antiserum to which 0.1% Photo-Flo had been added; ×7000. b, Labeling in the medial septal area of a rat of a target neuron by local injection of PHA-L, together with labeling of axon terminals of fornix fibers by means of the anterograde degeneration technique (transection of the fornix). A degenerating axon terminal (DB) forms a synaptic contact with a spine of a PHA-L-labeled dendrite (D); ×53,500.

Behan, M. (1981). Identification and distribution of retinocollicular terminals in the cat. An electron microscopic autoradiographic analysis. *J. Comp. Neurol.* **199,** 1.

Berod, A., Hartman, B. K., and Pujol, B. F. (1981). Importance of fixation in immunohistochemistry: Use of formaldehyde solutions at variable pH for the localization of tyrosine hydroxylase. *J. Histochem. Cytochem.* **29,** 844.

Björklund, A., and Hökfelt, T. (1983–1985). "Handbook of Chemical Neuroanatomy" Vols. 1–4. Elsevier, Amsterdam.

Blackstad, T. W. (1965). Mapping of experimental axon degeneration by electron microscopy of Golgi preparations. *Z. Zellforsch. Mikrosk. Anat.* **67,** 819.

Blackstad, T. W. (1975). Electron microscopy of experimental axonal degeneration in photochemically modified Golgi preparations: A procedure for precise mapping of nervous connections. *Brain Res.* **95,** 191.

Bolam, J. P., Somogyi, P., Totterdell, S., and Smith, A. D. (1981). A second type of striatonigral neuron: A comparison between retrogradely labelled and Golgi-stained neurons at the light and electron microscopic levels. *Neuroscience* **6,** 2141.

Bolam, J. P., Clarke, D. J., Smith, A. D., and Somogyi, P. (1983). A type of aspiny neuron in the rat neostriatum accumulates ^{3}H-γ-aminobutyric acid: Combination of Golgi-staining, autoradiography and electron microscopy. *J. Comp. Neurol.* **213,** 121.

Bolam, J. P., Ingham, C. A., and Smith, A. D. (1984a). The section-Golgi procedure. 3. Combination of Golgi impregnation with enzyme histochemistry and electron microscopy to characterize acetylcholinesterase-containing neurons in the rat neostriatum. *Neuroscience* **12,** 687.

Bolam, J. P., Wainer, B. H., and Smith, A. D. (1984b). Characterization of cholinergic neurons in the rat neostriatum. A combination of choline acetyltransferase immunocytochemistry, Golgi impregnation and electron microscopy. *Neuroscience* **12,** 711.

Bouyer, J. J., Joh, T. H., and Pickel, V. M. (1984). Ultrastructural localization of tyrosine hydroxylase in rat nucleus accumbens. *J. Comp. Neurol.* **227,** 92.

Buijs, R. M. (1982). Tissue treatment in immunocytochemistry. *In* "Immunocytochemistry and its Application in Brain Research," 2nd EMBO Practical Course, pp. 49–56. Netherlands Institute for Brain Research and the University of Amsterdam.

Buijs, R. M., and Swaab, D. F. (1979). Immuno-electron microscopical demonstration of vasopressin and oxytocin synapses in the limbic system of the rat. *Cell Tissue Res.* **204,** 355.

Carlsen, J., De Olmos, J., and Heimer, L. (1982). Tracing of two-neuron pathways in the olfactory system by the aid of transneuronal degeneration: Projections to the amygdaloid body and hippocampal formation. *J. Comp. Neurol.* **208,** 196.

Carson, K. A., and Mesulam, M.-M. (1982). Electron microscopic demonstration of neural connections using horseradish peroxidase: A comparison of the tetramethylbenzidine procedure with seven other histochemical methods. *J. Histochem. Cytochem.* **30,** 425.

Cesaro, P., Pollin, B., Nguyen-Legros, J., Azerad, J., and Laplante, S. (1982). Iron-dextran: An neuronal tracer without uptake by fibers of passage. *Neurosci. Lett., Suppl.* **10,** 108.

Colonnier, M. (1964). Experimental degeneration in the cerebral cortex. *J. Anat.* **98,** 47.

Connor, J. R., and Peters, A. (1984). Vasoactive intestinal polypeptide immunoreactive neurons in rat visual cortex. *Neuroscience* **12,** 1027.

Cuénod, M., Bagnoli, P., Beaudet, A., Rustioni, A., Wiklund, L., and Streit, P. (1982). Transmitter-specific retrograde labeling of neurons. *In* "Cytochemical Methods in Neuroanatomy" (V. Chan-Palay and S. L. Palay, eds.), pp. 17–44. Alan R. Liss, Inc., New York.

Descarries, L., Watkins, K. C., Garcia, S., and Beaudet, A. (1982). The serotonin neurons in the nucleus raphe dorsalis of adult rat: A light and electron microscopic radioautographic study. *J. Comp. Neurol.* **207,** 239.

DiFiglia, M., and Aronin, N. (1982). Ultrastructural features of immunoreactive somatostatin neurons in the rat caudate nucleus. *J. Neurosci.* **2,** 1267.

DiFiglia, M., Aronin, N., and Martin, J. B. (1982). Light and electron microscopic localization of immunoreactive leu-enkephalin in the monkey basal ganglia. *J. Neurosci.* **2,** 303.

Eldred, W. D., Zucker, C., Karten, H. J., and Yazulla, S. (1983). Comparison of fixation and penetration enhancement techniques for use in ultrastructural immunocytochemistry. *J. Histochem. Cytochem.* **31,** 285.

Fairén, A., Peters, A., and Saldanha, J. (1977). A new procedure for examining Golgi impregnated neurons by light and electron microscopy. *J. Neurocytol.* **6,** 311.

Fillenz, M., Gagnon, C., Stoeckel, K., and Thoenen, H. (1976). Selective uptake and retrograde axonal transport of dopamine-β-hydroxylase antibodies in peripheral adrenergic neurons. *Brain Res.* **114,** 293.

Freund, T. F., and Somogyi, P. (1983). The section-Golgi impregnation procedure. 1. Description of the method and its combination with histochemistry after intracellular iontophoresis or retrograde transport of horseradish peroxidase. *Neuroscience* **9,** 463.

Friedrich, V. L., Jr., and Mugnaini, E. (1981). Electron microscopy: Preparation of neural tissues for electron microscopy. *In* "Neuroanatomical Tract-Tracing Methods" (L. Heimer and M. J. Robards, eds.), pp. 345–375. Plenum, New York.

Frotscher, M., and Léránth, C. (1986). The cholinergic innervation of the rat fascia dentata: Identification of target structures on granule cells by combining choline acetyltransferase immunocytochemistry and Golgi impregnation. *J. Comp. Neurol.* **243,** 58–70.

Frotscher, M., and Zimmer, J. (1983a). Commissural fibers terminate on nonpyramidal neurons in the guinea pig hippocampus. A combined Golgi/EM degeneration study. *Brain Res.* **265,** 289.

Frotscher, M., and Zimmer, J. (1983b). Lesion-induced mossy fibers to the molecular layer of the rat fascia dentata: Identification of postsynaptic granule cells by the Golgi-EM technique. *J. Comp. Neurol.* **215,** 299.

Frotscher, M., Rinne, U., Hassler, R., and Wagner, A. (1981). Termination of cortical afferents on identified neurons in the caudate nucleus of the cat. A combined Golgi-EM degeneration study. *Exp. Brain Res.* **41,** 329.

Frotscher, M., Léránth, Cs., Lübbers, K., and Oertel, W. H. (1984). Commissural efferents innervate glutamate decarboxylase immunoreactive nonpyramidal neurons in the guinea pig hippocampus. *Neurosci. Lett.* **46,** 137.

Frotscher, M., Schlander, M., and Léránth, C. (1986). Cholinergic neurons in the hippocampus. A combined light and electron microscopic immunocytochemical study in the rat. *Cell Tissue Res.* **246,** 293–301.

Gabbott, P. L., and Somogyi, J. (1984). The 'single' section Golgi-impregnation procedure: Methodological description. *J. Neurosci. Methods* **11,** 221.

Gallyas, F., Görcs, T., and Merchenthaler, I. (1982). High grade intensification of the end-product of the diaminobenzidine reaction for peroxidase activity. *J. Histochem. Cytochem.* **30,** 183.

Gentschev, T., and Sotelo, C. (1973). Degeneration patterns in the ventral cochlear nucleus of the rat after primary deafferentation. An ultrastructural study. *Brain Res.* **62,** 37.

Gerfen, C. R., and Sawchenko, P. E. (1984). An anterograde neuroanatomical tracing method that shows the detailed morphology of neurons, their axons and terminals: Immunohistochemical localization of an axonally transported plant lectin, *Phaseolus vulgaris*-leucoagglutinin (PHA-L). *Brain Res.* **290,** 219.

Gerfen, C. R., and Sawchenko, P. E. (1985). A method for anterograde axonal tracing of chemically specified circuits in the central nervous system: Combined *Phaseolus vulgaris*-leucoagglutinin (PHA-L) tract tracing and immunohistochemistry. *Brain Res.* **343,** 144.

Gonatas, N. K., Harper, C., Mizutani, T., and Gonatas, J. O. (1979). Superior sensitivity of conjugates of horseradish peroxidase with wheat germ agglutinin for studies of retrograde axonal transport. *J. Histochem. Cytochem.* **27,** 728.

Graham, R. C., Jr., and Karnovsky, M. J. (1966). The early stages of absorption of injected horseradish peroxidase in the proximal tubules of mouse kidney: Ultrastructural cytochemistry by a new technique. *J. Histochem. Cytochem.* **14,** 291.

Gray, E. G., and Hamlyn, L. H. (1962). Electron microscopy of experimental degeneration in the avian optic tectum. *J. Anat.* **96,** 309.

Groenewegen, H. J., and Van Dijk, C. A. (1984). Efferent connections of the dorsal tegmental region in the rat, studied by means of anterograde transport of the lectin *Phaseolus vulgaris*-leucoagglutinin (PHA-L). *Brain Res.* **304,** 367.

Groenewegen, H. J., and Wouterlood, F. G. (1984). Light and electron microscopic identification of neuronal somata, dendrites, axons and axon terminals with *Phaseolus vulgaris* leucoagglutinin (PHA-L). *Abstr. Soc. Neurosci.* **10,** 421.

Harris, K. M., Cruce, W. L. R., Greenough, W. T., and Teyler, T. J. (1980). A Golgi impregnation technique for thin brain slices maintained in vitro. *J. Neurosci. Methods* **2,** 363.

Hattori, T., Singh, V. K., McGeer, E. G., and McGeer, P. L. (1976). Immunohistochemical localization of choline acetyltransferase containing neostriatal neurons and their relationship with dopaminergic synapses. *Brain Res.* **102,** 164.

Hayashi, H., Takagi, H., Takeda, N., Kubota, Y., Tohyama, M., Watanabe, T., and Wada, H. (1984). Fine structure of histaminergic neurons in the caudal magnocellular nucleus of the rat as demonstrated by immunocytochemistry using histidine decarboxylase as a marker. *J. Comp. Neurol.* **229,** 233.

Hendrickson, A. E. (1969). Electron microscopic radioautography: Identification of origin of synaptic terminals in normal nervous tissue. *Science* **165,** 194.

Hendry, S. H. C., and Jones, E. G. (1983). Thalamic inputs to identified commissural neurons in the monkey somatic sensory cortex. *J. Neurocytol.* **12,** 299.

Holländer, H., and Vaaland, J. L. (1968). A reliable staining method for semi-thin sections in experimental neuroanatomy. *Brain Res.* **10,** 120.

Houser, C. R., Crawford, G. D., Barber, R. P., Salvaterra, P. M., and Vaughn, J. E. (1983). Organization and morphological characteristics of cholinergic neurons: An immunocytochemical study with a monoclonal antibody to choline acetyltransferase. *Brain Res.* **266,** 97.

Houser, C. R., Crawford, G. D., Salvaterra, P. M., and Vaughn, J. E. (1985). Immunocytochemical localization of choline acetyltransferase in rat cerebral cortex: A study of cholinergic neurons and synapses. *J. Comp. Neurol.* **234,** 17.

Hunt, S. P., and Künzle, H. (1976). Selective uptake and transport of label within three identified neuronal systems after injection of ^{3}H-GABA into the pigeon optic tectum. An autoradiographic and Golgi study. *J. Comp. Neurol.* **170,** 173.

Ingham, C. A., Bolam, J. P., Wainer, B. H., and Smith, A. D. (1985). A correlated light and electron microscopic study of identified cholinergic basal forebrain neurons that project to the cortex in the rat. *J. Comp. Neurol.* **239,** 176.

Jonkers, B. W., Sterk, J. C., and Wouterlood, F. G. (1984). Transcardial perfusion fixation of the CNS by the use of a compressed-air-driven perfusion device. *J. Neurosci. Methods* **12,** 141.

Kiss, J., and Halász, B. (1985). Demonstration of serotoninergic axons terminating on luteinizing hormone-releasing hormone neurons in the preoptic area of the rat using a combination of immunocytochemistry and high resolution autoradiography. *Neuroscience* **14,** 69.

Köhler, C., Chan-Palay, V., and Wu, J.-Y. (1984). Septal neurons containing glutamic acid decarboxylase immunoreactivity project to the hippocampal region in the rat brain. *Anat. Embryol.* **169,** 41.

Kristensson, K., Olsson, Y., and Sjöstrand, J. (1971). Axonal uptake and retrograde transport of exogeneous proteins in the hypoglossal nerve. *Brain Res.* **32,** 399.

Kruger, L., and Maxwell, D. S. (1969). Wallerian degeneration in the optic nerve of a reptile: An electron microscopic study. *Am. J. Anat.* **125,** 247.

Kuypers, H. G. J. M., Catsman-Berrevoets, C. E., and Padt, R. E. (1977). Retrograde axonal transport of fluorescent substances in the rat's forebrain. *Neurosci. Lett.* **6,** 127.

Kuypers, H. G. J. M., Bentivoglio, M., Van der Kooy, D., and Catsman-Berrevoets, C. E. (1979). Retrograde transport of bisbenzimide and propidium iodide through axons to their parent cell bodies. *Neurosci. Lett.* **12,** 1.

Lakos, S., and Basbaum, A. I. (1986). Benzidine dihydrochloride as a chromogen for single- and double-label light and electron microscopic immunocytochemical studies. *J. Histochem. Cytochem.* **34,** 1047.

Landas, S., and Phillips, I. M. (1982). Staining of human and rat brain Vibratome sections by a new Golgi method. *J. Neurosci. Methods* **5,** 147.

LaVail, J. H., and LaVail, M. M. (1972). Retrograde axonal transport in the central nervous system. *Science* **176,** 1416.

Léránth, Cs., and Frotscher, M. (1983). Commissural afferents to the rat hippocampus terminate on vasoactive intestinal polypeptide-like immunoreactive non-pyramidal neurons. An EM immunocytochemical degenerative study. *Brain Res.* **276,** 357.

Léránth, Cs., Williams, T. H., Chrétien, M., and Palkovits, M. (1980). Ultrastructural investigation of ACTH immunoreactivity in arcuate and supraoptic nuclei of the rat. *Cell Tissue Res.* **210,** 11.

Léránth, Cs., Frotscher, M., Tömböl, T., and Palkovits, M. (1984). Ultrastructure and synaptic connections of vasoactive intestinal polypeptide-like immunoreactive non-pyramidal neurons and axon terminals in the rat hippocampus. *Neuroscience* **12,** 531.

Léránth, Cs., Sakamoto, H., MacLusky, N. J., Shanabrough, M., and Naftolin, F. (1985). Application of avidin-ferritin and peroxidase as contrasting electron-dense markers for simultaneous electron microscopic immunocytochemical labelling of glutamic acid decarboxylase and tyrosine hydroxylase in the rat arcuate nucleus. *Histochemistry* **82,** 165.

Liposits, Z., Lengvári, I., Vigh, S., Schally, A. V., and Flerkó, B. (1983). Immunohistological detection of degenerating CRF-immunoreactive nerve fibers in the median eminence after lesion of the paraventricular nucleus of the rat. A light and electron microscopic study. *Peptides (Fayetteville, N.Y.)* **4,** 941.

Liposits, Z., Görcs, T., and Trombitás, K. (1985). Ultrastructural analysis of central serotoninergic neurons immunolabeled by silver-gold intensified diaminobenzidin chromogen. Completion of immunocytochemistry with X-ray microanalysis. *J. Histochem. Cytochem.* **33,** 604.

Liposits, Zs., Phelix, C., and Paull, W. K. (1986). Adrenergic innervation of corticotropin releasing factor (CRF)-synthesizing neurons in the hypothalamic paraventricular nucleus of the rat. A combined light and electron microscopic immunocytochemical study. *Histochemistry* **84,** 201.

Ljungdahl, A., Hökfelt, T., Goldstein, M., and Park, D. (1975). Retrograde peroxidase tracing of neurons combined with transmitter histochemistry. *Brain Res.* **84,** 313.

McLean, I. W., and Nakane, P. K. (1974). Periodate-lysine-paraformaldehyde fixative. A new fixative for immunoelectron microscopy. *J. Histochem. Cytochem.* **22,** 1077.

Mesulam, M.-M. (1982). Principles of horseradish peroxidase neurochemistry and their applications for tracing neural pathways. Axonal transport, enzyme histochemistry and light microscopic analysis. *In* "Tracing Neural Connections with Horseradish Peroxidase" (M.-M. Mesulam, ed.), IBRO Handb. Ser., pp. 1–151. Wiley, Chichester.

Mihailoff, G. A., and Watt, C. B. (1981). Cerebral cortical afferent terminations on identified spiny basilar pontine neurons, a combined Golgi-EM degeneration study. *Brain Res. Bull.* **6,** 227.

Milner, T. A., and Pickel, V. M. (1986). Ultrastructural localization and afferent sources of substance P in the rat parabrachial region. *Neuroscience* **17,** 687.

Moore, R. Y. (1981). Methods for selective, restricted lesion placement in the central nervous system. *In* "Neuroanatomical Tract-Tracing Methods" (L. Heimer and M. J. Robards, eds.), pp. 55–89. Plenum, New York.

Mugnaini, E. (1985). GABA neurons in the superficial layers of the rat dorsal cochlear nucleus: Light and electron microscopic immunocytochemistry. *J. Comp. Neurol.* **235,** 61.

Mugnaini, E., and Friedrich, V. L., Jr. (1981). Electron microscopy: Identification and study of normal and degenerating neural elements by electron microscopy. *In* "Neuroanatomical Tract-Tracing Methods" (L. Heimer and M. J. Robards, eds.), pp. 377–406. Plenum, New York.

Nakane, P. K., and Pierce, G. B. (1967). Enzyme labelled antibodies for light and electron microscopic localization of tissue antigens. *J. Cell Biol.* **33,** 307.

Newman, G. R., and Jasani, B. (1984). Post-embedding immunoenzyme techniques. *In* "Immunolabeling for Electron Microscopy" (J. M. Polak, and I. M. Varndell, eds.), pp. 53–70. Elsevier, Amsterdam.

Oertel, W. H., Mugnaini, E., Schmechel, D. E., Tappaz, M. L., and Kopin, I. J. (1982a). The immunocytochemical demonstration of gamma aminobutyric acidergic neurons. Methods and application. *In* "Cytochemical Methods in Neuroanatomy" (V. Chan-Palay and S. L. Palay, eds.), pp. 297–329. Alan R. Liss, New York.

Oertel, W. H., Tappaz, M. L., Berod, A., and Mugnaini, E. (1982b). Two-color immunocytochemistry for dopamine and GABA neurons in rat substantia nigra and zona incerta. *Brain Res. Bull.* **9,** 463.

Olschowska, J. A., Molliver, M. E., Grzanna, R., Rice, F. L., and Coyle, J. T. (1981). Ultrastructural demonstration of noradrenergic synapses in the rat central nervous system by dopamine-β-hydroxylase immunocytochemistry. *J. Histochem. Cytochem.* **29,** 271.

Onteniente, B., Geffard, M., and Calas, A. (1984). Ultrastructural immunocytochemical study of the dopaminergic innervation of the rat lateral septum with anti-dopamine antibodies. *Neuroscience* **13,** 385.

Ordronneau, P., Lindström, P. B.-M., and Petrusz, P. (1981). Four unlabeled antibody bridge techniques: A comparison. *J. Histochem. Cytochem.* **29,** 1397.

Palkovits, M., Léránth, Cs., Jew, J. Y., and Williams, T. H. (1982). Simultaneous characterization of pre- and postsynaptic neuron contact sites in brain. *Proc. Natl. Acad. Sci. U.S.A.* **79,** 2705.

Peters, A., and Proskauer, C. C. (1980). Smooth or sparsely spined cells with myelinated axons in rat visual cortex. *Neuroscience* **5,** 2079.

Peters, A., Proskauer, C. C., Feldman, M. L., and Kimerer, L. (1979). The projection of the lateral geniculate nucleus to area 17 of the rat cerebral cortex. V. Degenerating axon terminals synapsing with Golgi impregnated neurons. *J. Neurocytol.* **8,** 331.

Phelps, P. E., and Vaughn, J. E. (1986). Immunocytochemical localization of choline acetyltransferase in rat ventral striatum: A light and electron microscopic study. *J. Neurocytol.* **15,** 595–617.

Phelps, P. E., Houser, C. R., and Vaughn, J. E. (1985). Immunocytochemical localization of choline acetyltransferase within the rat neostriatum. A correlated light and electron microscopic study of cholinergic neurons and synapses. *J. Comp. Neurol.* **238,** 286.

Pickel, V. M. (1981). Immunocytochemical methods. *In* "Neuroanatomical Tract-Tracing Methods" (L. Heimer and M. J. Robards, eds.), pp. 483–509. Plenum, New York.

Pickel, V. M., Joh, T. H., and Reis, D. J. (1975). Ultrastructural localization of tyrosine hydroxylase in noradrenergic neurons of brain. *Proc. Natl. Acad. Sci. U.S.A.* **72,** 659.

Pickel, V. M., Joh, T. H., and Reis, D. J. (1976). Monoamine-synthesizing enzymes in central dopaminergic, noradrenergic and serotonergic neurons. Immunocytochemical localization by light and electron microscopy. *J. Histochem. Cytochem.* **24,** 792.

Pickel, V. M., Reis, D. J., and Leeman, S. E. (1977). Ultrastructural localization of substance P in neurons of rat spinal cord. *Brain Res.* **122,** 534.

Pickel, V. M., Joh, T. H., Reis, D. J., Leeman, S. E., and Miller, R. J. (1979). Electron microscopic localization of substance P and enkephalin in axon terminals related to dendrites of catecholaminergic neurons. *Brain Res.* **160,** 387.

Pickel, V. M., Joh, T. H., Chan, J., and Beaudet, A. (1984). Serotoninergic terminals: Ultrastructure and synaptic interaction with catecholamine-containing neurons in the medial nuclei of the solitary tracts. *J. Comp. Neurol.* **225,** 291.

Pickel, V. M., Chan, J., and Milner, T. (1986a). Autoradiographic detection of ^{125}I-secondary antiserum: A sensitive light and electron microscopic labeling method compatible with peroxidase immunocytochemistry for dual localization of neuronal antigens. *J. Histochem. Cytochem.* **34,** 707–718.

Pickel, V. M., Chan, J., Park, D. H., Joh, T. H., and Milner, T. A. (1986b). Ultrastructural localization of phenylethanolamine N-methyltransferase in sensory and motor nuclei of the vagus nerve. *J. Neurosci. Res.* **15,** 439–456.

Piekut, D. T., and Casey, S. M. (1983). Penetration of immunoreagents in vibratome-sectioned brain: A light and electron microscope study. *J. Histochem. Cytochem.* **31,** 669.

Pinching, A. J. (1969). Persistence of post-synaptic thickenings after degeneration of olfactory nerves. *Brain Res.* **16,** 277.

Pool, C. W., Buijs, R. M., Swaab, D. F., Boer, G. J., and van Leeuwen, F. W. (1983). On the way to specific immunocytochemical localization. *In* "Immunocytochemistry" (A. C. Cuello, ed.), IBRO Hand. Ser., pp. 1–46. Wiley, Chichester.

Priestley, J. V. (1984). Pre-embedding ultrastructural immunocytochemistry: Immunoenzyme techniques. *In* "Immunolabeling for Electron Microscopy" (J. M. Polak and I. M. Varndell, eds.), pp. 37–52. Elsevier, Amsterdam.

Priestley, J. V., and Cuello, A. C. (1983). Electron microscopic immunocytochemistry for CNS transmitters and transmitter markers. *In* "Immunocytochemistry" (A. C. Cuello, ed.), IBRO Hand. Ser., Vol. 3, pp. 273–322. Wiley, Chichester.

Priestley, J. V., Somogyi, P., and Cuello, A. C. (1981). Neurotransmitter-specific projection neurons revealed by combining PAP-immunohistochemistry with retrograde transport of HRP. *Brain Res.* **220,** 231.

Puchtler, H., and Meloan, S. N. (1985). On the chemistry of formaldehyde fixation and its effects on immunohistochemical reactions. *Histochemistry* **82,** 201.

Ramón y Cajal, S. (1972). "Histologie du système nerveux de l'homme et des vertébrés," 2nd Printing. Consejo Superior de Investigaciones Cientificas, Madrid.

Repérant, J., and Rio, J.-P. (1976). Retinal projections in *Vipera aspis*. A reinvestigation using light radioautographic and electron microscopic techniques. *Brain Res.* **107,** 603.

Ribak, C. E. (1978). Aspinous and sparsely-spinous stellate neurons in the visual cortex of rats contain glutamic acid decarboxylase. *J. Neurocytol.* **7,** 461.

Ribak, C. E., Vaughn, J. E., and Saito, K. (1978). Immunocytochemical localization of glutamic acid decarboxylase in neuronal somata following colchicine inhibition of axonal transport. *Brain Res.* **140,** 315.

Ribak, C. E., Vaughn, J. E., and Barber, R. P. (1981). Immunocytochemical localization of GABAergic neurones at the electron microscopic level. *Histochem. J.* **13,** 555.

Rye, D. B., Saper, C. B., and Wainer, B. H. (1984). Stabilization of the tetramethylbenzidine (TMB) reaction product: Application for retrograde and anterograde tracing, and combination with immunohistochemistry. *J. Histochem. Cytochem.* **32,** 1145.

Schmued, L. C., and Fallon, J. H. (1986). Fluoro-Gold: A new fluorescent retrograde axonal tracer with numerous unique properties. *Brain Res.* **377,** 147.

Schönitzer, K., and Holländer, H. (1981). Anterograde tracing of horseradish peroxidase (HRP) with the electron microscope using the tetramethylbenzidine reaction. *J. Neurosci. Methods* **4,** 373.

Schwab, M. E., Javoy-Agid, F., and Agid, Y. (1978). Labeled wheat germ agglutinin (WGA) as a new, highly sensitive retrograde tracer in the rat brain hippocampal system. *Brain Res.* **152,** 145.

Seguela, P., Gamrani, H., Geffard, M., Calas, A., and Le Moal, M. (1985). Ultrastructural immunocytochemistry of γ-aminobutyrate in the cerebral and cerebellar cortex of the rat. *Neuroscience* **16,** 865.

Seress, L., and Ribak, C. E. (1984). Direct commissural connections to the basket cells of the hippocampal dentate gyrus: Anatomical evidence for feed-forward inhibition. *J. Neurocytol.* **13,** 215.

Skirboll, L., Hökfelt, T., Norell, G., Phillipson, O., Kuypers, H. G. J. M., Bentivoglio, M., Catsman-Berrevoets, C. E., Visser, T. J., Steinbusch, H., Verhofstad, A., Cuello, A. C., Goldstein, M., and Brownstein, M. (1984). A method for specific transmitter identification of retrogradely labeled neurons: Immunofluorescence combined with fluorescence tracing. *Brain Res. Rev.* **8,** 99.

Smith, L. M., Ebner, F. F., and Colonnier, M. (1980). The thalamocortical projection in *Pseudemys* turtles: A quantitative electron microscopic study. *J. Comp. Neurol.* **190,** 445.

Somogyi, P. (1978). The use of Golgi stained cells and of experimental degeneration under the electron microscope: A direct method for the identification in the visual cortex of three successive links in a neuron chain. *Neuroscience* **3,** 167.

Somogyi, P., and Hodgson, A. J. (1985). Antisera to γ-aminobutyric acid. III. Demonstration of GABA in Golgi-impregnated neurons and in conventional electron microscopic sections of cat striate cortex. *J. Histochem. Cytochem.* **33,** 249.

Somogyi, P., and Smith, A. D. (1979). Projection of neostriatal spiny neurons to the substantia nigra. Application of a combined Golgi-staining and horseradish peroxidase procedure at both light and electron microscopic levels. *Brain Res.* **178,** 3.

Somogyi, P., and Takagi, H. (1982). A note on the use of picric acid-paraformaldehyde-glutaraldehyde fixative for correlated light and electron microscopic immunocytochemistry. *Neuroscience* **7,** 1779.

Somogyi, P., Hodgson, A. J., and Smith, A. D. (1979). An approach to tracing neuron networks in the cerebral cortex and basal ganglia. Combination of Golgi staining, retrograde transport of horseradish peroxidase and anterograde degeneration of synaptic boutons in the same material. *Neuroscience* **4,** 1805.

Somogyi, P., Bolam, J. P., and Smith, A. D. (1981a). Monosynaptic cortical input and local axon collaterals of identified striatonigral neurons. A light and electron microscopic study using the Golgi-peroxidase transport-degeneration procedure. *J. Comp. Neurol.* **195,** 567.

Somogyi, P., Bolam, J. P., Totterdell, S., and Smith, A. D. (1981b). Monosynaptic input from the nucleus accumbens-ventral striatum region to retrogradely labelled nigrostriatal neurones. *Brain Res.* **217,** 245.

Somogyi, P., Freund, T. F., Halász, N., and Kisvárday, Z. F. (1981c). Selectivity of neuronal ^{3}H GABA accumulation in the visual cortex as revealed by Golgi staining of the labeled neurons. *Brain Res.* **225,** 431.

Somogyi, P., Priestley, J. V., Cuello, A. C., Smith, A. D., and Takagi, H. (1982). Synaptic connections of enkephalin-immunoreactive nerve terminals in the neostriatum: A correlated light and electron microscopic study. *J. Neurocytol.* **11,** 779.

Somogyi, P., Freund, T. F., and Kisvárday, Z. F. (1984). Different types of ^{3}H-GABA accumulating neurons in the visual cortex of the rat. Characterization by combined autoradiography and Golgi impregnation. *Exp. Brain Res.* **54,** 45.

Somogyi, P., Hodgson, A. J., Chubb, I. W., Penke, B., and Erdei, H. (1985). Antisera to γ-aminobutyric acid. II. Immunocytochemical application to the central nervous system. *J. Histochem. Cytochem.* **33,** 240.

Sotelo, C. (1971). The fine structural localization of norepinephrine-^{3}H in the sustantia nigra and area postrema of the rat. An autoradiographic study. *J. Ultrastruct. Res.* **36,** 824.

Spreafico, R., Cheema, S., Ellis, L. C., Jr., and Rustioni, A. (1982). On "comparison of horseradish peroxidase visualization methods." *J. Histochem. Cytochem.* **30,** 487.

Sternberger, L. A. (1979). "Immunocytochemistry," 2nd ed. Wiley, New York.

Stoeckel, K., Schwab, M., and Thoenen, H. (1977). Role of gangliosides in the uptake and retrograde axonal transport of cholera and tetanus toxin as compared to nerve growth factor and wheat germ agglutinin. *Brain Res.* **132,** 273.

Stürmer, C., Bielenberg, K., and Spatz, W. B. (1981). Electron microscopical identification of 3,3′,5,5′-tetramethylbenzidine-reacted horseradish peroxidase after retrograde axoplasmic transport. *Neurosci. Lett.* **23,** 1.

Takagi, H., Mizuta, H., Matsuda, T., Inagaki, S., Tateishi, K., and Hamaoka, T. (1984). The occurrence of cholecystokinin-like immunoreactive neurons in the rat neostriatum: Light and electron microscopic analysis. *Brain Res.* **309,** 346.

Takeda, N., Inagaki, S., Shiosaka, S., Taguchi, Y., Oertel, W. H., Tohyama, M., Watanabe, T., and Wada, H. (1984). Immunohistochemical evidence for the coexistence of histidine decarboxylase-like and glutamate decarboxylase-like immunoreactivities in nerve cells of the magnocellular nucleus of the posterior hypothalamus of rats. *Proc. Natl. Acad. Sci. U.S.A.* **81,** 7647.

Ter Horst, G. J., Groenewegen, H. J., Karst, H., and Luiten, P. G. M. (1984). *Phaseolus vulgaris* leuco-agglutinin immunohistochemistry. A comparison between autoradiographic and lectin tracing of neuronal efferents. *Brain Res.* **307,** 379.

Totterdell, S., Bolam, J. P., and Smith, A. D. (1984). Characterization of pallidonigral neurons in the rat by a combination of Golgi impregnation and retrograde transport of horseradish peroxidase: Their monosynaptic input from the neostriatum. *J. Neurocytol.* **13,** 593.

Van den Pol, A. N. (1985). Silver-intensified gold and peroxidase as dual ultrastructural immunolabels for pre- and postsynaptic neurotransmitters. *Science* **228,** 332.

Van den Pol, A. N., Herbst, R. S., and Powell, J. F. (1984). Tyrosine hydroxylase-immunoreactive neurons of the hypothalamus. A light and electron microscopic study. *Neuroscience* **13,** 1117.

Van den Pol, A. N., Smith, A. D., and Powell, J. F. (1985). GABA axons in synaptic contact with dopamine neurons in the substantia nigra: Double immunocytochemistry with biotin-peroxidase and protein A-colloidal gold. *Brain Res.* **348,** 146.

Varndell, I. M., and Polak, J. M. (1984). Double-immunostaining procedures: Techniques and applications. *In* "Immunolabelling for Electron Microscopy" (J. M. Polak and I. M. Varndell, eds.), pp. 155–178. Elsevier, Amsterdam.

Vinores, S. A., Herman, M. M., Rubinstein, L. J., and Marangos, P. J. (1984). Electron microscopic localization of neuron-specific enolase in rat and mouse brain. *J. Histochem. Cytochem.* **32,** 1295.

Voorn, P., and Buijs, R. M. (1983). An immuno-electronmicroscopical study comparing vasopressin, oxytocin, substance P and enkephalin containing nerve terminals in the nucleus of the solitary tract of the rat. *Brain Res.* **270,** 169.

Wainer, B. H., Bolam, J. P., Freund, T. F., Henderson, Z., Totterdell, S., and Smith, A. D. (1984). Cholinergic synapses in the rat brain: A correlated light and electron microscopic immunohistochemical study employing a monoclonal antibody against choline acetyltransferase. *Brain Res.* **308,** 69.

Warr, W. B., De Olmos, J. S., and Heimer, L. (1981). Horseradish peroxidase. The basic procedure. *In* "Neuroanatomical Tract-Tracing Methods" (L. Heimer and M. J. Robards, eds.), pp. 207–262. Plenum, New York.

Westrum, L. E. (1969). Electron microscopy of degeneration in the lateral olfactory tract and plexiform layer of the prepyriform cortex of the rat. *Z. Zellforsch. Mikrosk. Anat.* **98,** 157.

Westrum, L. E. (1973). Early forms of terminal degeneration in the spinal trigeminal nucleus following rhizotomy. *J. Neurocytol.* **2,** 189.

White, E. L. (1978). Identified neurons in the mouse SMI cortex which are postsynaptic to thalamocortical terminals: A combined Golgi-electron microscopic and degeneration study. *J. Comp. Neurol.* **181,** 627.

White, E. L., and Hersch, S. M. (1982). A quantitative study of thalamocortical and other synapses involving the apical dendrites of corticothalamic projection cells in mouse SmI cortex. *J. Neurocytol.* **11,** 137.

White, E. L., Hersch, S. M., and Rock, M. P. (1980). Synaptic sequences in mouse SmI cortex involving pyramidal cells labeled by retrograde filling with horseradish peroxidase. *Neurosci. Lett.* **19,** 149.

White, E. L., Benshalom, G., and Hersch, S. M. (1984). Thalamocortical and other synapses involving nonspiny multipolar cells of mouse SmI cortex. *J. Comp. Neurol.* **229,** 311.

Williams, M. N., and Faull, R. L. M. (1985). The striatonigral and nigrotectal neurons in the rat. A correlated light and electron microscopic study demonstrating a monosynaptic striatal input to identified nigrotectal neurons using a combined degeneration and horseradish peroxidase procedure. *Neuroscience* **14,** 991.

Willingham, M. C. (1983). An alternative fixation-processing method for preembedding ultrastructural immunocytochemistry of cytoplasmic antigens: The GBS (glutaraldehyde-borohydride-saponin) procedure. *J. Histochem. Cytochem.* **31,** 791.

Wiśniewski, H. M., Ghetti, B., and Horoupian, D. S. (1972). The fate of synaptic membranes of degenerating optic nerve terminals, and their role in the mechanism of transsynaptic changes. *J. Neurocytol.* **1,** 297.

Wouterlood, F. G. (1979). Light microscopic identification of Golgi impregnated CNS neurons during sectioning for electron microscopy. *Stain Technol.* **54,** 325.

Wouterlood, F. G., and Groenewegen, H. J. (1985). Neuroanatomical tracing by use of

Phaseolus vulgaris-leucoagglutinin (PHA-L): Electron microscopy of PHA-L-filled neuronal somata, dendrites, axons and axon terminals. *Brain Res.* **326,** 188.

Wouterlood, F. G., and Mugnaini, E. (1984). Cartwheel neurons of the dorsal cochlear nucleus. A Golgi-electron microscopic study in rat. *J. Comp. Neurol.* **227,** 136.

Wouterlood, F. G., and Nederlof, J. (1983). Terminations of olfactory afferents on layer II and III neurons in the entorhinal area: Degeneration-Golgi-electron microscopic study in the rat. *Neurosci. Lett.* **36,** 105.

Wouterlood, F. G., Alones, V. E., Elprana, D., and Lohman, A. H. M. (1982). Terminal degeneration in the mediodorsal cortex of the lizard *Agama agama:* Light and electron microscopy. *J. Morphol.* **172,** 45.

Wouterlood, F. G., Nederlof, J., and Paniry, S. (1983). Chemical reduction of silver chromate: A procedure for electron microscopical analysis of Golgi-impregnated neurons. *J. Neurosci. Methods* **7,** 295.

Wouterlood, F. G., Mugnaini, E., and Nederlof, J. (1985). Projection of olfactory bulb efferents to layer I GABA-ergic neurons in the entorhinal area. Combination of anterograde degeneration and immunoelectron microscopy in rat. *Brain Res.* **343,** 283.

Wouterlood, F. G., Sauren, Y. M. H. F., and Steinbusch, H. W. M. (1986a). Histaminergic neurons in rat brain: Correlative immunocytochemistry, Golgi impregnation and electron microscopy. *J. Comp. Neurol.* **252,** 227.

Wouterlood, F. G., Steinbusch, H. W. M., and J. G. J. M. Bol. (1986b). Anterograde tracing with *Phaseolus vulgaris*-leucoagglutinin (PHA-L), combined with labeling of the target neurons by peroxidase- or alkaline phosphatase immunocytochemistry. *Abstr. Soc. Neurosci.* **12,** 444.

Wouterlood, F. G., Bol, J. G. J. M. and Steinbusch, H. W. M. (1987a). Double-label immunocytochemistry: Combination of anterograde neuroanatomical tracing with *Phaseolus vulgaris*-leucoagglutinin and enzyme immunocytochemistry of target neurons. *J. Histochem. Cytochem.* (In press.)

Wouterlood, F. G., Sauren, Y. M. H. F., and Pattiselanno, A. (1987b). The compromise between penetration of antigens and preservation of ultrastructure: Application of the surfactant Kodak Photo-Flo in pre-embedding electron microscopic immunocytochemistry. (In preparation.)

Wouterlood, F. G., Paniry, S., and Pattiselanno, A. (1987c). Stabilization of silver chromate Golgi impregnation in rat CNS neurons by means of photographic developers. II. Electron microscopy. *Stain Technol.* (In press).

Wu, J.-Y., Lin, C.-T., Brandon, C., Chan, T.-S., Möhler, H., and Richards, J. G. (1982). Regulation and immunocytochemical characterization of glutamic acid decarboxylase. *In* "Cytochemical Methods in Neuroanatomy" (V. Chan-Palay and S. L. Palay, eds.), pp. 279–296. Alan R. Liss, Inc., New York.

Záborszky, L., Léránth, Cs., and Heimer, L. (1984). Ultrastructural evidence of amygdalofugal axons terminating on cholinergic cells of the rostral forebrain. *Neurosci. Lett.* **52,** 219.

Záborszky, L., Heimer, L., Eckenstein, F., and Léránth, C. (1986). GABAergic input to cholinergic forebrain neurons: An ultrastructural study using retrograde tracing of HRP and double immunolabeling. *J. Comp. Neurol.* **250,** 282.

Zahm, D. S., and Heimer, L. (1985). Synaptic contacts of ventral striatal cells in the olfactory tubercle of the rat: Correlated light and electron microscopy of anterogradely transported *Phaseolus vulgaris*-leucoagglutinin. *Neurosci. Lett.* **60,** 169.

5

Intracellular Staining of Neurons in the Central Nervous System: Combined Physiological, Morphological, and Cytological Analysis of Identified Neurons

GEORGIA A. BISHOP AND JAMES S. KING

Department of Anatomy and Neuroscience Program
The Ohio State University
Columbus, Ohio

INTRODUCTION

Chapter 4 in this volume describes procedures for correlating light and electron microscopic analyses of populations of neurons and Chapter 8 deals with visualizing individual neurons using two different techniques in an *in vitro* model. The intent of this chapter is to describe methods for: (1) intracellular recording of physiological data from single neurons in the central nervous system in an *in vivo* preparation; (2) injecting horseradish peroxidase (HRP) into the same neuron which, subsequent to appropriate histochemical processing, will allow for the visualization of the cell from which the physiological data were obtained; and (3) processing the tissue for electron microscopy so that the cytological features and synaptic relationships of the identified neuron can be analyzed.

The major advantages of this combined physiological–morphological–cytological technique are as follows. First the majority of nuclei in the central nervous system are composed of a heterogeneous population of neurons, often intermingled with each other. Thus when physiological data are obtained from a neuron located within a particular nucleus it is often difficult to determine the specific type of cell from which the data were obtained. Conversely, neuroanatomists have a variety of techniques at their disposal which allow them to determine the somadendritic morphology (e.g., Golgi technique), the neurochemical anatomy (e.g., immunohistochemistry), and the connections (e.g., anterograde and retrograde tracing techniques) of populations of neurons. However, they can only speculate on the physiological properties of the neurons of interest. Intracellular staining of individual neurons resolves the problems of both the physiologist and the neuroanatomist as this technique provides both physiological and morphological information for the same cell (e.g.,

Jankowska *et al.*, 1976; Kitai *et al.*, 1976; Light and Durkovic, 1976; McCrea *et al.*, 1976, 1977; Snow *et al.*, 1976; Kocsis and Kitai, 1977; Kocsis *et al.*, 1977; Christensen and Ebner, 1978; Gilbert and Wiesel, 1978; Chang *et al.*, 1979; Light and Perl, 1979; Light *et al.*, 1979; Preston *et al.*, 1980; VanderMaelen *et al.*, 1979; Bennett *et al.*, 1980; Deschenes and Landry, 1980; Landry and Labelle, 1980; Nahvi *et al.*, 1980; Ribak *et al.*, 1980; Spencer *et al.*, 1980; Bishop *et al.*, 1981; Bishop and King, 1986; Bishop *et al.*, 1987).

Another advantage of the intracellular staining technique is that HRP is easily visualized at the electron microscopic level (e.g., Jankowska *et al.*, 1976; Muller and McMahan, 1976; Cullheim *et al.*, 1977; Christensen and Ebner, 1978; King and McCrea, 1978; Bishop *et al.*, 1979a, 1980a,b; Chang *et al.*, 1979; Gobel *et al.*, 1980; King, 1980; Ribak *et al.*, 1980; King and Bishop, 1982; Bishop and King, 1986). Thus, the cytological features and synaptic relationships, both pre- and postsynaptic, of physiologically and morphologically characterized neurons can be analyzed. In ultrastructural studies it is often necessary to carry out tedious serial section reconstructions in order to confirm whether a process under investigation is continuous with a particular neuron. Analysis of intracellularly injected neurons obviates this procedure as all of the processes which are derived from a specific cell are labeled. Thus, even small discontinuous processes in the neuropil can be identified as belonging to a particular cell. If serial section reconstruction is required for a particular paradigm, it is facilitated by the presence of an opaque reaction product within the process of interest on each adjacent section. Finally, it is possible to correlate a specific process seen under a light microscope with its ultrastructural equivalent. Details of this procedure will be described in succeeding parts of this chapter.

Given the advantages of the procedure, what are the types of questions that can be addressed using this technique? Applications of the intracellular staining procedure include its use in: (1) developmental studies in which a correlative study of the physiological and morphological maturation of identified neurons can be carried out (Crepel *et al.*, 1980); (2) studies designed to trace the course and distribution of the axons of neurons (e.g., Cullheim and Kellereth, 1976; Snow *et al.*, 1976; McCrea *et al.*, 1976, 1977; Bishop *et al.*, 1979a,b; Mason and Robson, 1979; Robson and Mason, 1979; Preston *et al.*, 1980; Deniau *et al.*, 1980); (3) *in vitro* paradigms (e.g., Nelson *et al.*, 1978; Calvet and Calvet, 1979; Hendelman and Marshall, 1980; Llinás and Yarom, 1980; Prince *et al.*, 1978); and (4) in the analysis of local circuitry (e.g., Cullheim and Kellereth, 1976; Kitai *et al.*, 1976; McCrea *et al.*, 1976, 1977; Snow *et al.*, 1976; Cullheim *et al.*, 1977; Bishop *et al.*, 1979a, 1980a,b, 1986; Preston *et al.*, 1980; Chang *et al.*, 1979; Deniau *et al.*, 1980; Gobel *et al.*, 1980; Bishop, 1982). The use of the intracellular staining paradigm in the analysis of local circuitry will be used as the focus of this chapter as specific procedures for carrying out the experiments and analyzing data are discussed. The model system that will be used is the recurrent axonal collaterals of Purkinje cells in the feline cerebellum. The procedures described for carrying out an analysis of the distribution of these collaterals can be applied to address any question in the central nervous system in any species, both in the *in vivo* and *in vitro* paradigms.

Use of the Intracellular Staining Technique to Analyze Local Circuitry within the Cerebellar Cortex

Local circuits have been defined as connections that are wholly contained within the boundaries of a given area (Shepherd, 1979) whose function is to process information at the local level rather than to propagate it out of the given structure (Rakic, 1979). These local circuits may represent important substrates for information processing in the central nervous system as the number of local circuits appears to increase in more phylogenetically advanced species (Ramón y Cajál, 1899). Rakic (1976) postulated that involvement of local circuits in certain pathological conditions may be responsible for several poorly defined neurological syndromes. The introduction of intracellular markers such as horseradish peroxidase (HRP) allows for the detailed analysis of the physiological, morphological, and cytological characteristics of intrinsic circuits.

The precise laminar organization of the cerebellum and the well-defined physiological properties of the neuronal elements within the cortex make it an ideal model for analyzing local circuitry. In particular, the focus of this chapter will be on the local collaterals derived from the axons of Purkinje cells. These neurons have a dual role in cerebellar circuitry. First, they inhibit (Kitai *et al.*, 1977; Ito *et al.*, 1964, 1970) neurons in the deep cerebellar nuclei which are located within the white matter core of the cerebellum. In addition, Purkinje cells give rise to axonal collaterals that remain within the cerebellar cortex and play a role in the intrinsic circuitry of the cortical mantle. The circuit formed between Purkinje cells and their target neurons in the deep cerebellar nuclei has been defined as a corticonuclear complex (Oscarsson, 1979) which has been proposed to represent the basic structural–functional unit of the cerebellum. At present, little is known about the intrinsic organization of the corticonuclear complex. It is likely that the recurrent collaterals of Purkinje cells play a role in integrating information within and between these units. Specific questions related to local circuitry within the cerebellar cortex, which can be addressed by the intracellular staining technique, include the following.

1. Can the response characteristics of Purkinje cells, subsequent to activation of peripheral nerves, be correlated with the pattern of distribution of the collaterals? Various peripheral nerves may be activated to determine the areas of the body which provide an input to each impaled Purkinje cell. After histochemical processing the distribution of the collaterals from injected cells can be correlated with their physiological responses. Comparisons can be made between the patterns of collateral distribution for cells with similar and different physiological responses.
2. Do the local collaterals derived from Purkinje cells in all areas of the cerebellar cortex have the same pattern of distribution? Recent studies have shown that the cerebellar cortex is not a homogeneous structure. Differences exist with respect to the number and size of neurons in various areas of the cortex (Lange, 1982), the size and orientation of the dendritic trees of Purkinje cells (Fox *et al.*, 1967; Palay and Chan-Palay, 1974), the chemical neuroanatomy of the cortex (Takeuchi *et al.*, 1982; Bishop *et al.*, 1985; King *et al.*, 1987), and the pattern of distribution of Purkinje cell collaterals in the intermediate cortex of the anterior lobe versus the paramedian lobule (Bishop, 1982), as well as between Purkinje cells in zones *a* and *b* of the vermis (Blake *et al.*, 1986; Bishop *et al.*, 1986) and zone *c* laterally (O'Donoghue *et al.*, 1986). The question posed may be addressed by injecting Purkinje cells in different areas of the cerebellum and comparing the distribution of their local collaterals.
3. Do Purkinje cell collaterals remain within the confines of a cortical zone or do they extend to adjacent zones? The cerebellar cortex has been divided into eight parasagittal zones on the basis of physiological and anatomical studies (Ito, 1984). These zones are designated from medial to lateral, *a, x, b,* (located in the vermis), c_1, c_2, c_3 (comprising the intermediate cortex of the anterior lobe and the paramedian lobule) and d_1 and d_2 (which include the lateral hemispheres). Subsequent to serial section reconstruction of the axon collaterals of Purkinje cells localized to a particular zone, it may be possible to determine whether any of the collaterals cross the borders of the zone of origin, and, further, to define their terminal distribution in an adjacent zone.
4. What is the spatial distribution of the collateral? What is the density of distribution of varicosities within the area encompassed by the collateral? The questions can be addressed by serial section analysis of the distribution of varicosities located within the plexus.
5. What are the postsynaptic targets of the axon collaterals within their zone of origin as well as in adjacent zones? The HRP reaction product is

electron opaque; thus, the cytological characteristics of intracellularly injected neurons can be analyzed (King and Bishop, 1982). Specific axonal branches can first be identified at the light microscopic level and then examined with the electron microscope.

METHODS

Intracellular Recording

Electrode Preparation

Forming Microelectrodes. The recording microelectrodes are made from 2- or 3-mm glass tubing which has been cleaned in nitric acid and thoroughly rinsed. The clean glass blanks are placed in an electrode puller in order to form micropipettes with a gradual taper and a small tip diameter ($< 0.25\ \mu$m).

Beveling Microelectrodes. The HRP molecule is large (approximately 40,000 mol. wt.). Experience has shown that it is difficult to move the molecule through a fine micropipette tip. However, if the electrode tip is too large, or blunt, it will either be difficult to impale a neuron or, if it is impaled, it may severely damage the cell. As a compromise, beveled electrodes are used. In this way the tip size may be increased while still yielding a sharp electrode. Ideally, bevels of 30–45° with tip diameters of 0.7–0.8 μm are desirable.

HRP Solution and Filling Microelectrodes. A 4% solution of HRP is made in KCl diluted to 0.5 *M* with phosphate buffer (pH 7.4). About 3–4 μl of solution is used to fill an electrode formed from 2-mm glass tubing. Any of the procedures designed to fill microelectrodes can be followed (Purves, 1981). Beveled electrodes filled with the HRP solution described above will have impedances ranging from 50 to 80 Mohms. Filled electrodes may be stored in a humid environment for up to 1 week at 4°C. However, if electrodes are not used immediately the impedance should be checked and the tip examined microscopically to ensure that no major changes have occurred from the time they were originally beveled and/or filled.

Surgical Preparation

Animals are first anesthetized with an appropriate agent. The choice of animal and anesthesia will depend on the specific experimental paradigm. In the model presented in this chapter, the cat is the animal of choice and a combination of sodium thialmyl [Surital—35 mg/kg given intraperitoneally (IP)] and α-chlorolose (70 mg/kg, IP) is used as the anesthetic agent. The sciatic, ulnar, and radial nerves are exposed bilaterally and bipolar stimulating electrodes placed on each one. These electrodes are made from flexible Teflon-insulated wire bared at the end and wrapped around each nerve. In addition, bipolar stimulating electrodes are made of insulated "00" insect pins, bared 1 mm at the tip. These electrodes are stereotaxically inserted into the inferior cerebellar peduncle for orthodromic activation of Purkinje cells.

Physiological Recording and HRP Injections

Figure 1 is a schematic drawing that illustrates the experimental paradigm for recording intracellularly from Purkinje cells and for injecting HRP. In any intracellular recording experiment it is important to have

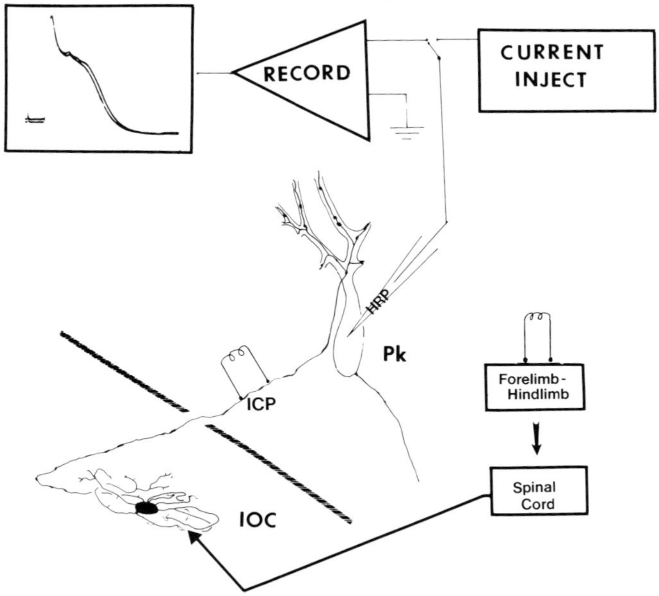

Fig. 1. Schematic drawing of the experimental paradigm used to record from and inject Purkinje cells in the cerebellar cortex. Purkinje cells (Pk) are activated via stimulation of the inferior cerebellar peduncle (ICP) which activates the axons of neurons located within the inferior olivary complex (IOC). This produces a characteristic response in Purkinje cells. In addition, cells are physiologically characterized as to whether they can be activated by forelimb and/or hindlimb stimulation. Subsequent to recording physiological data, the HRP is iontophoretically injected into the cell.

some physiological criteria for identifying the cell type which has been impaled. There are a number of different neuronal elements in the cerebellar cortex. However, only the Purkinje cell has a characteristic response to input from the inferior olive. The olivary afferents enter the cerebellum via the inferior cerebellar peduncle. Purkinje cells respond to this input (Fig. 1) with an initial action potential that is followed by a prolonged depolarization on which smaller secondary spike potentials are often superimposed. Impaled neurons that do not show this response are considered to be a different cell type and may or may not be injected.

After a cell has been identified as a Purkinje cell, the peripheral nerve electrodes are sequentially activated and the response, if any, to each site is recorded. After obtaining physiological data, cells which have a stable resting membrane potential (> 50 mV) are iontophoretically injected with HRP. Positive pulses (100 msec duration, at a frequency of 5/sec), 5–10 nA in amplitude are applied to the HRP filled microelectrode for 3–5 min. This is sufficient time to completely fill the soma, dendrites, initial portion of the axon, and axonal collaterals of all cells (Figs. 2–4). After the injection, additional physiological data can be obtained from the cell if desired (Fig. 5). The electrode is then withdrawn from the brain and moved to another area. Care should be taken to record the position of the penetrations relative to a zero point or to each other, as well as their depth from surface, in order to facilitate correlation of physiological responses with injected neurons subsequent to histochemical processing of the tissue.

Vascular Perfusion

At the conclusion of the experiment, the animal is perfused through the aorta with saline followed by a fixative made up of 2% paraformaldehyde and 2% glutaraldehyde in phosphate buffer. The brain is removed and stored in the same fixative for 4–6 hr and then transferred to phosphate buffer, to which 15% sucrose has been added, and stored overnight (or until the brain sinks) at 4°C.

TISSUE PROCESSING

Processing for Light Microscopy

The brain is cut on a freezing microtome at 60–80 μm in the desired plane. The sagittal plane is chosen for the cerebellar model used in this chapter. Serial sections are collected in phosphate buffer within individual wells and processed for HRP histochemistry as follows:

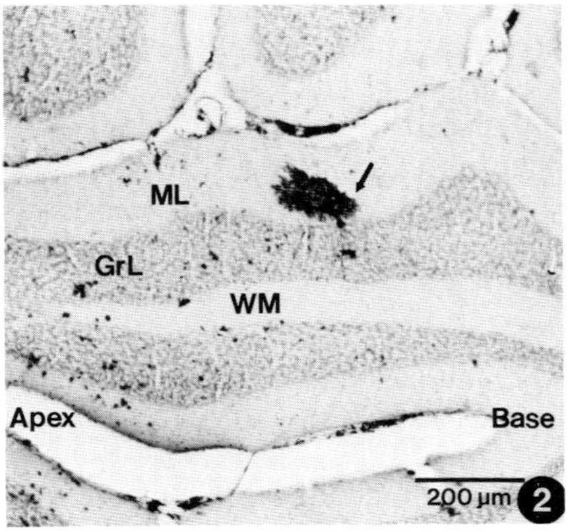

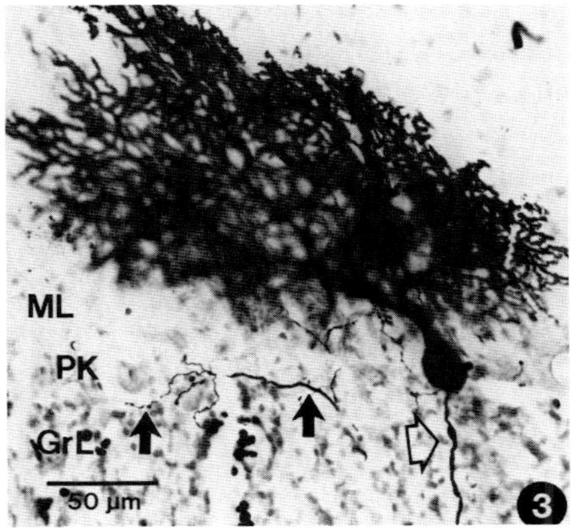

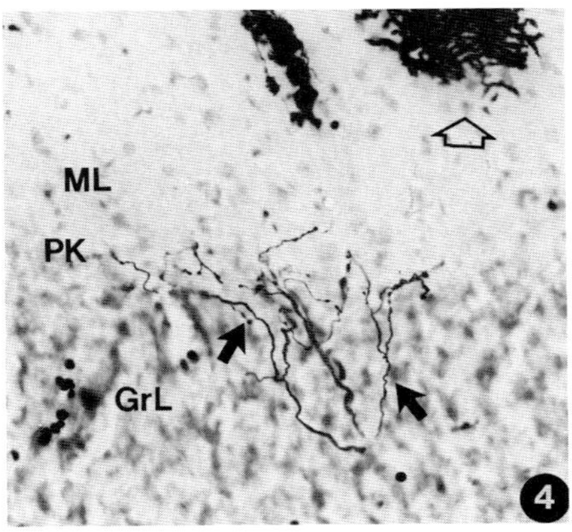

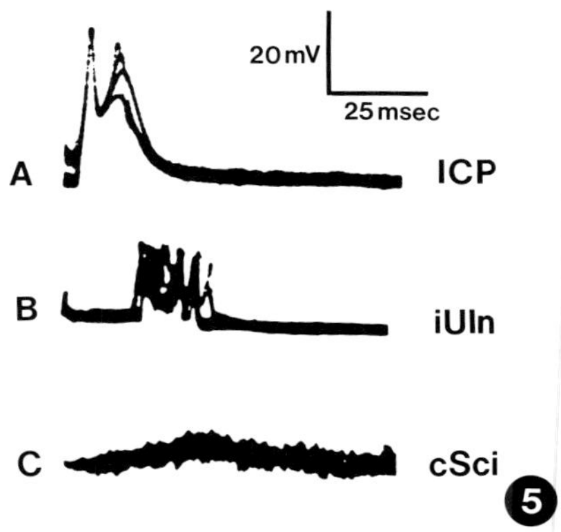

Fig. 2. Low power photomicrograph of an intracellularly injected Purkinje cell (arrow). Note it is the only stained element in the tissue. Apex refers to the top of the folium, and base to the bottom of the folium. ML, molecular layer; GrL, granule cell layer; WM, white matter.

Fig. 3. High power photomicrograph of the Purkinje cell shown in Fig. 2. The HRP reaction project completely fills the soma, dendrites, and parent axon (open block arrow). In addition, a portion of the recurrent collaterals derived from the axon of this cell are visible (closed arrows). ML, molecular layer; GrL, granule cell layer; PK, Purkinje cell layer. Calibration bar also applies to Fig. 4.

Fig. 4. High power photomicrograph illustrating a portion of the collateral plexus of the cell shown in Fig. 3 that was located on an adjacent section (closed arrows). The open block arrow indicates dendrites of the injected cell located in the molecular layer (ML). PK, Purkinje cell layer; GrL, granule cell layer.

Fig. 5. Physiological responses of the cell shown in Fig. 3 subsequent to activation of A, the inferior cerebellar peduncle (ICP); B, the ipsilateral ulnar nerve (iUln); and C, the contralateral sciatic nerve.

1. Rinse tissue three times in phosphate buffer (5 min each).

2. Incubate sections in a 1–3% cobalt chloride solution that is made in phosphate buffer and filtered. The tissue is placed in this solution for 10 min and agitated. This pretreatment intensifies the reaction, yielding a black reaction product in the cells (Adams, 1977).

3. Rinse tissue three times in phosphate buffer (5 min each).

4. Incubate tissue in 0.05% 3,3′-diaminobenzidine tetrahydrochloride (DAB; Baker Chemical Co., Phillipsburg, New Jersey) for 30 min at 37°C.

5. Add 4 ml of 0.006% H_2O_2 per 100 ml of DAB and continue incubation for another 30 min at room temperature with agitation.

6. Rinse tissue three times in phosphate buffer.

7. Mount sections on chrome alum-coated slides.

8. Counterstain with cresyl violet, dehydrate through a graded series of alcohols, clear in xylene and coverslip.

Processing for Electron Microscopy

Tissue to be processed for electron microscopic analysis is cut on a Vibratome rather than a freezing microtome. After collecting serial sections, the tissue is processed as described above through step 6. Sections are then mounted on clean glass slides and quickly examined, while still wet, to locate intracellularly injected neurons. Sections containing neurons, as well as sections adjacent to them, are collected in phosphate buffer. The area of the section containing the labeled neurons is blocked and processed for electron microscopy. The order of the sections must be maintained in order to ensure proper reconstruction at a later time (see the section on Data Analysis). The procedures to be followed for electron microscopy are as follows:

1. Postfix tissue for 20 min in osmium tetroxide diluted to 1% in phosphate buffer.

2. Rinse tissue three times in phosphate buffer (5 min each).

3. Dehydrate through a graded series of acetone (50, 70, 80, 95, 100, 100%). The sections should remain in each concentration of acetone for 10 min.

4. Place in a 1 : 1 mixture of 100% acetone and the resin (e.g., Spurrs, Epon, or Maraglas) to be used for embedding tissue for 1 hr.

5. Place in resin overnight (8–12 hr).

6. Place in fresh change of resin for 2 hr.

7. Flat embed tissue sections as follows:

a. Place a sheet of clear plastic (e.g., Aclar) on large piece of glass. Tape in place.

b. Place sections, in serial order, on Aclar.
c. Cover with a few drops of resin.
d. Place another piece of Aclar on top of sections (as in coverslipping).
e. Label sections with experiment and section number.
f. Place another sheet of glass on top to hold sections flat.
g. Cure resin in oven.
h. Remove Aclar from glass. Cut into strips of convenient size for storage. During storage, leave two pieces of Aclar around section. These will be removed just prior to the initiation of thin sectioning.

An alternative method for flat embedding is to use Teflon-coated slides and coverslips instead of the Aclar (Wilson and Groves, 1979).

DATA ANALYSIS

Correlation of Physiology and Morphology

In analyzing data, the first step is to examine every tissue section and to note the position of all injected neurons (Figs. 2–4) or electrode tracts (section number, location on section such as depth from surface, anterior–posterior relationships of each cell, etc.). This information is then correlated with the maps that were made at the time of recording to determine the probe on which a particular cell was impaled. Thus, the appropriate physiological data can be assigned to the cell from which they were obtained (Figs. 2–5).

Light Microscopic Analysis

In the model used in this chapter, the goal of the light microscopic analysis is to achieve a detailed picture of the local collaterals derived from the axons of Purkinje cells. In particular, interest is directed toward determining whether Purkinje cells in different parts of the cerebellum are uniform with respect to the laminar distribution, extent in the sagittal and transverse planes, and the postsynaptic targets of the collaterals.

Initially, the collateral plexus is reconstructed from serial sections (Fig. 6). This is carried out with the aid of a drawing tube attached to a light microscope. It must first be determined whether the collateral is completely filled. To assess this, note the intensity of labeling and whether branches fade without giving rise to some type of terminal specialization

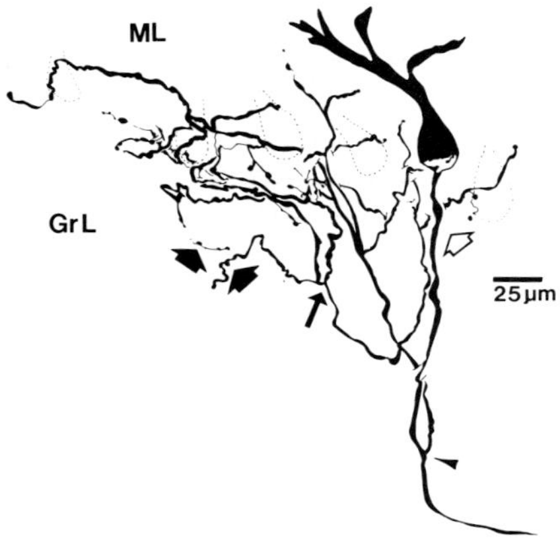

Fig. 6. Serial section reconstruction of the collateral plexus of the cell shown in Fig. 3. The dendritic tree of the cell has not been reconstructed in this illustration. The open block arrow indicates the parent axon. The arrowhead indicates the point where a single collateral arises from the parent axon. This single branch arborizes into three secondary branches (arrow) which further ramify to form a plexus that is located primarily in the Purkinje cell layer (PK). A few branches (closed block arrows) remain in the granule cell layer.

(e.g., bouton termineux). The selected plexuses are then drawn, maintaining the relationship of the branches to each other as well as to the parent cell (Fig. 6). Branches which continue to an adjacent section will be truncated at either the very top or bottom of a section. These points should be noted and marked on the drawing. When the next section is examined, points located at the top of the section just drawn should line up with points located at the bottom of the next section. Similarly, points located at the bottom of the original section should line up with points at the top of the section on the other side of the first section. This pattern of matching points on opposite sides of the tissue on adjacent sections is carried out until the entire plexus is drawn (Fig. 6). At this level of analysis several observations can be made. First, the lamina(e) to which the plexus distributes can be identified and compared with Purkinje cells located in different parts of the cerebellum. Second, the distance over which the plexus distributes in the sagittal plane can be measured and compared. In this type of reconstruction, however, the plexus is compressed into a two-dimensional drawing. Data with respect to the transverse distribution of the plexus are not apparent from the drawing. In order to provide this information a different type of analysis can be carried out.

Electron microscopic analysis (King and Bishop, 1982) has shown that axonal swellings or varicosities distributed along axonal branches correspond to presynaptic boutons. Thus, they represent points of information transfer between a Purkinje cell and the neuronal elements with which they interact. In a second type of analysis only varicosities are plotted. However, their relationship to the parent cell and to each other is maintained by use of the technique described above for lining up adjacent sections. Varicosities located on different sections are drawn with different symbols or colors. These data are still in a two-dimensional format at this point. They must be transferred to a coordinate system for final analysis. The varicosity drawing is divided into 25-μm bins, beginning at the parent cell body. This interval was chosen as it corresponds to the average diameter of a Purkinje cell body. The bins are oriented normal to the Purkinje cell layer regardless of its orientation in the folium. The number of varicosities on each section (as determined by its respective symbol or color) are counted and plotted on graph paper. The distance, from the cell of origin, both in the sagittal and transverse planes, is maintained. In the final plots of the varicosity distribution (Fig. 7), the star represents the position of the cell of origin and each dot represents the position of a varicosity relative to the parent cell in the sagittal plane (abscissa) and the transverse plane (ordinate). For further information the

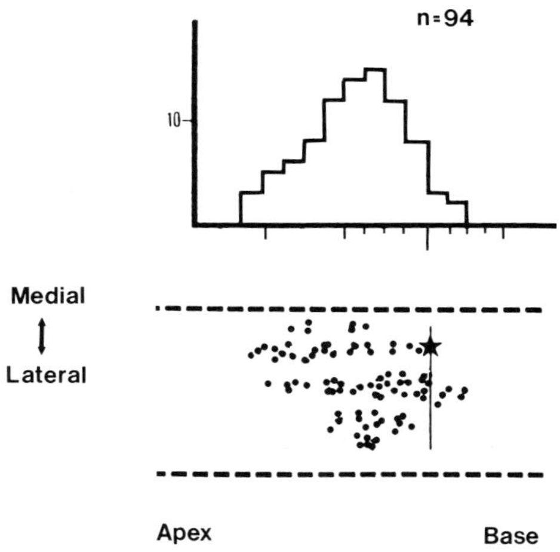

Fig. 7. Scatter plot and histogram illustrating the distribution of varicosities (dots) around the cell body (star) of the neuron shown in Fig. 6. The distance between the dashed lines equals 200 μm. The vertical line through the cell body (star) represents the linear extent of the plexus in the transverse plane. The histogram is in register with the scatter plot and quantifies the number of varicosities located within 25-μm transverse bins relative to the cell body. The total number of varicosities derived from this collateral plexus are shown at the top of the histogram. See text for further details.

total number of varicosities within each 25-μm bin are counted and illustrated in the histograms that accompany each scatter plot. This type of analysis allows for a graphic depiction of the collaterals in their two major planes of distribution. In Fig. 7, no data are provided with respect to the laminar distribution of varicosities. However, this information could be represented on the figure as different symbols or letters.

Electron Microscopic Analysis

The sections flat-embedded in Aclar can be placed on a microscopic slide, taped in place, and a light microscopic analysis, as described above, carried out prior to analyzing the tissue with the electron microscope. After reconstructing the collaterals (Fig. 8), sections with HRP-labeled varicosities are selected. The position of the varicosities are carefully noted with respect to laminar distribution, depth from the surface of the block, and relationship to other structures in the section (e.g., blood vessels, other injected processes). The Aclar is removed from both sides of the section; it is then blocked, and mounted on the top of a plastic block. The latter is formed by curing the resin used to embed the tissue in a BEEM capsule. The section is fixed to the top of the block with a small amount of cyanoacrylate (Super glue, etc.) and left to dry overnight. Thick (1 μm) and thin (800–900 nm) sections are cut from the block on an ultramicrotome. The number and thickness of sections cut should be recorded in order to determine depth of sectioning into a block. The thin sections are collected on Formvar-coated slot grids, stained with uranyl acetate and lead citrate, and examined with an electron microscope. Labeled somata (Fig. 9) and processes (Figs. 10–14) will stand out as dark profiles in the neuropil. The reaction product, which varies in density, fills the cytoplasm and surrounds membrane-bound organelles (e.g., synaptic vesicles and endoplasmic reticulum). One goal of this analysis is to correlate a process seen with the electron microscope with its light microscopic counterpart. This can be achieved by comparing the position of a profile

---

Fig. 8. Camera lucida drawing of an HRP-filled Purkinje cell and part of its collateral plexus. Electron micrographs of the cell body and branches of the collateral identified S and A are shown in subsequent figures: block arrow S in Figs. 10–12 and block arrows A in Figs. 13 and 14.

Fig. 9. The cell body of the Purkinje cell illustrated in Fig. 8. NUC, nucleus.

Fig. 10–12. Serial sections of an HRP-labeled terminal which contacts the apical portion of an adjacent Purkinje cell (PK) body at block arrow S. Fig. 10 is a higher magnification of the labeled terminal in Fig. 11. In this latter figure unlabeled terminals (open block arrows) with similar cytological features contact the same Purkinje cell. [Figs. 8–12 reprinted from *J. Neurocytol.* **11,** 867–880 (1982) with permission of Chapman and Hall Ltd.]

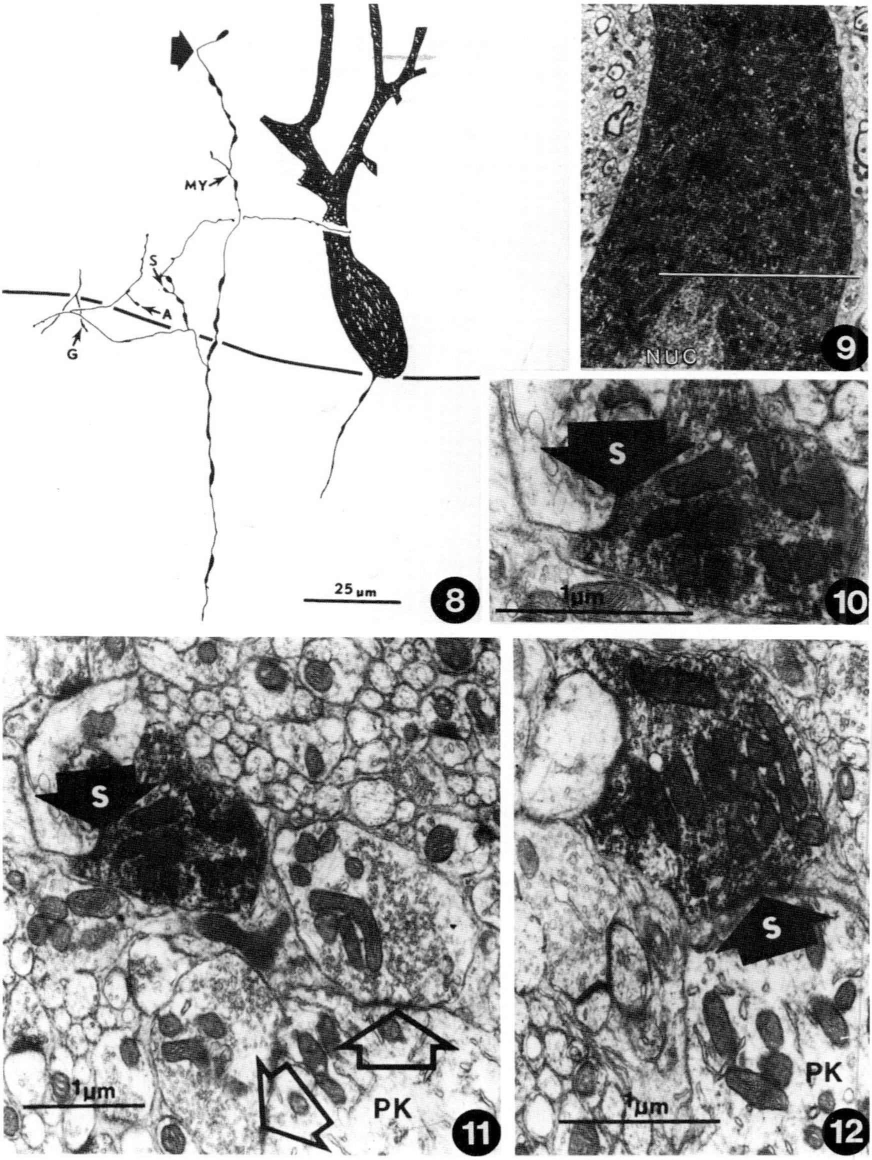
MY
S
A
G
25 µm
8
NUC
9
S
1µm
10
S
1µm
PK
11
S
PK
1µm
12

identified with the electron microscope with varicosities drawn under light microscopic observation. Correlation can be achieved by comparing depth of a profile from the surface, laminar distribution, continuity with other HRP-labeled neuronal elements, relationship to specific structures within the neuropil, or other landmarks.

Several types of information can be derived from the ultrastructural analysis. The synaptic relationships of specific varicosities can be identified. For varicosities, the number, size, and shape of synaptic vesicles can be determined, and the nature of the pre- and postsynaptic densities analyzed (Figs. 10-14). If the postsynaptic structure is not labeled, as will almost always be the case, its identity will require knowledge of distinguishing cytological characteristics of neuronal elements in the area being studied. If the postsynaptic process is labeled, and if no other neuronal elements in the area were injected with HRP, one has evidence for the existence of autaptic connections (Figs. 13 and 14).

The cytological characteristics of dendrites and somata can be determined. The number and type of boutons that form synaptic junctions with a single cell or a portion thereof could be quantified. The existence of specialized junctions such as gap junctions or dendro–dendritic interactions between labeled and unlabeled profiles could be identified and quantified. The interaction of labeled processes in specialized synaptic arrangements (e.g., glomeruli) could also be identified.

LIMITATIONS OF THE METHOD

One limitation of this procedure is that the investigator must have the equipment and expertise to carry out intracellular recording or to collaborate with an investigator who has these facilities. A few problems exist within the technique itself.

Physiological Recording

As noted above, the electrodes are relatively large as compared to the size normally used for intracellular recording, in order to eject the HRP. The size of the electrode, even though it is beveled, may damage the cell on impalement. This may lead to a low resting membrane potential, injury discharge which masks a response to an applied stimulus, or inability to hold the cell for any length of time. Thus, optimal physiological records may not be obtained for every cell. A second problem, inherent to all intracellular recording paradigms regardless of whether a marker is being used, is that the number of units from which data are obtained is relatively

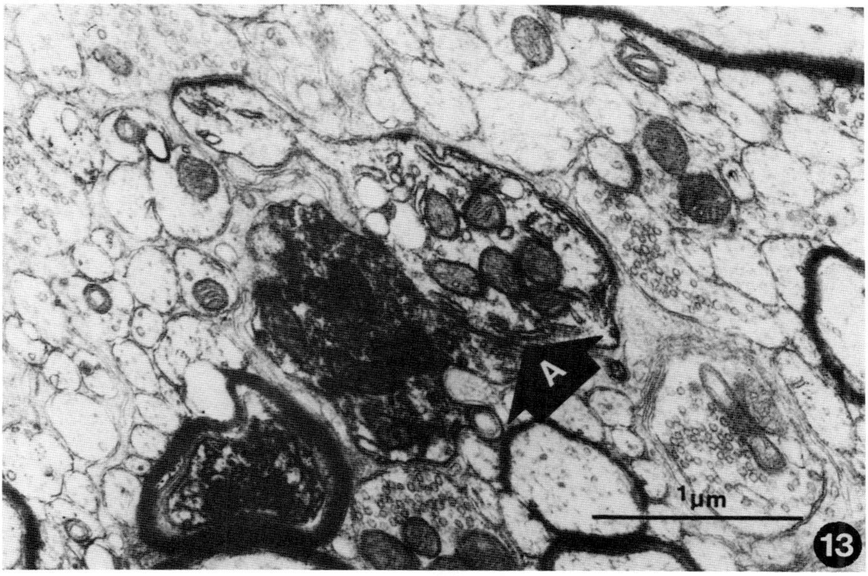

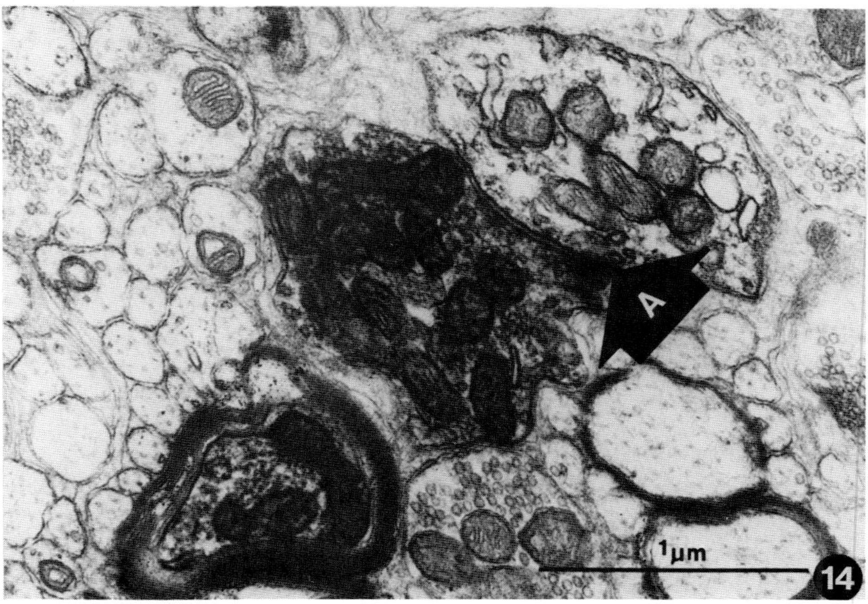

Figs. 13 and 14. Serial sections of an autapse (block arrow A). The myelinated axon that gives rise to the terminal is labeled by HRP as is the postsynaptic Purkinje cell dendrite. The location of the autapse is shown in Fig. 8 and is labeled A. [Figs. 13 and 14 reprinted from *J. Neurocytol.* **11,** 867–880 (1982) with permission of Chapman and Hall Ltd.]

small. Thus, generalizations about a population of neurons may be based on data obtained from 50 to 100 cells.

Light Microscopic Analysis

All of the neurons that are injected during the recording session may not be located. This may be due to the fact that the electrode was not in the cell throughout the injection, or that the electrode blocked and an insufficient amount of HRP was ejected. A second problem is that all of the recovered neurons may not be suitable for analysis. In some cases other neuronal elements in the area of the injected cell may also be labeled which may make reconstructions difficult. Some recovered neurons will be extensively damaged or only lightly labeled; both conditions preclude analysis.

Electron Microscopic Analysis

Prior to perfusion, the area from which data are recorded is subjected to several electrode penetrations, any of which may rupture blood vessels or damage tissue. Thus, optimal preservation of the tissue may not always be obtained, which may yield electron micrographs of less than perfect quality. In addition, the reaction product may be so dense that it obscures cytological detail. In contrast, if the process is only lightly labeled it may not be evident in the neuropil. As stated for physiological recordings, the number of cells that can be critically analyzed with the electron microscope is limited due to the tedious work involved in serial section reconstruction. Thus, generalizations may have to be based on a small sample size.

CONCLUSIONS

The goal of this volume is to present techniques which allow for correlative microscopy. The intracellular staining procedure is clearly one such technique. An identified process can be isolated under the light microscope, prepared for sectioning, and analyzed at the electron microscopic level. An additional advantage of the technique presented in this chapter is that the morphological and cytological characteristics of a given neuron can be correlated with its physiological properties. Thus, the investigator can learn something about the structure and function of identified populations of neurons.

REFERENCES

Adams, J. C. (1977). Technical considerations on the use of HRP as a neuronal marker. *Neuroscience* **2,** 141.

Bennett, G. J., Abdelmoumene, M., Hayashi, H., and Dubner, R. (1980). Physiology and morphology of substantia gelatinosa neurons intracellularly stained with horseradish peroxidase. *J. Comp. Neurol.* **194,** 809.

Bishop, G. A. (1982). The pattern of distribution of the local axonal collaterals of Purkinje cells in the intermediate cortex of the anterior lobe and paramedian lobule of the cat cerebellum. *J. Comp. Neurol.* **210,** 1.

Bishop, G. A., and King, J. S. (1986). Reticulo-olivary circuits: An intracellular HRP study in the rat. *Brain Res.* **371,** 133.

Bishop, G. A., Chang, H. T., and Kitai, S. T. (1979a). Light and electronmicroscopic analysis of various neostriatal neurons intracellularly labelled with HRP. II. Patterns of axonal distribution. *Soc. Neurosci. Abstr.* **5,** 214.

Bishop, G. A., McCrea, R. A., Lighthall, J. W., and Kitai, S. T. (1979b). An HRP and autoradiographic study of the projection from the cerebellar cortex to the nucleus interpositus anterior and nucleus interpositus posterior of the cat. *J. Comp. Neurol.* **185,** 735.

Bishop, G. A., King, J. S., and McCrea, R. A. (1980a). Light and electron microscopic analyses of Purkinje cell axon collaterals. An intracellular HRP study in the cat. *Soc. Neurosci. Abstr.* **6,** 470.

Bishop, G. A., McCrea, R. A., and King, J. S. (1980b). Analysis of the morphology and cytology of HRP labelled Purkinje cells. *Brain Res. Bull.* **5,** 563.

Bishop, G. A., Chang, H. T., and Kitai, S. T. (1981). Morphological and physiological properties of rat neostriatal neurons: An intracellular HRP study. *Neuroscience* **7,** 179.

Bishop, G. A., Ho, R. H., and King, J. S. (1985). Localization of serotonin immunoreactivity in the opossum cerebellum. *J. Comp. Neurol.* **235,** 301.

Bishop, G. A., Blake, T. L., and O'Donoghue, D. L. (1987). The distribution pattern of Purkinje cell axon collaterals: Variations on a theme. *In* "New Concepts in Cerebellar Neurobiology" (J. S. King, ed.), pp. 29–56. Alan R. Liss, Inc., New York.

Blake, T. L., O'Donoghue, D. L., King, J. S., and Bishop, G. A. (1986). An analysis of recurrent collaterals of Purkinje cells in zone a of the cat's vermis. *Anat. Rec.* **214,** 13A.

Calvet, M. C., and Calvet, J. (1979). Horseradish peroxidase iontophoretic intracellular labelling of cultured Purkinje cells. *Brain Res.* **173,** 527.

Chang, H. T., Bishop, G. A., and Kitai, S. T. (1979). Light and electron microscopic analysis of various neostriatal neurons intracellularly labelled with HRP. I. Soma dendritic morphology. *Soc. Neurosci. Abstr.* **5,** 220.

Christensen, B. N., and Ebner, F. E. (1978). The synaptic architecture of neurons in opossum somatic sensory-motor cortex: A combined anatomical and physiological study. *J. Neurocytol.* **7,** 39.

Crepel, F., Delhaye-Bouchaud, N., Dupont, J. L., and Sotelo, C. (1980). Dendritic and axonic fields of Purkinje cells in developing and x-irradiated rat cerebellum. A comparative study using intracellular staining with horseradish peroxidase. *Neuroscience* **5,** 333.

Cullheim, S., and Kellereth, J. O. (1976). Combined light and electron microscopic tracing of neurons, including axons and synaptic terminals, after intracellular injections of horseradish peroxidase. *Neurosci. Lett.* **2,** 307.

Cullheim, S., Kellereth, J. O., and Conradi, S. (1977). Evidence for direct synaptic interconnections between cat spinal alpha motoneurons via the recurrent axon collaterals: A morphological study using intracellular injection of horseradish peroxidase. *Brain Res.* **132,** 1.

Deniau, J. M., Grofova, I., Steindler, D., and Kitai, S. T. (1980). Axonal arborizations of nigrothalamic and nigrotectal cells: anatomical and electrophysiological studies. *Soc. Neurosci. Abstr.* **6,** 807.

Deschenes, M., and Landry, P. (1980). Axonal branch diameter and spacing of nodes in the terminal arborization of identified thalamic and cortical neurons. *Brain Res.* **191,** 538.

Fox, C. A., Hillman, D. E., Seigesmund, K. A., and Dutta, C. R. (1967). The primate cerebellar cortex: A Golgi and electron microscopic study. *Prog. Brain Res.* **25,** 174.

Gilbert, C. D., and Wiesel, T. N. (1978). Intracellular staining of physiologically identified cells and afferents in cat visual cortex. *Soc. Neurosci. Abstr.* **4,** 1235.

Gobel, S., Falls, W. M., Bennett, G. S., Abdelmoumene, M., and Hayashi, H. H. E. (1980). An EM analysis of the synaptic connections of horseradish peroxidase filled stalked cells and islet cells in the substantia gelatinosa of adult cat spinal cord. *J. Comp. Neurol.* **194,** 781.

Hendelman, W. J., and Marshall, K. C. (1980). Axonal projection patterns visualized with horseradish peroxidase in organized cultures of cerebellum. *Neuroscience* **5,** 1833.

Ito, M. (1984). "The Cerebellum and Neural Control." Raven Press, New York.

Ito, M., Yoshida, M., and Obata, K. (1964). Monosynaptic inhibition of the intracerebellar nuclei induced from the cerebellar cortex. *Experientia* **20,** 575.

Ito, M., Yoshida, M., Obata, K., Kawai, N., and Udo, M. (1970). Inhibitory control of intracerebellar nuclei by the Purkinje cell axons. *Exp. Brain Res.* **10,** 64.

Jankowska, E., Rastad, J., and Westman, J. (1976). Intracellular application of horseradish peroxidase and its light and electron microscopical appearance in spinocervical tract cells. *Brain Res.* **105,** 555.

King, J. S. (1980). The synaptic organization of the inferior olivary complex. *In* "The Inferior Olivary Nucleus: Anatomy and Physiology" (J. Courville, C. de Montigny, and Y. Lamarre, eds.), pp. 1–33. Raven Press, New York.

King, J. S., and Bishop, G. A. (1982). The synaptic features of horseradish peroxidase labelled recurrent collaterals in the ganglionic plexus of the cat cerebellar cortex. *J. Neurocytol.* **11,** 867.

King, J. S., and McCrea, R. A. (1978). A cytological analysis of inferior olivary neurons after intracellular injections of horseradish peroxidase. *Anat. Rec.* **190,** 446.

King, J. S., Ho, R. H., and Bishop, G. A. (1987). The origin and distribution of enkephalin-like immunoreactivity in the opossum's cerebellum. *In* "New Concepts in Cerebellar Neurobiology" (J. S. King, ed.), pp. 1–28. Alan R. Liss, Inc., New York.

Kitai, S. T., Kocsis, J. D., Preston, R. J., and Sugimori, M. (1976). Monosynaptic inputs to caudate neurons identified by intracellular injection of horseradish peroxidase. *Brain Res.* **109,** 601.

Kitai, S. T., McCrea, R. A., Preston, R. J., and Bishop, G. A. (1977). Electrophysiological and horseradish peroxidase studies of precerebellar afferents to the nucleus interpositus anterior. I. Climbing fiber system. *Brain Res.* **122,** 197.

Kocsis, J. D., and Kitai, S. T. (1977). Dual excitatory inputs to caudate spiny neurons from substantia nigra stimulation. *Brain Res.* **138,** 271.

Kocsis, J. D., Sugimori, M., and Kitai, S. T. (1977). Convergence of excitatory synaptic inputs to caudate spiny neurons. *Brain Res.* **124,** 403.

Landry, P., and LaBelle, A. (1980). Intracortical distribution of axonal collaterals of pyramidal tract cells in the cat motor cortex. *Brain Res.* **191,** 327.

Lange, W. (1982). Regional differences in the cytoarchitecture of the cerebellar cortex. *Exp. Brain Res., Supp.* **6,** 93–107.

Light, A. R., and Durkovic, R. G. (1976). Horseradish peroxidase: an improvement in intracellular staining of single, electrophysiologically characterized neurons. *Exp. Neurol.* **53,** 847.

Light, A. R., and Perl, E. R. (1979). Spinal termination of functionally identified primary afferent neurons with slowly conducting myelinated fibers. *J. Comp. Neurol.* **186,** 133.

Light, A. R., Trevino, D. L., and Perl, E. R. (1979). Morphological features of functionally defined neurons in the marginal zone and substantia gelatinosa of the spinal dorsal horn. *J. Comp. Neurol.* **186,** 151.

Llinás, R., and Yarom, Y. (1980). Electrophysiological properties of mammalian inferior olivary cells in vitro. *In* "The Inferior Olivary Nucleus: Anatomy and Physiology" (J. Courville, C. de Montigny, and Y. Lamarre, eds.), pp. 379–388. Raven Press, New York.

McCrea, R. A., Bishop, G. A., and Kitai, S. T. (1976). Intracellular staining of Purkinje cells and their axons with horseradish peroxidase. *Brain Res.* **118,** 132.

McCrea, R. A., Bishop, G. A., and Kitai, S. T. (1977). Morphological and electrophysiological characteristics of projection neurons in the nucleus interpositus of the cat cerebellum. *J. Comp. Neurol.* **181,** 397.

Mason, C. A., and Robson, J. A. (1979). Morphology of retino-geniculate axons in the cat. *Neuroscience* **4,** 79.

Muller, K. J., and McMahan, U. J. (1976). The shapes of sensory and motor neurons and the distribution of their synapses in ganglia of the leech: A study using intracellular injection of horseradish peroxidase. *Proc. R. Soc. London. Ser. B* **194,** 481.

Nahvi, M. J., Woody, C. D., Tzebelikos, E., and Ribak, C. E. (1980). Electrophysiologic characterization of morphologically identified neurons in the cerebellar cortex of awake cats. *Exp. Neurol.* **67,** 368.

Nelson, P. G., Neale, E. E., and MacDonald, R. L. (1978). Formation and modification of synapses in central nervous system cell cultures. *Fed. Proc., Fed. Am. Soc. Exp. Biol.* **37,** 2010.

O'Donoghue, D. L., Blake, T. L., King, J. S., and Bishop, G. A. (1986). Recurrent axon collaterals of Purkinje cells in zone cl of the cat cerebellum. *Anat. Rec.* **214,** 92A.

Oscarsson, O. (1979). Functional units of the cerebellum-sagittal zones and microzones. *Trends Neurosci.* **2,** 143.

Palay, S. L., and Chan-Palay, V. (1974). "The Cerebellar Cortex: Cytology and Organization." Springer-Verlag, Berlin and New York.

Preston, R. J., Bishop, G. A., and Kitai, S. T. (1980). Medium spiny neuron projection from the rat striatum: An intracellular horseradish peroxidase study. *Brain Res.* **183,** 253.

Prince, D. A., Wong, R. K. S., and Basbaum, A. I. (1978). Membrane properties of identified human cortical neurons. *Soc. Neurosci. Abstr.* **4,** 227.

Purves, R. D. (1981). "Microelectrode Methods for Intracellular Recording and Ionophoresis." Academic Press, London.

Rakic, P. (1976). "Local Circuit Neurons." MIT Press, Cambridge, Massachusetts.

Rakic, P. (1979). Genetic and epigenetic determinants of local neuron circuits in the mammalian central nervous system. *In* "The Neurosciences: Fourth Study Program" (F. O. Schmitt and F. G. Worden, eds.), pp. 109–127. MIT Press, Cambridge, Massachusetts.

Ramón y Cajál, S. (1899). Comparative study of the sensory areas of the human brain. "Clark University 1889–1899: Decennial Celebration."

Ribak, C. E., Woody, C. D., Nahvi, M. J., and Tzebelikos, E. (1980). Ultrastructural identification of physiologically recorded neurons in the cat cerebellum. *Exp. Neurol.* **67,** 372.

Robson, J. A., and Mason, C. A. (1979). The synaptic organization of terminals traced from individual labelled retino-geniculate axons in the cat. *Neuroscience* **4,** 99.

Shepherd, G. M. (1979). Functional analysis of local circuits in the olfactory bulb. *In* "The Neurosciences: Fourth Study Program" (F. O. Schmitt and F. G. Worden, eds.), pp. 129–143. MIT Press, Cambridge, Massachusetts.

Snow, P. J., Rose, P. K., and Brown, A. G. (1976). Tracing axons and axon collaterals of spinal neurons using intracellular injection of horseradish peroxidase. *Science* **191,** 312.

Spencer, R. F., Baker, R., and McCrea, R. A. (1980). Localization and morphology of cat retractor bulbi muscles. *J. Neurophysiol.* **43,** 754.

Takeuchi, Y., Kimura, H., and Sano, Y. (1982). Immunohistochemical demonstration of serotonin-containing nerve fibers in the cerebellum. *Cell Tissue Res.* **226,** 1.

VanderMaelen, C. P., Bonduki, A., and Kitai, S. T. (1979). Excitation of caudateputamen neurons following stimulation of the dorsal raphe nucleus in the rat. *Brain Res.* **175,** 356.

Wilson, C. W., and Groves, P. M. (1979). A simple and rapid section embedding technique for sequential light and electron microscopic examination of individually stained central neurons. *J. Neurosci. Methods* **1,** 383.

6

Correlative Microscopy of Catecholamines and Neuropeptides in Adrenal Medulla

MARKKU PELTO-HUIKKO

Department of Biomedical Sciences
University of Tampere
Tampere, Finland

INTRODUCTION

In most mammalian species the two major catecholamines, norepinephrine and epinephrine, are stored in separate chromaffin cells. These cells are innervated by preganglionic sympathetic nerve terminals. During the last decade many so-called regulatory- or neuropeptides have been demonstrated immunocytochemically in the adrenal chromaffin cells and in the nerve terminals innervating these cells.

To understand the function of adrenal medulla it has been elucidative to study the coexistence of catecholamines and neuropeptides in the chromaffin cells and to correlate peptidergic nerve terminals to the different types of chromaffin cells in adrenal medulla. This chapter describes a method for correlative microscopy of catecholamines and neuropeptides (Pelto-Huikko *et al.*, 1982, 1985a,b). It is based on the histochemical demonstration of catecholamines by fluorescence microscopy and subsequent immunocytochemistry of neuropeptides in the same Vibratome section. A brief description of other available methods is also given.

VIBRATOME SECTIONING AND FLUORESCENCE MICROSCOPY

Fixation and Generation of Fluorescence

According to the original fluorescence method of Eränkö (1955), the norepinephrine fluorescence was generated by immersion fixation with formaldehyde and the tissues were subsequently sectioned in cryostat and dried on glass slides. This allowed the visualization of norepinephrine in the adrenal medullary cells while epinephrine cells remained nonfluorescent. Hökfelt and Ljungdahl (1972) and Hökfelt *et al.* (1972) combined formaldehyde-induced fluorescence with Vibratome sectioning for demonstration of norepinephrine fluorescence in brain and adrenal medulla. Furness *et al.* (1977, 1978) introduced the mixture of 4% paraformaldehyde and 0.5–1.0% glutaraldehyde (Faglu) for fluorescence microscopy of catecholamines in whole mount preparations and Vibratome sections.

This method is more sensitive and also allows the visualization of adrenergic nerve terminals. In addition, the fluorescence is stable in aqueous solutions and the same sections can be used for electron microscopic studies after fluorescence microscopy.

We have used a mixture of 4% paraformaldehyde and 0.2% of glutaraldehyde in 0.1 *M* phosphate buffer (pH 7.3) for fixation and generation of fluorescence. Even the low, 0.2% concentration of glutaraldehyde increases the intensity of fluorescence compared to formaldehyde alone. It also hardens the tissues so that it is easier to cut sections from the tissue. Most of the antigens such as enkephalins, neurotensin, substance P, neuropeptide Y, somatostatin- and catecholamine-synthesizing enzymes tolerate this fixation without loosing their antigenicity. Kobayashi *et al.* (1983, 1984) suggested that fixation with aldehydes may mask the antigenicity of peptides stored in norepinephrine cells. However, we could not find differences in the distribution of enkephalin-like immunoreactivity in the hamster adrenal medulla after various fixatives tested. However, to exclude the possibility that fixation causes differences in the distribution of peptides between epinephrine and norepinephrine cells, it is important to compare immunocytochemical results after using different fixatives.

The adrenals of small rodents such as rat, hamster, and mouse can be properly fixed by immersion fixation. The animals are anesthesized and the adrenals are removed and freed of the surrounding adipose tissue. Adrenals are cut in half with a razor blade, taking care not to damage the tissues, and placed in ice-cold fixative for 6 hr. In these animal species perfusion fixation dilates the adrenal medullary sinusoids; this makes the sectioning difficult and should be avoided. The adrenals of larger species like cats and guinea pigs are fixed by perfusion via left ventricle for 20 min and about 1 l of fixative is perfused. The adrenals are sliced with a razor blade and further fixed by immersion for 1 hr.

After fixation the tissues are stored in phosphate buffer. According to Furness *et al.* (1977), the norepinephrine fluorescence is stable for weeks if the tissues are stored in Faglu solution but we have found that antigenicity of many peptides is lost if fixation time is longer than 6 hr. We have stored the adrenals in buffer for some days before sectioning and fluorescence microscopy, and the intensity of the fluorescence is decreased only slightly; however, longer storage in buffer causes fading of the fluorescence.

Vibratome Sectioning

A Lancer 1000 Vibratome (Lancer, St. Louis, Mo) has been used for cutting the sections. The Vibratome blade holder mounting rod has been modified by cutting a sharp groove into the rod. This allows moving the blade sideways without changing the height of the blade. Ordinary razor blades have been used instead of the blades supplied with the Vibratome. Johansson (1983) has recommended the use of a Ralph knife and this may be useful when very thin sections are required.

Before the adrenals are glued on the Vibratome chuck they are cut to about 2–3 mm thick with a razor blade. A small drop of syanoacrylic glue is put on a piece of glass slide which has been glued to the Vibratome chuck. Tissues are gently dried with paper, placed on the glue drop, and slightly compressed for 10 sec. After the tissue is glued, the adrenal can be trimmed to a square by removing the capsule and part of the cortex with a razor blade. This is often advantageous when cutting adrenals of larger animals. The Vibratome specimen chamber is filled with phosphate buffer which is kept at 4°C with noncorrosive metal bars which have been cooled in liquid nitrogen or in the mixture of dry ice and ethanol. The adjustments of the Vibratome are set so that sectioning speed is 0–2 and amplitude 5–7. The sectioning angle is set at 20°. The adjustments depend on the instrument and the tissue to be cut. We have been able to cut 5- to 10-μm-thick sections (on the Vibratome scale) of the adrenals of most species. Up to 25 consecutive 10-μm sections can be obtained before the blade has to be changed. About 7- to 10-μm-thick sections are suitable for fluorescence microscopy. In thicker sections clear overlapping of cells is evident and the sections thinner than 7 μm are fragile and difficult to handle. After cutting, the sections may be stored on porcelain test plates in 0.5-*M* Tris-saline buffer (pH 7.6) for some hours before fluorescence microscopy.

Fluorescence Microscopy

For fluorescence microscopy individual sections are embedded under a coverslip in Tris-saline buffer or in Tris-saline which contains 10–20% glycerol. The addition of glycerol to buffer makes it more viscous, which prevents wrapping of the sections when coverslipped. Glycerol also makes the fluorescence image clearer. A small drop of buffer is placed on the glass slide and the Vibratome section is placed in it and straightened. The section should not float on the buffer. A coverslip is carefully placed on the section. It is important to use little buffer and a long coverslip for embedding so that the coverslip adheres to the glass slide. If the coverslip moves after embedding it will wrap or break the section. If the section is wrapped the coverslip can be removed by placing the glass slide in buffer in a Petri dish and the section can be reembedded.

Vibratome sections can be dried on the glass slides and mounted in paraffin oil. According to Furness *et al.* (1977), fluorescence is stable in these preparations and the sections can be used for light microscopic immunocytochemistry if the antigens survive drying and deparaffinization.

A Leitz Orthoplan microscope equipped with Ploemtype II-epi-illumination optics and filter set D for catecholamines has been used for fluorescence microscopy. At low magnifications (objectives 40× and less) the optics are used dry and at high magnifications (objectives 63–100×) immersion oil is used. The norepinephrine cells and adrenergic nerve fibers and terminals exhibit bright fluorescence while epinephrine cells are virtually nonfluorescent (Figs. 1a, 2a, and 3a). At a high magnification the fluorescence is clearly granular. When the same section is labeled with antibodies to phenylethanol-*n*-methyltransferase (PNMT), the enzyme that converts norepinephrine to epinephrine, the fluorescing cells are not labeled (Fig. 1b), demonstrating the specificity of the Faglu-fluorescence.

IMMUNOCYTOCHEMISTRY

Processing of the Sections

The preembedding modification of Pickel *et al.* (1979) of the original peroxidase–antiperoxidase (PAP) method of Sternberger (1979) has been used. The sections are processed free-floating through the steps of the PAP method. The coverslip is removed over the sections by dipping the glass slide in Tris-saline buffer. The sections are transferred individually with a small brush or with a hair glued to a wooden stick to small test tubes. The incubation solutions are pipetted on the sections after the previous solution has been drawn away. Sections can also be incubated in the wells in porcelain test plates by transferring the section from solution to another with a hair but this easily breaks the sections. If the sections are used only for light microscopic immunocytochemistry, they can be dried on glass slides. Sections are easier to handle but in our experience the intensity of the immunoreaction is decreased either by the effect of drying or that penetration of antibodies is possible only from one side of the section.

The endogenous peroxidase activity can be blocked with treatment of 0.5% H_2O_2 in 100% methanol for 30 min. This easily causes curling up of the sections and treatment with 1% phenylhydrazine hydrochloride solution for 30 min may be used instead or the step can be omitted. To increase the antibody penetration the sections are treated with 0.25% solution of Triton X100 or saponin for 5–10 min. If the sections are used

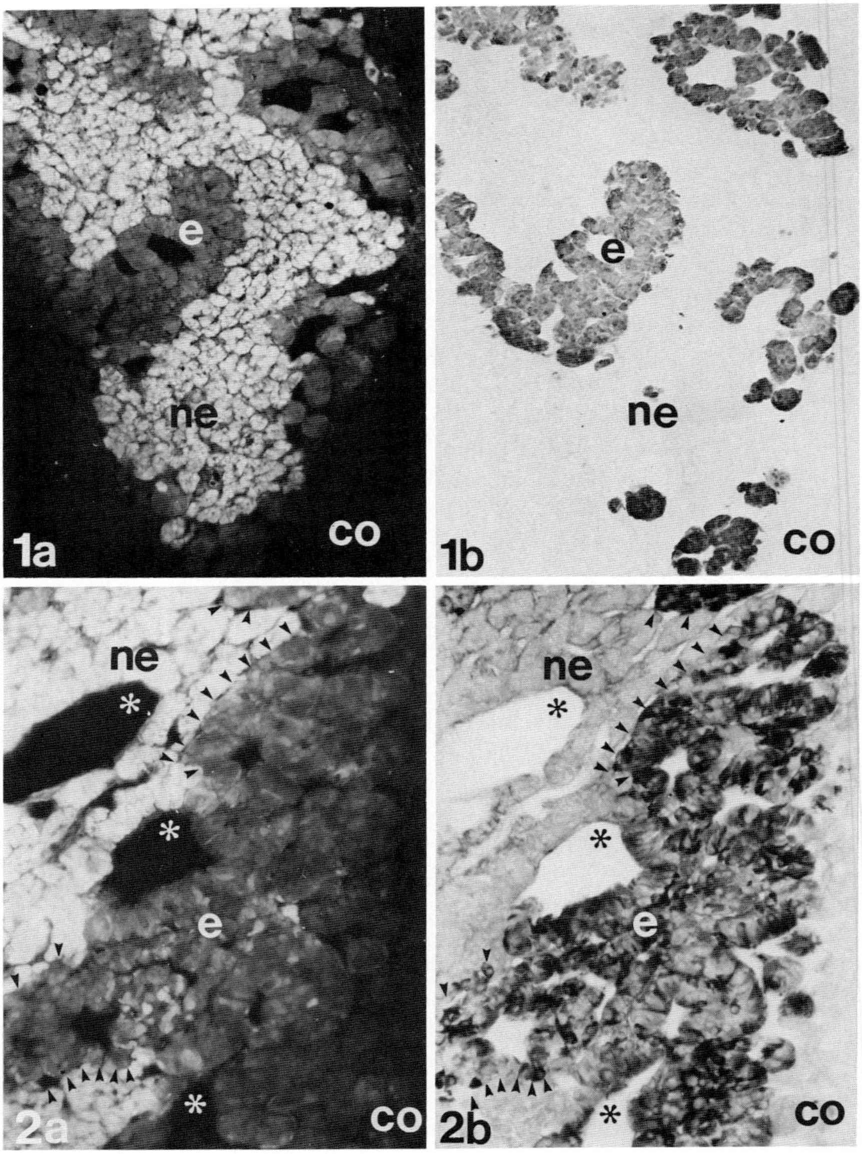

Fig. 1. Fluorescence micrograph of cat adrenal medulla. Norepinephrine cells (ne) exhibit bright fluorescence. Epinephrine cells (e) and adrenal cortex (co) are almost nonfluorescent (a). The same section is labeled with antibodies against PNMT. The epinephrine cells are immunoreactive while norepinephrine cells are not labeled (b); ×94.

Fig. 2. Fluorescence micrograph of cat adrenal medulla (a). Same section is labeled with antibodies to neuropeptide Y (NPY). Most of the epinephrine cells (e) marked with arrows are immunoreactive for NPY; norepinephrine cells (ne); ×188.

only for light microscopy these substances can be added to all solutions containing antibodies.

To block nonspecific binding of the antibodies, the sections are incubated in 10% normal serum (from the same species as the linking antiserum) for 30 min. The specific antisera are diluted between 1 : 500 and 1 : 5000 depending on the antigen and the species studied and incubation time. The sections are incubated in specific antiserum for 10–18 hr at 4°C for electron microscopic studies. If the sections are processed only for light microscopy the incubation time can be prolonged.

The goat anti-rabbit antiserum (Sternberger-Meyer Immunochemicals Inc. Jarretsville, MD) and the rabbit PAP complex (Dako, Copenhagen, Denmark) are used by diluting 1 : 40 for 30 min at room temperature. Sections are washed twice (10 min each) with Tris-saline buffer containing 1% normal serum between every incubation step.

The PAP complex is visualized using diaminobenzidine (DAB) as a chromogen. The reaction solution contains 0.02% DAB and 0.001% H_2O_2 in Tris-water (0.5 *M*, pH 7.6). The reaction time is 5–10 min. The reaction is terminated with distilled water.

Embedding for Light Microscopy

After the PAP reaction the sections are counterstained if desired before mounting. Aquamount (Gurr) or ordinary embedding media can be used. We usually use Aquamount for embedding because permanent preparations can be made without dehydration of the sections. The sections are placed in distilled water on glass slides and allowed to dry and attach to the glass. After drying, the sections are dipped in distilled water and embedded in Aquamount. Sections can also be dehydrated and embedded in ordinary embedding media (Entellan, Fluoromount).

Embedding and Processing for Electron Microscopy

We use two-stage flat-embedding procedure for embedding the sections for electron microscopy which has been described by Pickel (1981) and Priestley and Cuello (1983). This allows correlation between light- and electron microscopical observations (Figs. 3b, 4, and 5). The sections are first postfixed with 2% aqueous solution of OsO_4 for 1–2 hr and washed in phosphate buffer. Sections are dehydrated free floating through graded ethanol, propylene oxide, and propylene oxide–epon mixture in test tubes. After dehydration, they are placed in a drop of Epon 812 on the glass slide and a plastic coverslip is mounted on the section. A small

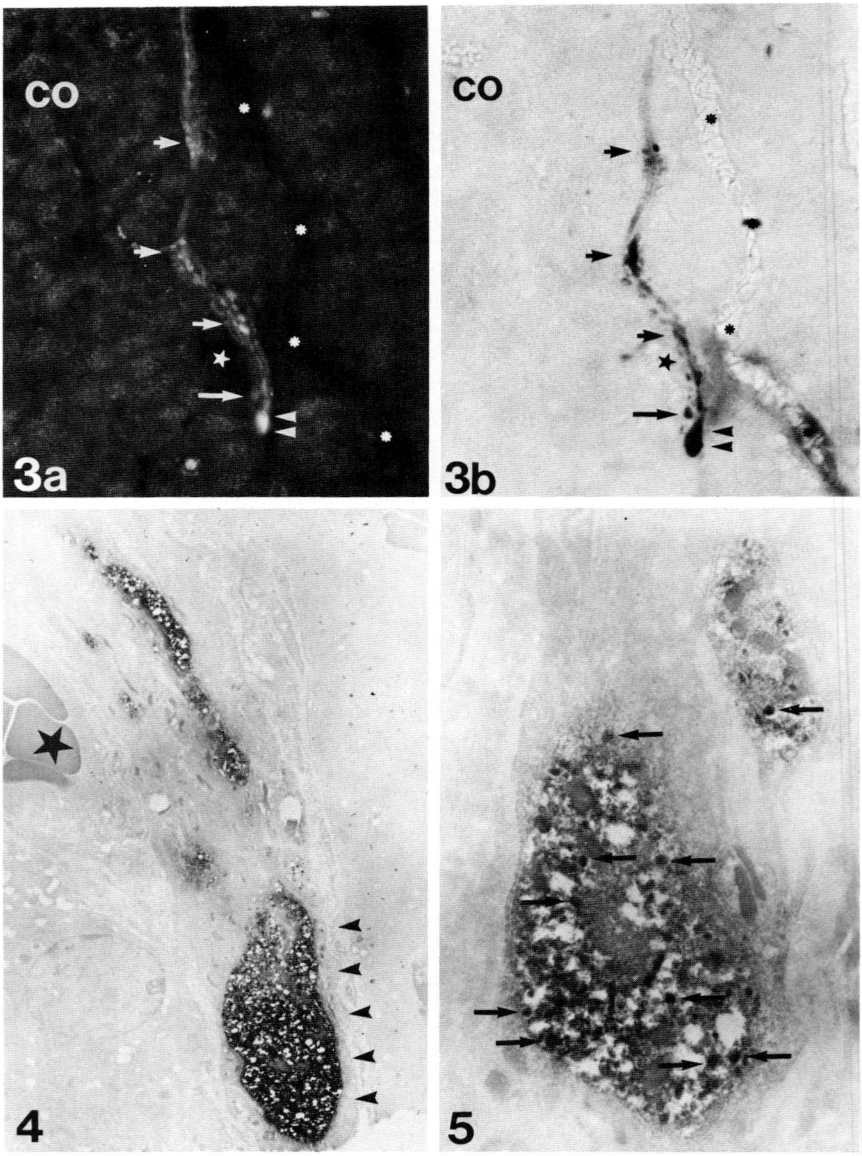
co
3a
co
3b
4
5

weight is placed on the coverslip to straighten the section. Epoxy resin is polymerized at 60°C for 48 hr.

Dehydration easily causes the sections to curl up and makes them fragile to handle. If good quality light micrographs are wanted from the sections before electron microscopical examination, the sections can be lightly dried and attached onto the glass slides and dehydrated on glass slides. Drying causes some shrinkage of the tissues which can be noted by electron microscopy.

After polymerization of the resin, the sections are photographed. The coverslip and the section are removed from the glass slide by heating it slightly and peeling the coverslip. Desired areas are cut out from the sections and glued on polymerized Epon blocks with the section facing upward. Semithin sections are cut from the surface and when the Epon has been removed above the Vibratome section, thin sections are cut from the block. Sections are viewed unstained in the electron microscope at 40 or 60 kV (Figs. 4 and 5).

POSSIBLE MODIFICATIONS

Fixation and Generation of Fluorescence

Livett *et al.* (1982) used formaldehyde alone as a fixative when demonstrating enkephalin-like immunoreactivity with indirect immunofluorescence method. This may be used if antigen or the immunohistochemical procedure does not tolerate the use of glutaraldehyde. Other methods for inducing catecholamine fluorescence such as glyoxylic acid or ALFA method may also prove useful although these are not as comparable fixatives as Faglu solution. For references, see Björklund (1983).

Fig. 3. Adrenergic nerve fibers (arrows) running along sinusoids (star, asterisks) in adrenal cortex of guinea pig. A large axonal swelling is seen (arrow heads) (a). Same section is labeled with antibodies against somatostatin. Many of the varicosities can be identified as in 3a (b); ×400.

Fig. 4. Electron micrograph of the same area as in Fig. 3. The large axonal swelling marked by arrowheads in Fig. 3 is seen at the bottom of the picture (arrowsheads). Star indicates the same sinusoid as in Fig. 3; ×4400.

Fig. 5. Electron micrograph of somatostatin immunoreactive varicosity marked with long arrow in Fig. 3. Immunoreactivity is localized mainly in the large, dense, cored vesicles (arrows) but some diffused PAP precipitates are also seen in axoplasma; ×13,333.

Sectioning

Frozen sections have been used by Livett *et al.* (1982) instead of Vibratome sections. These are easier to cut and serial sections are easier to obtain. In our experience the fluorescence is more diffuse in frozen sections and ice crystal artifacts are often seen in light microscopic examination. The ultrastructure is also inferior when compared to the Vibratome sections. If the tissues are frozen after fixation it is possible to store tissues without fading of the fluorescence.

Schöler and Armstrong (1982) combined Faglu method with polyethylene glycol embedding. This method allows cutting sections 2 μm thick with an ordinary microtome. The fluorescence is stable for several months after embedding. Mazurkiewicz and Nakane (1972) used polyethylene glycol (PEG)-embedded sections for light and electron microscopic immunocytochemistry. This method may well substitute the use of Vibratome sections if the antigens tolerate the PEG embedding.

Immunocytochemistry

Any other compatible immunocytochemical method can be used instead of the PAP method. If fluorescent markers are used, care must be taken that the catecholamine fluorescence is not mixed with fluorescence marker because some catecholamine fluorescence may survive despite extensive washing. Double-labeling can also be performed in a single section after fluorescence microscopy, making it possible to demonstrate two antigens in the same section.

DOUBLE-LABELING METHODS FOR STUDYING COEXISTENCE

Antibodies to Catecholamine-Synthesizing Enzymes and Catecholamines

Catecholamine-synthesizing cells can be demonstrated with antibodies against the enzymes involved in the synthesis of epinephrine and norepinephrine. The cells synthesizing norepinephrine possess the enzymes tyrosine hydroxylase (TH) and dopamine β-hydroxylase (DBH) but not phenylethanol-*n*-methyltransferase (PNMT). The epinephrine cells contain the enzymes TH and DBH and also the PNMT. Thus, antibodies to PNMT can be used to distinguish between norepinephrine and epinephrine cells (Fig. 1b). These antibodies are commercially available from

Eugene Technics International Inc. (Allendale, NJ). Antibodies against norepinephrine and epinephrine have been raised by some research groups. These can be used to demonstrate catecholamines by immunocytochemistry. Because norepinephrine is present in epinephrine cells as a precursor of epinephrine, the antibodies against norepinephrine also stain the epinephrine cells to some extent. Thus, only the antibodies to epinephrine can be used with certainty to distinguish between the adrenal medullary cell types. So far these antibodies have not been commercially available. The use of the antisera to catecholamines has been reviewed by Verhofstad *et al.* (1983).

Double-Labeling Methods

Several double-labeling methods have been developed to demonstrate two or more antigens in the same section at light and electron microscopic levels. Many of these methods can be combined with the use of antibodies to the catecholamine-synthesizing enzymes or to the catecholamines. These methods have been recently reviewed by Mason *et al.* (1983), Varndell and Polak (1984), and Wang and Larsson (1985). The choice of the method depends on the antigens studied and the localization of the antigen, whether it is localized in nerve terminals or chromaffin cells.

The major disadvantage of double-labeling methods is that the specificity of two immunohistochemical labeling procedures has to be verified. Contamination between the two staining steps may also occur. The antibody eluation used in some double-staining methods may destroy antigenicity of some peptides. The double-labeling methods performed at the electron microscopic level usually require resin embedding, which can destroy antigenicity.

ADVANTAGES AND LIMITATIONS OF COMBINED FAGLU-FLUORESCENCE AND IMMUNOCYTOCHEMISTRY METHOD

Advantages

The major advantage of the described method is that it combines a simple and reliable way for demonstration of catecholamines with fluorescence microscopy and preembedding immunocytochemistry. Because the same Vibratome section is used, it is possible to correlate the localization of neuropeptides and catecholamines both at light and electron microscopic levels. Because catecholamines are demonstrated with histochemical reaction, immunocytochemical localization can be carried out without possible interference of antibodies, which can occur in double-labeling methods. No eluation steps are needed which otherwise might destroy antigenicity of the peptides. It is also possible to combine fluorescence microscopy with double-labeling methods making it possible to demonstrate two antigens in addition to catecholamines in the same section.

The preembedding method using Vibratome sections preserves the antigenicity of most neuropeptides better than postembedding methods. It also offers acceptable quality of ultrastructure because osmication can be performed after immunocytochemical staining.

Limitations

Because the fluorescence is unstable, the tissues can not be stored for longer than a few days before Vibratome sectioning and fluorescent microscopy. The processing is rather laborious and time-consuming so it is difficult to process a large number of samples at the same time. This is a problem especially when large experimental series are processed. The problem may be solved by storing the tissues frozen or by embedding the tissues in polyethylene glycol.

Vibratome sectioning is time-consuming and considerable practice is needed to cut sections thinner than 10 μm. The embedding and processing of the Vibratome sections for electron microscopy is laborious. If antigens tolerate the use of postembedding methods, these are faster and easier to perform.

REFERENCES

Björklund, A. (1983). Fluorescence histochemistry of biogenic amines. *In* "Methods in Chemical Neuroanatomy" (A. Björklund and T. Hökfelt, eds.), Vol. 1, pp. 50–121. Elsevier, Amsterdam.

Eränkö, O. (1955). Distribution of fluorescing islets, adrenaline and noradrenaline in the adrenal medulla of the hamster. *Acta Endocrinol. (Copenhagen)* **18,** 174.

Furness, J. B., Costa, M., and Wilson, A. J. (1977). Water stable fluorophores, produced by reaction with aldehyde solutions, for histochemical localization of catechol- and indolethylamines. *Histochemistry* **52,** 159.

Furness, J. B., Heath, J. W., and Costa, M. (1978). Aqueous aldehyde (Faglu) methods for fluorescence histochemical localization of catecholamines and for ultrastructural studies of central nervous tissue. *Histochemistry* **57,** 285.

Hökfelt, T., and Ljungdahl, Å. (1972). Modification of the Falck-Hillard formaldehyde method using the Vibratome: Simple, rapid and sensitive localization of catecholamines in sections of unfixed or formalin fixed brain tissue. *Histochemie* **29,** 324.

Hökfelt, T., Fuxe, K., Goldstein, M., and Joh, T. (1972). Immunohistochemical localization of three catecholamine synthetizing enzymes: Aspects on methodology. *Histochemie* **33,** 231.

Johansson, O. (1983). The Vibratome-Ralph knife combination: A useful tool for immunohistochemistry. *Histochem. J.* **15,** 265.

Kobayashi, S., Ohashi, T., Fujita, T., Nakao, K., Yoshimasa, T., Imura, H., Mochizuki, T., Yanaihara, C., Yanaihara, N., and Verhofstad, A. (1983). An immunohistochemical study on the co-storage of met-enkephalin-arg6-gly7-leu8 and met-enkephalin-arg6-phe7 with adrenaline and/or noradrenaline in the adrenal chromaffin cells of the rat, dog and cat. *Biomed. Res.* **4,** 433.

Kobayashi, S., Ohashi, T., Uchida, T., Nakao, K., Imura, H., Yanaihara, N., and Verhofstad, A. A. J. (1984). Co-storage of adrenaline and noradrenaline with met-enkephalin-arg6-gly7-leu8 and met-enkephalin-arg6-phe7 in chromaffin cells of hamster adrenal medulla. *Arch. Histol. Jpn.* **47,** 319.

Livett, B., Day, R., Elde, R., and Howe, P. (1982). Co-storage of enkephalins and adrenaline in the bovine adrenal medulla. *Neuroscience* **7,** 1323.

Mason, D. Y., Abdulaziz, Z., Falini, B., and Stein, H. (1983). Double immunoenzymatic labelling. *In* "Immunocytochemistry" (J. M. Polak and S. Van Noorden, eds.), pp. 113–128. Wright, Bristol, England.

Mazurkiewicz, J. E., and Nakane, P. K. (1972). Light and electron microscopic localization of antigens in tissue embedded in polyethylene glycol with a peroxidase labelled antibody method. *J. Histochem. Cytochem.* **20,** 969.

Pelto-Huikko, M., Salminen, T., and Hervonen, A. (1982). Enkephalin-like immunoreactivity is restricted to the adrenaline cells in hamster adrenal medulla. *Histochemistry* **73,** 493.

Pelto-Huikko, M., Salminen, T., and Hervonen, A. (1985a). Localization of enkephalins in adrenaline cells and the nerves innervating adrenaline cells in rat adrenal medulla. *Histochemistry* **82,** 377.

Pelto-Huikko, M., Salminen, T., Toivanen, M., Partanen, M., and Hervonen, A. (1985b). Immunohistochemical localization of neurotensin in hamster adrenal medulla. *Anat. Rec.* **211,** 458.

Pickel, V. M. (1981). Immunocytochemical methods. *In* "Neuroanatomical Tract-Tracing Methods" (L. Heimer and M. J. Robards, eds.), pp. 483–509. Plenum, New York.

Pickel, V. M., Joh, T. H., Reis, D. J., Leeman, S. E., and Miller, R. J. (1979). Electron-microscopic localization of substance-P and enkephalin in axon terminals related to dendrites of catecholaminergic neurons. *Brain Res.* **160,** 387.

Priestley, J. V., and Cuello, A. C. (1983). Electron microscopic immunocytochemistry for CNS transmitters and transmitter markers. *In* "Immunohistochemistry" (A. C. Cuello, eds.), pp. 273–321. Wiley, Chichester.

Schöler, J., and Armstrong, W. E. (1982). Aqueous aldehyde (Faglu) histofluorescence for catecholamines in 2 μm sections using polyethylene glycol embedding. *Brain Res. Bull.* **9,** 27.

Sternberger, L. A. (1979). "Immunocytochemistry," 2nd ed. Wiley, New York.

Varndell, I. M., and Polak, J. M. (1984). Double immunostaining procedures: Techniques and applications. *In* "Immunolabelling for Electron Microscopy" (J. M. Polak and I. M. Varndell, eds.), pp. 155–177. Elsevier, Amsterdam.

Verhofstad, A. A. J., Steinbusch, H. W. M., Joosten, H. W. J., Penke, B., Varga, J., and Goldstein, M. (1983). Immunocytochemical localization of noradrenaline, adrenaline and serotonin. *In* "Immunocytochemistry" (J. M. Polak and S. Van Noorden, eds.), pp. 143–168. Wright, Bristol, England.

Wang, B.-L., and Larsson, L.-I. (1985). Simultaneous demonstration of multiple antigens by indirect immunofluorescence or immunogold staining. *Histochemistry* **83,** 47.

7

Sequential Immunocytochemical and Silver Staining of Neuroendocrine Cells in the Same Section

ERIK WILANDER AND MONALILL LUNDQVIST

Endocrine Laboratory
Department of Pathology
University Hospital
Uppsala, Sweden

INTRODUCTION

The application of silver stains and immunocytochemistry with hormone specific antisera is a reliable tool for the identification and characterization of peptide and/or biogenic amine-producing neuroendocrine cells in histological sections of normal and neoplastic human tissues (Grimelius and Wilander, 1980; Facer *et al.*, 1980). In addition, new markers for the visualization of neuroendocrine differentiated cells like neuron-specific enolase, prealbumin, and neurofilament have recently been introduced (Schmechel *et al.*, 1978; Miller *et al.*, 1984; Lehto *et al.*, 1983). The presence of these substances is easily demonstrated immunocytochemically using the respective specific antiserum. A sequential staining technique has been introduced in which an immunocytochemical stain and a silver stain are applied to the same section, and thus both the silver staining properties and an immunoreaction can be evaluated in the same cell (Lundqvist and Wilander, 1983). The method facilitates the simultaneous identification of various peptide hormones, biogenic amines, and other neuroendocrine markers in specific neuroendocrine cells.

DESCRIPTION OF SEQUENTIAL STAINING TECHNIQUE

Immunocytochemical Staining

1. Fix the tissue specimens in 10% buffered formaldehyde solution.
2. Cut in 4-μm-thick sections on chrome-gelatinated glass slides.
3. Dry overnight at 37°C.
4. Deparaffinize in two changes of xylene for 5 min each.
5. Rehydrate in descending ethanol concentrations: 100% ethanol (two changes), 95% ethanol (two changes), 70% ethanol, and distilled water for 1 min each.
6. Rinse three times in Tris (hydroxymethyl)-amino-methan-saline buffer (1 : 9) with a final dilution of 0.05 *M* Tris–0.15 *M* NaCl (pH 7.6) (Tris-saline) for 5 min in each.
7. Incubate overnight with monoclonal antibody against serotonin (Sera Lab., Crawley Down, Sussex, England; Lot Mas 055 clone YC5/HLK), diluted 1 : 800 in Tris-saline buffer at room temperature in a moist chamber.
8. Rinse three times in Tris-saline buffer for 5 min in each.
9. Incubate with rabbit anti-mouse IgG (Cappel Laboratories, Malvern, Pennsylvania; code No. 15109), diluted 1 : 800 in Tris-saline buffer for 30 min in a moist chamber.

10. Rinse three times in Tris-saline buffer for 5 min in each.

11. Incubate with sheep anti-rabbit IgG (diluted 1:30 in Tris-saline buffer) for 30 min in a moist chamber.

12. Rinse three times in Tris-saline buffer for 5 min in each.

13. Incubate with the rabbit peroxidase–antiperoxidase (PAP) complex (Dakopatts; Santa Barbara, California; code Z1132) for 30 min in a moist chamber.

14. Rinse three times in Tris-saline buffer for 5 min in each.

15. Visualize the antibody complex with 0.00002% 3,amino-9-ethyl-carbazole (A-5754 Sigma Chemical Company, St. Louis, Missouri) dissolved in 0.02 *M* sodium acetate buffer containing 0.000006% H_2O_2 for 10 min.

16. Rinse 10 min in tap water.

17. Mount in glycerine-gel.

18. Photograph selected areas.

Destaining

19. Heat the slides at 60°C until the coverslip is separated from the slides.

20. Immerse the slides in two changes of 100% ethanol for a total period of 10 min.

21. Check under the microscope that no positive cells remain in the section.

22. Pass the section through descending ethanol concentrations: 95% (two changes), and 70% for 1 min in each.

23. Rinse in double-distilled water five times for 2 min in each.

Restaining with Grimelius Technique: Impregnation

24. Immerse the slides in a solution of 60 mg silver nitrate (Merck 1512) dissolved in 90 ml double-distilled water and 10 ml Walpol's buffer (pH 5.6) at room temperature.

25. Place the slides at 60°C for $3\frac{1}{2}$–4 hr.

26. Reduce the section in a freshly prepared solution containing 1 g Qu'nol (hydroquinone) (BDH Chemicals Ltd., Poole, England) and 5 g sodium sulfite (anhydrous) (Merck 6657) for 1–3 min.

27. Rinse in distilled water.

28. Dehydrate in increasing ethanol concentrations: 70% (two changes), 95% (two changes), 100%, and xylene (two changes), and mount in Canada balsam.

Buffer Solutions

Tris-buffer

Dissolve 60.57 g of Tris in 500 ml of distilled water.

Adjust the pH to 7.6 with 1 *N* HCl, and make up the volume to 1000 ml with distilled water.

0.15 M *sodium chloride*

Dissolve 8.766 g of NaCl in 1000 ml of distilled water.

Working solution: Tris-saline

0.05 *M* Tris	1 part
0.15 *M* NaCl	9 parts

Acetate-buffer

Solution A: 0.02 *M* sodium acetate; 1.64 g/1000 ml distilled water.

Solution B: 0.02 *M* acetic acid; 1.2 ml (37%) acetic acid/988 ml distilled water.

Working solution: A, 282 ml + B, 118 ml

Walpol's buffer

Solution A: 0.2 *M* acetic acid; mix 12 ml of acetic acid with 1000 ml of distilled water.

Solution B: 0.2 *M* sodium acetate; dissolve 27.2 g of sodium acetate in 1000 ml of distilled water.

Working solution: A, 48 ml; B, 952 ml

3-amino-9-ethylcarbazol incubation solution

3-Amino-9-ethylcarbazol	10 mg
Dimethyl sulfoxide	6 ml
Acetate buffer (0.02 *M*, pH 5.0)	50 ml
H_2O_2 (30%)	10 μl

COMMENTS ON METHODOLOGY

Although a monoclonal antibody against serotonin for the immunocytochemical procedure and the argyrophil reaction of Grimelius as the silver stain are described in this protocol for the sequential staining method, a great variety of antisera can be applied for the initial stain as well as different types of silver stains, such as the argentaffin reaction of Masson (Singh, 1964) and the argyrophil reaction of Sevier-Munger (Sevier and Munger, 1965) for the secondary stain. A slight modification of the Grimelius argyrophil stain is used which is particularly suitable for staining intestinal endocrine cells and neuroendocrine tumors (Grimelius and Wilander, 1980).

It is an advantage if both the immunocytochemical reaction and the silver stain individually show a clearly positive reaction in the tissue sections before the sequential staining is attempted. The specimens must be fixed in solutions allowing adequate silver staining and immunostaining. Formaldehyde is preferable but glutaraldehyde and glutaraldehyde–picric acid mixture are also suitable for Masson's argentaffin reaction, while Bouin's fluid results in a relatively weak or unsatisfactory staining. The best fixatives for the Grimelius argyrophil reaction are formaldehyde–glutaraldehyde mixture or Bouin's fluid, while solutions containing ethanol, mercuric chloride, biochromate and, OsO_4 are unsatisfactory. Cryostat section material stains weakly or is found to be unreactive. For the Sevier-Munger staining of neuroendocrine cells, formaldehyde or Bouin's fluid are recommended (Grimelius and Wilander, 1980). It is worth noting that at no time is the initial immunocytochemical staining seen to interfere with the silver staining that follows.

APPLICATIONS

Neuroendocrine cells in both normal and neoplastic human tissues can be identifed in several ways, e.g., electron microscopically after the demonstration of neuroendocrine secretory granules, immunocytochemically after the application of hormone-specific antisera (Facer *et al.*, 1980), or other antisera directed against specific neuroendocrine markers (Schmechel *et al.*, 1978; Miller *et al.*, 1984), and with various silver staining techniques (Grimelius and Wilander, 1980). Since the chemical background of several of the silver methods is not completely understood, the identification of neuroendocrine cells with these stains is often complemented by ultrastructural and immunocytochemical studies. The present immunocytochemical and silver-restaining technique for neuroendocrine cells in the same section represents a simple method for the histopathological characterization of these cells using different techniques.

In the silver stains described, the silver particles are accumulated and confined to the hormonal secretory granules and thus the silver technique visualizes at the light microscopic level these granules which otherwise could only be observed by electron microscopy (Grimelius and Wilander, 1980). Although several silver stains do not discriminate between different neuroendocrine cell types, they have the advantage, at the light microscopic level, of aiding the identification of neuroendocrine cells and neuroendocrine tumor cells irrespective of hormone product and are, for that reason, still in current use despite the development of immunocytochemistry. Furthermore, several neuroendocrine cell types lack an identifiable hormone and the visualization of these cells is achieved using silver stains.

In principle, the staining and restaining method could be adapted to all neuroendocrine cell systems exhibiting neurohormonal granulation; it has already been applied to the pancreatic islet A_2 cells and intestinal neuro-

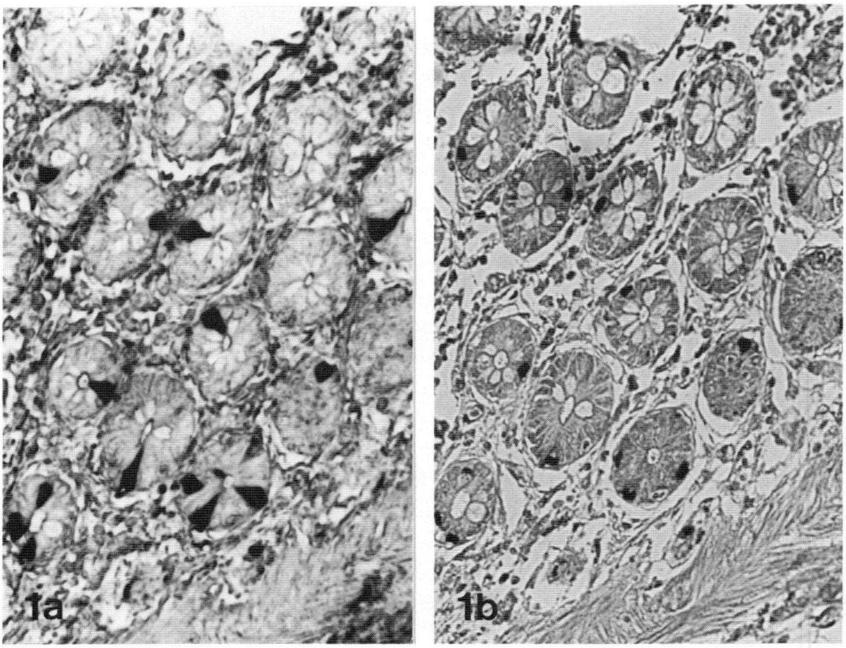
1a
1b

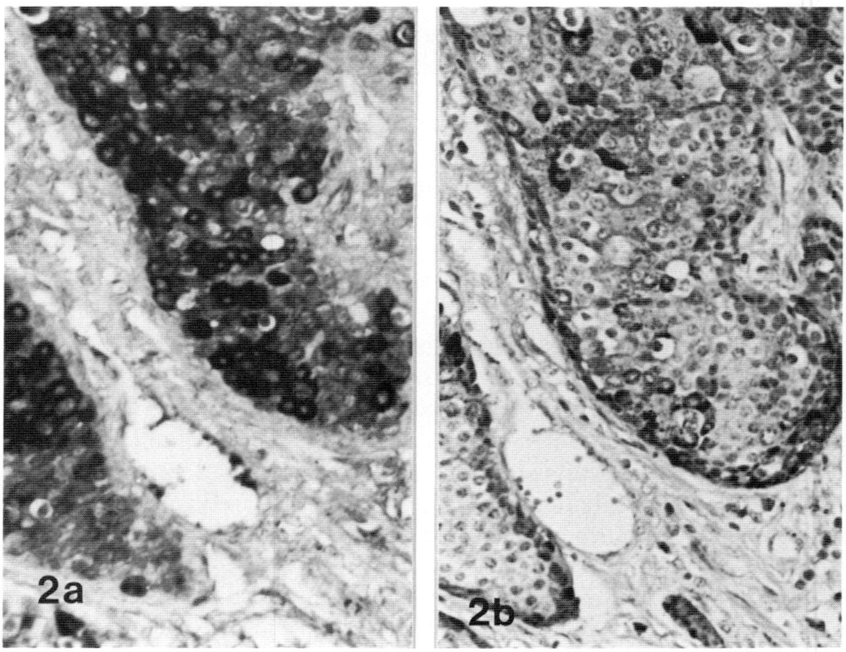
2a
2b

endocrine cells (Lundqvist and Wilander, 1983, 1984; Lundqvist *et al.*, 1985). Using this method, it has been possible to present evidence for neuron-specific enolase content and serotonin immunoreactivity in the silver stained neuroendocrine cells of the intestinal mucosa (Fig. 1).

A comparative study of the neuron-specific enolase immunoreactivity and the argentaffin reaction in intestinal carcinoids has shown that, at the cellular level, the two staining methods do not completely correlate. Thus, the neuron-specific enolase reactive tumor cell may be non-argentaffin and vice versa (Fig. 2) (Lundqvist *et al.*, 1985).

This work was supported by grants from the Swedish Medical Research Council (Project No. 6817).

REFERENCES

Facer, P., Bishop, A. E., and Polak, J. M. (1980). Immunocytochemistry: Its applications and drawbacks for the study of gut neuroendocrinology. *Invest. Cell Pathol.* **3,** 13.

Grimelius, L, and Wilander, E. (1980). Silver stains in the study of endocrine cells of the gut and pancreas. *Invest. Cell Pathol.* **3,** 3.

Lehto, V. P., Stenman, S., Miettinen, M., Dahl, D., and Virtanen, I. (1983). Expression of a neural type of intermediate filament as a distinguishing feature between oat-cell carcinoma and other lung cancers. *Am. J. Pathol.* **110,** 113.

Lundqvist, M., and Wilander, E. (1983). A simple procedure for immunocytochemical and silver staining of endocrine cells in the same section. *Acta Pathol. Microbiol. Immunol. Scand., Sect. A* **91,** 493.

Lundqvist, M., and Wilander, E. (1984). Small intestinal chromaffin cells and carcinoid tumours: A study with silver stains formalin-induced fluorescence and monoclonal antibodies to serotinin. *Histochem. J.* **16,** 1247.

Lundqvist, M., Wilander, E., Esscher, T., and Pahlman, S. (1985). Neuron-specific enolase in mucosal endocrine cells and carcinoid tumors of the small intestine: A comparative study with neuron-specific enolase immunocytochemistry and silver stains. *Histochem. J.* **17,** 323.

Miller, I. D., Reid, W. A., Liddle, C. N., and Horne, C. H. (1984). Immunolocalization of prealbumin as a marker for carcinoid tumors. *J. Pathol.* **143,** 199.

Schmechel, D., Marangos, P. J., and Brightman, M. (1978). Neuron-specific enolase is a molecular marker for peripheral and central neuroendocrine cells. *Nature (London)* **276,** 285.

Sevier, A. C., and Munger, B. L. (1965). A silver method for paraffin sections of neural tissues. *J. Neuropathol. Exp. Neurol.* **24,** 130.

Fig. 1. Human small intestinal mucosa. Neuroendocrine cells in the mucosal crypts (a) as revealed immunocytochemically after application of monoclonal serotonin antibodies, and (b) after staining with the argyrophil reaction of Grimelius; ×250.

Fig. 2. An illustration of human small intestinal carcinoid. (a) Some tumor cells are neuron-specific enolase immunoreactive, and (b) some cells, but not identical, are argentaffin; ×250.

8

Correlative Light and Electron Microscopic Method for the Visualization of the Same *in Vitro* Cell Using Radioautography and Serial Sectioning

JACQUELINE VUILLET

Centre de Microcopie Electronique
Faculté de Médecine
Marseille, France

CORRELATIVE MICROSCOPY IN BIOLOGY:
INSTRUMENTATION AND METHODS

INTRODUCTION

It is very difficult to study a minority cell type among large heterogenous cell cultures directly with the electron microscope. Light microscopy is necessary for preliminary cell identification, and labeling permits fine visualization and identification of typical cells. But it has not been possible until now, for technical reasons, to identify with certainty the same cell labeled *in vitro* by a radioactive compound with light and electron microscopy. With the method described here, it is possible to study with the electron microscope the same cell with a specific tritiated uptake, that has been first visualized by radioautography using the light microscope. It is thus possible to choose cells of interest with the light microscope, and then to section them into serial ultrathin sections. A detailed analysis of morphological differentiation can be carried out within the whole culture thickness with the electron microscope.

METHODOLOGY

Culture Preparation, Tritiated Uptake, Fixation, and Embedding (Vuillet *et al.*, 1984)

1. Mesencephalic cells (1.5×10^6) of mouse embryos are dissociated according to the method of Prochiantz *et al.* (1979), and then directly plated on polyornithin-coated plastic Falcon dishes (35 mm in diameter) in serum-free medium to prevent the non-neuronal cells from proliferating (Di Porzio *et al.*, 1980; Prochiantz *et al.*, 1982).

The following conditions must be met to obtain marked neurons directly on polyornithin layer. Collagen, glial cells, or other products present between the cells and the bottom of the dish stop tritium radiation and prevent successful visualization on photographic emulsion. The main condition for satisfactory visualization of marked cells in light microscopy on the whole cultures is that the least possible distance should exist between marked cells and the bottom of the dish. Indeed, the substratum choice is very important. Satisfactory preparations are obtained after a week of culture with good morphological differentiation. No apparent artefacts are observed during this time and under these culture conditions. The culture latency depends on its nature.

2. After 6 days, the mesencephalic cultures are incubated at 37°C for 1 hr with tritiated dopamine (5.10^{-7} *M*, 12.5 Ci/mmol NEN) in 1 ml phosphate-buffered saline (PBS). The uptake is stopped by three brief washes

(1 min each) and one thorough wash (5 min) in complete PBS at room temperature. Thorough washes are necessary to avoid background noise.

3. The cells are fixed for 1 hr at 4°C with 3% glutaraldehyde in PBS, rinsed in cold PBS, and postfixed in 2% OsO_4 in PBS for 1 hr at 4°C.

4. Alcohol dehydration is necessary, instead of acetone and propylene oxide, because of the plastic dishes, according to the following schedule:

25% ethanol	two changes for 7 min each at room temperature
50% ethanol	two changes for 7 min each at room temperature
60% ethanol	two changes for 7 min each at room temperature
80% ethanol	two changes for 7 min each at room temperature
90% ethanol	two changes for 7 min each at room temperature
100% ethanol	two changes for 10 min each at room temperature

5. The embedding is carried out in Epon as follows:

Epon: ethanol (1 : 1)	30 min
Epon	3 hr or overnight at room temperature

Epon polymerization is carried out for 3 days in a 60°C oven.

6. After polymerization, the cultures are separated from Falcon dishes. The separation must be done carefully to avoid the deposition of plastic fragments on the culture surface. The lateral walls of the box are cut out with scissors and completely removed. The plastic support with the culture is immersed in liquid nitrogen, and then warmed by placing it in a warm waterbath. By successive retraction and dilation, the two walls are separated. The lower face of the culture is smooth, and the desired cells are found at the surface.

Visualization and Identification of Labeled Neurons in the Light Microscope

After separation from plastic dishes, large resin blocks with cells (35 mm in diameter) are obtained. Marked neurons are spread over the bottom among other heterogenous cells. These blocks are turned over so that the cells are now at the top. They are coated directly with liquid photographic emulsion. For coating, we use K5 Ilford emulsion or NTB2 (Kodak) diluted 1 : 1 with water and mixed for 1 hr at 40°C in waterbath under safe light. After coating and drying overnight at room temperature, the blocks are stored for 3–4 weeks in a light-proof box with dessicant at 4°C.

The blocks are developed in Kodak Dektol for 4 min at 17°C and rinsed for 30 sec at 17°C in distilled water. This is followed by fixation with 30% sodium thiosulfate for 6 min at 17°C, and two washes in distilled water for 5 min each at 17°C. During development the emulsion coating is kept upward. One should avoid touching the emulsion side with fingers or any other object to prevent its tearing.

After drying overnight at room temperature, the blocks are ready for light microscopy, the emulsion lying at the top of the block directly under the microscope lens. Labeled neurons and their processes are clearly revealed on coated emulsion by silver grains, while pale, noncontrasted nuclei of unlabeled cells are scarcely visible (Fig. 1).

Selection and Mounting of Small Blocks for Semithin and Ultrathin Sectioning

After observing, as described above, labeled neurons are selected, marked by circling them with a felt-tip pen, numbered, and photographed. They are then cut out in small blocks with a punch and mounted on prepolymerized Epon blocks. Trapezoid-shaped small blocks centered on labeled cells are prepared. Under the emulsion coating, the labeled cells lie directly on the surface of the blocks. The culture thickness is as little as 4–5 μm. Since they contain the richest part of the cell culture, the first serial sections are the most important. It is here that most of the cytoplasms, nuclei, and processes are found. This area is the preferential site of intercellular exchanges, especially synaptic connections.

Radioautography of Semithin Sections

The radioautography of semithin sections is necessary to locate more precisely the cell culture organization. Because of the small thickness of cultures, semithin sections are not cut from the same block as that chosen for electron microscopy, but from neighboring blocks cut out from the large culture blocks. These semithin serial sections (1 μm thick) are treated according to the methods of Larra and Droz (1970) and Boyenval and Fischer (1976).

Radioautography of Serial Ultrathin Sections

As stated above, an electron microscopic control by radioautography is necessary to recognize labeled neurons among the whole cell population. This can be carried out either by the classical Larra and Droz method (1970) or by the LIGOP method as described by Bouteille *et al.* (1975) and Bouteille (1976). When the former is used, a modification is necessary since single-holed grids are used. The grids are first coated with Formvar, and then placed on collodion surface to collect serial sections. After isoamyl acetate dissolves the collodion, the sections remain on the Formvar film.

However, for this type of work, a modification of the LIGOP method of

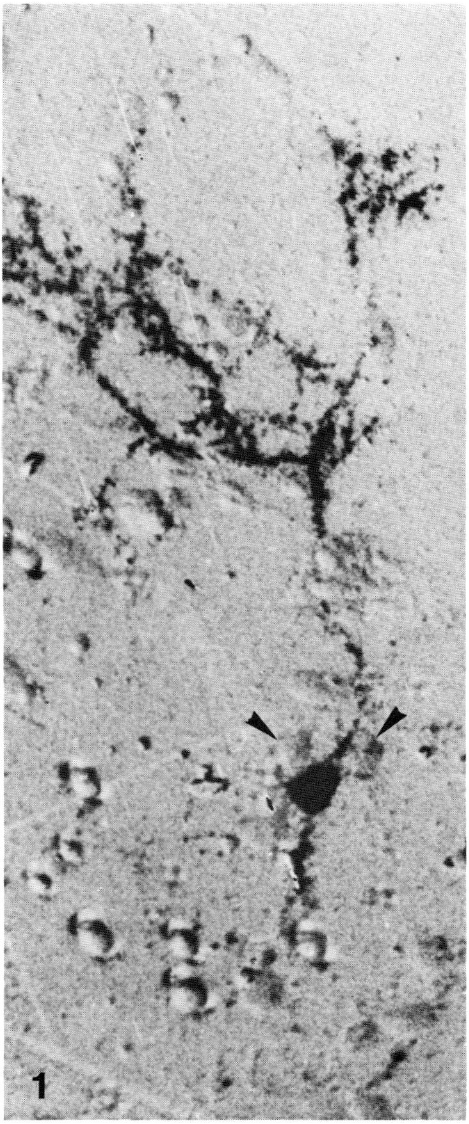

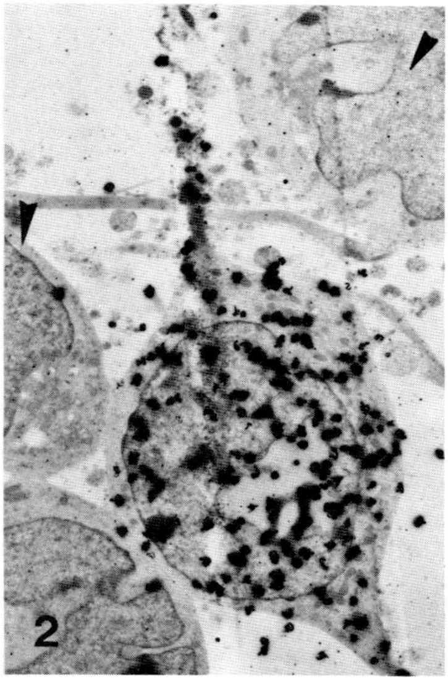

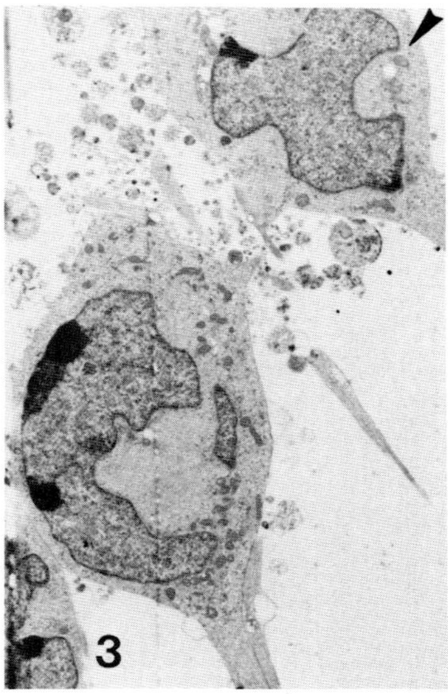

Fig. 1. Light microscopic visualization of a labeled neuron on a large Epon block after Epon embedding, emulsion coating, and development. Note the unlabeled nuclei (arrowheads) surrounding the marked perikaryon; ×500 (Vuillet *et al.*, 1984).

Fig. 2. The same neuron as in Fig. 1 after serial ultrathin sectioning. A radioautographic treatment permits the study of neuronal relationships between labeled and unlabeled (arrowheads) neurons and their processes; ×5000 (Vuillet *et al.*, 1984).

Fig. 3. The same neuron as in Fig. 1. An adjacent section of the neuron without radioautography, permitting a higher resolution and finer observation of cytological details, which are concealed by silver grains in Fig. 2; ×5000.

Bouteille (1976) seems to be more suitable (Vuillet and Seite, 1982; Vuillet *et al.*, 1984). Serial sections showing yellow to silver interference colors are cut through the whole thickness of the block with a diamond knife. As stated earlier, the first sections are the most important and also the most difficult to obtain because the block surface is rough and the emulsion coating elastic. To facilitate sectioning, the block should not be placed parallel to the knife cutting edge but at a slight angle. In this way the culture surface is gradually cut with the Epon and not wasted. The sections are collected on Formvar-coated grids (3 mm in diameter with a 2- × 0.6-mm single hole), and can be stored in LKB boxes.

Serial sections are stained with uranyl acetate and lead citrate. After carbon layering, a few grids are examined to locate the richest part of the block to study with radioautography. Some grids are coated as described by Bouteille (1976) with L4 Ilford emulsion and stored for one month in a light-proof box with dessicant at 4°C. When ready, the grids are developed as follows. They are developed in Kodak D19 for 4 min at 17°C, rinsed for 30 sec with distilled water at 17°C, fixed for 6 min with 30% sodium thiosulfate solution at 17°C, and then rinsed twice (5 min each) with distilled water at the same temperature. The grids are dried overnight at room temperature and then examined (Fig. 2). If needed, a second staining with uranyl acetate and lead citrate can be carried out after development. The LKB Ultrostainer can be used at 20°C for automatic staining.

Since the grid hole is necessarily large in size to achieve a clear view of the cells, it requires strengthening by a carbon film. The Formvar film when combined with serial sections and hydrated emulsion (during development) becomes quite heavy. Therefore, the grid should be dried gently at room temperature, otherwise the support film may show tearing, especially in the center of the hole or around the sections. The majority of the grids are not subjected to radioautography because improved staining and resolution are achieved without radioautographic emulsion (Fig. 3). However, if needed, the remaining grids can be used for radioautography. This approach is not possible with the Larra and Droz method (1970). Also, this method may result in wasted blocks because the control used for radioautography is not effective.

APPLICATIONS

A technical improvement is reported, which makes it possible to study, by a combined light and electron microscopic method, the same *in vitro* cell (among an heterogenous culture) by radioautography and serial sectioning. As previously described, this technique has a wide range of applications. It is possible, after light microscopic radioautography on entire cultures, to choose the cells to be investigated by electron microscopy. Also, it is feasible to treat, by electron radioautography, some serial sections and put aside other sections for a subsequent treatment of sections from another plane of sectioning. It is thus possible to proceed step-by-step, accompanied by a control if necessary. This approach avoids the loss of the entire experiment or the block should some technical error occur. This aspect of the method is particularly important when cultures of human cells are used, where each case is unique (e.g., tumors).

We have used this method for mouse mesencephalon cultures to study the dopaminergic neurons as an *in vitro* model. It can be employed for other tritiated-labeled neuromediator cells such as GABA-ergic or serotoninergic neurons. It can be used for other nerves or cell cultures where a tritiated-labeled component can be fixed at any moment of the cell cycle. Beside the ease and reliable results of this method, it allows a quantitative assessment made in the same way. An easier version of this method can be used if all serial sections can be collected on standard grids (150–200 mesh) with or without Formvar coating. This makes localization of labeled cells far easier and allows a random quantitative assessment.

I thank M. A. Hayat for his kind invitation to contribute. I am deeply indebted to R. Seïte (Faculté de Médecine, Marseille, in whose laboratory this work was made), for teaching me the skills on which I have relied in this work, and to J. Glowinski (College de France, Paris) for entrusting me with the completion of this method. All culture preparations were made by A. Prochiantz and associates (College de France, Paris), without whom the application of this method would not have been possible. I thank A. Autillo-Touati for advice and encouragement and D. Araud, H. Debroas, and E. Canale for their skillful technical and secretarial assistance. This work was supported, in part, by a grant from MRI No. DN 84/P.58.

REFERENCES

Bouteille, M. (1976). The "LIGOP" method for routine ultrastructural autoradiography. A combined of single grid coating, gold latensification and phenidon development. *J. Microsc. Biol. Cell.* **27,** 121.

Bouteille, M., Dupuy-Coin, A. M., and Moyne, G. (1975). Methods for localization of proteins and nucleo-proteins in the cell nucleus by high resolution autoradiography and cytochemistry. *In* "Methods in Enzymology" (B. W. O'Malley and J. G. Hardmann eds.), Vol. 40, p. 3. Academic Press, New York.

Boyenval, J., and Fischer, J. (1976). Dipping technique. *J. Microsc. Biol. Cell.* **27,** 115.

Di Porzio, U., Daguet, M. C., Glowinski, J., and Prochiantz, A. (1980). Effect of striatal cells on in vitro maturation of mesencephalic dopaminergic neurons grown in serum-free conditions. *Nature* (*London*) **288,** 370.

Larra, F., and Droz, B. (1970). Techniques radioautographiques et leur application à l'étude du renouvellement des constituants cellulaires. *J. Microsc.* (*Paris*) **9,** 845.

Prochiantz, A., Di Porzio, U., Kato, A., Berger, B., and Glowinski, J. (1979). *In vitro* maturation of mesencephalic dopaminergic neurons from mouse embryos is enhanced in presence of their striatal target cells. *Proc. Natl. Acad. Sci. U.S.A.* **76,** 5387.

Prochiantz, A., Delacourte, A., Daguet, M. C., and Paulin, D. (1982). Intermediate filament proteins in mouse brain cells cultured in the presence or absence of fetal calf serum. *Exp. Cell Res.* **139,** 404.

Vuillet, J., and Seïte, R. (1982). Technique de mise en évidence d'un même neurone en culture en microscopie photonique et électronique par radioautographie. *Biol. Cell.* **43,** 25a.

Vuillet, J., Daguet-De Montety, M. C., Autillo-Touati, A., Glowinski, J., Prochiantz, A., and Seïte, R. (1984). A combined light and electron microscopic method for the visualization of the same *in vitro* neuron by radioautography and serial sections. *J. Microsc.* (*Oxford*) **133,** 171.

9

Correlative Light and Transmission Electron Microscopy of Mitotic Cells in the Nervous System

R. R. STURROCK

Department of Anatomy
University of Dundee
Dundee, Scotland

INTRODUCTION

The technical problems involved in studying mitotic cells in the central nervous system (CNS) differ with the stage of development. In the case of embryos and fetuses the main problem is inadequate fixation, particularly for electron microscopy. Since large numbers of mitotic cells are present in the embryonic nervous system, finding mitotic cells is no problem whereas in later stages of development, when ventricular mitotic activity has virtually ceased, mitotic cells are much more difficult to find. Finally, in the adult and aged CNS mitotic cells are very infrequent, for example, in adult white matter only about one glial cell in 10,000 is undergoing mitosis at any time. The morphology of mitotic cells in the CNS has been the subject of a recent review (Sturrock, 1982a) and will not be described in detail here.

The first question to be asked when beginning a study of embryos or fetuses from small laboratory animals such as mice and rats is whether immersion fixation or perfusion is preferable and if perfusion fixation is the method of choice how does one go about perfusing a small embryo? The advantage of immersion fixation is that it is quick and simple and in small embryos is probably as effective as perfusion. Rickmann and Wolff (1985) have recently claimed that they obtained better fixation of neopallial tissue in rat embryos younger than E17 by immersion fixation, because perfusion fixation took too long and removal of blood was incomplete. This has not been our own experience. We have found that for electron microscopy of brain and spinal cord, perfusion gives excellent results in mouse embryos from E11 (Sturrock, 1978, 1981b). The successful perfusion fixation of very young mouse embryos was first carried out by Henrikson and Vaughn (1974) and the method described below was based on their technique.

One advantage of using perfusion fixation throughout is that if an age-related series of animals is being examined there will be no transition period between perfusion and immersion fixation since perfusion fixation is necessary in older animals for preservation of mitotic figures (Cavanagh and Lewis, 1969). The choice of fixative will firstly depend on whether the tissue is being prepared for conventional light microscopy using paraffin wax embedding or whether the tissue is to be resin-embedded for semithin section light microscopy and/or electron microscopy.

MITOTIC CELLS IN THE EMBRYONIC CNS

A high-quality operating microscope is essential for successful perfusion of small embryos; in our experience the Zeiss operating microscope is ideal due to the range of magnification, foot controls, and above all, depth of focus. A separate fiber optic lighting system is a useful adjunct to the lighting system of the microscope as oblique lighting can eliminate glare which is often present due to light reflecting from either the embryo or the perfusion fluid. For very small embryos cannulae can be improvised from 30G disposable dental needles (Fig. 1) or fine glass cannulae

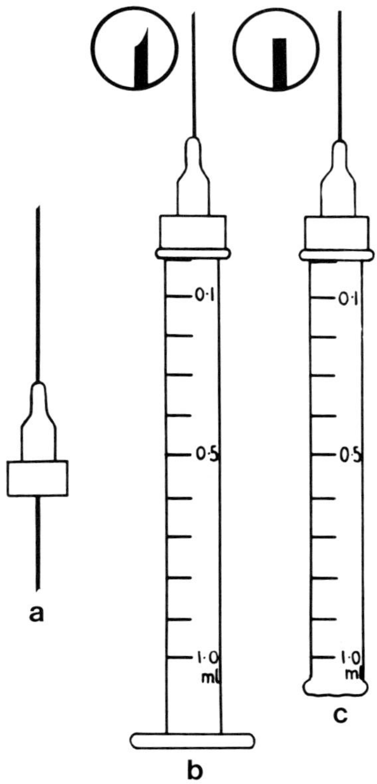

Fig. 1. A typical 30G dental needle (a) which is attached to a 1-ml syringe by sawing off most of the shank and heating the remaining stump to melting point before attaching the needle to it (b). The pointed needle is then converted to a cannula using a whetstone, the wings of the syringe are removed, and the barrel is heated and widened slightly to accept a connector (c).

can be made in the laboratory. The cardinal rule is to use the largest cannula which will pass into the ascending aorta. A method of holding the jellylike embryo in place is essential. Henrikson and Vaughn (1974) used a V-shaped clay trough but a simple, robust, and inexpensive holder can be manufactured from a Gillette GII razor cartridge and holder. The cartridge, in its holder, is filled with epoxy resin up to the level of the bottom of the V-shaped trough and the resin is polymerized. The transverse plastic bars are trimmed so that when the embryo is placed in the holder with its thorax over the bar its head is extended (Fig. 2). This position gives ideal access to the thoracic contents.

The degree of difficulty in exposing the heart varies with the stage of development. In very young embryos in which the cartilage of the tho-

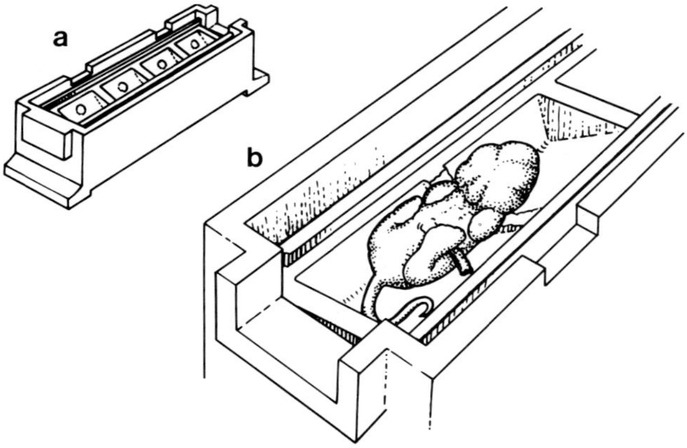

Fig. 2. a, Razor cartridge. b, Embryo in position. Note the head placed over a trimmed transverse bar.

racic cage has not yet developed, the heart can be easily exposed by slitting the skin with a fine needle. Dr. Seymour, formerly of our department, recommended cactus needles mounted with glue on orange sticks, but 30G dental needles are just as effective. Later in development the developing rib cage has to be excised with ophthalmology scissors. Insertion of the cannula into the aorta via the left ventricle may be done manually or with the aid of micromanipulators. Provided the embryo holder is placed on a low block so that the ulnar border of the operator's hand rests on a firm surface, it is usually easier to insert the cannula manually as the hand can move in a greater variety of planes than micromanipulators; the disadvantage is that the operator has to hold the cannula in position for the duration of perfusion. In very young embryos before septation of the ventricles has begun the aorta is approached via the common ventricle. As perfusion commences the right atrium or common atrium is incised with either a needle or ophthalmology scissors.

If a highly coagulative fixative such as Bouin's solution is used, a prerinse with physiological saline or Ringer's solution is essential to flush out the vascular system. This necessitates the use of a three-way tap (Fig. 3). A prerinse is not usually necessary in prenatal animals when glutaraldehyde or an aldehyde mixture is used.

For electron microscopy the osmolarity of the fixative is important. The amount of extracellular space seen in electron microscopy depends on the osmolarity of the fixative and the stage of development of the nervous system (Sumi, 1969). If the same fixative is used in sequential studies of the mouse, large extracellular spaces appear in the ependymal layer of the

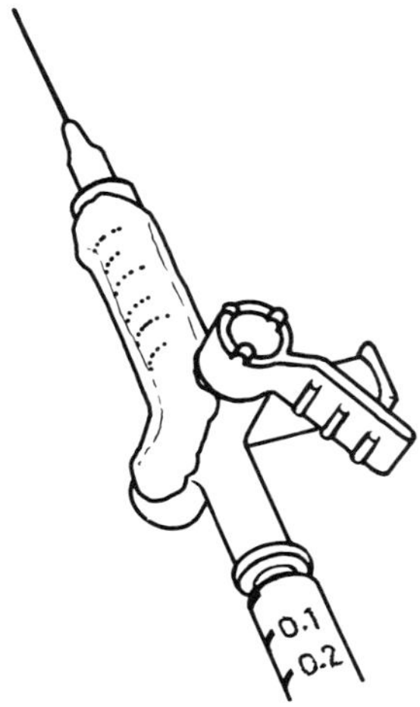

Fig. 3. Three-way tap attached to a shortened cannula–syringe combination. When this is used it is necessary to employ micromanipulators to hold the needle in position in the aorta while the tap is being switched from saline to perfusion fluid.

lateral ventricles from E17 until a few days after birth (Sturrock and Smart, 1980). Similar spaces were described in the neonatal rabbit brain by Stensaas and Gilson (1972). These are almost certainly artifactual as they are absent from parts of the ependyma which has choroid plexus adhering to it. Identical extracellular spaces appear in the ependyma of the central canal of the mouse spinal cord at E14 but are absent at E13 and E15 (Sturrock, 1981a). By varying the osmolarity of the fixative and the percentage of aldehyde, these spaces can be prevented but this can only be done by trial and error. One argument against varying the fixative is that reduction of the extracellular space to the "normal" adult level may in itself be creating artifacts. Cragg (1979, 1980) has suggested that the generally accepted narrow intracellular spaces in electron micrographs of the CNS are artifactual since the extracellular space thus displayed is much less than that measured physiologically. The size of the extracellular space also depends on the integrity of the blood-brain barrier. Kalimo (1976) demonstrated that the same fixative will give rise to a "normal" appearance in parts of the CNS with tight-junctioned capillaries but causes large extracellular spaces in regions such as the median eminence which contain fenestrated capillaries.

The composition of mixed aldehyde fixatives may be altered to preserve specific organelles. Rickmann and Wolff (1985) found that a three-aldehyde mixture was ideal for membrane preservation whereas a two-aldehyde fixative was better for preserving microtubules. It is therefore beyond the scope of this chapter to advise an ideal fixative for perfusion fixation of embryonic CNS for electron microscopy and one can only recommend a process of trial and error. Likewise the temperature at which the fixative is used varies widely from laboratory to laboratory. Most authors recommend that the flushing fluid, if used, should be at 37°C but successful results appear to have been obtained using a fixative at 37°C, room temperature, or 4°C although most laboratories tend to use fixatives at room temperature or 4°C for electron microscopy. Sodium cacodylate is still widely used as a buffer but it appears to have little advantage over phosphate buffers and may have hidden dangers because of its arsenic content (Weakley, 1977).

The number and position of mitotic cells in the developing nervous system varies with the stage of development and different regions develop at different times. In the early stages of development the neural tube consists of a pseudostratified columnar epithelium. As development proceeds this ventricular layer becomes surrounded by a mantle layer of cells and a marginal layer of fibers. In the developing spinal cord the ventricular layer is the site of most mitotic activity during the relatively short period of neurogenesis (Smart, 1972a). The spinal cord does not develop a secondary proliferative zone such as the subependymal layer of the telencephalon (Sturrock, 1982b) and in the spinal cord the most rapid increase in glial number is brought about by proliferation of glial precursors already situated in the grey and presumptive white matter (Bensted *et al.*, 1957; Sturrock, 1982b). Even during the phase of high mitotic activity in the ventricular layer the level of activity is not uniform even in closely anatomically related areas (Smart, 1972a,b, 1973, 1976). In certain regions of the CNS, nonventricular areas of mitotic activity are present. Some, such as the rhombic lip and external granular layer of the cerebellum, are transient, neuron-producing zones but the subventricular layer around the lateral ventricles persists throughout life as the glial-producing, subependymal layer, although the number of mitotic cells present in adult animals is small (Allen, 1912; Bryans, 1959; Smart, 1961; Sturrock, 1985).

At the stages of development during which mitotic activity is high, no other special techniques are necessary as mitotic cells can be found easily even in small blocks of tissue. It is possible to stain selectively for mitotic figures in paraffin wax embedded material fixed with modified Bouin's solution (Fraser, 1982). This technique has the advantage that the mitotic nuclei are stained blue while interphase nuclei are stained red; in areas such as the ventricular layer where nuclei are closely packed together, the mitotic cells stand out clearly which allows accurate identification and rapid counting. Such a stain would obviously be invaluable if semiautomated image analysis was being used to count mitotic figures.

MITOTIC CELLS IN THE CNS DURING MYELINATION

Immediately before the onset of myelination and during the phase of rapid myelination there is a substantial increase in the number of glial cells in white matter (Roback and Scherer, 1935; Hillebrand, 1966; Schonbach *et al.*, 1968; Schmidt, 1969; Matthews and Duncan, 1971; Lucas, 1972; Sturrock, 1974a, 1982b,c). Analysis of mitotic activity in the rabbit spinal cord (Sturrock, 1982c) and mouse forebrain (Sturrock, 1983) indicates that while some of this increase can be attributed to migration of glial cells, most of it (~75%) is due to proliferation of glial cells already present in the presumptive white matter. The percentage of mitotic cells in the presumptive white matter during this myelination gliosis is remarkably similar in both the anterior commissure of the mouse forebrain (Sturrock, 1974b) and tracts of the rabbit spinal cord. The rapid increase in glial number begins at E20 in the rabbit cord and at postnatal day 5 in the mouse anterior commissure; in both cases the increase begins to tail off 8–10 days later. During this period the percentage of mitotic cells falls from 1.5 to 0.5% in both species.

With a mitotic index of around 1% it is a practical proposition to look for mitotic cells in ultrathin sections provided areas such as the corpus callosum, optic nerve (Fig. 4), or complete spinal cords are examined so that the sections contain 100–200 glial cells. This, however, is a tedious and time-consuming procedure. An alternative method is to examine serial semithin and ultrathin sections (Fig. 5). The semithin sections (0.5–1 μm) are stained with 1% toluidine blue and examined under oil immersion light microscopy. In our experience coverslipping is not necessary, although unless the immersion oil is removed from the slide after examination myelin staining is gradually lost. The oil can be easily removed by rapidly flushing with toluene.

This technique has the advantage that large areas can be scanned quickly to find mitotic cells and if none are present the serial ultrathin sections can be discarded immediately. When mitotic cells are identified, their position in the section relative to obvious landmarks, such as blood vessels, can be noted thus making the finding of the same cell in serial ultrathin sections relatively easy. The disadvantage of this technique is that it is quite time-consuming. Ideally, each semithin section should be examined as it is cut to decide whether it is worth cutting serial ultrathin sections. It is also possible (some would say probable) that the vital feature in the semithin section, e.g., a mitotic cell process in contact with a myelin sheath, will not be present in the serial section. This problem can be minimized by cutting the semithin sections at 0.5 μm.

An alternative method is to reembed semithin sections for electron

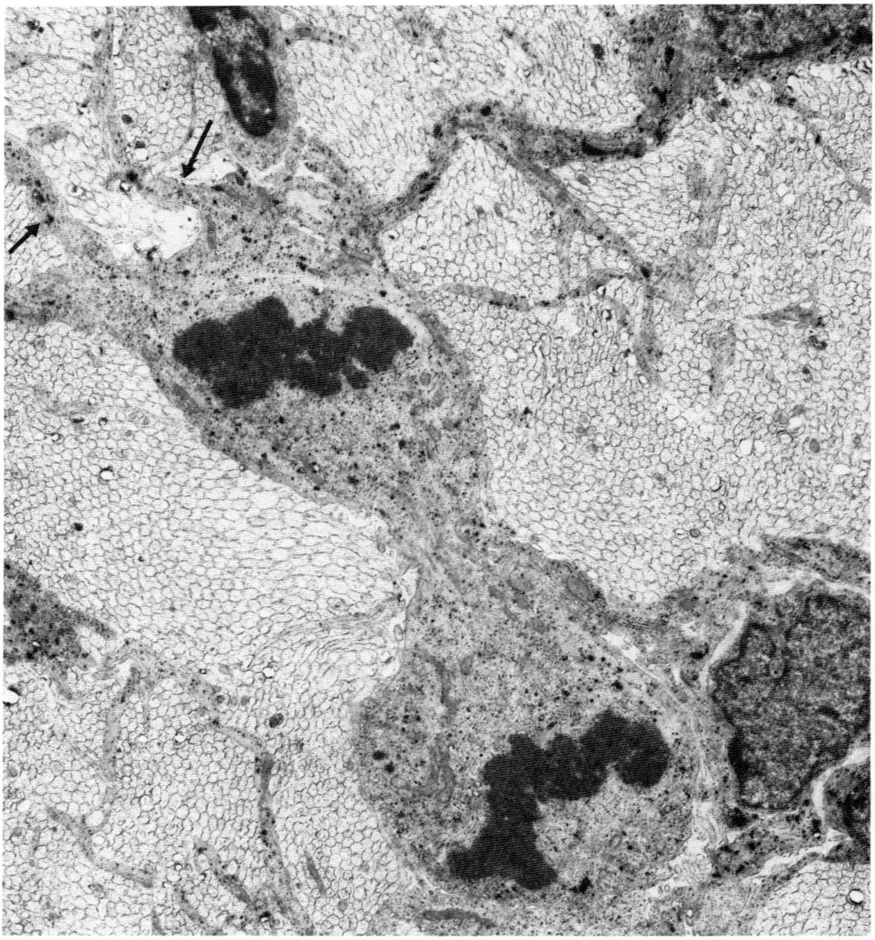

Fig. 4. Mitotic astrocyte from an 18-week human fetal optic nerve. This is an example of a mitotic cell found by scanning ultrathin sections in the electron microscope. At this stage mitotic cells are relatively common. This micrograph demonstrates that astrocytes retain their processes (arrows) during division; ×7500.

microscopy. The introduction of the Ralph knife (Bennett *et al.*, 1976) has made it possible to cut semithin sections from large blocks of tissue. For example, it is possible to cut complete coronal sections of adult rodent brains.

A number of different methods of reembedding semithin light microscopic sections for electron microscopy have been published in the literature. This indicates that there is a general need for such a technique and

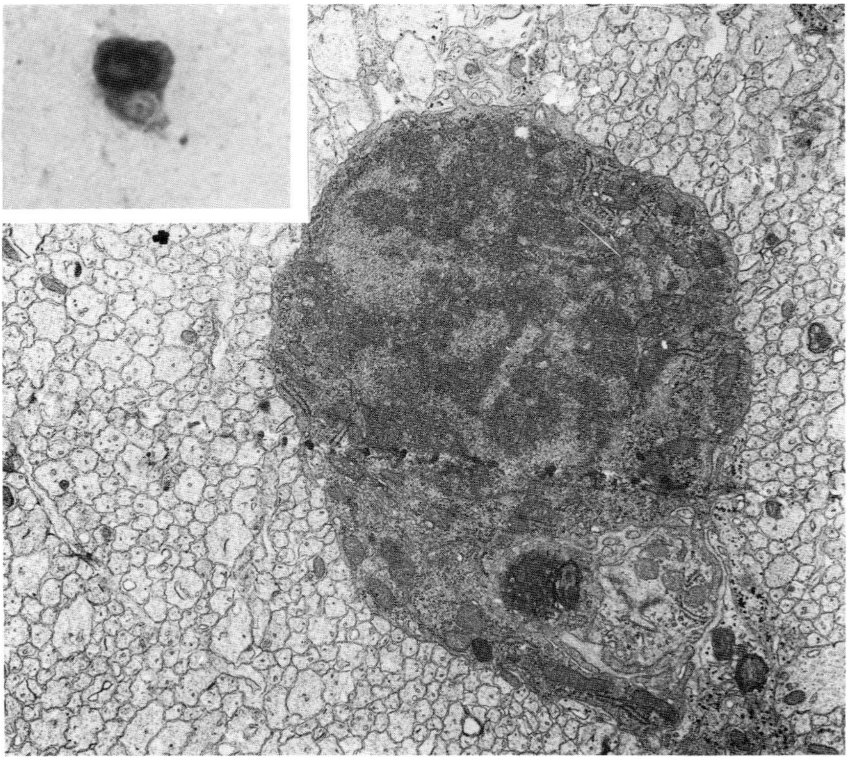

Fig. 5. Five-day postnatal mouse corpus callosum with a mitotic microglial cell. The process of this cell partially surrounds an area of pale cytoplasm with a lipofuscin granule. This is an example of serial semithin/ultrathin sections; ×15,740. Inset: semithin, toluidine blue-stained sections; ×2400.

that no one method is entirely satisfactory. Some have involved the manufacture of apparatus (Grimley, 1965; Woodcock and Bell, 1967) or mounting slides on plastic dishes (Springer, 1973) or acrylic slides (Davidowitz *et al.*, 1976). The removal of sections from glass slides with a razor blade prior to reembedding has also been recommended (Campbell and Hermans, 1972). Schabtach and Parkening (1974) recommended covering the slide with a capsule full of resin, polymerizing the resin and snapping the capsule, complete with section, off the slide. Kaplan and Hinds (1977) varied this technique by breaking off most of the slide from around the resin cylinder and dissolving most of the remaining glass in hydrofluoric acid. The thin sliver of glass remaining was then lifted off with fine forceps or the cylinder plus glass was plunged into liquid nitrogen for 30

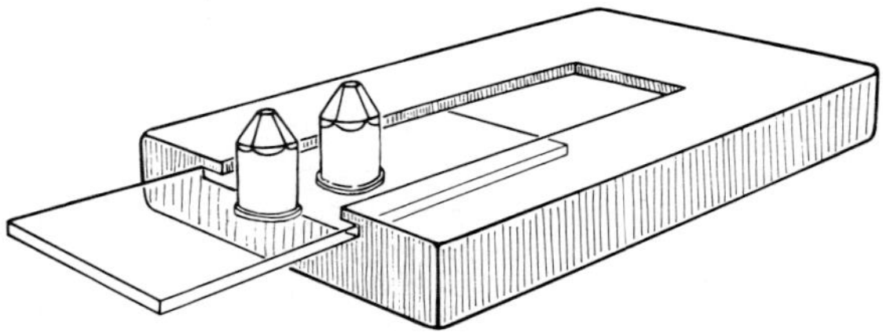

Fig. 6. Slide and capsules being placed in holder.

sec before being ultrasonically cleaned. A variation of this technique is to mount the section on a coverslip to avoid having to dissolve glass away.

King *et al.* (1982) developed a slide holder to enable the resin-filled capsule to be snapped off the slide without breaking the glass. A simplified version of their holder was designed by Donald Cathcart of Dundee University and is shown in Fig. 6. We found that the following procedure is extremely reliable with tissue embedded in Spurr's resin provided each step is followed. The material is fixed for electron microscopy and embedded in Spurr's resin in the usual manner except that with large blocks instead of going directly from absolute alcohol to resin the block is put in propylene oxide, then propylene oxide resin mixture, (50 : 50) followed by 2 hr in resin before being put in fresh resin for embedding. Failure to do this with large blocks often leads to inadequate penetration of resin.

Procedure

1. Cut 1- to 3-μm sections from the resin block in the normal manner.
2. Mount six to eight sections equal distances apart on a glass slide, placing each section on an individual drop of distilled water. Place the slide on a hot plate at 90°C for 5 min to weld the sections to the glass and to prevent wrinkling of the sections.
3. Stain with 1% toluidine blue and examine at 400× or 1000× under oil immersion without a coverslip.
4. Photograph the section at low power.
5. Photograph the cell or structure of interest at high power.
6. Remove the immersion oil by flushing with toluene and gently blotting with tissue paper.
7. Reheat the slide on a hot plate at 120°C for 2 min.
8. Place the slide, section side down, over a vial(s) filled with freshly prepared Spurr's resin; one vial beneath each section to be reembedded.
9. Invert the slide and vial(s). There is little or no spillage of resin due to its viscosity.
10. Place the slide on a *flat* surface in an oven at 70°C overnight.

11. The following morning remove the slide from the oven and place on a hot plate at 90°C for 2 min.
12. Place the slide in a Perspex holder (Fig. 6).
13. Snap the vial off the slide using a pair of pliers (Fig. 7).
14. The section can be clearly seen on the surface of the block and can be accurately trimmed by using features such as blood vessels as a guide and the low-power photomicrograph.
15. Ultrathin sections are cut and stained in the usual manner and examined in the electron microscope.

The critical points in our experience are steps 7 and 8. If reheating is omitted, the sections partly, or completely, float off into the vial and cannot be resectioned. If the surface in the oven is not absolutely level the vial slowly slides off the section before the resin hardens.

Even with a well-trimmed block it is not easy to find the cell one is seeking. It is usually quicker to scan the section systematically than to try to find the cell using landmarks such as blood vessels. The stain contrast with uranyl acetate and lead citrate is less than normal and occasionally a fine precipitate is present, especially over cell somata. The edges of the tissue tend to wrinkle slightly during reembedding so cells in this region should be avoided if possible.

One practical problem we have found with the reembedding technique is that occasionally very small pieces of glass remain adherent to the

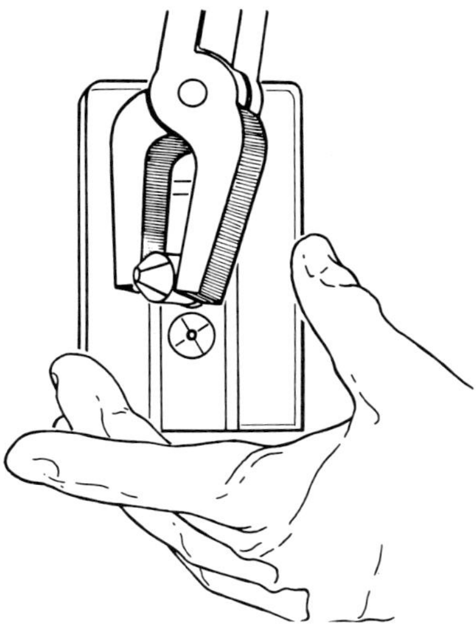

Fig. 7. Capsule snapped off slide with the aid of pliers.

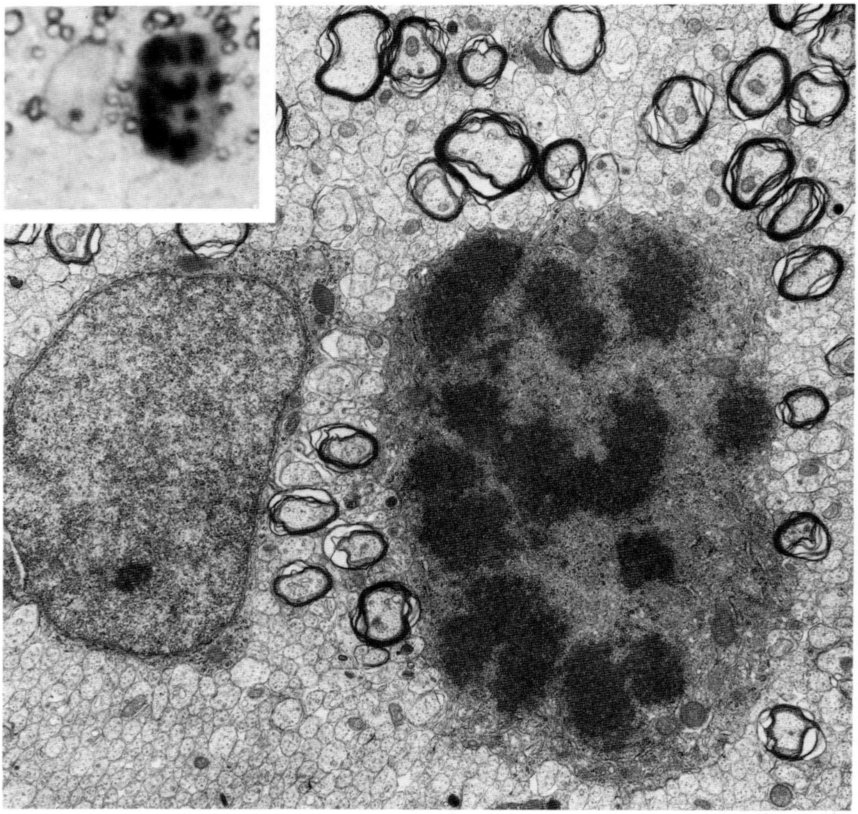

Fig. 8. A glioblast and mitotic oligodendrocyte from a 17-day postnatal mouse corpus callosum. This is an example of an ultrathin section of a reembedded semithin section. Note the disrupted myelin sheaths; ×12,000. Inset: semithin, toluidine blue-stained section; ×2400.

capsule. These are not visible to the naked eye or to routine examination of the capsule microscopically but do damage the knife. For this reason glass knives should be used to section reembedded material to avoid expensive damage to diamond knives.

The extra processing involved in the reembedding technique appears to have some effect on the material. We found that myelin sheaths tended to be disrupted (Sturrock, 1984) and the staining properties with uranyl acetate and lead citrate were affected, resulting in a reduction in contrast (Fig. 8). It was necessary to examine sections at 40 kV rather than the usual 60 kV. This meant that the resultant micrographs were not as sharply contrasted as one would have wished. On balance the semithin-ultrathin serial sectioning technique results in better electron micrographs and may be the better method in the early stages of myelination.

IDENTIFICATION OF MITOTIC CELLS IN THE MATURE AND AGED CNS

A small number (two to four) of mitotic cells are present in the subependymal layer in each coronal section of mouse brain from 6 to 31 months of age (Sturrock, 1985) but mitotic cells are rare in the rest of the forebrain (1 : 10,000). The only practical technique for finding such cells and identifying them electron microscopically is to use the reembedded semithin section method. In this case, 3-μm sections have the advantage over 1-μm sections of reducing the number of sections examined.

APPLICATIONS

These methods can be used to study mitosis of cells in the CNS at different stages of development in both normal and pathological tissues. With care they should help to resolve current controversies on exactly which types of differentiated glial cells retain the capacity to divide under normal and pathological circumstances. The method for perfusion fixation of embryos can be used for other tissues apart from the CNS. Indeed in a well-perfused embryo most tissues are ideally fixed for electron microscopy. The semithin/ultrathin and reembedding techniques are applicable to any tissue where the subject of interest is a rare component and where the number of ultrathin sections which would have to be scanned would otherwise make investigation of the structure impractical.

REFERENCES

Allen, E. (1912). Cessation of mitosis in the central nervous system of the albino rat. *J. Comp. Neurol.* **22,** 547.

Bennett, H. S., Wyrick, A. D., Lee, S. W., and McNeil, J. H., Jr. (1976). Science and art of preparing tissues embedded in plastic for light microscopy with special reference to glycol methacrylate, glass knives and simple stains. *Stain Technol.* **5,** 71.

Bensted, J. P. M., Dobbing, J., Morgan, R. S., Reid, R. T. W., and Payling Wright, G. (1957). Neuroglial development and myelination in the spinal cord of the chick embryo. *J. Embryol. Exp. Morphol.* **5,** 428.

Bryans, W. A. (1959). Mitotic activity in the brain of the adult rat. *Anat. Rec.* **133,** 65.

Campbell, R. D., and Hermans, C. O. (1972). A rapid method for resectioning 0.5–4 μ epoxy sections for electron microscopy. *Stain Technol.* **47,** 115.

Cavanagh, J. B., and Lewis, P. D. (1969). Perfusion-fixation, colchicine and mitotic activity in the adult rat brain. *J. Anat.* **104,** 341.

Cragg, B. (1979). Brain extracellular-space fixed for electron microscopy. *Neurosci. Lett.* **15,** 301.

Cragg, B. (1980). Preservation of extracellular space during fixation for electron-microscopy. *Tissue Cell* **12,** 63.

Davidowitz, J., Pachter, B. R., and Breinin, G. M. (1976). Acrylic slides for the clearing and re-embedding of thick ultramicrotome sections. *Stain Technol.* **51,** 247.

Fraser, F. J. (1982). A selective stain for mitotic figures, particularly in the developing brain. *Stain Technol.* **57,** 219.

Grimley, P. M. (1965). Selection for electron microscopy of specific areas in large epoxy tissue sections. *Stain Technol.* **40,** 259.

Henrikson, C. K., and Vaughn, J. E. (1974). Fine structural relationships between neurites and radial glial processes in developing mouse spinal cord. *J. Neurocytol.* **3,** 659.

Hillebrand, H. (1966). Quantitative Untersuchungen Über Postnatale Veränderungen der Glia im Corpus Callosum der Katze. *Z. Zellforsch. Mikrosk. Anat.* **73,** 303.

Kalimo, H. (1976). The role of the blood-brain barrier in perfusion-fixation of the brain for electron microscopy. *Histochem. J.* **8,** 1.

Kaplan, M. S., and Hinds, J. W. (1977). Neurogenesis in the adult rat: Electron microscopic analysis of light radioautographs. *Science* **197,** 1092.

King, D. G., Kammlade, N., and Murphy, J. (1982). A simple device to help re-embed thick plastic sections. *Stain Technol.* **57,** 307.

Lucas, D. (1972). Über die postnatale Entwicklung der Neuroglia im Corpus callosum des Kaninchens. *Z. Mikrosk.-Anat. Forsch.* **85,** 341.

Matthews, M. A., and Duncan, D. (1971). A quantitative study of morphological changes accompanying the initiation and progress of myelin production in the dorsal funiculus of the rat spinal cord. *J. Comp. Neurol.* **142,** 1.

Rickmann, M., and Wolff, J. R. (1985). Prenatal gliogenesis in the neopallium of the rat. *Adv. Anat., Embryol. Cell Biol.* **93,** 1.

Roback, H. N., and Scherer, H. J. (1935). Über die feinere Morphologie des frühkindlichen Hirnes unter besonderer Berücksichtigung der Gliaenlwicklung. *Virchows Arch. Pathol. Anat. Physiol.* **294,** 365.

Schabtach, E., and Parkening, T. A. (1974). A method for sequential high resolution light and electron microscopy of selected areas of the same material. *J. Cell Biol.* **61,** 261.

Schmidt, W. (1969). Über die postnatale Entwicklung der Neuroglia im Corpus callosum des Menschen. *Z. Anat. Entwicklungsgesch.* **128,** 18.

Schonbach, J., Hu, K. H., and Friede, R. L. (1968). Cellular and chemical changes during myelination: Histologic, autoradiographic, histochemical and biochemical data on myelination in the pyramidal tract and corpus callosum of rat. *J. Comp. Neurol.* **134,** 21.

Smart, I. (1961). The subependymal layer of the mouse brain and its cell production as shown by radioautography after thymidine-H^3 injection. *J. Comp. Neurol.* **116,** 325.

Smart, I. H. M. (1972a). Proliferative characteristics of the ependymal layer during the early development of the spinal cord in the mouse. *J. Anat.* **111,** 365.

Smart, I. H. M. (1972b). Proliferative characteristics of the ependymal layer during the early development of the mouse diencephalon, as revealed by recording the number, location, and plane of cleavage of mitotic figures. *J. Anat.* **113,** 109.

Smart, I. H. M. (1973). Proliferative characteristics of the ependymal layer during the early development of the mouse neocortex: A pilot study based on recording the number, location and plane of cleavage of mitotic figures. *J. Anat.* **116,** 67.

Smart, I. H. M. (1976). A pilot study of cell production by the ganglionic eminences of the developing mouse brain. *J. Anat.* **121,** 71.

Springer, M. (1973). Casting epon slides containing serial 3–5 μm sections for scanning prior to electron microscopy. *Stain Technol.* **48,** 45.

Stensaas, L. J., and Gilson, B. C. (1972). Ependymal and subependymal cells of the caudato-pallial junction in the lateral ventricle of the neonatal rabbit. *Z. Zellforsch. Mikrosk. Anat.* **132,** 297.

Sturrock, R. R. (1974a). Histogenesis of the anterior limb of the anterior commissure of the mouse brain. A quantitative study of changes in the glial population with age. *J. Anat.* **117,** 17.

Sturrock, R. R. (1974b). Histogenesis of the anterior limb of the anterior commissure of the mouse brain. II. A quantitative study of pre- and postnatal mitosis. *J. Anat.* **117,** 27.

Sturrock, R. R. (1978). A developmental study of epiplexus cells and supraependymal cells and their possible relationship to microglia. *Neuropathol. Appl. Neurobiol.* **4,** 307.

Sturrock, R. R. (1981a). An electron microscopic study of the development of the ependyma of the central canal of the mouse spinal cord. *J. Anat.* **132,** 119.

Sturrock, R. R. (1981b). A quantitative and morphological study of vascularisation of the developing mouse spinal cord. *J. Anat.* **132,** 203.

Sturrock, R. R. (1982a). Cell division in the normal central nervous system. *Adv. Cell. Neurobiol.* **3,** 3.

Sturrock, R. R. (1982b). Gliogenesis in the prenatal rabbit spinal cord. *J. Anat.* **134,** 771.

Sturrock, R. R. (1982c). Changes in cell number in the central canal ependyma and in the dorsal grey matter of the rabbit thoracic spinal cord during fetal development. *J. Anat.* **135,** 635.

Sturrock, R. R. (1983). Cellular proliferation in the central nervous system. *J. Anat.* **136,** 446.

Sturrock, R. R. (1984). Identification of mitotic cells in the central nervous system by electron microscopy of reembedded semithin sections. *J. Anat.* **138,** 657.

Sturrock, R. R. (1985). Age related changes in cellularity, mitotic activity and pyknotic cell number in the mouse subependymal layer. *J. Anat.* **141,** 19.

Sturrock, R. R., and Smart, I. H. M. (1980). A morphological study of the mouse subependymal layer from embryonic life to old age. *J. Anat.* **130,** 391.

Sumi, S. M. (1969). The extracellular space in developing brain: Its variation with changes in osmolarity of the fixative, method of fixation and maturation. *J. Ultrastruct. Res.* **29,** 398.

Weakley, B. S. (1977). How dangerous is sodium cacodylate? *J. Microsc.* (*Oxford*) **109,** 249.

Woodcock, C. L. F., and Bell, P. R. (1967). A method for mounting 4 μ resin sections routinely for ultra-thin sectioning. *J. R. Microsc. Soc.* **87,** 485.

10

Light and Scanning Electron Microscopy of the Same Metaphase Chromosomes

CHRISTINE J. HARRISON, ELSPETH M. JACK, AND TERENCE D. ALLEN

Departments of Cytogenetics and Ultrastructure
Paterson Laboratories
Christie Hospital and Holt Radium Institute
Manchester, England

CORRELATIVE MICROSCOPY IN BIOLOGY:
INSTRUMENTATION AND METHODS

INTRODUCTION

Clinical cytogenetics is a rapidly expanding field, particularly in prenatal diagnosis and cancer. The aim of the cytogenetic approach is to identify the chromosome abnormalities involved in a particular syndrome or malignancy by examination of metaphases derived from the appropriate tissues. The results in a large proportion of cases now provides the clinician with the definitive diagnosis.

Cytogenetic analysis is carried out routinely using light microscopy. The limited resolving power of the light microscope restricts the precise location of many small chromosome abnormalities. Recent improvements have been made in cytogenetic techniques for light microscopy by using "high-resolution" procedures (Yunis *et al.*, 1978). These have improved the accuracy of clinical cytogenetics and made it possible to visualize previously undetectable chromosome defects in relation to clinical syndromes (Buhler *et al.*, 1983; Fryns *et al.*, 1983) and cancer (Yunis *et al.*, 1981, 1982).

In spite of these innovations, cytogenetic interpretation continues to be restricted by the limited resolving power of the light microscope. Electron microscopy, with the inherent capacity for high resolution, has the scope for further definition of minute chromosome abnormalities, by taking over when light microscope resolution runs out. Sections of metaphase chromosomes viewed in the transmission electron microscope (TEM) provide too small an area to determine the presence of chromosome abnormalities, although Hozier *et al.* (1981) successfully used this difficult procedure on complete metaphases to examine chromatin fiber organization. Whole mount chromosomes prepared for TEM may be observed at a higher magnification than in the light microscope (Burkholder, 1981; Goyanes and Mendez, 1982; Xu and Wu, 1983). Metaphase chromosomes are highly electron dense and as a result no additional information may be gained relating to small chromosome defects. Therefore, the scanning electron microscope (SEM) is the best tool for this purpose.

Extensive studies of metaphase chromosomes have been carried out in the SEM (Brecher, 1977; Cervenka *et al.,* 1973; Christenhus *et al.,* 1967; Daskal *et al.,* 1976; Filip *et al.,* 1975; Golomb, 1976; Golomb and Bahr, 1971, 1974; Kingsley-Smith, 1970; Korf and Diacumakos, 1978; Mace *et al.,* 1978; Scheid and Tarut, 1971). The range of protocols for chromosome preparation used in these investigations differed significantly from the light microscope methods routinely used in clinical cytogenetics. Therefore, the results are not directly comparable. To undertake cytogenetics in the SEM, a procedure closely following the light microscope method must be used. Recently, several authors have successfully produced complete human metaphases, prepared initially as for light microscopy, in the SEM (Harrison *et al.,* 1981; Sweney *et al.,* 1979; Utsumi, 1982). Harrison *et al.* (1981) presented the first three-dimensional human karyotype which led to examination of the same metaphases in the light microscope and the SEM (Harrison *et al.,* 1985a). The advantage of this procedure is that small chromosome abnormalities identified initially in the light microscope can be transferred to the SEM and the same defective chromosome examined. This has provided a breakthrough in clinical cytogenetics by opening up the powers of magnification when the light microscope has reached its limit.

This clinical application for correlative light microscopy and SEM of human chromosomes will be discussed in detail in this chapter. In addition, use of the technique has provided insight toward the understanding of some of the mechanisms involved in chromosome banding. This area will also be covered in this chapter.

METHODOLOGY

Light Microscope Procedures

Correlative light microscopy and SEM of the same chromosomes has been mainly directed toward human clinical cytogenetics. It is of help in those cases where small or ill-defined chromosome abnormalities are present. Standard light microscope preparative procedures are employed in the clinical situation to extract metaphases from a wide variety of human material, ranging from blood to tumor tissue. SEM has frequently been carried out on material which has already been prepared for light microscopy by one of these routine methods. Therefore, although this material is not always the best to undertake SEM processing on, modifications have been made to the protocol, after the light microscope fixation stages, to ensure a reliably good chromosome morphology in the SEM.

It is beyond the scope of this chapter to enter into detailed discussion of culture methods for individual tissues. The aim in each case is to produce a fixed sample with sufficient metaphases for cytogenetic analysis. The majority of correlative light microscropy and SEM of chromosomes has been undertaken on blood and other mammalian cell lines. Therefore, the complete light microscope preparative procedures for these tissues only are presented here, with special emphasis on those modifications required for the best SEM result.

Chromosome Preparation from Peripheral Blood

The most convenient tissue for production of the human constitutional karyotype is peripheral blood. Clinical cytogenetic laboratories undertake the preparation of blood samples in a similar manner, with minor variations, for example, in the choice of medium used for culture, the amount of serum added to the cultures, and the type of culture vessel. These variables appear to have no effect on the success of the final result in the SEM. The critical steps in the procedure are:

1. To add only a small amount of heparinized peripheral blood to each culture, for example 0.25 ml blood into 5-ml medium.
2. To add phytohemagglutinin to stimulate division of the lymphocytes, approximately 0.1 ml into a 5-ml culture.
3. To incubate the cultures for 72 hr at 37°C.
4. After this time to add colcemid to the cultures at a final concentration of approximately 0.05 μg/ml and to reincubate for an additional 1–2 hr, to arrest those cells entering cell division during this period in metaphase. Metaphase chromosomes continue to condense in the presence of colcemid and become very short and as a result the details of chromosome banding may be obscured. Therefore, incubation time in colcemid should be kept to the minimum for optimum chromosome morphology in the SEM without reducing the mitotic index. This has been found to be 1–2 hr.

It is during the blood culture stage that the chromosomes are manipulated for "high-resolution" light microscope observation. The definition of high resolution according to these light microscope procedures does not refer to the usual method of increasing magnification, by increasing the size of the image, but to improving the resolution by changing the size of the object. In the case of metaphase chromosomes this is achieved by producing chromosomes of increased length at prometaphase. Successful methods have made use of agents which synchronize the cells in the early S phase of the cell cycle, for example, amethopterin (Yunis *et al.*, 1978), fluorodeoxyuridine (Webber and Garson, 1983), and excess thymidine

(Viegos-Pequignot and Dutrillaux, 1978). After release from this block, cells are arrested in prometaphase by a short pulse of colcemid. These cultures may also be processed for SEM following the same protocol.

After exposure to colcemid the cells in suspension in the medium are transferred to centrifuge tubes and spun at 200 *g* for 5 min. The pellet of cells is then resuspended in freshly made 0.075 *M* potassium chloride (KCl) for 8 min at room temperature. The time in this hypotonic may be varied, but in most cases it is recommended that the cells be exposed to hypotonic for no longer than 20 min, since a prolonged hypotonic treatment results in an ill-defined chromosome morphology in the SEM. The temperature of the hypotonic does not appear to be critical. Incubation of cells in KCl at either room temperature or 37°C produces the same result. However, it is advisable not to heat the KCl to a temperature greater than 37°C or to use chilled hypotonic, as drastic changes in temperature may affect the physiological functions of the cells, resulting in inefficient action of the hypotonic effect.

The function of hypotonic treatment is to separate the chromosomes and swell the cytoplasm of the lymphocytes to weaken the plasma membrane for ease of metaphase spreading. In addition, it produces lysis of the red blood cells to provide a clean preparation of lymphocytes and other white blood cells. The cells are then centrifuged in KCl at 200 *g* for 5 min and the pellet of cells is resuspended in 1–2 ml of freshly prepared 3 : 1 methanol : acetic acid fixative. This should be added over a few seconds, drop by drop, with constant agitation of the tube, to ensure proper and fast fixation. The volume is then adjusted to 10 ml and the suspension left at room temperature for 20–30 min. The cells are washed an additional two to three times, resuspending the cells each time and stored at 4°C for a minimum of 1 hr. Overnight storage appears to be a favorite option. At this stage the cells can be stored at −20°C for periods of up to a year. Also, at this stage, cells can be collected for further SEM processing.

For light microscope cytogenetic investigations, in the routine manner, the cells are resuspended in fresh fixative, dropped onto slides, stained, and exposed to a range of banding procedures prior to light microscope analysis. These routine methods are not given in this chapter. Included herein are the modified procedures utilized for correlate light microscope SEM observation only.

Chromosome Preparation of Mammalian Cell Lines

Correlative light microscopy and SEM may also be undertaken on chromosome preparations from monolayer cell cultures of human or other mammalian species. For example, cells from human amniotic fluids, skin biopsies, and tumor biopsies are routinely cultured in clinical cytogenetics laboratories. A range of mammalian tissues are also grown in culture in many research laboratories. As described for lymphocyte cultures, the choice of medium and amount of serum used may be highly variable and equally successful. The critical steps necessary to achieve the best chromosome preparations are:

1. To use logarithmically growing cultures with the medium changed 24 hr prior to harvesting.
2. To add colcemid to the cultures at a final concentration of 0.05 μg/ml, as for lymphocytes cultures, but to experiment with the incubation time in colcemid to achieve a high mitotic index. More slowly dividing cells will require a longer incubation in colcemid.

Prior to harvesting the metaphases, the cells must be removed from the surface of the culture vessel. The medium is collected first to prevent the loss of metaphases. Cells are removed by incubating the cultures in 0.1% trypsin at 37°C, until all the cells are detached. This takes only a few minutes with trypsin at this concentration. The cells and the collected medium are transferred to a centrifuge tube and the surface of the flask rinsed again to ensure that all cells are collected.

To achieve a higher mitotic index, the "metaphase shake off" technique of Stubblefield and Klevecz (1965) may be used. This technique makes use of the property of most mammalian cell lines, that the mitotic cells are easily detached from the surface of the culture vessel. Metaphases accumulated by 0.05 μg/ml colcemid are collected by physically tapping the culture flask with a flat hand and agitating the medium to selectively dislodge the metaphases from the surface of the vessel. The metaphases from several flasks may be pooled and concentrated by centrifugation. This ensures a consistently high mitotic index. Chromosome preparation of the cells from the complete monolayer and the selected metaphases uses the same method as that described for the peripheral blood procedure with minor modifications to prevent premature rupture of these more fragile cell membranes. These cells are exposed to 0.075 *M* KCl for 10 min in suspension as described for blood cultures. Three drops of 3 : 1 methanol : acetic acid fixative are added prior to centrifugation and the resultant pellet is fixed, as a pellet, for 20 min in fresh 3 : 1 fixative and maintained at 4°C in a refrigerator for a minimum of 1 hr. This fixation of a pellet prevents premature cell bursting, which leads to chromosome loss. As described for blood cultures, at this stage the cells may be stored or slides may be made or the cells may be collected for further SEM processing. From this point lymphocyte and mammalian somatic cell chromosomes are treated in the same manner.

Metaphase Spreading

The first part of the correlative light microscope and SEM preparation involves examination of the metaphases in the light microscope. For subsequent SEM observation several modifications of the standard light microscope slide-making procedure are carried out, which have been found to be essential for preservation of good chromosome morphology in the SEM.

The first variation is that metaphases are spread onto 15-mm circular glass coverslips instead of light microscope slides. To facilitate rapid relocation of the same metaphases in the SEM after observation in the light microscope, it is useful to have a point of reference on the coverslip. The most convenient method is to divide the coverslip into segments, using an engraving pen, and to number each segment as demonstrated in Fig. 1a. The numbers are clearly visible in the SEM when viewed at a low magnification (Fig. 1b). Care must be taken not to engrave too many markings onto the coverslip, as it becomes very fragile and metaphase spreading may be inhibited.

The fixed cell suspension, after removal from storage, is washed in fresh fixative an additional three times to eliminate cell debris and cytoplasmic background. At the final wash the pellet of cells is resuspended in sufficient cold fixative to provide an optimum cell density. The main consideration of cell density is to permit the chromosomes within individual metaphases to be well spread, without the cells themselves being too far apart. An example of optimum cell density and cell spreading is shown in Fig. 2. The best method of achieving this is to experiment on one or two glass slides and adjust the density accordingly.

For SEM investigation the aim is to produce optimally spread metaphases with evenly distributed chromosomes. Individual laboratories employ variation in their metaphase spreading procedures, which have all been found to be equally successful when making spreads onto coverslips. The normal procedure involves dropping a small drop of the adjusted, fixed cell suspension onto the surface of the slide, allowing the drop to spread and then to air dry. When making spreads onto coverslips, a few minor modifications have been found useful:

1. The coverslips are easier to handle if adhered onto a glass slide by placing a small drop of water underneath each coverslip. In this way two coverslips may be treated at the same time (Fig. 3a).

2. To facilitate cell spreading, the coverslips should be lightly moistened by allowing breath to condense onto the surface, immediately prior to dropping the suspension. This controls the amount of moisture without flooding the coverslips and allows the drop to be accurately placed onto the coverslip.

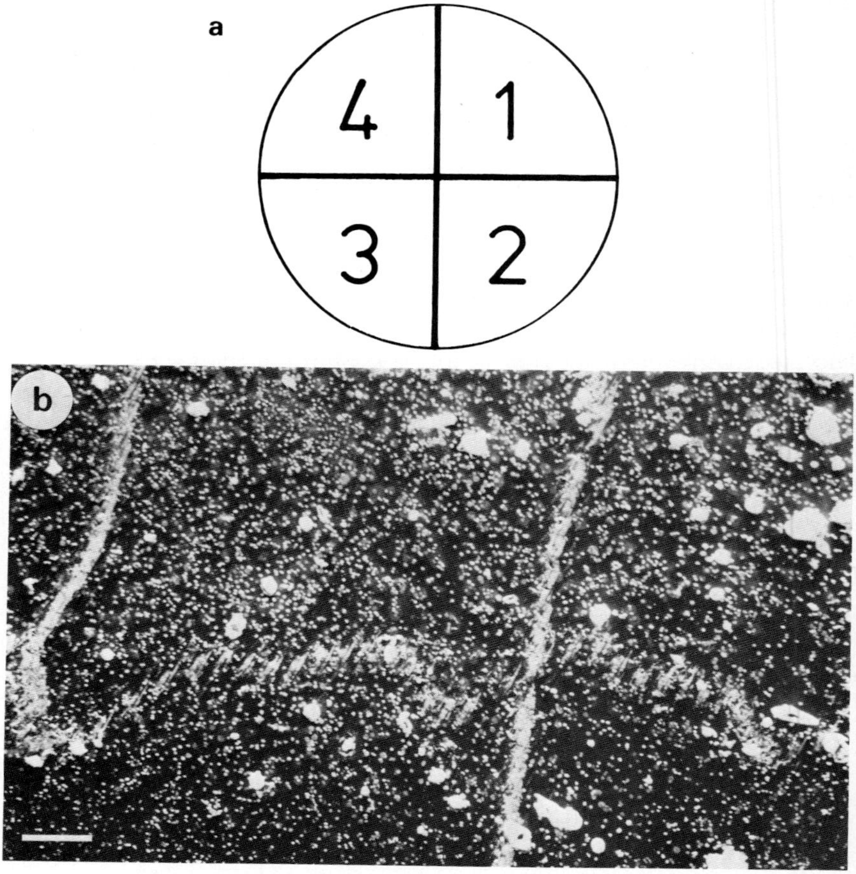

Fig. 1. a, Diagram of a circular coverslip engraved with suitable markings to allow relocation in the SEM of metaphases previously photographed in the light microscope. b, An engraving shown at low magnification in the SEM. Metaphases are relocated by their position in relation to the engravings; bar = 300 μm.

3. Deposit only a tiny drop of suspension onto each coverslip to permit the main area of spreading to occur over the surface of the coverslip (Fig. 3b). If too much suspension is allowed to overspill onto the supporting slide the most suitable metaphases are to be found on the slide and not on the coverslip.

4. To encourage the drop to spread over the coverslip it may be necessary to tilt the slide and drain the excess suspension onto blotting paper (Fig. 3c).

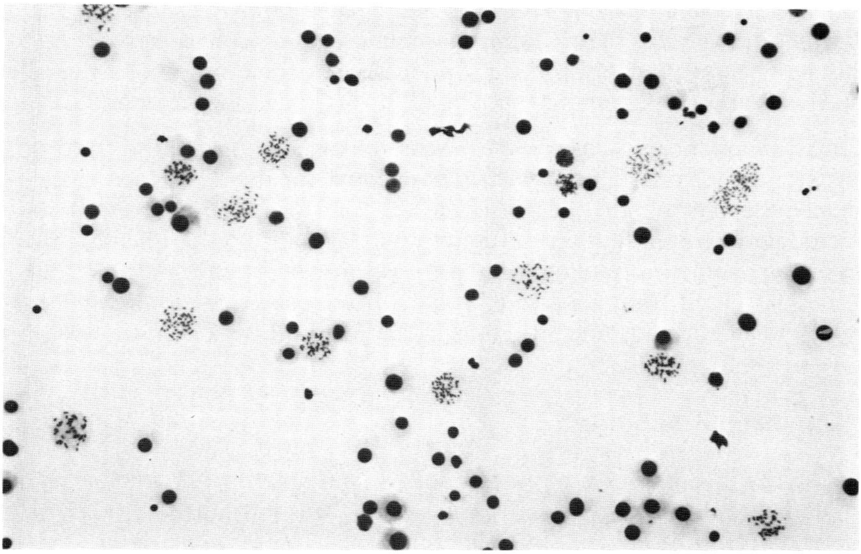

Fig. 2. Low-power field of view in the light microscope of a fixed lymphocyte suspension showing the optimum cell density to achieve the best metaphase spreading. Note that this preparation has a relatively high mitotic index.

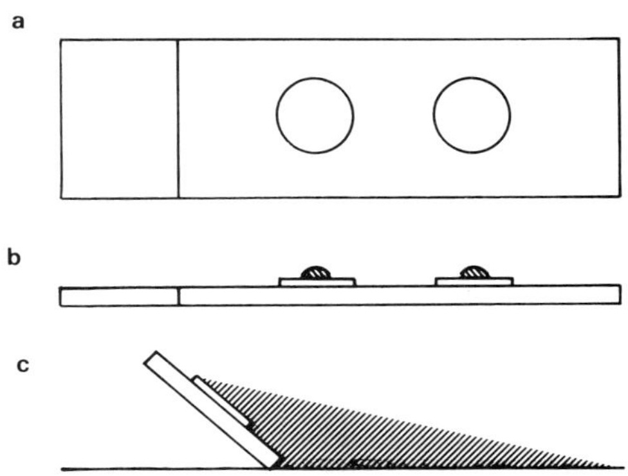

Fig. 3. Diagramatic representation of the metaphase spreading procedure. a, Two coverslips are adhered onto the surface of a microscope slide by placing a small drop of water underneath each coverslip. b, a tiny drop of suspension is deposited onto each coverslip to permit the main area of spreading to occur over the surface of the coverslip. c, The slide is tilted onto blotting paper to drain the excess fluid and encourage metaphase spreading.

5. After draining do not be tempted to use intensive heat for rapid drying of coverslips in an attempt to encourage metaphase spreading, as this has been found to produce a poor chromosome morphology in the SEM. Ideally the coverslips should be air-dried at room temperature.

It is uncommon to air-dry specimens for examination in the SEM but this step is necessary, as in light microscopy, for the adherence of complete metaphases onto the glass surface for karyotype analysis. It has been demonstrated that air-drying at this stage has no detrimental effect on chromosome morphology in the SEM. However, chromosomes need to be prevented from air-drying at all subsequent stages to prevent collapse of the chromosome structure in the SEM (Harrison *et al.*, 1982).

Chromosome Banding

Karyotypic analysis is now regarded as incomplete without analysis of banded chromosomes. Chromosome banding was introduced by Caspersson *et al.* (1969, 1970a,b). Since that time a wide range of procedures have been used to produce the highly characteristic differential staining along the length of the chromosome arms (Drets and Shaw, 1971; Kato and Yosida, 1972; Patil *et al.*, 1971; Schnedl, 1971; Seabright, 1971; Sumner *et al.*, 1971; Utakoji, 1972; Wang and Federoff, 1972).

G-Banding

The most popular of the banding techniques used in cytogenetics is Giemsa (G)-banding. The individual bands of human metaphase and prometaphase chromosomes are well characterized using a standard nomenclature as described by the International System for Human Cytogenetic Nomenclature (ISCN, 1985). Prometaphase chromosome analysis, achieved by the high-resolution techniques described previously, have improved the accuracy of cytogenetics. The chromosomes of these metaphases are elongated and show an increased number of G-bands along the chromatids. Yunis *et al.* (1978) identified these bands and demonstrated that in human prometaphase chromosomes, the increased number of G-bands were subdivisions of the bands normally observed in metaphase chromosomes. The most successful G-banding procedure for SEM of human and other mammalian species has been found to be the method described by Seabright (1971), which involves the use of a dilute trypsin solution. The details of this technique, with the modifications employed for SEM observation of the chromosomes, are described here.

Once suitable metaphase spreads have been achieved on the coverslips they are stored at room temperature for 2–3 days. Storage for longer periods has been found to produce a deterioration in chromosome structure when observed in the SEM. The metaphases are then G-banded using a modification of the Seabright (1971) method. This involves the use of 0.025% trypsin in 0.85% saline at room temperature. The coverslips are immersed in this solution for 5–10 sec. Individual variation exists between laboratories in the choice of manufacturer for the trypsin and the buffer used. These changes may require the concentration of the trypsin and time of exposure to be altered; therefore, a few coverslips should be sacrificed to establish the correct conditions. Largely all the variations known to produce G-bands when observed in the light microscope are suitable for SEM investigation.

The most important factor to maintain the best chromosome morphology and visualize the "G-banding pattern" in the SEM is to produce "just-banded" metaphases in the light microscope, as described by Harrison *et al.* (1983). The changes in chromosome morphology in response to the G-banding treatment, monitored by timed trypsinization, are totally reproducible. It appears that with increasing exposure to trypsin a progressive disruption of chromosome structure occurs. The various orders of chromosome organization are revealed, without changing the form of the chromosomes. Nonbanded human metaphases, observed by SEM, show chromosomes of three-dimensional cylindrical profile, with well-defined chromatids and centromeres (Fig. 4b). The surface appears smooth with occasional indentations along the chromatids. In "just-banded" chromosomes the major light microscope G-positive bands, only, were revealed, after 5–10 sec in the above concentration of trypsin (Fig. 4c). "Well-banded" chromosomes showed the entire G-banding pattern after 15–25 sec trypsinization when observed in the light microscope (Fig. 4e). These banded chromosomes show more pronounced indentations than the nonbanded chromosomes when viewed in the SEM. (Fig. 4d,f). The indentations in the chromatids resemble circumferential grooves subdividing the short (p) and long (q) arms of the chromosomes into segments. The segmentation appears to result from a coiling of the chromatids in a manner corresponding to the quaternary coiling of metaphase chromosomes, as shown in Fig. 5. This represents the final stages of chromosome condensation into metaphase chromosomes as described by Du Praw (1965, 1966). A characteristic pattern of quarternary coiling is consistently observed for each chromosome pair in all metaphases examined after G-banding. A positional correlation also exists between the circumferential grooves in the SEM and the light microscope G-banding pattern as shown in chromosome 2 in Fig. 5. This finding led to the production of the first three-dimensional human karyotype in the SEM, as illustrated in Fig. 6 (Harrison *et al.*, 1981).

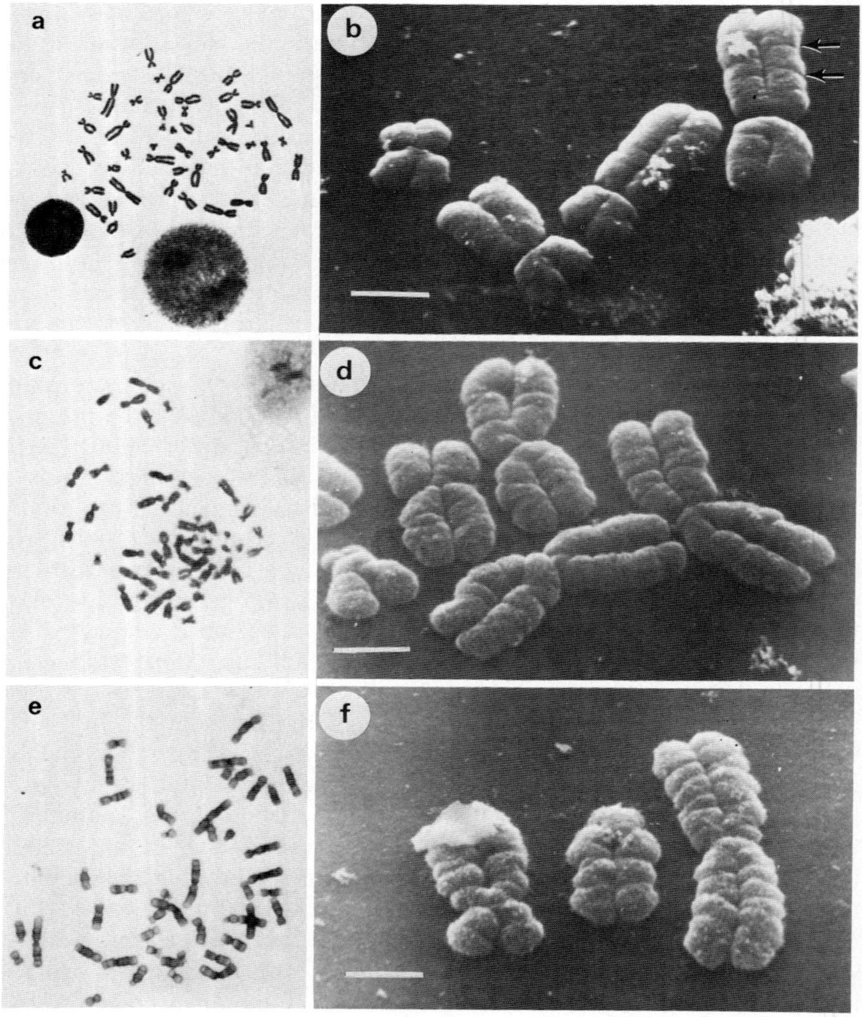

Fig. 4. Parallel preparations of human metaphase chromosomes observed in the light microscope (left) and SEM, at a tilt angle of 45° (right). a, Nonbanded metaphase chromosomes by light microscopy. b, Nonbanded chromosomes in the SEM. Arrows show occasional indentations along the chromatids. c, Just-banded metaphase chromosomes. d, Just-banded chromosomes in the SEM. e, Well-banded metaphase in the light microscope. f, Well-banded chromosomes in the SEM; SEM photographs bars = 2 μm. (Reprinted with permission from Harrison *et al.*, 1983; © S. Karger AG, Basel.)

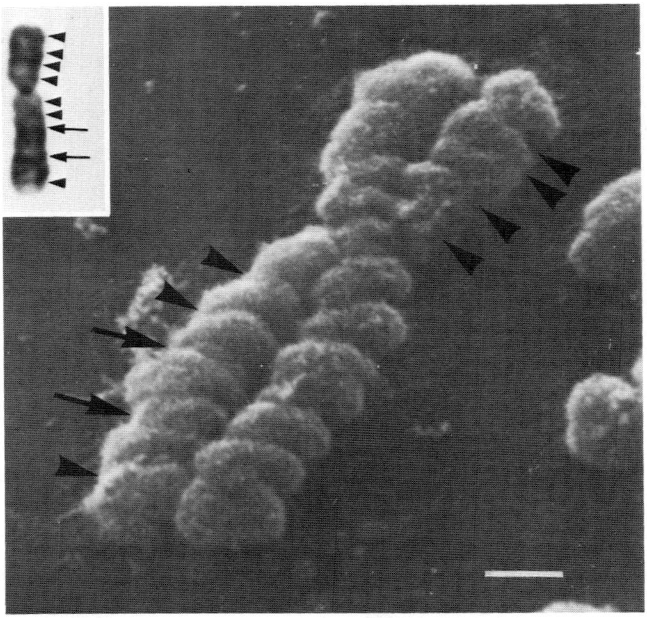

Fig. 5. A No. 2 human chromosome observed in the SEM. The positive microscope G-bands (black arrows and arrowhead, inset) and the most prominent circumferential grooves of the SEM chromosomes (black arrows and arrowheads in SEM photograph) are indicated. The smaller, positive G-bands correspond to the less prominent circumferential grooves (arrowheads). The circumferential grooves divide the chromatids into segments. From examination of the nearside q arm, the segments appear to represent coiling. The pattern of coiling is identical along the sister chromatid with parallel alignment of grooves along the length of the p and q arms; bar = 1 μm.

At the lower SEM magnifications, just-banded chromosomes show a well-defined outline and tightly packed chromosome structure in the SEM (Figs. 4d and 7a,b). In contrast, well-banded chromosomes show a clearly filamentous surface morphology (Fig. 8b) in association with a more diffuse chromosome boundary and a loss of sister chromatid separation (Figs. 4f and 8a). This appears to result from an expansion of the chromosome fine structure beyond the chromosome periphery, with an associated intermingling of fibers from adjacent chromatids. The circumferential grooves also lose definition in well-banded chromosomes due to the expansion and intermingling of fibers across them. Therefore, for study of the G-banding pattern in the SEM the more compact just-banded chromosome provide the best results.

For subsequent SEM viewing, the coverslips are removed from trypsin

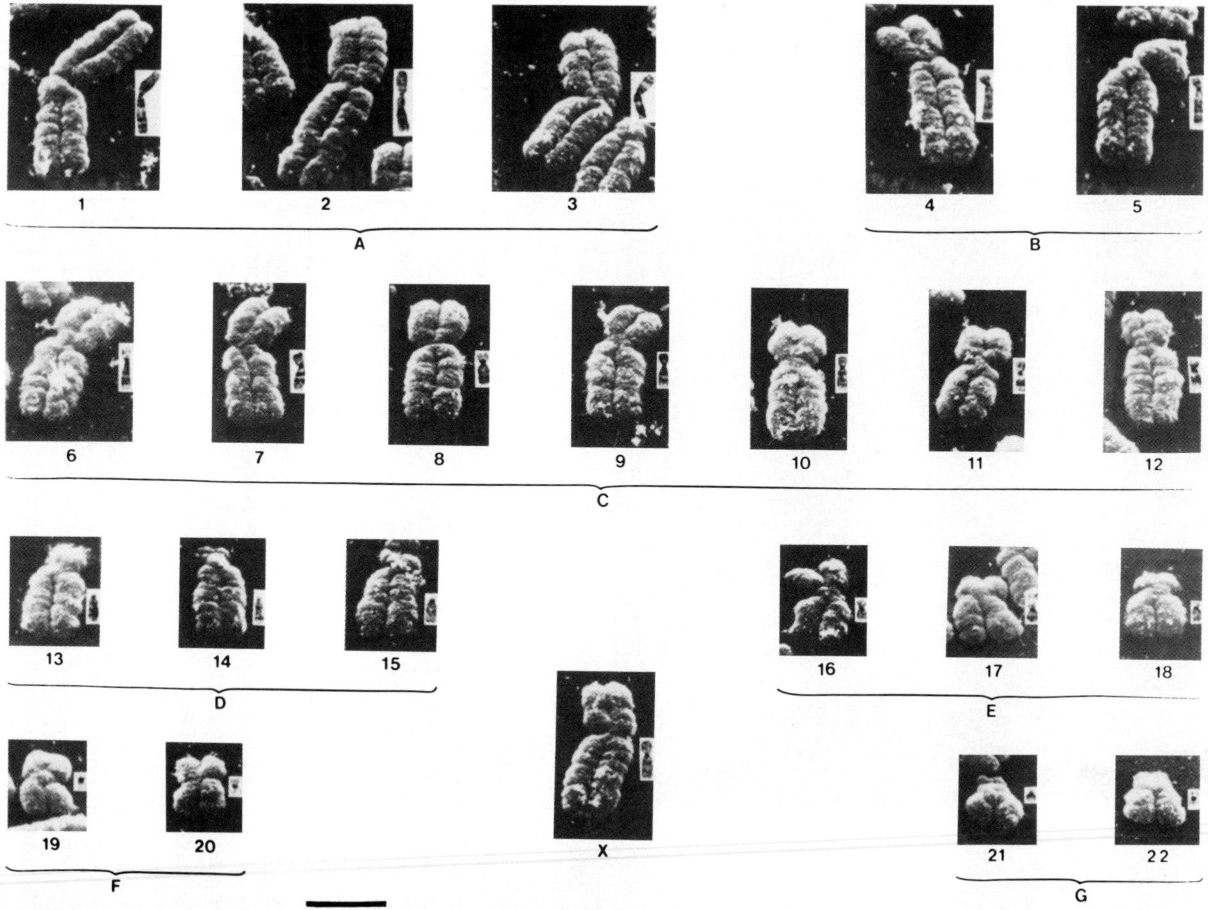

Fig. 6. A human karyotype showing SEM and light microscope homologues from parallel preparations. The chromosomes 1–22 plus the X are numbered according to the Paris convention (ISCN, 1985); bar = 4 μm.

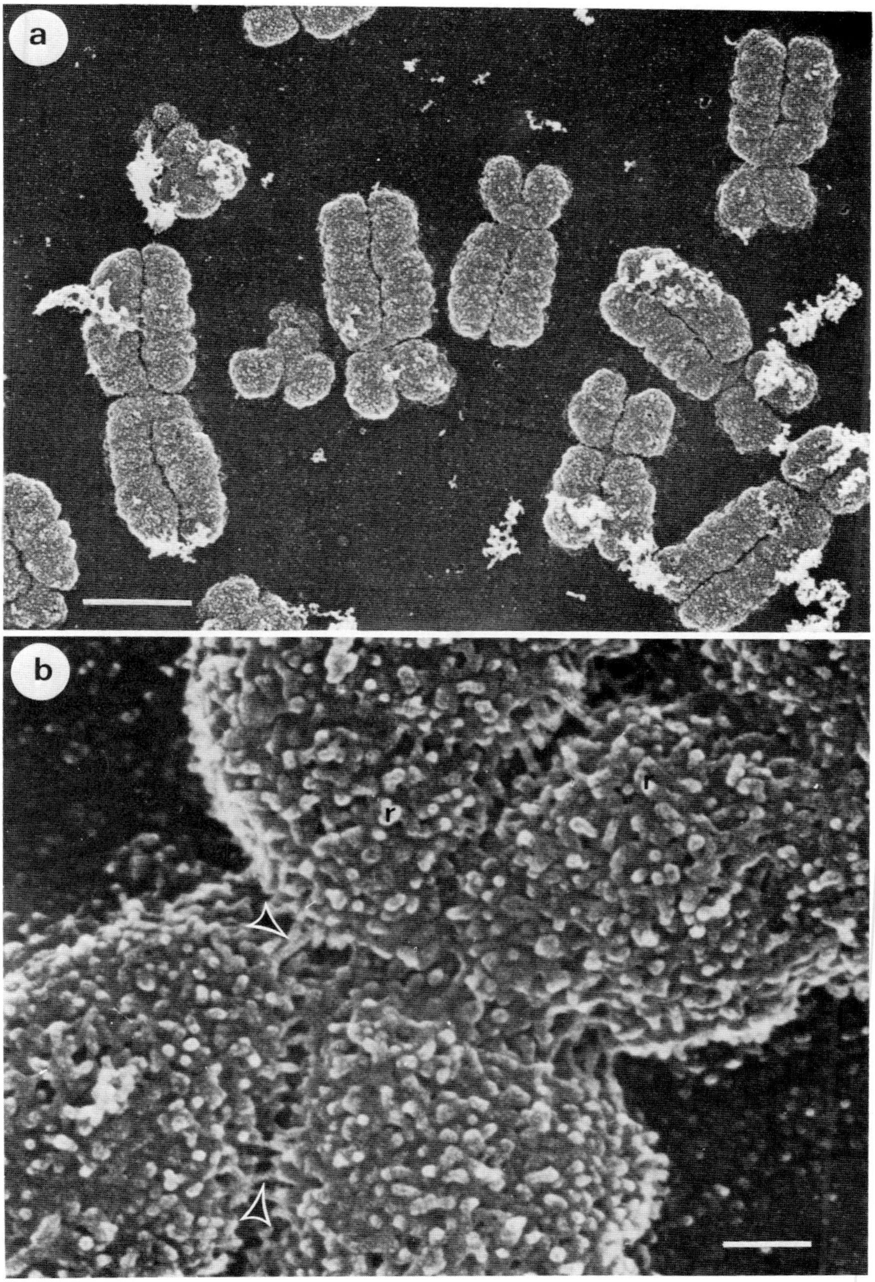

Fig. 7. Just-banded chromosomes viewed at 0° tilt in the SEM. a, Low-power view of part of a metaphase spread showing the well defined outline; bar = 2 μm. b, High-power view of the centromeric region of one of the above chromosomes, showing the tightly packed chromosome structure in which rounded projections (r) are evident, with occasional individual fibers of chromatin (arrows); bar = 0.2 μm. (Reprinted with permission from Harrison *et al.*, 1983; © S. Karger AB, Basel.)

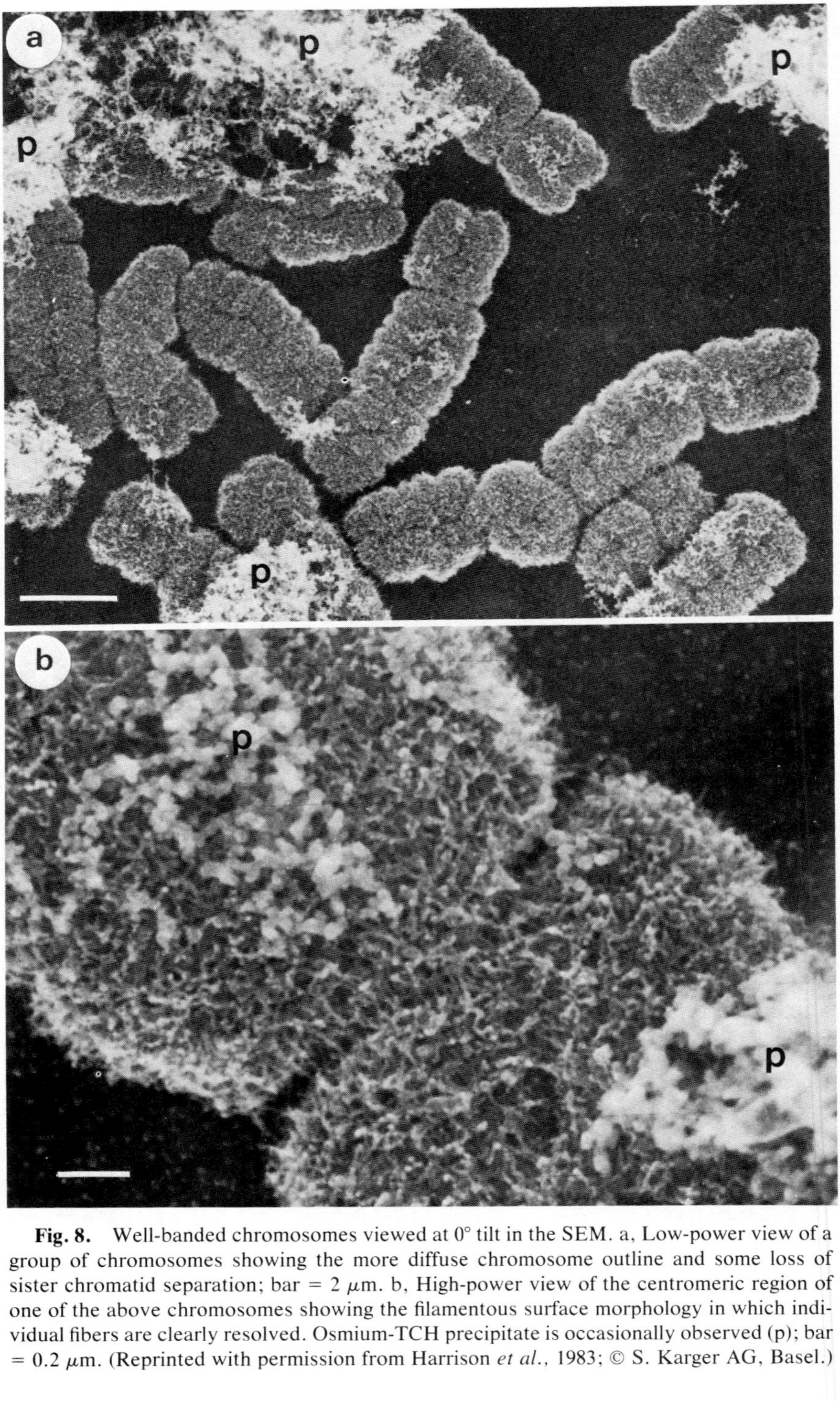

Fig. 8. Well-banded chromosomes viewed at 0° tilt in the SEM. a, Low-power view of a group of chromosomes showing the more diffuse chromosome outline and some loss of sister chromatid separation; bar = 2 μm. b, High-power view of the centromeric region of one of the above chromosomes showing the filamentous surface morphology in which individual fibers are clearly resolved. Osmium-TCH precipitate is occasionally observed (p); bar = 0.2 μm. (Reprinted with permission from Harrison *et al.*, 1983; © S. Karger AG, Basel.)

and placed immediately into 3% glutaraldehyde in 0.1 *M* Sorensen's phosphate buffer (pH 7.4) for 30 min. The methods of rinsing in Gurr's buffer (pH 6.8) or 0.85% saline after trypsinization, as applied in routine light microscope G-banding protocols, are insufficient to inactivate the trypsin totally. Gradual enzymatic breakdown of chromosome structure continues, resulting in a variable morphology in the SEM.

At this stage, if light microscope investigations are not required prior to SEM, the coverslips may be processed directly for SEM. It is unnecessary, for SEM viewing alone, to stain the chromosomes with Giemsa. The chromosome morphology is unchanged as a result of staining (Harrison *et al.*, 1982). However, Giemsa staining is essential for light microscopy. Therefore, those coverslips to be examined in the light microscope are stained at this stage.

From the glutaraldehyde the coverslips are rinsed twice in Sorensen's buffer, then stained in 2% Giemsa in Gurr's buffer for 8 min. Leishman's stain may be used as an alternative to Giemsa. There is a slight reduction in the uptake of stain after glutaraldehyde fixation (Fig. 14). A further rinse in Gurr's buffer is given. The coverslips are then mounted, in the buffer, with the chromosomes uppermost, onto a slide. A large coverslip is placed over the circular coverslip and sealed around the edge with nail varnish (Fig. 9). This creates a large chamber of buffer to prevent air-drying of the chromosomes and protect the circular coverslip from contamination by the nail varnish. Thus, throughout the G-banding procedure the coverslips are prevented from air-drying. The chromosomes are then observed in the light microscope. Suitable just-banded metaphases are selected and photographed. Their location, in relation to the engravings on the surface of the coverslip, is noted before progressing to the next metaphase. In the light microscope photographs, a reduction in the degree of contrast between G-positive and G-negative bands is produced as a result of undertaking photography through an air/buffer interface. However, this sacrifice allows the chromosome structure to be consistently preserved to a sufficiently high standard for detailed cytogenetic analysis to be subsequently carried out in the SEM.

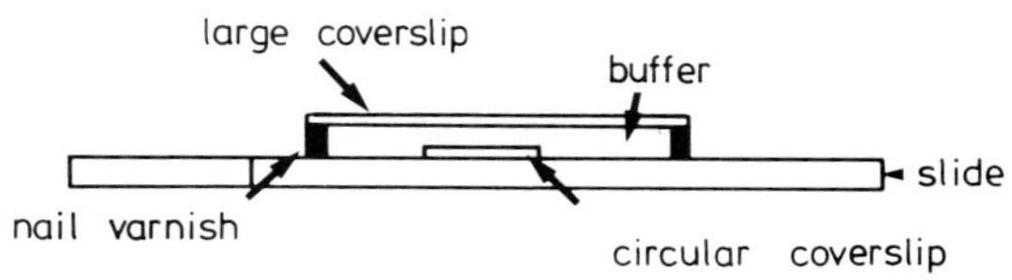

Fig. 9. Diagram to show the mounting of the circular coverslip in buffer for light microscope observation.

At the end of the light microscope investigation the top coverslip is removed, and the circular coverslip is dropped immediately into fresh 3% glutaraldehyde in Sorensen's buffer, without allowing to air-dry.

C-Banding

The SEM relocation procedure need not be restricted to G-banded chromosomes. Other banding techniques may also be investigated in a similar way. This includes centromeric (C)-banding. C-banding selectively stains heterochromatin. This chromatin consists of highly repetitive sequences of DNA (Hsu, 1975). In the human karyotype C-bands are observed as constitutive heterochromatin at three main locations: centromeres, the secondary constriction sites of chromosomes 1, 9, and 16, and the distal end of the Y chromosome. These regions stain positively with Giemsa (ISCN, 1985).

A current view of the mechanism of C-banding is that selective extraction of DNA is taking place from the chromosome arms, with a greater proportion of DNA being retained in the C-bands. That is, the constitutive heterochromatin remains resistant to the extraction procedure and largely intact. This interpretation is based on several studies using biochemical techniques (Burkholder and Duczek, 1982; Comings *et al.,* 1973; Holmquist, 1979; Pathak and Arrighi, 1973), light microscope observations (McKay, 1973; Comings *et al.,* 1973), and whole mount TEM (Burkholder, 1975; Comings *et al.,* 1973). The mechanisms of the C-banding technique and its effect on chromosome structure are still not fully understood. In order to investigate these problems further, the same C-banded chromosomes may be examined in the light microscope followed by more detailed study, at higher resolution in the SEM, in a similar way to that described for G-banded chromosomes. Using this approach, changes in both gross morphology and fine structure may be investigated and compared with the effects of routine C-banding techniques.

Current techniques of C-banding are based on treating chromosomes with a hot alkali solution, then a hot saline solution, followed by Giemsa staining (Arrighi and Hsu, 1971; Sumner, 1972; Yunis *et al.,* 1971). A variation of this procedure is used for light microscope and subsequent SEM study. Two days after metaphase spreading, coverslips may be C-banded, using a modification of the technique described by Sumner (1972). In particular, metaphases from individuals with known centromeric polymorphisms are more appropriate for this investigation. The coverslips are treated consecutively with 5% barium hydroxide for 10 min, followed by two washes in Gurr's buffer (pH 6.8). They are then incubated in 2 × SSC (0.3 *M* sodium chloride, 0.03 *M* sodium citrate) at 55°C for 17 min and then given two further washes in Gurr's buffer. The solutions are heated and accurately maintained at the correct temperature in a water bath. It may be necessary to experiment with a few coverslips to determine the optimum incubation times. At all stages air-drying is prevented.

In the same manner as the G-banding procedure the coverslips are fixed prior to staining in 3% glutaraldehyde in 0.1 *M* Sorensens buffer (pH 7.4) for 30 min. This step is necessary to completely arrest the action of the barium hydroxide and 2 × SSC. Without this fixation the C-banding process continues throughout the subsequent stages, resulting in a poor chromosome morphology in the SEM. The coverslips are then removed from the glutaraldehyde, rinsed twice in Gurr's buffer, and stained in 2% Giemsa in Gurr's buffer for 3–5 min at room temperature. Another rinse in Gurr's buffer is given and the coverslips are then mounted in the same buffer onto a slide and sealed around the edge with nail varnish, to prevent air-drying as described for G-banded chromosomes (Fig. 9). Suitable metaphase spreads, with well C-banded chromosomes, are selected and photographed in the light microscope. As expected the buffer does not provide the best medium for photography and contrast is reduced. Reference is made of their position in relation to the engravings previously made on the surface of the coverslips for later examination in the SEM. Once sufficient metaphases have been photographed, the coverslip is immediately transferred to fresh 3% glutaraldehyde.

R-Banding

The technique for reverse (R)-banding of human metaphase chromosomes involves a hot acidic sodium phosphate solution (NaH_2PO_4), followed by Giemsa staining (Dutrillaux and Lejeune, 1971). It produces a characteristic pattern of light and dark bands in the light microscope, which is the reverse of the G-banding pattern. This has been defined by ISCN (1985).

The mechanism of R-banding remains unknown, although several suggestions have been made; for example, that differences in nucleotide composition may play an important role in the appearance of R-bands (Comings, 1978; Goyanes, 1985). Holmquist *et al.* (1982) showed that the R-interbands are A-T rich while the R-bands are G-C rich. It has been demonstrated by biochemical, light microscope, and TEM studies that the R-banding process totally removes only a small amount of protein, but causes substantial aggregation of protein in the R-banded regions (Burkholder, 1981; Burkholder and Duczek, 1982; Comings *et al.*, 1973).

The R-banding method used for correlative light microscopy and SEM is a modification of the technique described by Sehested (1974). After metaphase spreading on circular coverslips, as previously described, they are stored overnight at room temperature. They are then incubated in a solution of 1 *M* $NaH_2PO_4 \cdot 2H_2O$ (pH 4.0–4.5) at 83°C for 10 min. This time may be variable; therefore, on each occasion it is necessary to experiment with different incubation times. The high temperature is maintained in a waterbath. The coverslips are rinsed briefly in Gurr's buffer (pH 6.8) and immediately fixed in 3% glutaraldehyde in Sorensen's buffer (pH 7.4) at room temperature. As described for the other banding techniques, this instantly arrests the action of the salt solution and prevents continuation of the banding process to preserve the chromosome morphology for SEM.

After, at least, a 30-min fixation the coverslips are rinsed briefly in Gurr's buffer and stained for 2 min in 4% Giemsa in Gurr's buffer. The coverslips are then mounted onto a slide in buffer for light microscope observation, as described for G-banded chromosomes (Fig. 9). Suitable R-banded metaphases are selected and photographed. Their position is noted in relation to the engravings on the coverslip for subsequent relocation of the same metaphases in the SEM. The coverslip is removed and transferred to fresh 3% glutaraldehyde in Sorensen's buffer prior to processing for SEM.

T-Banding

Telomeric (T)-banding is produced by an extension of the R-banding technique, either by increasing the length of time in NaH_2PO_4 or increasing the temperature (Dutrillaux and Covic, 1974). T-banded metaphase chromosomes observed in the light microscope are uniformly pale along the length of the chromatids with a positively staining region at the telomeres. This difference in staining is often difficult to detect in the light microscope because the region involved is often quite small.

The method found to produce the best chromosome morphology in the SEM is a modification of that of Dutrillaux (1973), involving an increase in temperature of NaH_2PO_4. The coverslips, 1 day after metaphase spreading, are incubated in 1 *M* Na_2PO_4 at an increased temperature of 88–93°C at pH 4.0–4.5 for 10 min. This time may vary with fluctuation in temperature, which is difficult to maintain, even in a waterbath. Therefore, a few experimental coverslips are required to determine the optimum incubation time to produce the best T-banding pattern. The R-banding procedure is followed from this point.

Harlequin Staining for the Study of Sister Chromatid Exchanges

Preparations in which sister chromatid exchanges (SCEs) are to be studied have 5 μg/ml of the thymidine analogue 5-bromo-2′-deoxyuridine

(BrdUrd) introduced into the cell culture over two successive cell cycles prior to the time of fixation (Perry and Wolff, 1974). In the case of human lymphocytes this is approximately 64 hr, and in Chinese hamsters is about 28 hr. Since BrdUrd-substituted DNA is susceptible to breaks, leading to SCE induction, upon exposure to light (Ikushima and Wolff, 1974), light is excluded from these cultures by placing the flasks in loosely sealed black plastic bags prior to placing in the incubator. The remainder of the procedure continues as described for routine chromosome prepartions until after the metaphase spreading step. The coverslips are stored at room temperature. It is better to use coverslips between 1 and 3 days old for subsequent observation in the SEM. To achieve Harlequin staining these coverslips are placed face up in phosphate-buffered saline 8 cm away from a 30-W Hanovia bactericidal ultraviolet lamp for 12 min, then transferred to 2 × SSC at 60°C for 1 hr. As described for the other banding procedures, the coverslips are immediately placed in 3% glutaraldehyde in 0.1 *M* Sorensen's buffer (pH 7.4) for a minimum of 30 min to rapidly fix the chromosomes and retain the optimum chromosome morphology in the SEM. From the glutaraldehyde the coverslips are rinsed twice in Sorensen's buffer, then stained in 2% Giemsa in Gurr's buffer (pH 6.8) for 10 min. The coverslips are then examined in the light microscope as previously described (Fig. 9). SCEs are examined in suitable second-division metaphases. After photography the location of the metaphases is noted in relation to the engravings on the coverslips. The coverslips are then transferred to fresh 3% glutaraldehyde and processed for SEM.

Scanning Electron Microscope Procedures and Observations

SEM Processing

The coverslips are processed for SEM using an osmium tetroxide-thiocarbohydrazide (osmium-TCH) fixation. This technique was first described by Ip and Fischman (1979) for SEM observation of extracted cytoskeletons. It was modified to observe metaphase chromosomes by Harrison *et al.* (1981). The major advantage of this procedure is that it allows direct SEM visualization of uncoated chromosomes. The osmium-TCH provides conductivity by a mordanting process to allow extensive impregnation of the specimens with osmium tetroxide, and thus reduces specimen deterioration and charging at high kilovolts. In addition, this impregnation provides supplementary rigidity to the chromosomes, preventing specimen collapse. The method is described below:

1. The coverslips are fixed in the second batch of 3% glutaraldehyde in Sorensen's buffer (pH 7.4) for a minimum of 30 min at room temperature.

2. They are removed from the glutaraldehyde and rinsed twice in Sorensen's buffer.

3. A fixation in 1% osmium tetroxide in Sorensen's buffer follows for 10 min.

4. The coverslips are then rinsed three times in distilled water.

5. This is followed by a 5-min incubation at room temperature in freshly prepared and filtered, saturated TCH in distilled water. It has been found that the TCH must be reasonably fresh. Bottles of TCH which have been in use for some time show deterioration and the crystals turn from white to pink. Pink crystals should be discarded, as only white crystals give results of reproducible quality. Also, the TCH must be fully saturated. The best method to ensure this, is to dissolve the TCH to saturation in hot distilled water and allow to cool to room temperature. It may then be filtered at each step, as it is required. Any unused TCH must be discarded as it rapidly deteriorates once in solution.

6. Three rinses in distilled water are given.

7. This is followed by an additional fixation in 1% osmium tetroxide in distilled water.

8. Steps 6 and 7 are repeated.

9. The specimens are then dehydrated through a graded acetone series (20–100%).

10. They are rinsed twice in absolute acetone.

11. This is followed by two changes of trichlorotrifluroethane (arklone).

12. Then the coverslips are critical-point dried from liquid carbon dioxide with arklone as the transitional fluid. For details of critical-point drying the reader is referred to Hayat (1976).

13. The coverslips may now be examined in the SEM without sputter coating.

The chromosomes consistently show a markedly cylindrical profile with well-defined chromatids and centromeres (Fig. 4b,d,e). To demonstrate that the osmium-TCH provides supplementary rigidity to the chromosomes, preparations for comparison may be made using conventional SEM processing. This eliminates steps 4–8 and after critical-point drying the coverslips are coated with 10–20 nm of gold, necessary to prevent specimen charging.

Chromosomes prepared in this way show some loss of three-dimensional relief, although definition of chromatids and centromeres is retained (Fig. 10). In the absence of osmium-TCH support, specimen collapse may occur as a result of the heat generated during the sputter coating process (Filip *et al.*, 1975).

In the majority of cases the osmium-TCH eliminates the necessity for

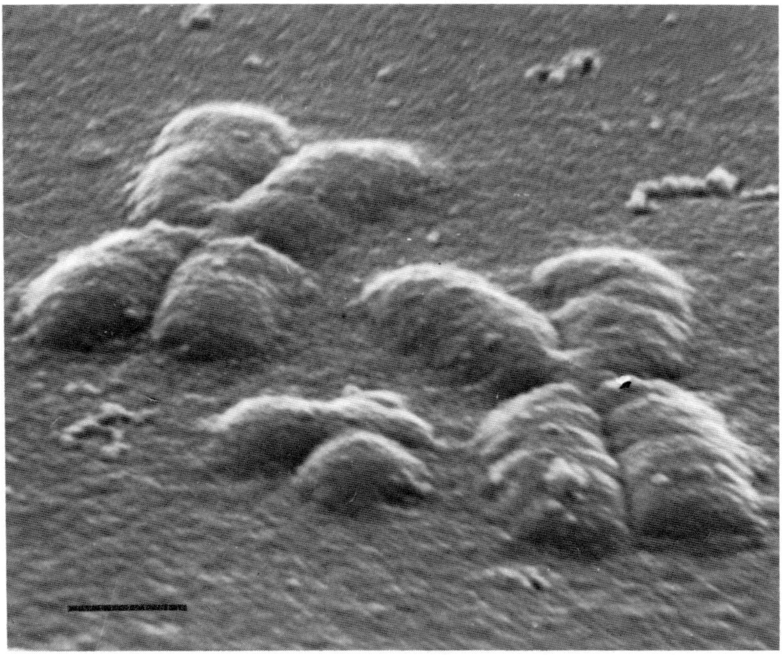

Fig. 10. Human metaphase chromosomes prepared by conventional SEM techniques and sputter-coated with 10–20 nm of gold. Chromatid and centromeres are evident, although some loss of relief is shown; bar = 1 μm. (Reprinted with permission from Harrison *et al.*, 1982.)

sputter coating, although no deterioration of chromosome structure is produced when specimens prepared by this method are subsequently coated with gold. This is demonstrated by examination of the same osmium-impregnated chromosome both with and without the deposition of 5–10 nm of gold. At high magnification, filament resolution is heightened and contrast is enhanced (Fig.11). Therefore, for examination of chromosomes at high magnification, minimal sputter coating may provide enhanced resolution. For optimum visualization of the banding patterns in the SEM the stage should be tilted to an angle of approximately 30° tilt.

Examination at 0° tilt provides increased resolution of fine structural detail in osmium-TCH prepared chromosomes (Fig. 12). Individual and folded fibers of chromatin are observed at the chromosome surface, providing insight into the structural organization of metaphase chromosomes (Harrison *et al.*, 1982, 1983). However, the reduced topographical aspect of imaging at 0° tilt considerably reduces definition of the circumferential

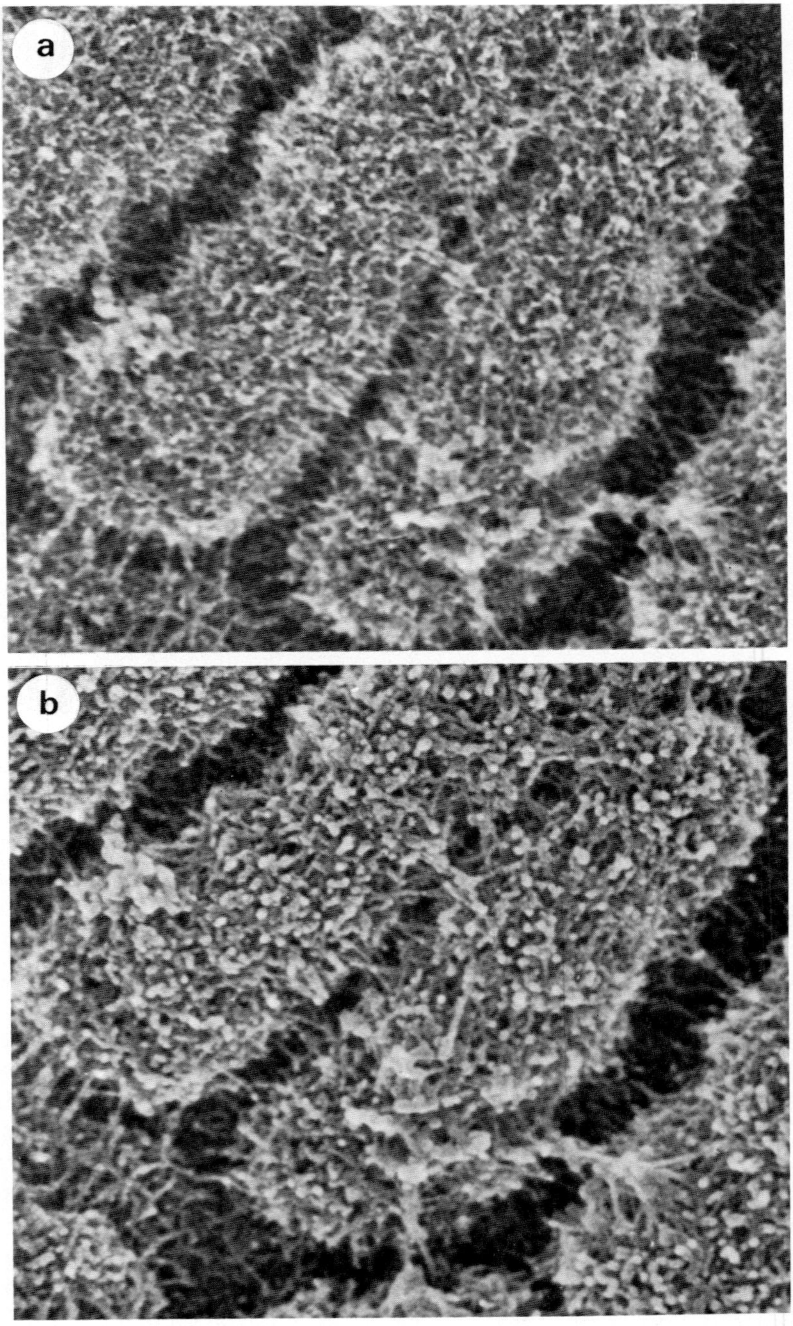
a
b

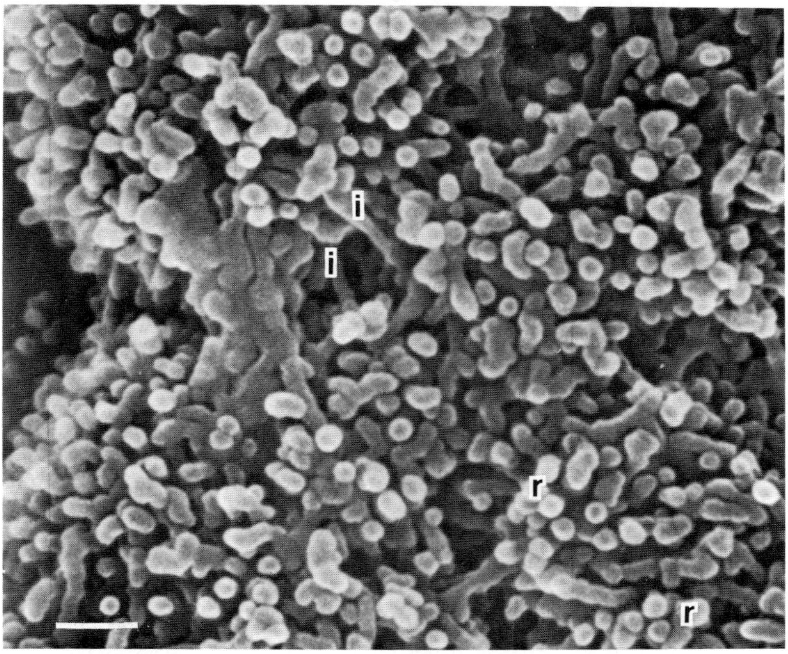

Fig. 12. Detail from the centromeric region of an osmium-TCH prepared and sputter-coated (5–10 nm gold) chromosome from the same metaphase as the chromosome in Fig. 11, demonstrating the increase resolution provided by this technique. Rounded projectings (r) of varied diameter are distributed over the entire surface of the chromosome. Individual fibers (i) are shown underlying these projections (i); bar = 0.1 μm. (Reprinted with permission from Harrison *et al.*, 1982.)

grooves of G-banded chromosomes (Fig. 13). The entire light microscope and SEM procedure is summarized in Table I.

SEM of G-Banded Chromosomes

The same G-banded metaphases observed in the light microscope are relocated in the SEM using the recognizable markings previously engraved on the coverslips. Figure 14 shows a relocated human metaphase and Fig. 15 a metaphase from a Chinese hamster ovary (CHO) cell line. In the SEM the chromosomes of both species show a series of indentations

---

Fig. 11. a, A G-banded chromosome after osmium-TCH preparation. b, The same chromosome sputter-coated with 5–10 nm gold. At this magnification, resolution of the individual fibers is heightened and contrast is enhanced by the additional gold coating. Note the increase in size of the fibers; bar = 0.5 μm.

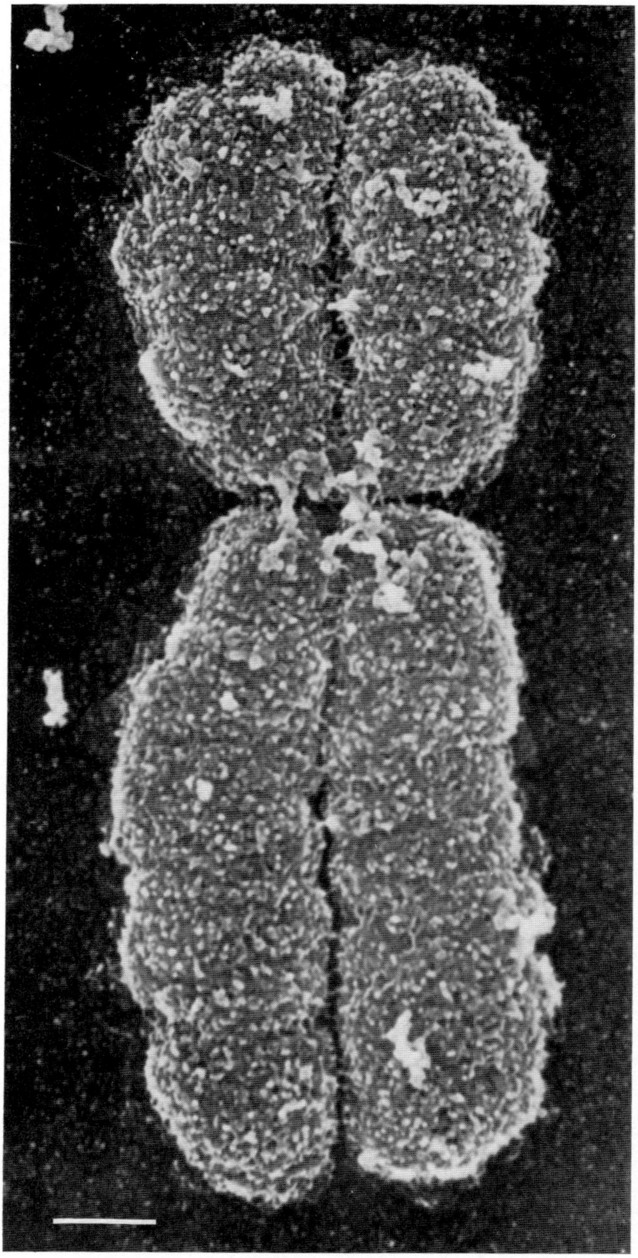

Fig. 13. A G-banded, osmium-TCH-prepared, uncoated human chromosome 2, viewed at 0° tilt in the SEM. Resolution of surface detail is enhanced at this angle. The definition of the circumferential grooves is reduced in comparison to tilted specimens (see Fig. 5); bar = 0.5 μm. (Reprinted with permission from Harrison *et al.*, 1982.)

TABLE I

Summary of light miscroscope and SEM procedure

CELL CULTURE

↓

COLCEMID

↓

HYPOTONIC

↓

3:1 METHANOL : ACETIC ACID FIXATION

↓

FIXED SUSPENSION DROPPED ONTO COVERSLIPS

↓

BANDING TREATMENT

↓

3% GLUTARALDEHYDE FIXATION

↓

RINSE SORENSENS BUFFER

↓

2% GIEMSA STAIN IN GURR'S BUFFER

↓

LIGHT MICROSOCOPE ← MOUNT IN GURR'S BUFFER ON SLIDE

↓

3% GLUTARALDEHYDE FIXATION

↓

1% OSMIUM TETROXIDE IN SORENSENS BUFFER

↓

TCH IN DISTILLED WATER

↓ ↑ ↓

1% OSMIUM TETROXIDE IN DISTILLED WATER

↓

ACETONE DEHYDRATION

↓

SEM ← CRITICAL POINT DRYING

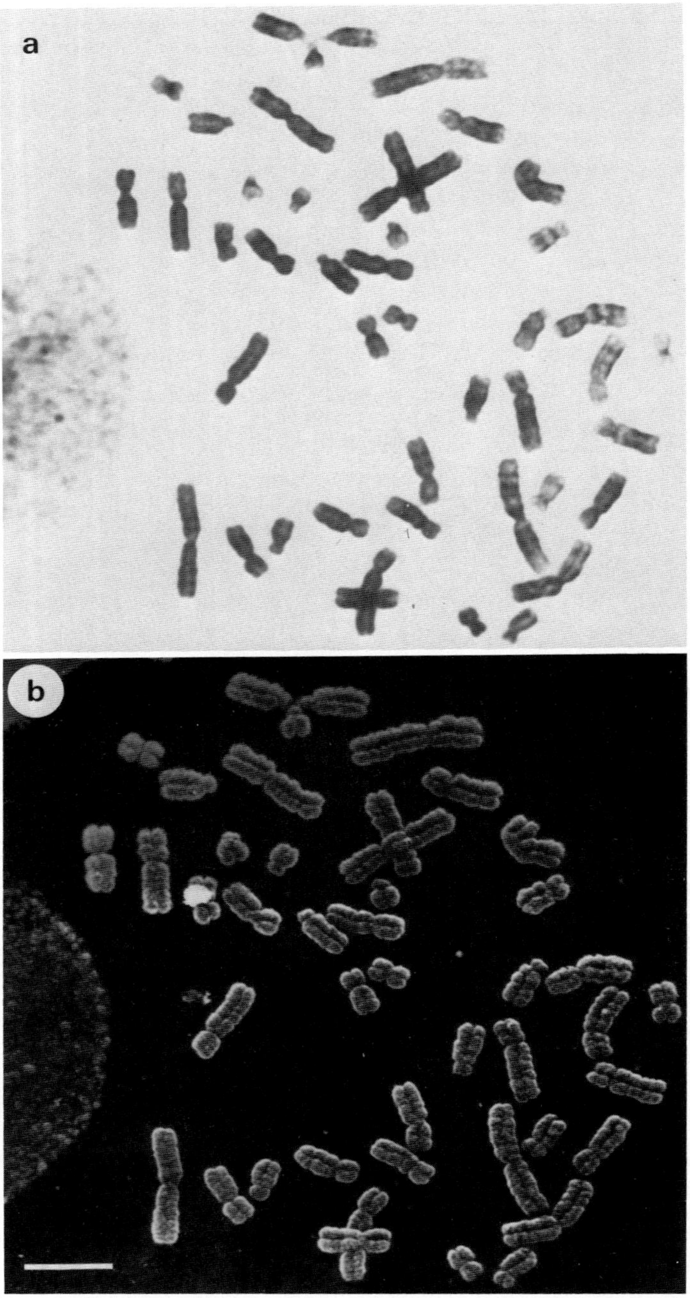
a
b

in the surface of the chromatids (Figs. 14b and 15b), as previously observed in parallel light microscope and SEM preparations (Harrison *et al.*, 1981, 1982). The metaphases may be reorientated by rotation of the specimen stage of the ISI SS-40 SEM. This facilitates correlation with the individual chromosomes observed in the light microscope. Using this facility selected chromosomes from each metaphase may be examined in detail at a higher magnification (Figs. 16 and 17). This confirms the previous findings from parallel preparations of human metaphases (see Fig. 5) of a correlation between the G-positive light microscope bands (black arrows, Fig. 16, insets) and the circumferential grooves along the chromatids, when observed in the SEM (white arrows, Fig. 16; Harrison *et al.*, 1985a). Human chromosome 1 from the metaphase in Fig. 14 has a large number of G-bands which clearly demonstrates this relationship (Fig. 16a). Certain G-positive bands correspond in position to two circumferential grooves, indicated by connected white arrows (Fig. 16a). This feature is also observed in the smaller chromosomes, as demonstrated in a human chromosome 15 from this metaphase (Fig. 16b), with one prominent G-positive band (black arrows, Fig. 16b, inset).

Generally, the circumferential grooves are situated at parallel positions on sister chromatids. Distortion of the chromosomes shifts the grooves from this precise alignment, as observed in chromosome 15, leaning to one side (Fig. 16b) and human chromosome 5 (Fig. 16c). The light microscope G-positive bands are shifted an equal distance in the same direction (Fig. 16b,c, insets).

Frequently in contracted metaphase chromosomes, only the major G-positive bands may be observed in the light microscope. For example, chromosome 5 is characterized by one large G-positive band in the center of the q arm in contracted metaphase chromosomes (connected black arrows, Fig. 16c, inset). In the SEM this region corresponds to four circumferential grooves in both chromatids (connected white arrows, Fig. 16c). Four G-positive bands, 5q12, 5q14, 5q21, and 5q23, are characteristic of this region when elongated metaphases are observed in the light microscope (black arrowheads, Fig. 18a,b, insets), as described by the International System for Human Cytogenetic Nomenclature (ISCN, 1985), shown diagrammatically in Fig. 18c(i). The white arrows (Fig. 16c) demonstrate the positional relationship between the circumferential

---

Fig. 14. The same G-banded human metaphase spread observed by light microscopy (a) and then by SEM (b). The chromosomes show reduced contrast between G-positive and G-negative bands (a) as a result of the modifications used to preserve the best chromosome morphology in the SEM (b); bar = 10 μm. (Reprinted with permission from Harrison *et al.*, 1985a.)

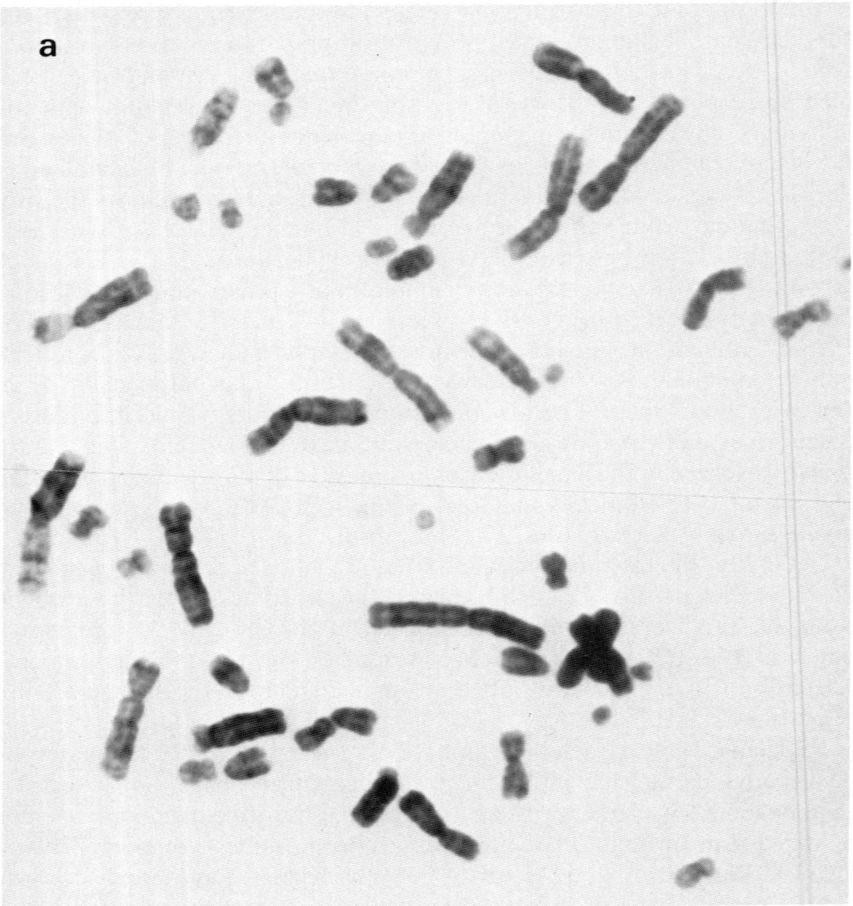

Fig. 15. The same G-banded CHO metaphase spread observed by light microscopy (a) and SEM (b). Note the similarity to the human G-banded metaphase (Fig. 14); bar = 10 μm.

grooves and the ISCN (1985) mid-metaphase banding pattern [black arrowheads, Fig. 18c(i)]. Therefore, the details of mid-metaphase chromosome banding are observed in contracted chromosomes in the SEM. This correlation may be applied to those clinical cases in which contracted chromosomes predominate, for example, leukemic bone marrows, preparations from solid tumors, or chorionic villi samples, to provide increased details of the metaphase G-banding pattern.

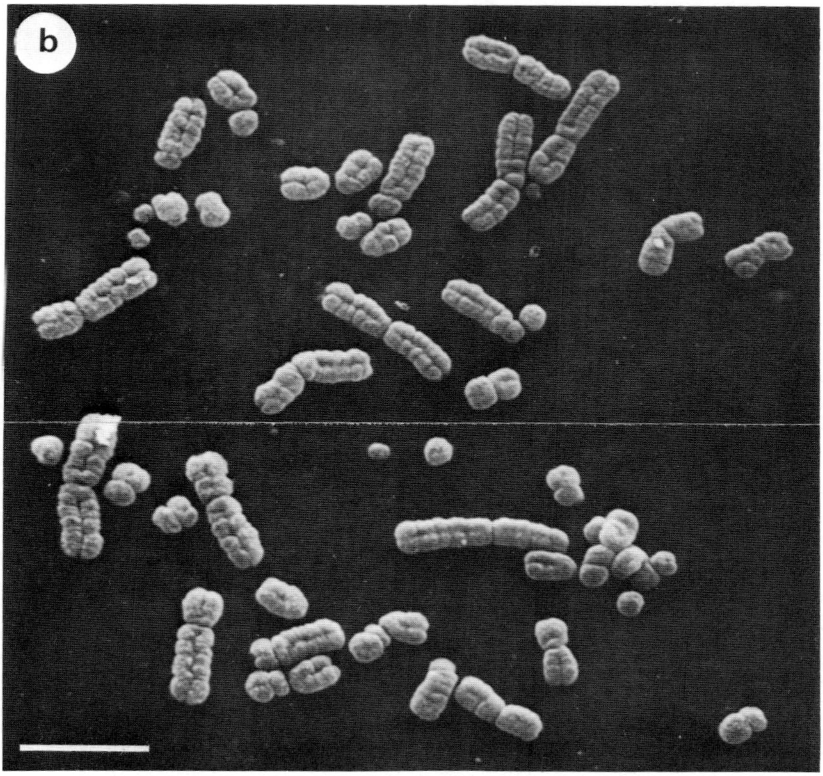

Fig. 15. (b).

The rotation of individual chromosomes in the SEM highlights different regions of the chromatids from different angles. Certain grooves appear to be more pronounced when observed from alternative directions. Figure 16d shows the same No. 5 chromosome as Fig. 16c viewed upside down. From this angle the circumferential groove at the distal end of the q arm appears more prominent (Fig. 16d) than when observed in the upright position (Fig. 16c). As a result any discrepancy in number and location of individual grooves may be accurately resolved in all chromosomes.

Similar correlations are shown in the CHO chromosomes. As an example, one of the large CHO chromosomes has been selected from the metaphase shown in Fig. 15. Enlargement of the light microscope photograph (Fig. 17b) to the same size as the SEM photograph (Fig. 17a) has produced a rather ill-defined chromosome, but allows a precise alignment to be shown between the light microscope G-bands and the grooves ob-

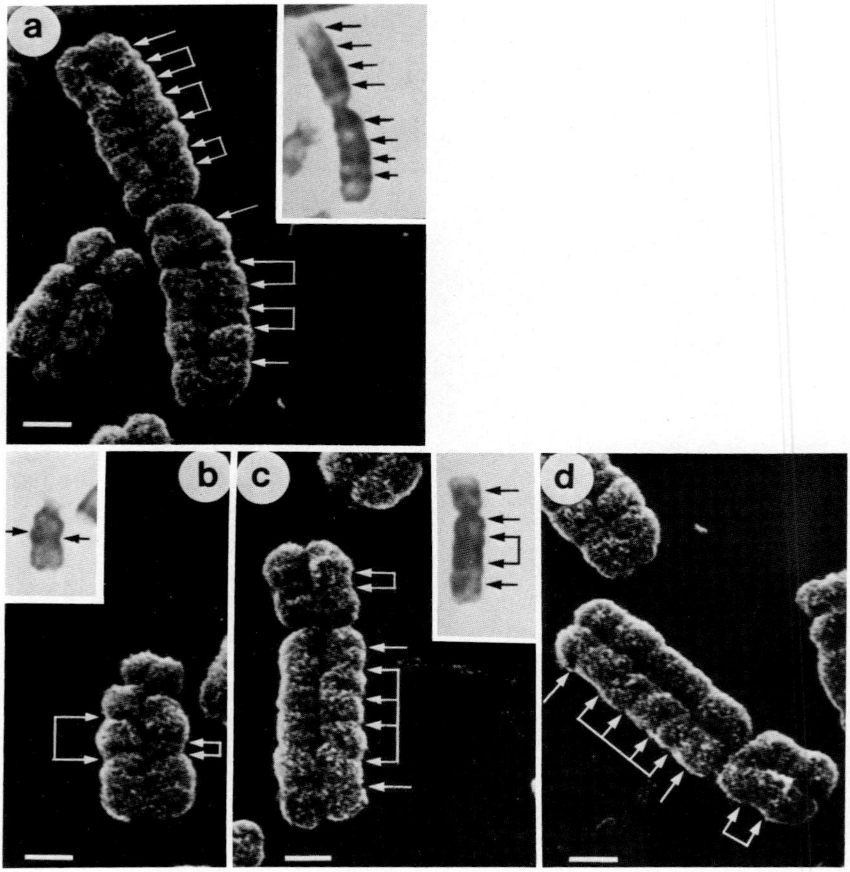

Fig. 16. Human chromosomes from the metaphase in Fig. 14. G-positive light microscope bands (black arrows, insets) correspond to circumferential grooves in the chromatids (white arrows). a, Chromosome 1 shows a large number of grooves. Certain G-positive bands correspond in position to two circumferential grooves, indicated by connected white arrows. b, Chromosome 15 shows one prominent G-band (black arrows, inset). This corresponds to two circumferential grooves in each chromatid (connected white arrows). The chromosome is leaning to one side and the circumferential grooves are shifted from parallel alignment. The G-band is shifted an equal distance in the same direction. c, Chromosome 5 is characterized by one large G-positive band in the center of the q arm (connected black arrows, inset). This corresponds to four circumferential grooves, in both chromatids, in the SEM (connected white arrows). The white arrowheads demonstrate the positional relationship between the grooves and the ISCN midmetaphase banding pattern [black arrowheads, Fig. 18c(i)]. d, The same No. 5 chromosome viewed upside down. From this angle the groove at the distal end of the q arm is more pronounced (white arrow) than when observed in the upright position (see c); bar = 2 μm. (Reprinted with permission from Harrison *et al.*, 1985a.)

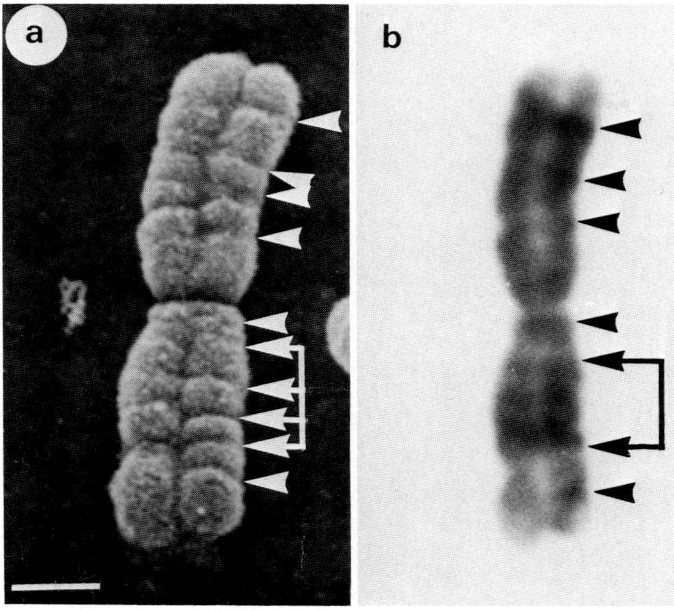

Fig. 17. Chinese hamster chromosome from the metaphase in Fig. 15, viewed upside down. The chromosome photographed in the light microscope (b) is enlarged to the same size as the SEM photograph (a) which produces a poor image. The advantage is that the correlation between the G-positive light microscope bands (black arrows, b) and the circumferential grooves (white arrows, a) is clearly demonstrated; bar = 2 μm.

served in the SEM. (Compare the black and white arrows, Fig. 17.) Therefore, the chromosome structural relationship to G-banding is not species specific and, although not presented here, has been demonstrated to exist in a large number of mammalian species. From this point the observations will refer to human chromosomes.

Figure 18 shows a homologous pair of elongated No. 5 chromosomes from a different human metaphase. These chromosomes, when compared with the more contracted No. 5 chromosomes in Fig. 16c, show an increased number of G-positive bands in the light microscope (black arrowheads, Fig. 18a,b, insets) as described above. A corresponding increase in the number of circumferential grooves is also observed in these extended chromosomes, when viewed in the SEM (white arrows, Fig. 18a,b). An identical pattern of coiling is present in both members of the homologous pair (compare grooves in Fig. 18a and b). This close correlation is shown in all homologues, in this and all other metaphases examined.

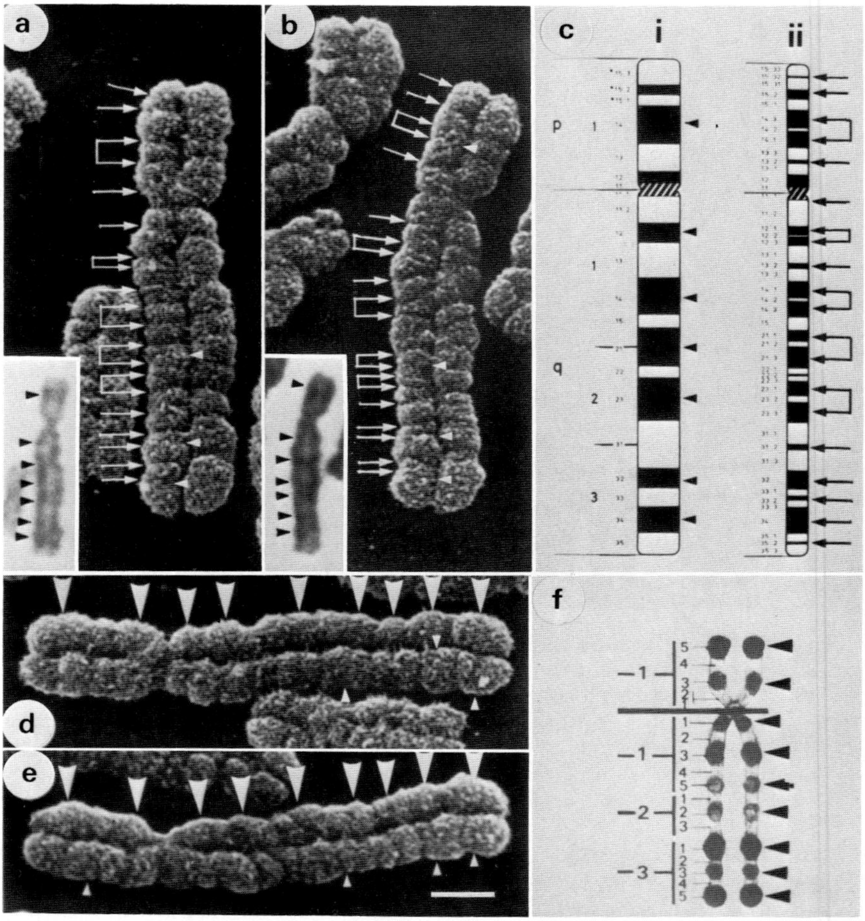

Fig. 18. a and b, A homologous pair of elongated No. 5 chromosomes from a different metaphase. An increased number of G-positive bands in the light microscope (black arrowheads, insets) and a corresponding increase in number of circumferential grooves in the SEM (white arrows) are observed, when compared with the more contracted chromosome 5 (Fig. 16c). Identical patterns of coiling are demonstrated in both members of the homologous pair; compare position and number of grooves in a and b. Each light microscope, G-positive midmetaphase band (black arrowheads, insets) corresponds to more than one circumferential groove in the SEM (connected white arrows). For example, G-bands 5p14, 5q12, 5q14, 5q21, and 5q23, indicated in the light microscope by black arrowheads (see insets), are represented by two circumferential grooves (white arrows connected in pairs), and correspond to two sub-bands in prometaphase (see Fig. 16c, black arrows connected in pairs). G-bands 5q32 and 5q34 are visible in the light microscope photographs (black arrowheads, distal portion of q arm, insets). In the SEM, circumferential grooves are present corresponding to these bands and also the sub-bands 5q31.2, 5q33.2, and 5q35.2, which are not visible in

The diagrammatic representation of chromosome 5 (Fig. 18c) clearly demonstrates the relationship between the midmetaphase [chromosome (i)] and prometaphase [chromosome (ii)] banding patterns, which are described in detail elsewhere (ISCN, 1985). That is, the G-bands observed at midmetaphase are made up of sub-bands in prometaphase. Each light microscope G-positive midmetaphase band, observed in the extended chromosomes [black arrowheads, Fig. 18a,b, insets; and chromosome (i), Fig. 18c] corresponds in position to more than one circumferential groove, when viewed in the SEM (connected white arrows, Fig. 18a,b). These in turn are correlated with the sub-bands of the prometaphase chromosomes [similarly connected black arrows, chromosome (ii), Fig. 18c]. For example, G-bands 5p14, 5q12, 5q14, 5q21, and 5q23, visible in the light microscope, are represented by two circumferential grooves in the SEM (white arrows connected in pairs, Fig. 18a,b) and correspond to two sub-bands in prometaphase (black arrows connected in pairs, Fig. 18c). G-bands 5q32 and 5q34 are visible in the light microscope photographs (Fig. 18a,b, insets). However, in the SEM circumferential grooves are demonstrated within the 5q3 region (single white arrows at distal portion of the q arm, Fig. 18a,b) corresponding to 5q32 and 5q34 and also to the sub-bands 5q31.2, 5q33.2, and 5q.35.2 not visible in the light microscope. [Compare single white arrows, Fig. 18 a and b and single black

---

the light microscope (compare single white arrows in these figures and single black arrows, in Fig. 16c, distal portion of q arm). Circumferential grooves corresponding in position to sub-bands 5p15.32, 5p15.2, 5p13.2, and 5q13.2 are also present in the SEM (compare black, in c, and white arrows, as above), which are not observed in the light microscope photographs (inset). These elongated chromosomes are prone to bending and at several points along the sister chromatids the grooves are shifted from positions directly parallel to each other, reflected by an equivalent distortion in the G-bands in the light microscope (insets).

c, Diagramatic representation of chromosome 5. Chromosome (i) shows the banding pattern at midmetaphase and chromosome (ii) represents the banding pattern at prometaphase. The black arrowheads on the left-hand chromosome correspond to the G-bands observed in the light microscope (black arrowheads, Fig. 16a,b, insets). The connected black arrows on the right-hand chromosome indicate those sub-bands correlated with each midmetaphase G-band (black arrowheads, left-hand chromosome) and correspond in position and number to those circumferential grooves on the chromosomes in a and b shown by connected white arrows.

d and e, Chromosomes from a and b viewed from one side. Those grooves shown by a small white arrowhead (see a and b) are resolved more clearly from this angle. The raised regions between the grooves, indicated by large white arrowheads, correspond in position to the G-negative or R-bands (see f); bar = 2 μm.

f, Diagramatic representation of an R-banded chromosome 5 (ISCN, 1985). The black arrowheads, indicating the positions of the R-bands, correlate with the large white arrowheads on d and e.

(Reprinted with permission from Harrison *et al.*, 1985a.)

arrows, Fig. 18 c(ii) at distal end of q arm.] Circumferential grooves corresponding in position to sub-bands 5p15.32, 5p13.2, and 5q13.2 are also observed in the SEM (compare single arrows as above) that are not resolved in the light microscope photographs. Therefore, the details of prometaphase banding are observed in midmetaphase chromosomes in the SEM.

Elongated chromosomes are more prone to bending of the chromatids than the more contracted chromosomes and, as a result, at several points along the sister chromatids the grooves are slightly shifted from positions directly parallel to each other (Fig. 18a,b). This is reflected by an equivalent distortion in the G-bands in the light microscope (Fig. 18a,b, insets). This precise correlation provides good evidence for the involvement of chromosome structure in the production of light microscope banding patterns. Those grooves, indicated by a small white arrowhead (Fig. 18a,b) are resolved more clearly when the chromosomes are viewed from one side (small white arrowheads, Fig. 18d,e). Discrepancies in number and precise position of the grooves between sister chromatids and homologues are excluded by rotational viewing in the SEM.

In support of the conclusion that the circumferential grooves correspond to the light microscope G-positive bands, it can also be demonstrated that the regions between the grooves are related to the G-negative or reverse bands (R-bands) (Harrison *et al.*, 1985a). Fig. 18f demonstrates the R-banding pattern as described by ISCN (1985). The R-positive bands, indicated by black arrowheads correspond in position to the raised regions of the chromatids between the circumferential grooves (large white arrowheads, Fig. 18d,e). Identical patterns are demonstrated in both members of this homologous pair (compare large arrowheads in Figs. 18d and e). A similar correlation is also observed in other homologous chromosome pairs taken from different metaphases.

These results demonstrate that using the SEM it is possible to show the details of the prometaphase G-banding pattern in metaphase chromosomes and to overcome some of the technical difficulties associated with the analysis of prometaphase chromosomes in the light microscope. These include the problems of spreading the prometaphases to prevent overlapping and the associated difficulties of individual chromosome identification. By detailed comparison of homologous chromosomes in both the light microscope and the SEM, this technique potentially allows the study of minute chromosome aberrations, and the precise location of breakpoints involved in structural chromosome rearrangements. Examples of the clinical applications of this procedure are presented later in this chapter.

SEM of C-Banded Chromosomes

Correlative light microscopy and SEM of C-banded chromosomes shows that in the light microscope chromosomes 1, 9, and 16, with paracentric secondary constrictions, are positively stained at the centromere. The remainder of the chromosome is pale in color. The staining is well illustrated in the homologous pair of number 1 chromosomes shown in Fig. 19a (inset). These chromosomes are closely associated at the centromeric region. The distal end of the Y chromosome in the male karyotype also stains positively (Fig. 20, inset). In the SEM the areas corresponding in position to these positively stained C-bands are of constricted appearance with a smooth surface morphology and a highly compact fine structure (Figs. 19a,c and 20). When viewed at a higher magnification, the C-banded regions appear to be coated with a "blanket" of material, beneath which fibrillar projections are observed. The details of fibrillar organization cannot be resolved, giving the appearance of a smooth surface morphology (Fig. 19b).

The areas corresponding in position to the non-C-banded regions show, in striking contrast, a very loosely packed chromosome structure, in which the individual fibers are easily resolved. These observations are described in detail by Jack *et al.* (1985). This difference in gross morphology between the positively staining heterochromatin and the nonstaining euchromatic regions may be produced as a result of alternative methods of coiling during compaction into the metaphase chromosome. This has been implicated by several authors from light microscope observations (Balicek *et al.*, 1977; Madan and Bobrow, 1974; Schmiady and Sperling, 1976; Selezneva *et al.*, 1977).

The boundary between the C-banded and non-C-banded regions of the same chromosome is well defined in the light microscope (Fig. 19a, inset). The same boundary between the smooth and fibrillar morphology in the SEM is of clearer definition (Fig. 19). However, variations are observed, between metaphases on the same coverslips, in the clarity of this division between heterochromatic regions and euchromatic regions in both light microscope and SEM photographs. It appears to be dependent on the degree of chromosome contraction. The C-banded region of chromosome No. 1 (Fig. 19c) is very tightly packed in structure, while the non-C-banded region is very loosely packed. A very marked delineation between the two areas is observed. In comparison, a more reduced definition is shown in another chromosome No. 1 (Fig. 19a). In spite of this variation, SEM provides an accurate assessment of the extent of the heterochromatic regions between homologous chromosome pairs and between individuals.

Sequential examination of the same C-banded chromosomes in the light microscope and the SEM demonstrates the precision of the correlation between the compact structure and heterochromatin and the fibrillar morphology and euchromatin. This finding provides direct evidence for a structural basis for C-banding.

SEM of R-Banded Chromosomes

In the light microscope a pattern of light and dark staining regions is observed along the chromosome arms of R-banded metaphases. The R-bands stain positively and the R-interbands remain pale, which is the reverse of the G-banding pattern. This is shown in the chromosome 11 in Fig. 21 (inset). As demonstrated in the other banding procedures, the photography of the chromosomes in buffer, which is necessary to retain the best chromosome morphology in the SEM, produces a reduction in contrast in the light microscope photographs.

When the same chromosomes are examined in the SEM, the darkly stained R-bands clearly correspond to highly fibrous regions and the pale staining R-interbands to flattened regions in between. Figure 21a shows a R-banded chromosome 11 viewed in the light microscope (inset) and the SEM. The fibrous regions correlate in size and position to positively stained R-bands in the light microscope at the bands p15, p11, q13, and q23. The flattened regions correspond to the pale staining R-interbands. A similar correlation between chromosome structure and light microscope staining is observed in all homologous pairs and is consistent from metaphase to metaphase (Jack *et al.*, 1986).

The ability to relocate the same R-banded chromosomes in the SEM after light microscopy enables detailed study to be undertaken on well-R-banded chromosomes, since the results may be highly variable. Observation of R-banded regions at higher magnification in the SEM reveals discrete fibers which form individual loops and projections (Fig. 21b). High-resolution observation of the R-interband region shows that these areas are covered uniformly with short, rounded, closely opposed protu-

Fig. 19. C-banded human chromosome 1. a, A homologous pair of No. 1 chromosomes closely associated at the centromeric region viewed in the SEM and light microscope (inset). The paracentric secondary constrictions stain positively in the light microscope. This C-banded region is represented in the SEM by a smooth, tightly compact area; bar = 1 μm. b, C-banded area at a higher magnification. Note the smooth highly compact morphology of the C-banded region in contrast to the loosely packed fibrous nature of the non-C-banded; bar = 0.5 μm. c, Chromosome 1 from a different metaphase spread showing a more marked distinction between the C-banded and non-C-banded region. The single arrow marks the extent of the C-banded region distal to the centromere (c); bar = 0.5 μm. (Reprinted with permission from Jack *et al.*, 1985.)

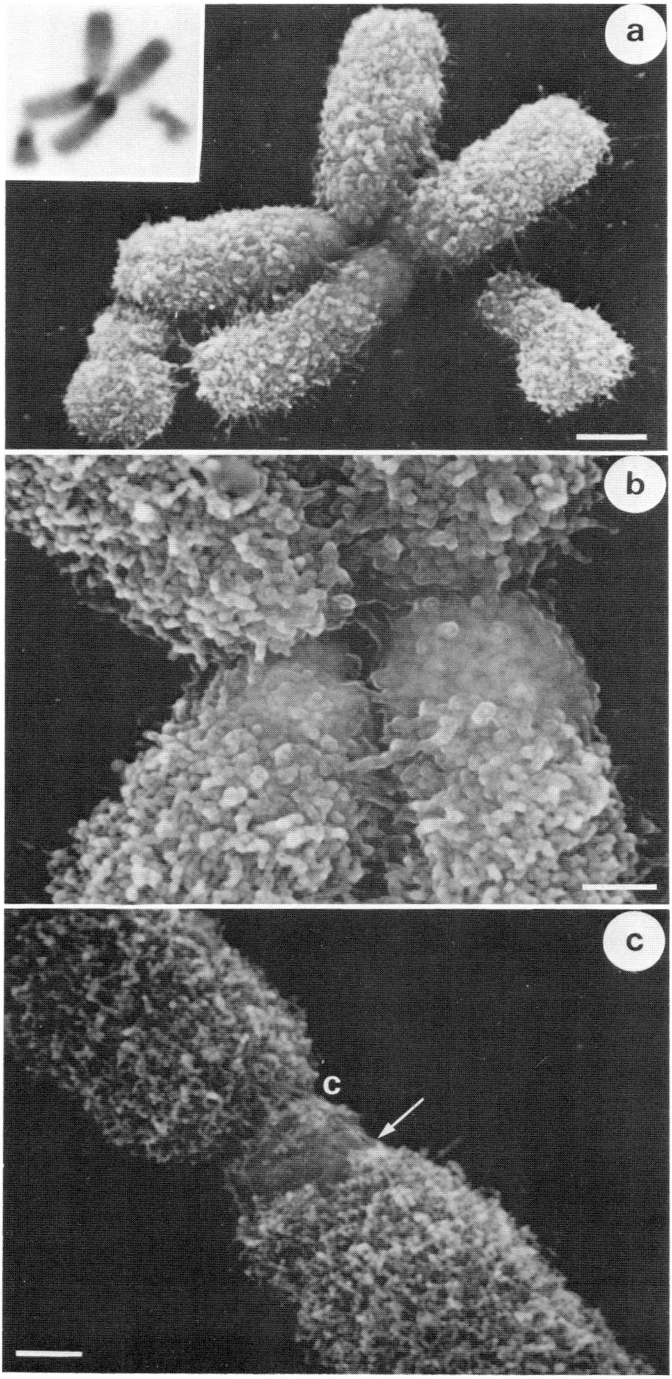
a
b
c
C

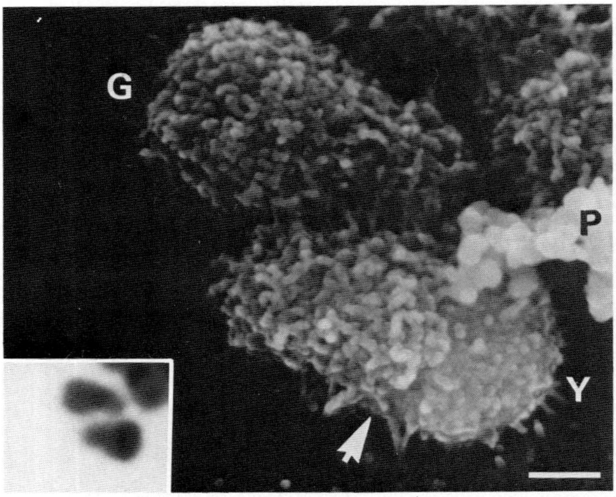

Fig. 20. C-banded human Y chromosome. The same Y chromosome (Y) viewed from a lateral position in both the light microscope (inset) and SEM, lying adjacent to a G group (G). The distal portion of the long arm stains positively with Giemsa (inset). This C-banded region in the SEM has a tightly packed fibrous morphology with a smooth surface appearance, in contrast to the loosely packed fibrous nature of the non-C-banded region. The arrow marks the extent of the C-banded region. A precipitate of osmium-TCH(p) lies over part of the Y chromosome; bar = 0.5 μm. (Reprinted with permission from Jack *et al.*, 1985.)

berances and are devoid of individual fibers. A substantial loss of chromosomal material appears to have occurred from these areas. Short, rounded projections, and occasionally short lengths of fibers, are observed forming a "blanket" of background material surrounding the chromosomes (Fig. 21a). This may represent a redistribution of the material extracted from the R-interbands (Jack *et al.*, 1986).

SEM of T-Banded Chromosomes

T-banded chromosomes are often pale with little distinction between the T-positive and T-negative bands (Fig. 22, inset). When the same T-banded chromosome is examined in the SEM, after light microscopy, a striking structural difference is demonstrated between the light microscope, positively staining telomere, and the remainder of the chromosome in all chromosomes examined. This demonstrates the advantage of SEM observation. The positively staining T-band in the light microscope corresponds to an area of large, looped fibers in the SEM (Fig. 22b). The pale staining region of the chromosomes, along the length of the p and q arms, show no structural variation. This corresponds to a flattened area, made

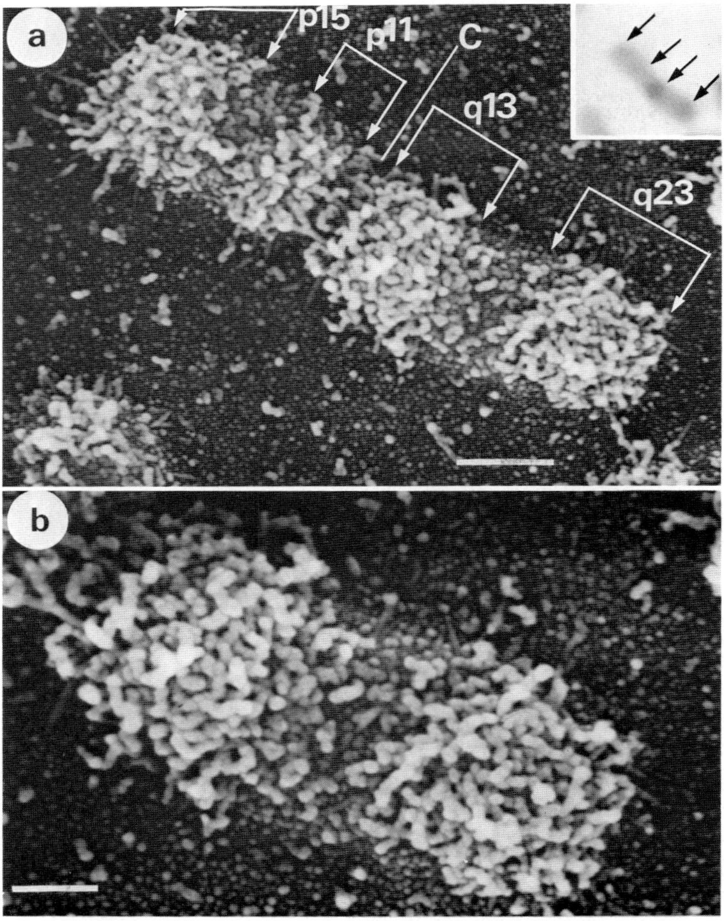

Fig. 21. a, An R-banded chromosome 11 viewed in the light microscope (inset) and SEM. The positively stained R-bands (black arrows, inset) correlate in position to highly fibrous regions in the SEM (connected white arrows). The bands are labeled according to ISCN (1985). The pale staining R-interbands (inset) correlate with flattened regions in between. Short, rounded projections and short lengths of fibers are observed forming a blanket of background material surrounding the chromosome; bar = 1 μm. b, Higher magnification of the q arm of the chromosome 11 in a. The R-banded regions are formed of individual loops and projections. The R-interband regions appear to have been flattened. The area is covered uniformly with short, rounded, closely opposed protuberances and appears to be devoid of individual fibers; bar = 0.5 μm. (Reprinted with permission from Jack *et al.*, 1986).

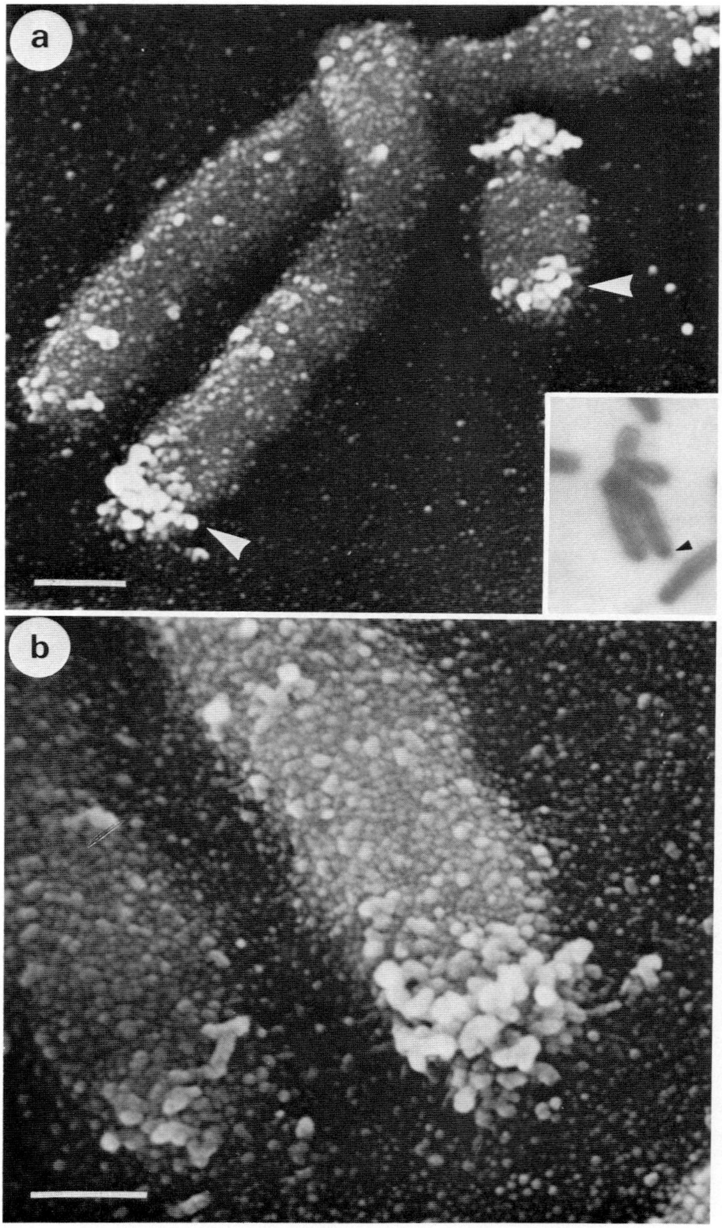

Fig. 22. a, Light microscope (inset) and SEM of the same T-banded C- and G-group chromosome. T-banded chromosomes are often pale in the light microscope (inset) with little distinction between T-positive and T-negative bands. A striking structural difference is shown in the SEM between the telomeres (arrows) and the remainder of the chromosome; bar = 1 μm. b. Higher magnification of the telomere of the C-group chromosome. This T-

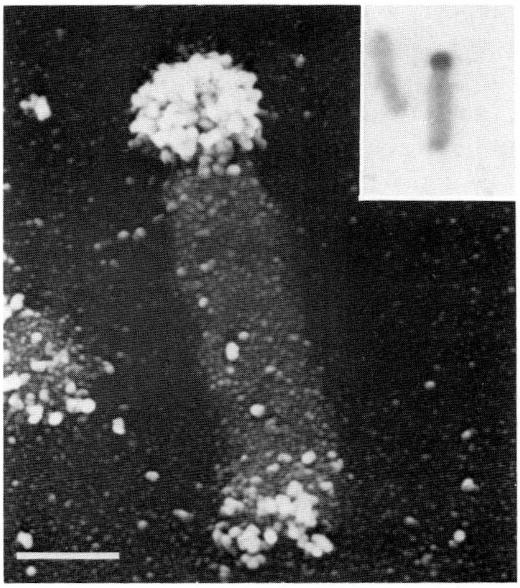

Fig. 23. A T-banded D-group chromosome in the light microscope (inset) and SEM. This acrocentric chromosome shows a highly fibrous structure within the satellite region on the p arm in addition to the telomere at the end of the q arm. The highly fibrous region corresponds to the positively stained satellite region in the light microscope (inset); bar = 1 μm. (Reprinted with permission from Jack *et al.*, 1986.)

up of closely apposed, short, rounded protuberances which appear to be devoid of any fibrous organization. The telomeres and the T-negative regions show a close resemblance in fine structural detail to the R-positive and R-negative regions, respectively (Jack *et al.*, 1986).

Acrocentric chromosomes show a similar highly fibrous structure in the satellite region on the p arm in addition to telomere at the end of the q arm (Figs. 22a and 23). This highly fibrous region corresponds to the positively stained satellite region in the light microscope photograph of the same chromosome (Fig. 23, inset). A large amount of fibrous background material is also observed scattered around the T-banded chromosomes (Fig. 22a) as seen in R-banded preparations (Jack *et al.*, 1986).

---

band shows an area of large looped fibers. The remainder of the chromosome appears flattened and is composed of closely opposed short, rounded protuberances, which appear to be devoid of any fibrous organization. A large amount of fibrous background material is observed scattered around the chromosomes; bar = 0.5 μm. (Reprinted with permission from Jack *et al.*, 1986.)

These results suggest that by increasing the temperature of the acid–salt solution, a progressive breakdown in structure is caused with eventual denaturation of the entire chromosome, except the telomere. This supports the hypothesis of Dutrillaux (1973) that T-banding is a an extension of the R-banding process.

APPLICATIONS OF CORRELATIVE LIGHT MICROSCOPY AND SCANNING ELECTRON MICROSCOPY

Ring Chromosomes

The correlative light microscopy and SEM procedure for cytogenetic analysis of human metaphase chromosomes may be successfully applied to clinical examples. One area where it has been found particularly to be of use is in those cases where obvious structural changes are present, for example, in identification of the form of ring chromosomes. Usually, ring chromosomes are formed by chromosome breakage at two points on opposite sides of the centromere. The ends of the centric fragment fuse to form a ring, or occasionally a figure 8. This is usually associated with the loss of chromosomal material. Figure 24 shows a complete ring, in the figure 8 formation, derived from a chromosome 14. In the SEM, observation of ring chromosomes from a series of different angles in a number of cells confirms the form of the ring.

A complex ring chromosome formation was observed in the leukemic cells of a patient with acute myelomonocytic leukemia (AMMoL). In the light microscope the form of the ring was difficult to interpret as it arose after a translocation event between chromosomes 5 and 11. From a series of light microscope photographs it was difficult to ascertain whether one or two centromeres were involved and whether the ring was completely or incompletely formed. Figure 25c shows one of the light microscope photographs, before relocation in the SEM, indicating the difficulties of interpretation. When viewed in the SEM (Fig. 25a) it is apparent that two complete chromosomes (5 and 11) are involved; therefore the ring has two centromeres (Fig. 25a, arrows). The p arms are fused completely together at or close to the telomeres (Fig. 25b, arrowhead), whereas the telomeres of the q arms are incompletely joined by a lateral fusion of a single chromatid from each chromosome (Fig. 25b, small arrow). This accuracy of information is dependent on SEM structural information, guided by light microscope karyotyping.

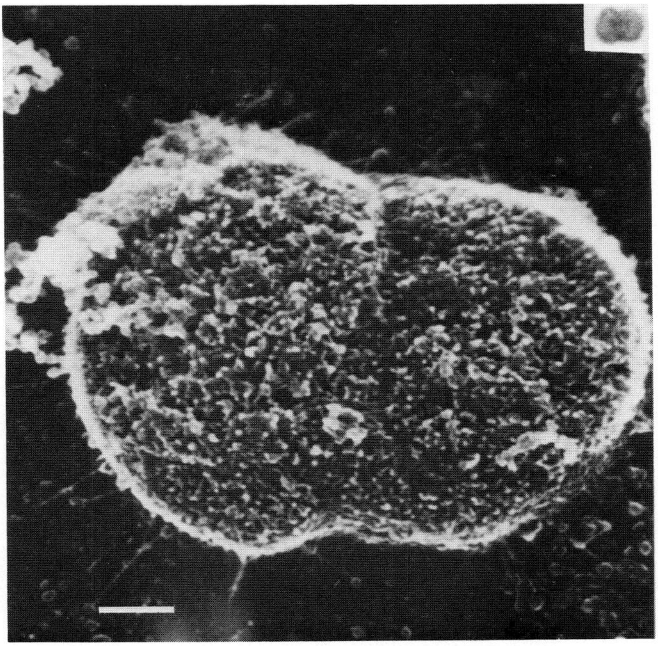

Fig. 24. A ring chromosome 14 observed in the light microscope (inset) and the SEM. The ring is complete and forms a figure 8; bar = 0.5 μm.

Double Minutes

Double minutes (DMs) are of interest since they have been described in a number of malignant and drug-resistant somatic mammalian cells. They are known to be sites of gene amplification. The origin of DMs cannot be explained by classical mechanisms of structural chromosome change. Although they lack centromeres they persist in cell populations. For detailed reviews of the biology of DMs refer to Barker (1982) and Cowell (1982).

For sequential light microscopy and SEM observation a human neuroendocrine tumor cell line was prepared containing a large number of DMs. Light microscopy of DMs is of limited value due to the small size, variable number, and pale staining of DMs after banding. G-banded metaphases, when photographed for light microscopy, require longer exposure times during printing of photographs to reveal the DMs. As a result the G-banded chromosomes are overexposed (Fig. 26a).

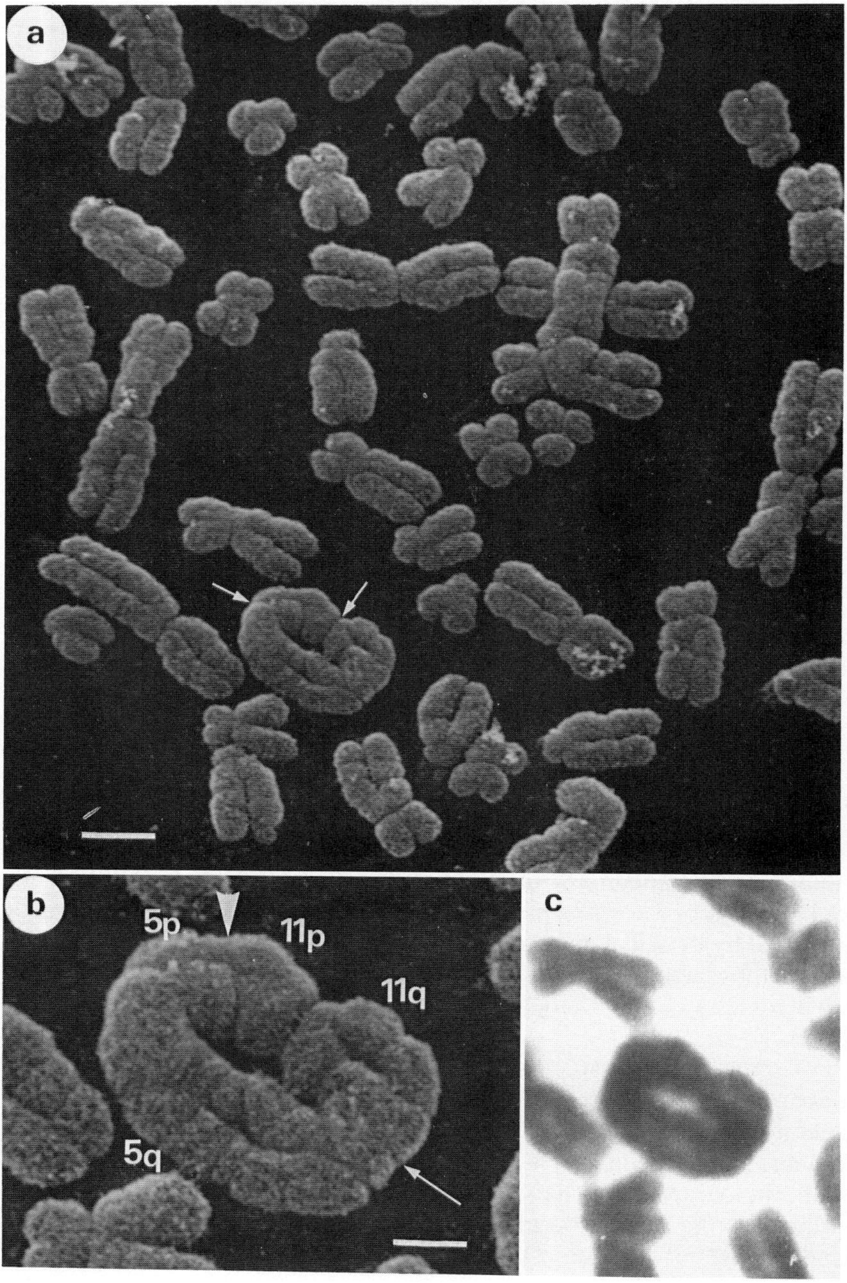
a
b
5p
11p
11q
5q
c

In the light microscope photographs the DMs are found to be highly variable in size and closely associated with the metaphase chromosomes (Fig. 26a). SEM observation of the same cells as observed in the light microscope confirms the three-dimensional structure of DMs (Fig. 26b). The sister minutes appear as two spherical chromatin bodies interconnected by chromatin fibers (Fig. 26c) and separated spacially from each other to varying degrees (Fig. 26b; Jack *et al.*, 1987).

In the SEM the close association between DMs and chromosomes is shown to be due to attachment by chromatin fibers (Fig. 26b, arrows). This observation indicates one possible mechanism of distribution of DMs to daughter cells, by adherence to the metaphase chromosomes themselves. The DMs show similar morphological features to the chromosomes within the same metaphases, although no centromeric constriction is observed in DMs. They show the same degree of compaction of chromatin fibers and a similar chromatin fiber diameter (Fig. 26c). The variation in size of the DMs is therefore not due to differential compaction of chromatin, in support of an earlier observation by Bahr *et al.* (1983). SEM confirms that the partners of the DMs are consistently of a similar size. The number of DMs per cell is known to be highly variable (Cowell, 1982). This relocation technique provides an accurate method for studying the distribution in number of DMs within a cell population (Jack *et al.*, 1987).

Chromosome Polymorphisms

Chromosome polymorphisms are structural variants of chromosomes displayed most readily at specific chromosome sites, including the highly variable centromeric regions of chromosomes 1, 9, and 16, the distal end of the Y, and the satellites of acrocentric chromosomes. These regions show a wide variation in size between individuals. They are generally stable, present in all cells in a person, and inherited from one generation to the next in a Mendelian fashion. Therefore, they have been shown to be informative genetic markers, for example, in establishing parental origin of chromosome abnormalities (Chamberlin and Magenis, 1980; Kajii and

Fig. 25. A complex ring chromosome observed in the light microscope (c) and SEM (a) and (b), from the leukemic cells of a patient with AMMoL. a, In the SEM it is apparent that two complete chromosomes are involved as two centromeres are present (arrows); bar = 2 μm. b, At higher magnification in the SEM the p arms of chromosomes 5 and 11 are fused completely together, at or close to the telomeres (arrowhead). The telomeres of 5q and 11q are incompletely joined by a lateral fusion of a single chromatid from each chromosome (small arrow); bar = 0.5 μm. c, In the light microscope the form of the ring is difficult to interpret.

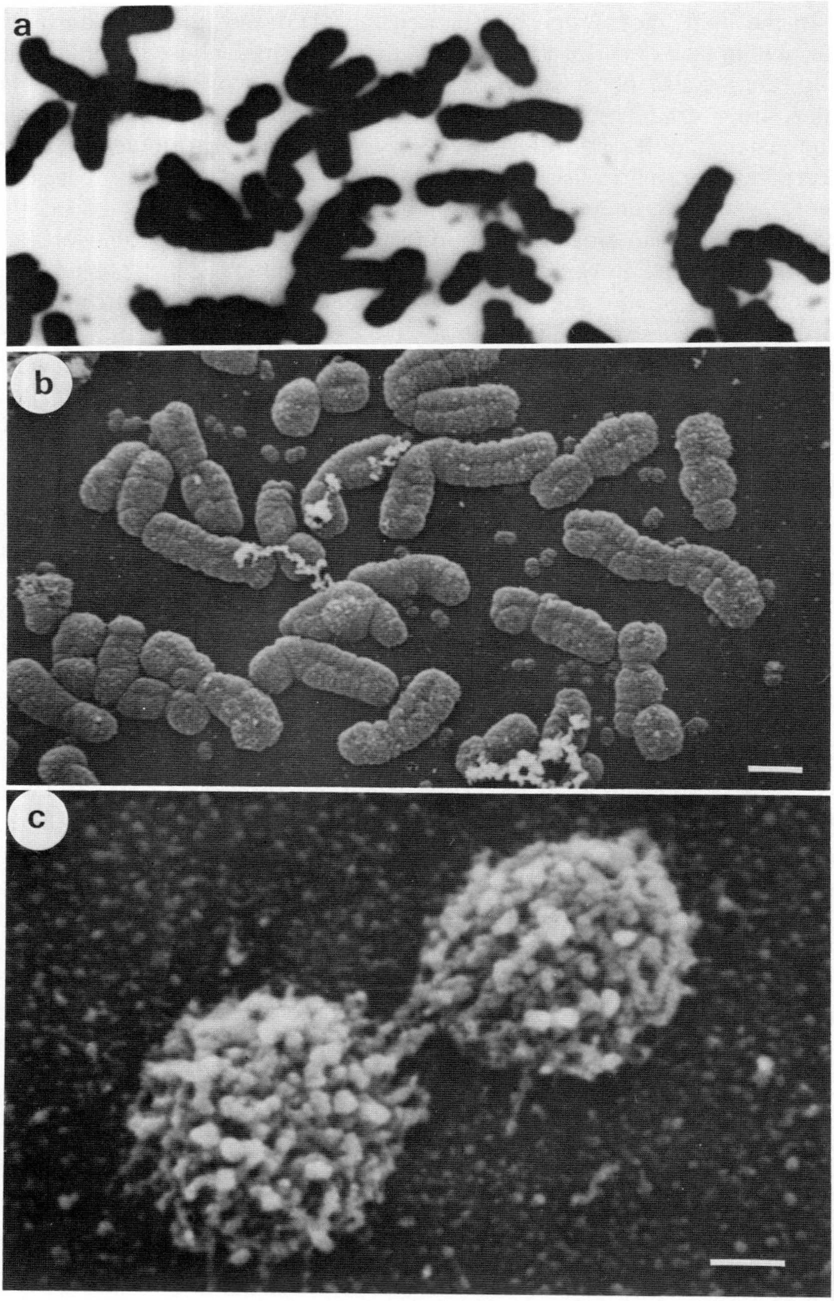
a
b
c

Niikawa, 1977), particularly in the origin of the extra chromosome 21 in Down's syndrome (Jacobs and Mayer, 1981; Magenis *et al.*, 1977; Mikkelsen *et al.*, 1980; Robinson, 1973; Uchida, 1973) and for exclusion of paternity in cases of disputed parentage (de la Chapelle *et al.*, 1967; Olson *et al.*, 1983). In light of these suggestions, methods to provide reliable comparison between polymorphisms are essential. Correlative light microscopy and SEM demonstrates small variations not readily detected by light microscopy alone. Polymorphisms may be subdivided into two categories: variable satellite regions and heterogeneic C-bands.

Variable Satellite Regions

The acrocentric chromosomes, in which the centromere is situated close to one end of the chromosome, have the highest incidence of minor variants within the human karyotype, occurring exclusively in the satellite region. The most common of these variations is the size of satellites. To demonstrate these differences in the light microscope, sequential and diverse staining is required (Balicek *et al.*, 1977; Chen *et al.*, 1981; Wachtler and Musil, 1980). These techniques may reduce the definition of the satellites. Therefore, only gross changes in morphology may be accurately assessed in the light microscope. After G-banding the satellites are readily identified in the SEM (Fig. 27), compared to the reduced definition when observed in the light microscope. Also, in the SEM the morphology of the satellites is not variable from metaphase to metaphase, since this observation does not depend on uptake of stain.

Figure 27 shows a group of acrocentric chromosomes with satellites of various sizes. The light microscope photographs demonstrate the difficulties involved in interpreting the size of the satellites (inset). When the same chromosomes are examined in the SEM the shape and size is clearly defined. For example, the chromosome 21 on the left-hand side shows enlarged satellites, of which one is displaced. The homologous 21 has, comparatively, smaller satellites. The chromosome 14 shows satellites of similar size to the second 21, and the 22 has smaller satellites. Therefore, the SEM allows accurate detection of satellite polymorphisms. In addition, the ability to rotate the specimen stage in the SEM and to observe the satellites from different angles, over a range of magnifications, allows

---

Fig. 26. a and b, Part of the same metaphase spread observed in the light microscope (a) and SEM (b). The light micrograph (a) is overexposed to enhance the DMs. In the SEM (b) the DMs are three-dimensional. Note the variation in size and separation of the DMs; bar = 2 μm. c, Enlargement of a pair of DMs from b to show the highly compact chromatin fibers and interconnecting fibers between sister minutes; bar = 0.2 μm. (Reprinted with permission from Jack *et al.*, 1987; © S. Karger AG, Basel.)

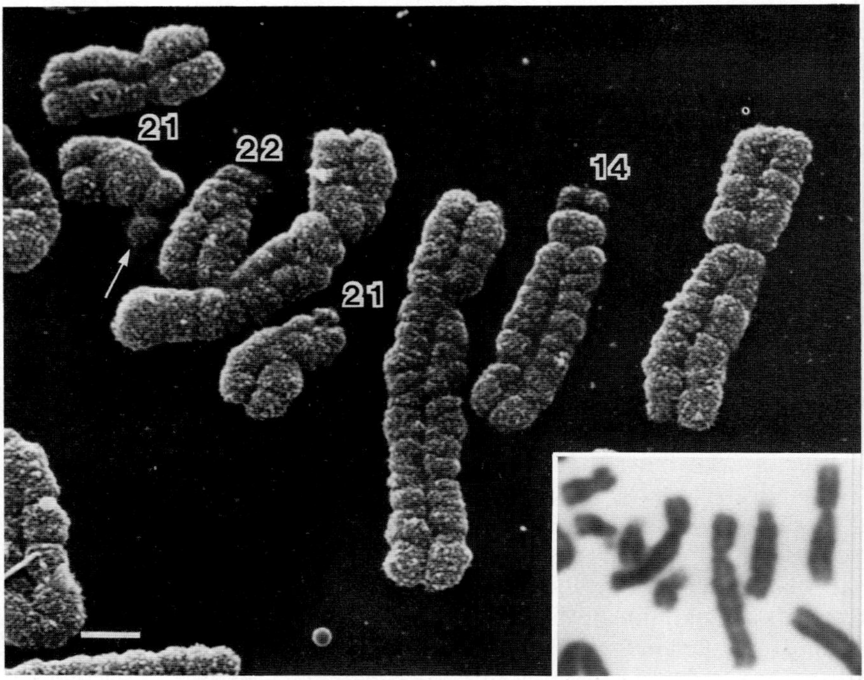

Fig. 27. A group of acrocentric chromosomes viewed in the light microscope (inset) and SEM. In the light microscope it is difficult to interpret the size of the satellites (inset). When the same chromosomes are observed in the SEM the size is clearly defined. The chromosome 21 on the left-hand side shows enlarged satellites, of which the lower one is displaced (small arrow). The homologous 21 has smaller satellites. The chromosome 14 shows satellites of similar size to the second 21 and the 22 has smaller satellites; bar = 1 μm.

accurate measurements to be made of satellite volumes (Harrison *et al.*, 1985b). Therefore, the increased detail of these polymorphisms observed by high-resolution SEM reveals the scope of this technique for application to clinical cases.

Heterogeneic C-Bands

Certain polymorphic human chromosomes show variations in the amount of heterochromatin present at specific regions, for example, the sites of the paracentric secondary constrictions of chromosomes 1, 9, and 16 and the distal end of the Y. Variations may be demonstrated between homologues by correlative light microscopy and SEM of C-banded chromosomes, as previously described. These variations may also be shown by examination of G-banded chromosomes. The number 9 chromosomes provide a good example for study since the paracentric secondary constriction is highly polymorphic (Harrison *et al.*, 1985b; Jack *et al.*, 1985). The C-banding technique reveals this region as a dense heterochromatic block showing considerable variation between persons. Figure 28a and b (upper insets) shows a homologous pair of C-banded chromosomes 9 in the light microscope of which one member exhibits an increased amount of heterochromatin at the centromere (9qh+) (Fig. 28a, upper inset) when compared to the other chromosome 9 (Fig. 28b), upper inset). The heterochromatic secondary constriction of chromosome 9 remains pale and rather poorly stained in all G-banded metaphases examined in the light microscope (Figs. 28a,b, lower insets). In these light microscope photographs the extent of the secondary constriction is demarcated by the proximal boundary of G-band 9q21 (ISCN, 1985) (Figs. 28a,b, black arrowheads, lower insets) which correlates with a prominent circumferential groove in both homologues when these chromosomes are

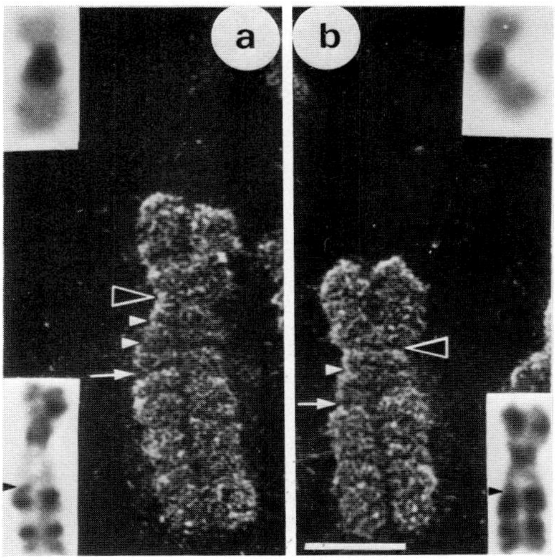

Fig. 28. A homologous pair of No. 9 chromosomes viewed in the light microscope after G-banding (lower insets) and then in the SEM. A light microscope C-banded pair (prepared in parallel) is shown (upper insets) in which the 9qh + (a) shows extra C-banded material at the centromere when compared to the smaller 9 (b). The centromeric region stains negatively in the light microscope after G-banding (lower insets) and the extent of this region is demarcated by the proximal boundary of G-band 9q21 (black arrowheads) corresponding to a prominent groove in the SEM (white arrows). The centromeres are clearly defined in the SEM (black on white arrowheads). Within the heterochromatic region circumferential grooves are observed (white arrowheads); bar = 2 μm. (Reprinted with permission from Harrison *et al.*, 1985b.)

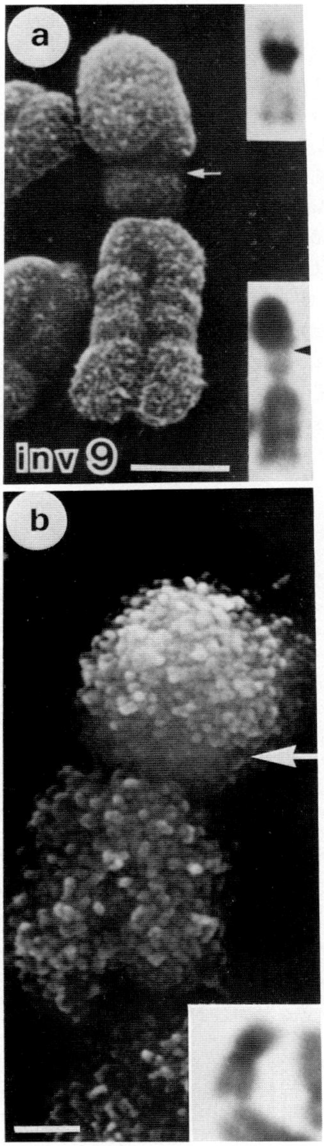

Fig. 29. a, This shows inv(9) detected by C-banding (upper inset). This C-banded region stains negatively after G-banding in the light microscope (lower inset). When the same G-banded inv(9) chromosome is viewed in the SEM this region is constricted in appearance and demarcated by a prominent circumferential groove (white arrow). This inversion is complete as no chromatin of constricted morphology is observed on the long arm of this chromosome; bar = 2 μm. (Reprinted with permission from Harrison *et al.*, 1985b.) b, C-

examined in the SEM (Fig. 28a,b, white arrows). The centromere is more clearly defined in the SEM (black on white arrows) than the light microscope. Therefore, in the SEM the extent of the heterochromatic region between the centromere and the boundary of the secondary constriction are more precisely defined, compared to the reduced definition provided by light microscope C- or G-banded preparations.

Also, when observed in the SEM, both chromosomes show additional circumferential grooves within the heterochromatic region. The 9qh+ has two grooves within this region (Fig. 28a, white arrowheads) compared to the homologue, in which only one groove is observed (Fig. 28b, white arrowhead). This distribution of grooves is consistently observed in numerous metaphases from this person. No variable staining regions which might relate to this pattern of intraheterochromatic grooves have been observed in the light microscope (insets). Therefore, this observation of variable numbers of grooves within the heterochromatin, observed by relocation of metaphases in the SEM, may be used as a reliable evaluation of the relative amounts of heterochromatin in homologous pairs of No. 9 chromosomes. To date, these grooves have not been observed in polymorphic chromosomes 1 or 16.

Pericentric inversions of the heterochromatin of chromosome 9 [inv(9)], form another polymorphism observed in this chromosome (Fig. 29). The chromosome appears metacentric with heterochromatin transferred from q to p arm. Inv(9) is readily detected in the light microscope by C-banding (Fig. 29a, upper inset, and 29b, inset). After G-banding the inverted heterochromatin remains poorly stained, even when the euchromatin is heavily stained (Fig. 29, lower inset). This G-banded chromosome, when observed in the SEM, shows a prominent circumferential groove demarcating the boundary of the secondary constriction (Fig. 29a, white arrow). Also, the heterochromatin shows a constricted morphology, as observed in the noninverted chromosomes 9 (Figs. 29a,b).

Inv(9) may show partial or complete inversion of the heterochromatin. Light microscope C-banding studies may be unclear. Variable staining is shown between inv(9) chromosomes from different metaphases of this same individual (Fig. 29a, upper inset and 29b, inset). However, SEM studies of the same G-banded (Fig. 29a) and C-banded chromosomes (Fig. 29b) confirm a complete inversion. In G-banded chromosomes, the chro-

---

banded inv(9) chromosome from a different metaphase spread of the same individual as a, viewed in the light microscope (inset) and SEM. The entire secondary constriction region has been transferred to the p arm from the q arm. A distinct line of division (arrow) between the smooth, tightly packed area and the loosely packed fibrous area marks the extent of the C-banded region; bar = 1 μ. (Reprinted with permission from Jack *et al.*, 1985.)

matin of constricted morphology, relating to the heterochromatin, is observed and is completely absent from the q arm in all metaphases examined (Fig. 29a). In C-banded chromosomes, the heterochromatic region appears as a smooth, tightly packed area in the SEM, as described previously. This morphology representing the heterochromatin is located in the p arm of all metaphases examined. Therefore, correlative light microscopy and SEM provides a useful method of determination of partially or completely inverted polymorphic chromosomes.

Sister Chromatid Exchanges

Sister chromatid exchanges (SCEs) have been used largely in the past as a measure of carcinogenicity or mutagenicity of a particular compound (Perry and Evans, 1975), and as an environmental monitor, for example, to detect the effects of cigarette smoking (Lambert *et al.*, 1982). Unusual SCE properties have also been reported for a number of disease states in humans (Evans, 1982) of which the most important is Bloom's syndrome.

Chromosomes visualized at metaphase have already undergone replication and consist of two sister chromatids, each containing a single identical DNA duplex. An SCE is a reciprocal exchange between strands of similar polarity, involving all four DNA strands of these two sister chromatids. SCE studies differentiate the two sisters, by harlequin staining, allowing SCEs to be observed in the light microscope. The chromatids stain lightly or intensely dependent on the extent of BrdUrd incorporation into the DNA, BrdUrd substitutes for thymidine. At the second cell division with BrdUrd in the culture, one chromatid has doubly BrdUrd-substituted DNA. This chromatid is unable to retain Giemsa stain and is, therefore, pale. The second chromatid has only singly substituted DNA and stains intensely (Michalova *et al.*, 1978; Speit, 1984). Therefore, in the light microscope the two chromatids may be readily identified (Fig. 30, inset). This allows SCEs to be detected.

For SEM investigation of chromosome structural changes associated with the formation of SCEs, it is essential to know which of the chromatids is doubly substituted with BrdUrd. The only way to determine this is by making use of the light microscope staining properties. Therefore, correlative light microscopy and SEM clearly defines the two chromatids. Using this procedure a pronounced structural difference is shown between the two chromatids in the SEM (Fig. 30). The doubly BrdUrd-substituted chromatid, which remains pale after Giemsa staining (inset), corresponds to a smooth chromosome surface morphology, in which rounded protuberances are observed (Fig. 30). The intensely staining chromatid, with only singly substituted DNA, corresponds to a highly

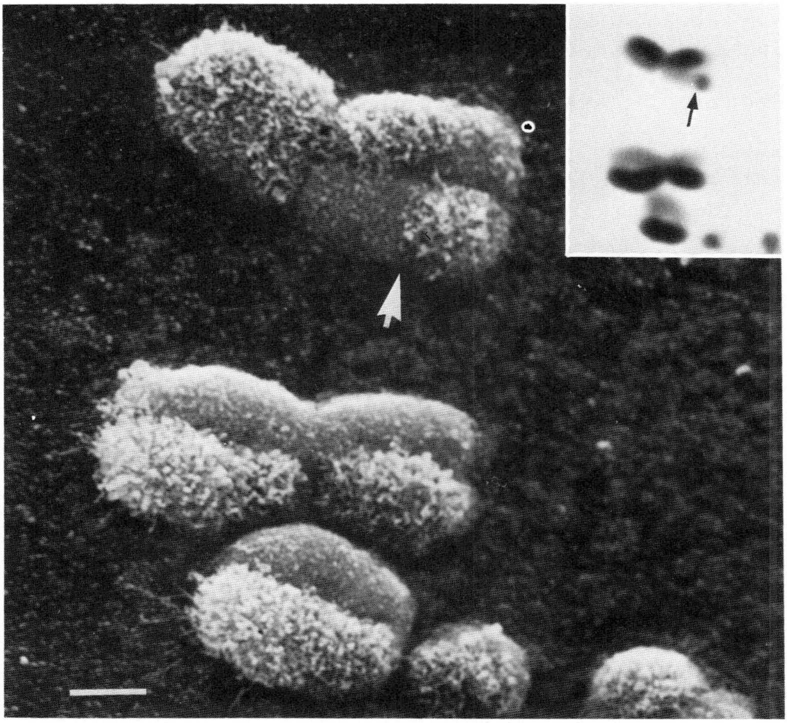

Fig. 30. Harlequin stained chromosomes observed in the light microscope (inset) and the SEM. The chromatid with doubly BrdUrd-substituted DNA stains pale in the light microscope, and the singly substituted chromatid stains intensely (inset). SCEs may be detected (black arrow, inset). When the same chromosomes are observed in the SEM the pale chromatid in the light microscope (inset) corresponds to a smooth chromosome surface in the SEM, in which rounded protuberances are observed. The intensely staining chromatid (inset) corresponds to a highly fibrous chromosome morphology in the SEM. SCE's may be readily detected in the SEM (white arrow); bar = 1 μm.

fibrous chromosome morphology (Fig. 30). The structural difference is very striking and accurately demonstrates the points of exchange between chromatids, where the SCEs are formed (arrow). These findings have enabled detailed studies to be made of the structural differences in chromatin between the two chromatids. Also, with the high-resolution available in the SEM, small SCEs may be detected which are not readily visible in the light microscope.

These applications of the correlative procedures for light microscopy and SEM are by no means exclusive. In addition to the scope of studying the whole range of chromosomal abnormalities in humans and other mammals, it is hoped that this technique may be successfully applied to chromosomes of all the available species in the animal and plant kingdom.

The work presented in this chapter was supported by grants from the North Western Regional Health Authority and the Cancer Research Campaign.

REFERENCES

Arrighi, F. E., and Hsu, T. C. (1971). Localization of heterochromatin in human chromosomes. *Cytogenetics* **10,** 81.

Bahr, G., Gilbert, F., Baaban, G., and Engler, W. (1983). Homogeneously staining regions and double minutes in a human cell line: Chromatin organization and DNA content. *JNCI, J. Natl. Cancer Inst.* **71,** 657.

Balicek, P., Zizka, J., and Skalska, H. (1977). Length of human constitutive heterochromatin in relation to chromosomal contraction. *Hum. Genet.* **38,** 189.

Barker, P. E. (1982). Double minutes in human tumour cells. *Cancer Genet. Cytogenet.* **5,** 81.

Brecher, S. (1977). Ultrastructure of observations of X-ray induced chromatid gaps. *Mutat. Res.* **42,** 249.

Buhler, E. M., Buhler, U. K., and Christen, R. (1983). Terminal or interstitial deletion in chromosome 8 long arm in Langer-Giedion syndrome (TRP11 syndrome)? *Hum. Genet.* **64,** 163.

Burkholder, G. D. (1975). The ultrastructure of G- and C-banded chromosomes. *Exp. Cell Res.* **90,** 269.

Burkholder, G. D. (1981). The ultrastructure of R-banded chromosomes. *Chromosoma* **83,** 473.

Burkholder, G. D., and Duczek, L. L. (1982). The effect of chromosome banding techniques on the proteins of isolated chromosomes. *Chromosoma* **87,** 425.

Caspersson, T., Zech, L., Modest, E. J., Foley, G. E., Wagh, U., and Simonsson, E. (1969). Chemical differentiation with fluorescent alkylating agents in *Vicia fibia* metaphase chromosomes. *Exp. Cell Res.* **58,** 128.

Caspersson, T., Zech, L., and Johansson, C. (1970a). Differential banding of alkylating fluorochromes in human chromosomes. *Exp. Cell Res.* **60,** 315.

Caspersson, T., Zech, L., Johansson, C., and Modest, E. J. (1970b). Identification of human chromosomes by DNA-binding fluorescent agents. *Chromosoma* **30,** 215.

Cervenka, J., Thorn, H. L., and Gorlin, R. J. (1973). Structural basis of banding patterns of human chromosomes. *Cytogenet. Cell Genet.* **12,** 81.

Chamberlin, J., and Magenis, R. E. (1980). Parental origin of de novo chromosome rearrangements. *Hum. Genet.* **53,** 343.

Chen, T. R., Kao, M. L., Marks, J., and Chen, Y. Y. (1981). Polymorphic variants in human chromosome 15. *Am. J. Med. Genet.* **9,** 61.

Christenhus, R., Buchner, T., and Pfeiffer, R. A. (1967). Visualization of human somatic chromosomes by scanning electron microscopy. *Nature (London)* **216,** 379.

Comings, D. E. (1978). Mechanisms of chromosome banding and implications for chromosome structure. *Annu. Rev. Genet.* **12,** 25.

Comings, D. E., Avelino, E., Okado, T. A., and Wyandt, H. E. (1973). The mechanism of C- and G-banding of chromosomes. *Exp. Cell Res.* **77,** 469.

Cowell, J. K. (1982). Double minutes and homogeneously staining regions: Gene amplification in mammalian cells. *Annu. Rev. Genet.* **16,** 21.

Daskal, Y., Mace, M. L., Jr., Wray, W., and Busch, H. (1976). Use of direct current sputtering for improved visualization of chromosome topology by scanning electron microscopy. *Exp. Cell Res.* **100,** 204.

de la Chapelle, A., Fellman, J., and Unnerus, V. (1967). Determination of human paternity from the length of the Y chromosome. *Ann. Genet.* **10,** 60.

Drets, M. E., and Shaw, M. W. (1971). Specific banding pattern of human chromosomes. *Proc. Natl. Acad. Sci. U.S.A.* **68,** 2073.

DuPraw, E. J. (1965). Macromolecular organization of nuclei and chromosomes: A "folder-fiber" model based on whole mount electron microscopy. *Nature (London)* **206,** 338.

DuPraw, E. J. (1966). Evidence for a folder fiber organization of human chromosomes. *Nature (London)* **209,** 577.

Dutrillaux, B. (1973). Nouveau systeme de marquage chromosomique: Les bandes T. *Chromosoma* **41,** 395.

Dutrillaux, B., and Covic, M. (1974). Etude de facteurs influencant la denaturation thermique menagee des chromosomes. *Exp. Cell Res.* **85,** 143.

Dutrillaux, B., and Lejeune, J. (1971). Sur une nouvelle technique d'analyse due caryotype humain. *Ct. R. Hebd. Seances Acad. Sci.* **272,** 2635.

Evans, H. J. (1982). Induction of sister chromatid exchanges by chemical agents. *In* "Sister Chromatid Exchanges" (S. Wolff, ed.). p. 87. Wiley, New York.

Filip, D. A., Gilly, C., and Mouriquand, C. (1975). The metaphase chromosome ultrastructure. 1. Acute angle metal deposition technique as an appropriate use of shadow casting in chromosome structure organization. *Exp. Cell Res.* **92,** 245.

Fryns, J. P., Heremans, G., Marien, J., and Van den Berge, H. (1983). Langer-Giedion syndrome and deletion of the long arm of chromosome 8. Confirmation of the critical segment of 8q33. *Hum. Genet.* **64,** 194.

Golomb, H. M. (1976). Human chromatin and chromosomes studied by scanning electron microscopy: Progress and perspectives. *J. Reprod.* **17,** 1024.

Golomb, H. M., and Bahr, G. F. (1971). Scanning electron microscopic observations of surface structure of isolated human chromosomes. *Science* **171,** 1024.

Golomb, H. M., and Bahr, G. F. (1974). Human chromatin from interphase to metaphase. A scanning electron microscope study. *Exp. Cell Res.* **84,** 79.

Goyanes, V. J. (1985). Electron microscopy of chromosomes. Towards an ultrastructural cytogenetics? *Cancer Genet. Cytogenet.* **15,** 349.

Goyanes, V. J., and Mendez, J. (1982). Karyotyping chromosomes by electron microscopy. II. A method for sequential examination of spread and banded metaphases by light and electron microscopy. *Hum. Genet.* **62,** 355.

Harrison, C. J., Britch, M., Allen, T. D., and Harris, R. (1981). Scanning electron microscopy of the G-banded human karyotype. *Exp. Cell Res.* **134,** 141.

Harrison, C. J., Allen, T. D., Britch, M., and Harris, R. (1982). High resolution scanning electron microscopy of human metaphase chromosomes. *J. Cell Sci.* **56,** 409.

Harrison, C. J., Allen, T. D., and Harris, R. (1983). Scanning electron microscopy of variations in human metaphase chromosome structure revealed by Giemsa banding. *Cytogenet. Cell Genet.* **35,** 21.

Harrison, C. J., Jack, E. M., Allen, T. D., and Harris, R. (1985a). Light and scanning electron microscopy of the same human metaphase chromosomes. *J. Cell Sci.* **77,** 143.

Harrison, C. J., Jack, E. M., Allen, T. D., and Harris, R. (1985b). Investigation of human chromosome polymorphisms by scanning electron microscopy. *J. Med. Genet.* **22,** 16.

Hayat, M. A. (1976). Critical point drying. *In* "Principles and Techniques of Electron Microscopy" (M. A. Hayat, ed.), Vol. 3, Chapter 6. Van Nostrand-Reinhold, New York.

Holmquist, G. (1979). The mechanism of C-banding. Depurination and B-elimination. *Chromosoma* **72,** 203.

Holmquist, G., Gray, M., Porter, T., and Jordan, J. (1982). Characterization of Giemsa dark and light band DNA. *Cell (Cambridge, Mass.)* **31,** 121.

Hozier, J. C., Furcht, L. T., and Wendelshafer-Grabb, G. (1981). Structure of human chromosomes visualized at the electron microscopic level. *Chromosoma* **82,** 55.

Hsu, T. C. (1975). A possible function of constitutive heterochromatin: The body guard hypothesis. *Genetics* **79,** 137.

Ikushima, T., and Wolff, S. (1974). Sister chromatid exchanges induced by light flashes to 5-bromodeoxyuridine- and 5-iododeoxyuridine substituted Chinese hamster chromosomes. *Exp. Cell Res.* **87,** 15.

Ip, W., and Fischman, D. A. (1979). High resolution scanning electron microscopy of isolated *in situ* cytoskeletal elements. *J. Cell Biol.* **83,** 249.

ISCN. (1985). An international system for human cytogenetic nomenclature (D.G. Harnden and H. P. Klinger, eds.), published in collaboration with *Cytogenet. Cell Genet.* Karger, Basel.

Jack, E. M., Harrison, C. J., Allen, T. D., and Harris, R. (1985). The structural basis for C-banding. A scanning electron microscope study. *Chromosoma* **91,** 363.

Jack, E. M., Harrison, C. J., Allen, T. D., and Harris, R. (1986). A structural basis for R- and T-banding: A scanning electron microscopy study. *Chromosoma (Berl.)* **94,** 395.

Jack, E. M., Waters, J. J., and Harrison, C. J. (1987). A scanning electron microscopy study of double minutes from a human tumour cell line. *Cytogenet. Cell Genet.* (in press).

Jacobs, P. A., and Mayer, M. (1981). The origin of human trisomy. A study of heteromorphisms and satellite associations. *Ann. Hum. Genet.* **45,** 357.

Kajii, T., and Niikawa, N. (1977). Origin of triploidy and tetraploidy in man: 11 cases with chromosome markers. *Cytogenet. Cell Genet.* **18,** 109.

Kato, H., and Yosida, T. H. (1972). Banding patterns of Chinese hamster chromosomes revealed by new techniques. *Chromosoma* **36,** 272.

Kingsley-Smith, B. V. (1970). The application of scanning and transmission electron microscopy to a study of whole chromosomes. *Micron* **2,** 39.

Korf, B. R., and Diacumakos, E. G. (1978). Microsurgically extracted metaphase chromosomes of the Indian muntjac examined with phase contrast and scanning electron microscopy. *Exp. Cell Res.* **111,** 83.

Lambert, B., Lindblad, A., Holmberg, K., and Francesconi, D. (1982). The use of sister chromatid exchange to monitor human populations for exposure to toxicologically harmful agents. *In* "Sister Chromatid Exchange" (S. Wolff, ed.), p. 149. Wiley, New York.

Mace, M. L., Jr., Daskal, Y., and Wray, W. (1978). Scanning electron microscopy of chromosome aberrations. *Mutat. Res.* **52,** 199.

McKay, R. D. G. (1973). The mechanism of G- and C-banding in mammalian metaphase chromosomes. *Chromosoma* **44,** 1.

Madan, K., and Bobrow, M. (1974). Structural variation in chromosome No. 9. *Ann. Genet.* **17,** 81.

Magenis, R. E., Overton, K. M., Chamberlin, J., Brady, T., and Lovrien, E. (1977). Parental origin of the extra chromosome in Downs syndrome. *Hum. Genet.* **37,** 7.

Michalova, K., Jami, J., Rousset, J. P., Ritz, E., and Bucchini, D. (1978). Differential staining of chromatids reveals substitution of BUdR for thymine in "old" DNA strands. *Exp. Cell Res.* **114,** 293.

Mikkelsen, M., Pousen, H., Grinsted, J., and Lange, A. (1980). Non-disjunction in trisomy 21: Study of chromosomal heteromorphisms in 110 families. *Ann. Hum. Genet.* **44,** 17.

Olson, S. B., Magenis, R. E., Rowe, S. I., and Lovrien, E. W. (1983). Chromosome heteromorphism analysis in cases of disputed paternity. *Am. J. Med. Genet.* **15,** 47.

Pathak, S., and Arrighi, F. E. (1973). Loss of DNA following C-banding procedures. *Cytogenet. Cell Genet.* **12,** 414.

Patil, S., Merrick, S., and Lubs, H. A. (1971). Identification of each human chromosome with a modified Giemsa stain. *Science* **173,** 821.

Perry, P., and Evans, H. J. (1975). Cytological detection of mutagen-carcinogen exposure by sister chromatid exchange. *Nature (London)* **258,** 121.

Perry, P., and Wolff, S. (1974). New giemsa method for the differential staining of sister-chromatids. *Nature (London)* **251,** 156.

Robinson, J. A. (1973). Origin of the extra chromosome in trisomy 21. *Lancet* **1,** 131.

Scheid, W., and Tarut, H. (1971). Visualization by scanning electron microscopy of achromatic lesion ("gap") induced by X-rays in chromosomes *Vicia faba. Mutat. Res.* **11,** 253.

Schmiady, H., and Sperling, K. (1976). Length of human C-bands in relation to the degree of chromosome condensation. *Hum. Genet.* **49,** 243.

Schnedl, W. (1971). Banding patterns of human chromosomes. *Nature (London), New Biol.* **233,** 93.

Seabright, M. (1971). A rapid banding technique for human chromosomes. *Lancet* **2,** 971.

Sehested, J. (1974). A simple method for R-banding of human chromosomes showing a pH dependent connection between R and G-bands. *Humangenetik* **21,** 55.

Selezneva, T. G., Deryaguin, G. V., Badaev, N. S., and Prokof'jeva-Bel'govskaya, A. A. (1977). The analysis of eu- and heterochromatin spiralization dynamics in heteromorphic homologous chromosomes no. 1 in man. *Cytologia* **3,** 298.

Speit, G. (1984). Considerations on the mechanism of differential Giemsa staining of BrdU-substituted chromosomes. *Hum. Genet.* **67,** 264.

Stubblefield, E. J., and Klevecz, R. (1965). Synchronization of Chinese hamster cells by reversal of colcemid treatment. *Exp. Cell Res.* **40,** 660.

Sumner, A. T. (1972). A simple technique for demonstrating centromeric heterochromatin. *Exp. Cell Res.* **75,** 304.

Sumner, A. T., Evans, H. J., and Buckland, R. A. (1971). New technique for distinguishing between human chromosomes. *Nature (London) New Biol.* **232,** 31.

Sweney, L. R., Lam, L. F. H., and Shapiro, B. L. (1979). Scanning electron microscopy of uncoated human metaphase chromosomes. *J. Microsc. (Oxford)* **115,** 151.

Uchida, I. A. (1973). Paternal origin of the extra chromosome in Downs syndrome. *Lancet* **2,** 1258.

Utakoji, T. (1972). Differential staining patterns of human chromosomes treated with potassium permanganage. *Nature (London)* **239,** 168.

Utsumi, K. R. (1982). Scanning electron microscopy of Giemsa-stained chromosomes and surface spread chromosomes. *Chromosoma* **86,** 683.

Viegos-Pequignot, E., and Dutrillaux, B. (1978). Une methode simple pour obtenir des prophases et des prometaphases. *Ann. Genet.* **21,** 122.

Wachtler, F., and Musil, R. (1980). On the structure and polymorphism of the human chromsome 15. *Hum. Genet.* **56,** 115.

Wang, H. C., and Federoff, S. (1972). Banding in human chromosomes treated with trypsin. *Nature (London), New Biol.* **235,** 52.

Webber, L. M., and Garson, O. M. (1983). Fluorodeoxyuridine synchronization of bone marrow cultures. *Cancer Genet. Cytogenet.* **8,** 123.

Xu, S., and Wu, M. (1983). Electron microscopy of G-banded mitotic chromosomes. *Chromosoma* **88,** 237.

Yunis, J. J., Roldan, L., Yasmineh, W. G., and Lee, J. C. (1971). Staining of satellite DNA in metaphase chromosomes. *Nature (London)* **231,** 532.

Yunis, J. J., Sawyer, J. R., and Ball, D. W. (1978). The characterization of high-resolution G-banded chromosomes of man. *Chromosoma* **67,** 293.

Yunis, J. J., Bloomfield, C. D., and Ensrud, K. M. (1981). All patients with acute non-lymphocytic leukaemia may have a chromosomal defect. *N. Engl. J. Med.* **305,** 135.

Yunis, J. J., Oken, M. M., Kaplan, M. E., Ensrud, K. M., Howe, R. R., and Theologides, A. A. (1982). Distinctive chromosomal abnormalities in histological sub-types of non-Hodgkins lymphoma. *N. Engl. J. Med.* **307,** 1231.

11

Correlative Light and Electron Microscopy on the Same Epoxy Section

CONLY L. RIEDER AND SAMUEL S. BOWSER

Wadsworth Center for Laboratories and Research
New York State Department of Health
Albany, New York
and
School of Public Health Sciences
State University of New York
Albany, New York

CORRELATIVE MICROSCOPY IN BIOLOGY:
INSTRUMENTATION AND METHODS

INTRODUCTION

The image quality obtained with the light microscope (LM) from sections of cells and tissues is dependent on the section thickness.* In general, the thinner the section the better the detail and resolution in the image because structures are less likely to be hidden by constituent overlap within the section thickness (e.g., Bennett *et al.*, 1976; Franklin, 1982, 1983; Rieder and Bowser, 1983). Thus, structural details and relationships that are not apparent in sections thicker than 2.0 μm may be visible in thinner sections prior to or after staining. Such sections can also be used in conjunction with immunocytochemical (e.g., DeMey, 1983; Lucocq and Roth, 1984; Kersken *et al.,* 1984) or histochemical (McGee-Russell, 1968; Aparicio and Marsden, 1968; see also Glauert, 1965) staining techniques to determine the composition of cellular components at a high resolution LM level.

In the past, the LM data obtained from epoxy sections thicker than 0.25 μm was corroborated, at the electron microscope (EM) level, by cutting adjacent 50- to 80-nm sections (e.g., McNary *et al.,* 1964; Rubin and Schwartz, 1984; Llewellyn-Smith *et al.,* 1985) or by reembedding the section for subsequent thin sectioning (e.g., Larramendi, 1985; Milroy, 1985). Many transmission electron microscopes are now available, however, which can be used to examine biological material directly within 0.25- to 1.0-μm sections.

When compared to traditional 50- to 80-nm sections, 0.25- to 1.0-μm sections possess a number of inherent advantages for ultrastructural studies (reviewed in Rieder *et al.,* 1985). Structures and structural relationships between cellular constituents may be clearly visible in thicker sections which are not visible or go unnoticed in thinner sections. The thicker sections also facilitate studies requiring an analysis of (or a reconstruction

* The terminology used to categorize a particular range of section thicknesses is confusing. The "semithin" 1- to 2-μm-thick epoxy section to the light microscopist (e.g., Franklin, 1983; Lucocq and Roth, 1984) is clearly a "thick" section to the electron microscopist (e.g., King *et al.*, 1980; Rieder, 1981; Bergeron and Thiéry, 1981). The same-section correlative light and electron microscopic methods outlined in this chapter are applicable to sections of various thicknesses. To avoid a semantic tangle we will describe these sections according to their nominal thickness, as determined from the microtome setting, not by the imprecise terms in current use (see Bowser and Rieder, 1986).

from) serial sections since fewer sections are needed to identify or reconstruct the structure of interest. In addition, conventionally (uranyl and lead) stained 0.25- to 1.0-μm sections are frequently the optimum thickness for stereomicroscopy. Finally, as noted above, biological material within these sections can be visualized with excellent resolution of detail by LM.

The ability to study biological material within 0.25- to 1.0-μm sections by high resolution LM, prior to examining the same section in the EM, provides two distinct advantages for the electron microscopist. First, since each section can be rapidly prescreened for content by LM, nonuseful sections need not be examined in the EM. In some cases this can provide enough of a timesaving advantage that a study can be conducted which may not otherwise be practical. Second, sections of cells which have been stained with fluorescent probes or treated to reveal endogenous fluorescent moieties, prior to epoxy embedding (Peppe and Fink, 1961; Hokfelt, 1965; Rubin and Coslo, 1978) or after sectioning (Kersken *et al.*, 1984), can be examined by fluorescence LM (FLM). Using this approach, the distribution of fluorescence within sections as thin as 50–80 nm can then be correlated with the ultrastructure of the same section (Rieder and Bowser, 1985). This same-section correlative FLM–EM procedure provides a rapid and simple first approach to many problems which require the ultrastructural localization of specific fluorescent probes.

The purpose of this chapter is to outline the methods involved in, and to illustrate the advantages of, same-section LM–EM. It is hoped that the information presented here may provide the reader with new insight concerning how best to approach a particular biological problem in which ultrastructural information is required.

METHODOLOGY

Same-Section Correlative Light and Electron Microscopy

Once mounted on Formvar-coated slot grids, the biological material in 0.25- to 0.75-μm plastic sections can be visualized, with excellent resolution of detail and without special coloring stains, by phase-contrast, differential-interference contrast, and in some cases darkfield LM. Under dry 40× phase objectives it is relatively simple to distinguish cell boundaries, nucleoli, nuclei, mitochondria, and even the centrosomes comprising the poles of mitotic spindles in uranyl and lead stained sections (Figs.

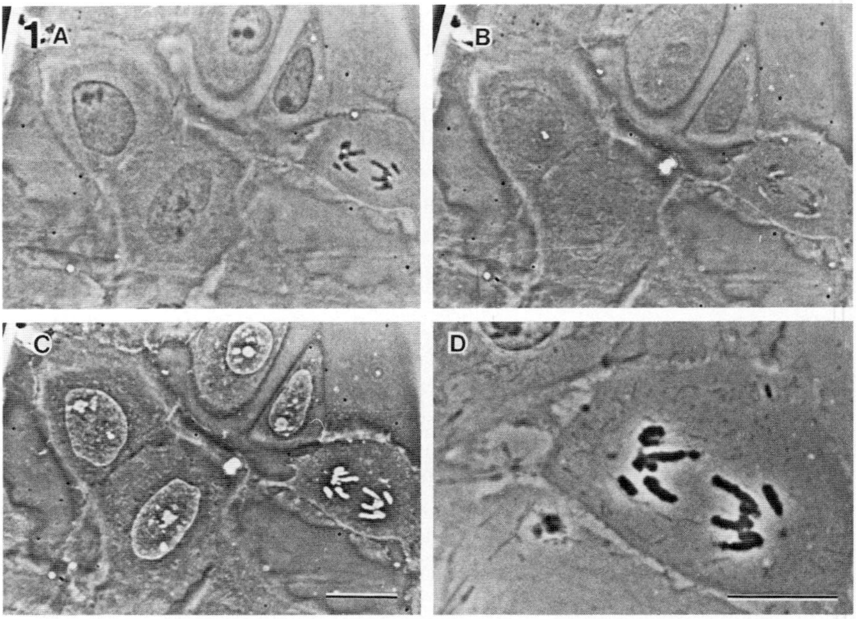
1 A
B
C
D

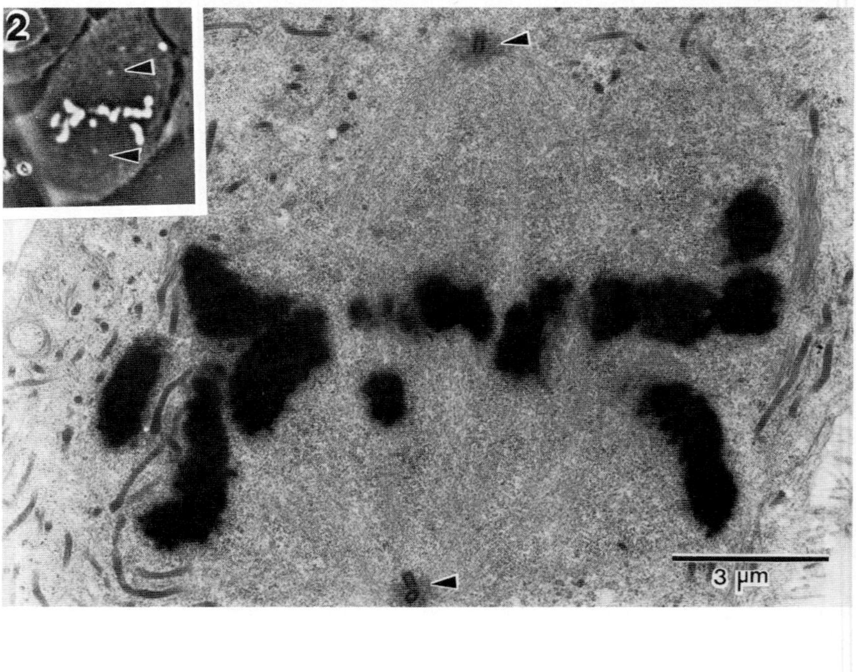
2
3 μm

1 and 2). Higher resolution is obtained by examining these sections, mounted on slot grids and immersed in glycerol, with 100× oil-immersion objectives (Fig. 1D). Finally, a further augmentation of resolution can be achieved by applying video image processing techniques (Allen, 1985; Inoué, 1986).

The approaches outlined in this chapter are particularly useful for single cell correlative LM–EM studies (e.g., observing the same cell *in vivo* and then in the EM). Thus, most of the methods and illustrations presented will be of material prepared on coverslips. However, by no means is the method limited to such studies, and the advantages of the technique (see section on Applications) can be fully appreciated in tissue samples as well.

Fixation and Embedding

Specimens can be fixed by any method proven suitable by previous thin section studies. Depending on the application, one may wish to consider the influence of the fixative on the final image contrast. For example, in *unstained* sections, material fixed only in glutaraldehyde shows better contrast than does material postfixed with OsO_4 (Rieder and Bowser, 1983). However, the addition of metals with a high Z number during section staining changes the refractive index of the biologic material relative to the embedding resin. Thus, for most purposes the fixation protocol is not critical for LM since both osmicated and nonosmicated material show excellent contrast after uranyl and lead staining (Figs. 1–5).

Fixed preparations are dehydrated in a graded series of ethanol or acetone. Depending on the application, they are then embedded by standard procedures in Araldite, Epon, Epon-Araldite, Spurr's, or Lowicryl

---

Fig. 1. A–C, negative (anoptral) phase-contrast light micrographs of the same 0.25-μm section from a monolayer of PtK_1 cells, fixed only in 3% glutaraldehyde (in 0.1 *M* phosphate buffer, pH 7.1) and embedded in Polybed 812. A, unstained. B, stained with uranyl acetate. C, further stained with lead citrate. Note that the phase-dense cell components lose their contrast in response to uranyl staining but undergo a dramatic reversal in contrast in response to lead staining. (D) 100× oil-immersion positive phase-contrast light micrograph of the section adjacent to that shown in A–C. This section has been stained with uranyl and lead. Note the visibility of mitochondria, mitotic spindle poles, and cell boundaries. Bar in C, 25 μm; D, 10 μm. (From Rieder and Bowser, 1983; reproduced by permission of Oxford Univ. Press.)

Fig. 2. Electron micrograph of a 0.25-μm section through the spindle of a mitotic PtK_1 cell fixed as in Fig. 1 and embedded in Polybed 812. This section has been stained in uranyl and lead. Both spindle poles (arrows) are contained in this section, a fact which was determined initially by negative phase-contrast light microscopy (inset). (From Rieder and Bowser, 1983; reproduced by permission of Oxford Univ. Press.)

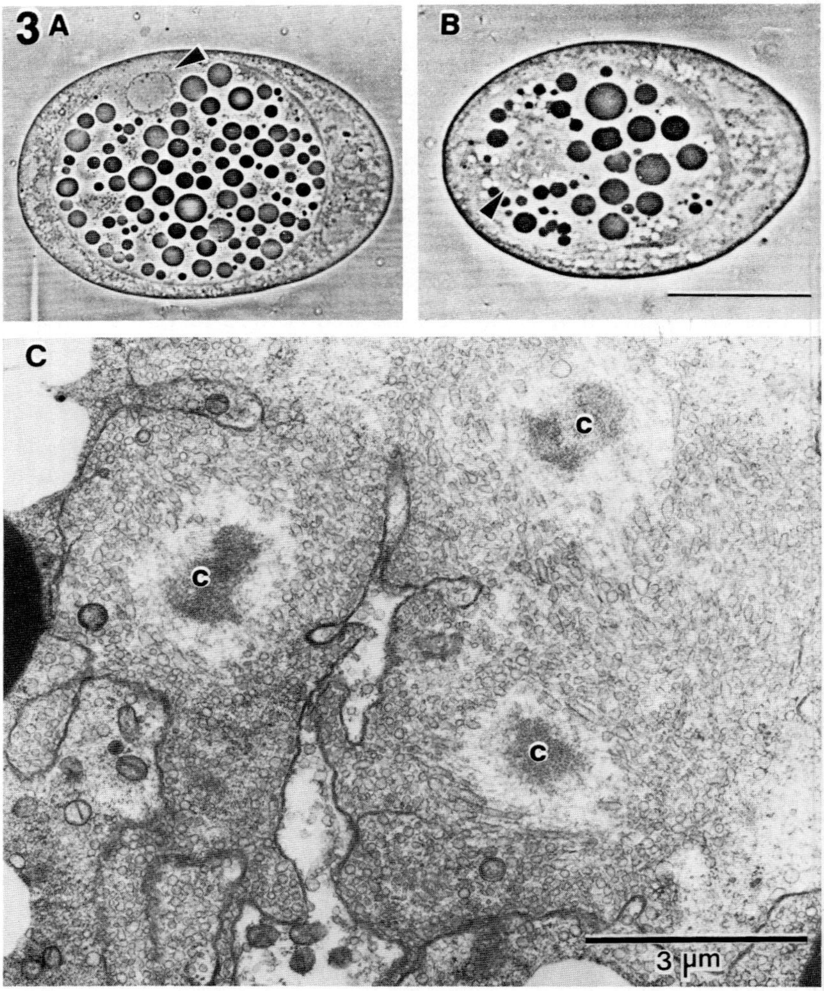

Fig. 3. A, B, phase-contrast light micrographs of 0.25-μm Epon-Araldite sections, stained with uranyl and lead, from interphase (A) and meiotic (B) *Xenos* oocytes. These cells were fixed in a 50 : 50 mixture of 4.5% glutaraldehyde and 2% OsO_4, each in 0.1 *M* phosphate buffer (pH 7.1). The interphase and meiotic nuclei are noted by arrowheads. C, electron micrograph of the early prometaphase meiotic apparatus from the section pictures in B. See text for details. c, Chromosome; bar in B, 25 μm. (From Rieder *et al.*, 1983; reproduced by permission of San Francisco Press, Inc., Box 6800, San Francisco, CA 94101.)

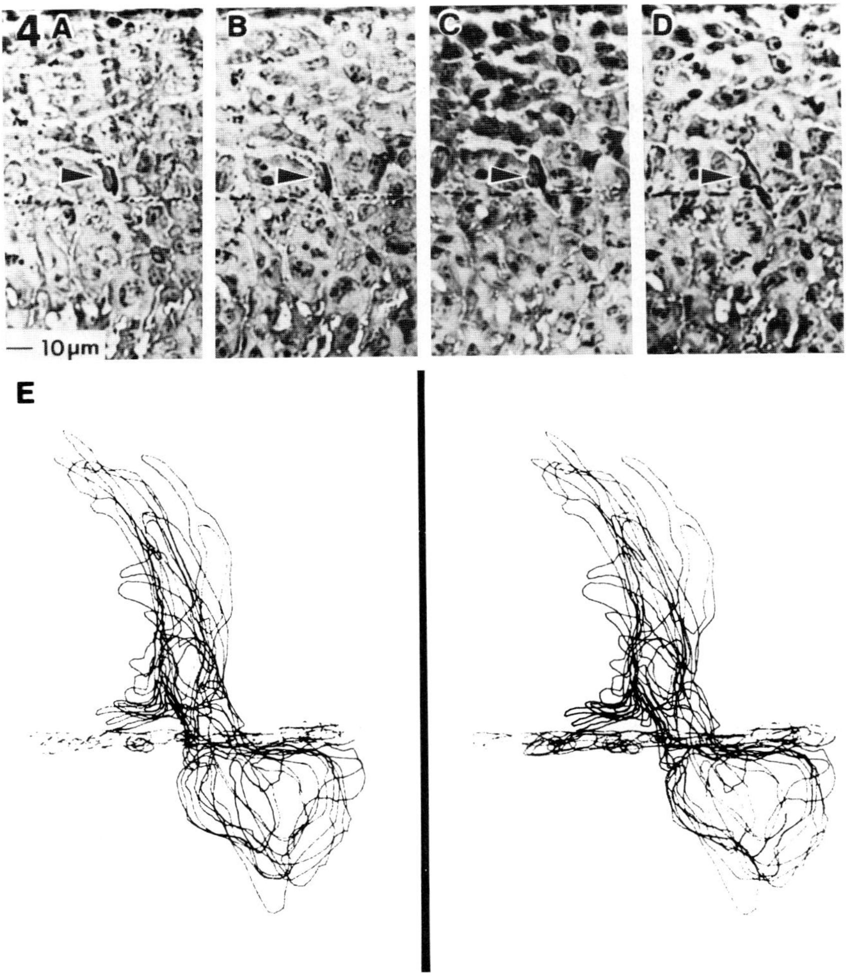

Fig. 4. A–D, phase-contrast light micrographs of serial 0.75-μm sections, stained in uranyl and lead, from rat peritoneum invaded by Ascites carcinoma. This material was fixed in 2.5% glutaraldehyde/2% paraformaldehyde, postfixed in 1% OsO_4, and embedded in Polybed 812. A very dense elongated cell is associated with a less dense cell (arrow). In consecutive sections it is clear that the dense cell is traversing the medial surface of the elastin–collagen reticulum (dense line, from left to right). E, stereo pair of a graphic reconstruction made from serial electron micrographs of the dense cell pictured in A–D. (From Parsons *et al.*, 1985; reproduced by permission of Longman Group Ltd.)

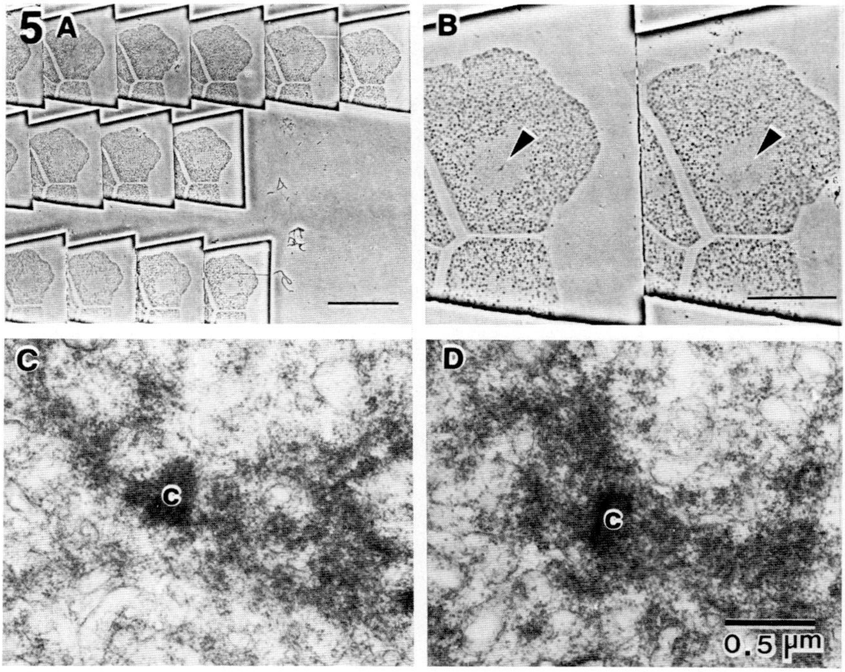

Fig. 5. A, low magnification phase-contrast light micrograph of serial 0.25-μm sections, stained in uranyl and lead, from a late anaphase *Lytechinus* blastomere which has undergone a monopolar–bipolar transition. This egg was fixed with 1% OsO_4 in 0.4 *M* Na-acetate buffer (pH 6.1) and embedded in Epon-Araldite. Scanning these sections at higher magnification reveals centrosomes (arrowheads in B). C and D are serial electron micrographs of the centriole-containing centrosome pictured in B. c, Centriole; bar in A, 0.1 mm; B, 50 μm. (From Rieder *et al.*, 1983; reproduced by permission of San Francisco Press, Inc., Box 6800, San Francisco, CA 94101.)

resins. For routine studies we use a fairly soft Epon-Araldite formulation described in detail by Rieder *et al.* (1985). After dehydration, we typically incubate the specimens overnight in a 1 : 1 (v/v) mixture of solvent (absolute ethanol or acetone) : Epon-Araldite lacking catalyst. Most of the solvent has evaporated by the next day, and the preparation is then further infiltrated with two changes (4 hr each) of uncatalyzed resin followed by two changes of freshly catalyzed resin (1–2 hr each). Alternatively, the more common method of clearing in a transitional fluid (i.e., propylene oxide) prior to infiltration can be employed.

We flat-embed most specimens on No. 1-1/2 glass coverslips, which can be easily removed after the epoxy has hardened. The advantage of this flat-embedment is that thin wafers can be produced which have excellent optical properties. In addition, the specimen usually rests against the coverslip as the resin cures and, as a result, it is found on the surface of the plastic (i.e., immediately accessible to cutting once the coverslip is removed). Fully infiltrated specimens, which were not initially prepared on coverslips (e.g., eggs, tissue chunks, etc.), are simply placed on a coverslip during the final embedding stage. In all cases the coverslip preparations are maintained on a *level* plane within the 60°C curing oven by placing them on a suitable holder, e.g., one made by gluing a cluster of three BEEM capsules to the bottom of a disposable Petri dish.

Trimming and Sectioning

After the resin has hardened, the coverslips are removed from the flat embedments by etching in cold hydrofluoric acid (Moore, 1975). Once the coverslip is etched away (typically 2–10 min) the plastic wafer is rinsed thoroughly with water and then dried by placing the embedment into a 60°C oven for several hours. Alternatively, the wafer can be separated from the coverslip by taking advantage of the difference in expansion coefficients between glass and plastic, e.g., by plunging the preparation into liquid nitrogen or by placing it on dry ice. However, these methods often crack the wafer and can ruin the area of interest. We therefore use the thermal separation method only on specimens embedded in water-soluble resins, e.g., Lowicryl.

After removing the coverslip the resin wafer containing the specimen is taped to a slide, specimen side up, and placed on the stage of a phase-contrast microscope. Specimens previously followed *in vivo* are selected from the embedment, circled with a diamond objective scribe (Carl Zeiss, West Germany), excised from the embedment, and glued with a cyanoacrylate-based cement onto Epon pegs (e.g., Collins, 1979). The pegs are placed in a Silastic holder and illuminated by a combination of epi- and transmitted light directed from a fiber-optical source (Rieder and Bowser, 1984). Using this system one can easily visualize with a dissecting microscope embedded objects as small as individual mammalian cells (and mitotic chromosomes) within a monolayer. The block containing the circled specimen is then trimmed using an acetone-cleaned, high quality, single-edge razor blade (e.g., GEM super stainless; American Safety Razor Co., Staunton, VA). Two factors should be considered during trimming. First, it is important to keep the blockface as small as practical to avoid unnecessary section compression during microtomy. Second, the leading and trailing edge of the block face should be parallel so that straight ribbons of sections are produced. When these two factors are optimized, many serial sections can be mounted on a single slot grid.

When three-dimensional reconstructions are being made from serial sections, the sections should be in a linear ribbon since the perspective of a specimen within a nonlinear ribbon of sections changes from section to section if the specimen is not perpendicular to the beam (e.g., when tilted—see Rieder, 1981). Finally, the block face can be sonicated briefly in detergent, rinsed in distilled H_2O, and then air-dried prior to sectioning in order to eliminate possible sources of contamination (e.g., from oils or other adhering debris).

We cut serial 0.25- to 0.75-μm sections with a diamond knife using a Sorvall MT 2B or MT 6000 ultramicrotome. (In this respect it should be noted that we have cut thousands of 0.25- to 0.75-μm sections on the same diamond knife over the past 4 years without a resharpening and without a noticeable change in section quality.) Compression of 0.25- to 0.75-μm sections, introduced during cutting, is more severe than in thin sections, and chatter results if the plastic is too hard and/or if the block face is too large or if the cutting speed is excessive.

Ribbons of 0.25-μm sections display a pale blue interference color after expansion with a heat pen, or $CHCl_3$, while 0.50- to 0.75-μm sections appear bright red (Locke and Krishnan, 1971). These ribbons are mounted on Formvar-coated 1 × 2 mm slot grids by methods previously outlined in Rieder (1981). With intermediate- or high-voltage EMs, thicker [cast from 0.5 to 0.8% (w/v) Formvar in ethylene dichloride; gold interference color] support films can be used without compromising resolution. This extra durability allows for the routine use of slot grids which is essential in obtaining high-resolution LM images of biological material within 0.25- to 0.75-μm sections (see below).

Staining for Electron Microscopy

In some cases it may be advantageous to stain the material *en bloc* before infiltrating with the embedding resin. The most commonly used stains are uranyl acetate and phosphotungstic acid dissolved in H_2O or alcohol. Aqueous stains can be applied after fixation but before dehydration, whereas alcoholic stains are used during dehydration. *En bloc* staining appears to act synergistically with section staining to yield additional specimen contrast (at both the LM and EM levels). It may also help to stabilize delicate structures against the deleterious effects of dehydration and infiltration. *En bloc* staining should *not* be used, however, when specimens are prepared for same-section correlative FLM–EM, since heavy metals may quench the fluorescence of the fluorochrome-coupled probe.

As previously noted, unstained material, whether or not it has been postfixed in OsO_4, is frequently visible by LM in 0.25- to 0.75-μm sections. This can prove particularly advantageous for monitoring progress toward a desired region or structure within the embedded material during microtomy. For this reason we have an inverted phase-contrast microscope immediately adjacent to our microtome. The best contrast is, however, obtained after uranyl and lead staining (Fig. 1; Rieder and Bowser, 1983). Methods for staining 0.25- to 0.75-μm sections have recently been reviewed (Rieder *et al.,* 1985) and will not be detailed here. In general, 0.25- to 0.75-μm Epon or Epon-Araldite sections are adequately stained by incubating the sections in 3–4% aqueous uranyl acetate for 2–4 hr at 60°C. After rinsing in distilled H_2O, the sections can be poststained in lead citrate for 25–30 min at room temperature.

Light Microscopy and Photomicroscopy

Slot grids containing ribbons of stained or unstained sections should be placed section-side up on a clean, dry microscope slide. The sections can be quickly screened, without placing a coverslip over the grid, by phase-contrast or fluorescence microscopy. For more detailed examination, oil-immersion objectives are typically employed. The following procedure is used to mount slot grids and examine sections with oil-immersion objectives. A small drop (ca. 10 μl) of glycerin is placed on a clean microscope slide and the slot grid is carefully placed, section-side up, within or on top of this drop. A second drop (ca. 10 μl) of glycerin is then placed on a clean oversized (e.g., 24 × 40 mm) coverslip, which is quickly inverted and placed on top of the grid on the microscope slide. The grid is thus tightly sandwiched in a small volume of glycerin between the coverslip and microscope slide. The condenser and objective can then be oiled, using the *minimum* amount of immersion oil, to this preparation to eliminate any glass/air interface. After examination and photography, the excess oil is blotted from the preparation and then the coverslip is gently pushed laterally off the slide. The grid, which may remain attached to either the slide or the coverslip, is flooded with distilled H_2O and removed with jeweler's forceps. It is then freed of glycerin by repeated distilled H_2O rinses. Care must be used when removing the grid to prevent immersion oil from mixing with the glycerin. Should this occur, the grid can be rinsed briefly in 1% Triton X-100 in 95% ethanol to remove the oil and then rinsed in distilled H_2O.

For maximum contrast and resolution the microscope must be properly aligned for Köhler illumination. We use a Nikon Optiphot microscope equipped with phase-contrast and epifluorescence optics. For recording images we use a Nikon UFX 35-mm automatic exposure system or a Dage MTI-66 SIT camera which is coupled to a digital image processor (Nippon Avionics) and a Panasonic 1/2″ time-lapse VCR. The video system is useful in those cases where contrast enhancement is necessary (e.g., for sections thinner than 0.25 μm), or where simultaneous viewing by others is desired. We photograph phase-contrast and darkfield images of 0.25- to 0.75-μm sections with green-filtered light on Kodak Technical Pan film (ASA 64), which is developed in Rodinal (1 : 50; 9 min). Fluorescence images are photographed on Kodak Tri-X film using an ASA setting of 3200. This film is push-processed in Perfection XR-1 developer (Perfection Photographic Products, Inc., Sun Valley, CA) for 16 min using the temperature drift method at an initial temperature of 30°C. For maximum contrast it is essential that the film be agitated continuously during the first minute of development, and periodically thereafter as described by the manufacturer.

Electron Microscopy

Serial 0.25- to 0.75-μm sections can be viewed and photographed, with good resolution and contrast, in a high-voltage EM, in an intermediate-voltage EM, or in one of the new generation 100-kV EMs (e.g., Philips 420) equipped with the so-called achromatic imaging lens system. In some cases 0.25-μm sections can also be examined and photographed in many of the older 100-kV EMs (e.g., Philips 300, 301; Siemens 102), but the adequacy of the data obtained from these instruments will depend on the type of specimen and on the specific information required. The older microscopes, for example, frequently give an adequate image of 0.25-μm sections from preparations lysed for FLM, but fine details such as cytoskeletal elements are typically lost in unlysed, dense cells. However, by optimizing the condenser lenses and apertures to maximize the illumination, and by using a small (10 μm) objective aperture, we have frequently been able to obtain adequate resolution of even thin cytoskeletal elements (e.g., microtubule protofilaments) in 0.25-μm sections of unlysed cells using the older instruments.

Same-Section Correlative Fluorescent-Light and Electron Microscopy

Various methods have been developed to stain cells and tissues, at both the LM and EM level, with specific molecular probes. All of these approaches are subject to the conflicting demands of good probe penetration into the specimen and optimum ultrastructural preservation. Postembedding methods, for example, provide good ultrastructural preservation, but they are plagued by very poor probe penetration since they only stain the surface of the section (e.g., Bendayan, 1984; Kersken *et al.*, 1984; Hearn *et al.*, 1985). Alternatively, reversible embedment methods (e.g., Parysek *et al.*, 1984; Gorbsky and Borisy, 1985; Coulter and Elda, 1986), in which the embedding medium is removed after sectioning and before staining with the probe, provide good probe penetration but relatively poor ultrastructural preservation. The preembedding approach utilized extensively in our work essentially represents a compromise between postembedding and reversible embedding approaches. In general this method requires that the material be gently lysed by detergents prior to, during, or after glutaraldehyde fixation, or extracted in alcohol after fixation, and then stained with the probe prior to embedding (e.g., DeBrabander *et al.*, 1979, 1981; DeMey *et al.*, 1981; Rieder and Bowser, 1985). When fluorescent probes are used it also requires that the material *not* be postfixed in OsO_4. In addition to its simplicity, one of the major advantages of staining material with fluorescent probes, prior to embedding, is that the potential disruption or masking of specific target sites during the embedding (and de-embedding) procedure is avoided (Eneström, 1986).

To enhance probe penetration, we typically lyse single cells and monolayers for 1 min in cytoskeleton buffer (PHEM; see Schliwa and van Blerkom, 1981) containing 0.15% Triton X-100. The specimens are then immediately placed in 1% glutaraldehyde in PHEM for 2–3 min followed by a 20-min fixation in 0.1% glutaraldehyde in PHEM (see Rieder and Bowser, 1985). For some individual cells and monolayers it is feasible to fix with glutaraldehyde first, and then to permeabilize and label with specific probes (e.g., Weber *et al.*, 1978; DeMey *et al.*, 1981; Rupp *et al.*, 1986). When possible this approach is highly desirable since the distribution of the target material will be closer to the *in vivo* situation, i.e., it is more likely that the target has not been redistributed or lost during the lysis step. Regardless of the exact protocol, free aldehydes in the extracted and fixed cells must be reduced, e.g., by treatment with $NaBH_4$, glycine, or ammonium chloride. These preparations are then labeled with the specific fluorescent probe as described below.

Immunolabeling

Specimens which have been made permeable and then fixed as described above are incubated for 30 min at 37°C in phosphate- or Tris-buffered saline containing goat serum (or any other suitable treatment, e.g., Tween-20, which blocks the nonspecific binding of antibodies). These specimens are then incubated in the primary antibody (e.g., tubulin, WTV outer-coat protein, keratin, etc.) for 45 min at 37°C, thoroughly rinsed in buffer, followed by a 30-min incubation in the appropriate fluorescein-, rhodamine-, or Texas red-conjugated secondary antibody. Alternatively, the primary antibody can be detected using fluorescently tagged protein A. However, if this method is employed it is important *not* to use nonimmune serum in the blocking step! When desired the secondary antibody can be conjugated to colloidal gold. Under these conditions a tertiary fluorescently tagged antibody, directed against the secondary antibody, can be used to reveal the distribution of gold at the LM level (see Alexander *et al.,* 1985). After immunolabeling, the cells are rinsed in buffer, dehydrated, embedded, and examined as described in the previous section on LM–EM.

Other Fluorescent Probes

In the preembedment labeling approach described above, fluorescein-, rhodamine-, or Texas red-conjugated antibodies remain bound to their antigenic targets and remain fluorescent after dehydration and embedding in Spurrs, Araldite, Epon, Epon-Araldite, or Lowicryl (Peppe and Fink, 1961; Eneström and Hed, 1982; Rieder and Bowser, 1985; unpublished). Similarly, the aldehyde-induced yellow fluorescence of certain amines is also stable to embedding in Epon and/or Araldite (Hokfelt, 1965; Rubin and Cosio, 1978; Rubin and Schwartz, 1984). Unfortunately, these successes cannot be automatically extrapolated to other fluorochromes or fluorescent probes. For example, the fluorescence of R-phycoerythrin-conjugated antibodies is irreversibly lost during dehydration in acetone or ethanol or after mounting in media containing glycerol or polyvinyl alcohol (G. Rupp and C. L. Rieder, unpublished). In addition, the actin-specific staining of rhodamine–phalloidin is lost during Epon-Araldite embedding. At present it is not clear whether the phalloidin loses its affinity for actin or whether the rhodamine becomes dissociated from the phalloidin–actin complex. However, we have recently found that rhodamine–phalloidin remains bound to filamentous actin during low-temperature embedding in Lowicryl K4M (Polysciences Inc., Warrington, PA). More importantly, the long wavelength UV light used during the photopolymerization of this resin does not quench the rhodamine fluorescence (S. S. Bowser, unpublished). Thus, low-denaturation and water-soluble embedding resins may prove more successful when standard epoxy embedding fails, and we are currently exploring the utility of such low-denaturation techniques with a variety of fluorescent probes.

APPLICATIONS

Sections as thick as 10 μm can be imaged, with good resolution of detail, with the high-voltage EM. However, all the details within the thickness of a conventionally stained section are brought into focus as a superimposed image. This overlapping of cell constituents in sections thicker than 0.5–1.0 μm often hides or obscures the structure of complicated cell components, even when viewed in stereo (see Rieder, 1981). The confusion arising from specimen overlap is a serious problem and can currently be overcome either by selectively staining the structure of interest (reviewed in Yamada and Ishikawa, 1981), by computer-aided tomography (e.g., McEwen *et al.*, 1986), or by using a thinner section. In this respect it has become apparent over the past few years that conventionally stained 0.25- to 0.75-μm sections are the optimal thickness for a wide variety of ultrastructural studies (reviewed in Rieder *et al.*, 1985). These sections can also be viewed, with adequate resolution, in many of the 100-kV EMs currently in use. Since these same sections can also be viewed with high resolution in the LM, the LM image of the section can be correlated with the underlying ultrastructure of the same section. Some practical applications of this same-section LM–EM approach are detailed in this section.

Screening Sections for Content with the Light Microscope

The fact that the biological material in 0.25- to 0.75-μm sections can be studied by the LM, with excellent detail and contrast (and frequently prior to uranyl and lead staining), enables one to use the LM to preliminarily screen these sections for content. In many cases this results in a significant timesaving advantage since nonuseful sections need not be examined in the EM. *This approach is especially useful for studies requiring an ultrastructural examination of a selected area or structure which is large enough to be visualized by LM, but which comprises only a small volume of the embedded material.*

Rieder and Nowogrodzki (1983; see also Rieder *et al.*, 1983) used this same-section LM–EM approach, for example, to study the ultrastructure of the meiotic spindle in oocytes of *Xenos peckii*. The eggs of this unusual insect are approximately 100 μm in diameter, and the ratio of the volume of the meiotic apparatus to the egg is 1 : 10. Thus, 90% of the sections cut from an egg were not useful to the study. However, since the egg nucleus was visible in *unstained* 0.25-μm sections, those eggs which were not in meiosis (the great majority; cf. Figs. 3A,B) could be quickly abandoned, as could those eggs in which the nucleus had already been completely sectioned. Furthermore, the ultrastructure of the egg nucleus, as well as its relationship to other structures within the egg, could be clearly documented using minimal publication space by illustrating both the LM image of the section and the electron micrograph of the nucleus (Figs. 3B,C).

Bowser *et al.* (1985) recently used this same approach to study the ultrastructure and cytochemistry of the oral zone in the foraminiferan protist *Allogromia*. Since the oral zone comprises only a small volume of the organism, many sections which were not useful for the study needed to be cut before the area of interest was reached. Prescreening ribbons of unstained 0.25-μm sections by *darkfield* LM allowed these investigators to pinpoint and examine by EM only those grids containing sections through the oral zone (not shown; see Rieder and Bowser, 1983).

The same-section LM–EM approach can also be used to facilitate the study of cell–cell interactions. For example, Rieder and Shain (1982) applied this method to examine the ultrastructure of contacts formed *in vitro* between horseradish peroxidase (HRP)-injected neurons and pinealocytes. HRP was used as an electron- (and light) opaque marker to distinguish neuronal processes from those of pineal origin. This study required extensive serial sectioning, since the neurons and their processes were found on the dorsal surfaces of pineal cells. Approximately 100 0.25-μm sections were obtained from each block, but from these only those sections containing suitably cut HRP-labeled processes, as determined by LM, were selected for examination in the EM (not shown; see Rieder and Bowser, 1983).

Investigations on the migration of inflamatory cells, embryonic cells, and tumor cells through tissue can also profit substantially from same-section LM–EM. Parsons *et al.* (1985) used this approach, for example, to study the migration of carcinoma cells through the rat peritoneum. To gain an adequate sample size in a reasonable amount of time, serial 0.75-μm sections were screened with the LM for the presence of invading cells (Fig. 4A–D). Once located, invading cells were photographed in the high-voltage electron microscope throughout the serial series. These micrographs were then used to reconstruct the cell profile graphically in three dimensions (Fig. 4E).

In some instances the ability to prescreen 0.25- to 0.75-μm sections for content with the LM allows a study to be conducted which would not otherwise be feasible. Sluder and Rieder (1985; see also Rieder *et al.*, 1983), for example, used serial 0.25-μm sections to analyze the polar (i.e., centrosomal) regions in sea urchin eggs during the unusual mitotic divisions induced by mercaptoethanol. The major goal of this correlative LM and EM study of eggs previously followed *in vivo* was to determine the number of centrioles at each pole during tetrapolar, monopolar, and bipolarized-monopolar divisions. Each centriole in the sea urchin egg is 0.25 μm wide and about 0.50 μm long, and these are contained within the volume of an egg which is approximately 100 μm in diameter (the appropriate calculations give the centriole-to-egg volume ratio as $1 : 1.25 \times 10^6$). Thus, the major technical difficulty in this study was to reproducibly locate and reconstruct a few needles (i.e., centrioles) within a haystack (i.e., egg) and to do this in a significant number of different haystacks (i.e., tetrapolar and monopolar spindles). A study using conventional 50- to 80-nm sections would require that 1000 serial sections be cut from each sea urchin egg and that most of these be examined in the EM for centrioles. The same task using 0.25-μm sections still required 250–300 sections. However, each of these sections could be rapidly screened for the presence of cenriole-containing centrosomes with the LM (Fig. 5A,B). Thus, only those sections containing centrosomes needed to be examined in the EM (Fig. 5C). Rieder *et al.* (1985) estimated that this approach eliminated two-thirds or more of the sectioning time, and more than 90% of the EM time which normally would have been required in a study using 50- to 80-nm sections.

Same-Section Correlative Fluorescence-Light and Electron Microscopy

The ultrastructural localization of a particular antigen usually involves an initial characterization of antigen distribution at the LM level after staining the cell (directly or indirectly) with the appropriate fluorescently-conjugated antibody. The antigenic site in question is then generally revealed, at the EM level, by preembedding or postembedding immunolabeling techniques that involve the use of colloidal gold, ferritin, or peroxidase reaction product as electron-opaque markers. Although these methods may provide the information desired, they are more complex than immunofluorescent staining techniques and each possesses limitations that restrict its usefulness, e.g., the peroxidase reaction product frequently masks the details of the labeled structure and is diffusable, ferritin is clearly visible only in sections thinner than 80 nm, and colloidal gold penetrates poorly into well fixed tissues (see DeMey, 1983; Horisberger, 1979).

In 1961, Peppe and Finck demonstrated that the fluorescent staining of chick muscle by fluorescently-conjugated antimyosin is not quenched after Araldite embedding and that Araldite is not autofluorescent (see also McNary *et al.,* 1964). Indeed, it has since become evident that other commonly used epoxy embedding resins (e.g., Epon, Spurr's) are not significantly autofluorescent and that immunofluorescent staining is not quenched during standard embedding procedures (Eneström and Hed, 1982; Rieder and Bowser, 1985). Thus, sections of cells, stained *prior* to embedding with fluorescein- (Fig. 6), rhodamine- (Fig. 7), or Texas red- (not shown) conjugated antibodies, show high-resolution fluorescent patterns. That these patterns are specific for the distribution of the antigen

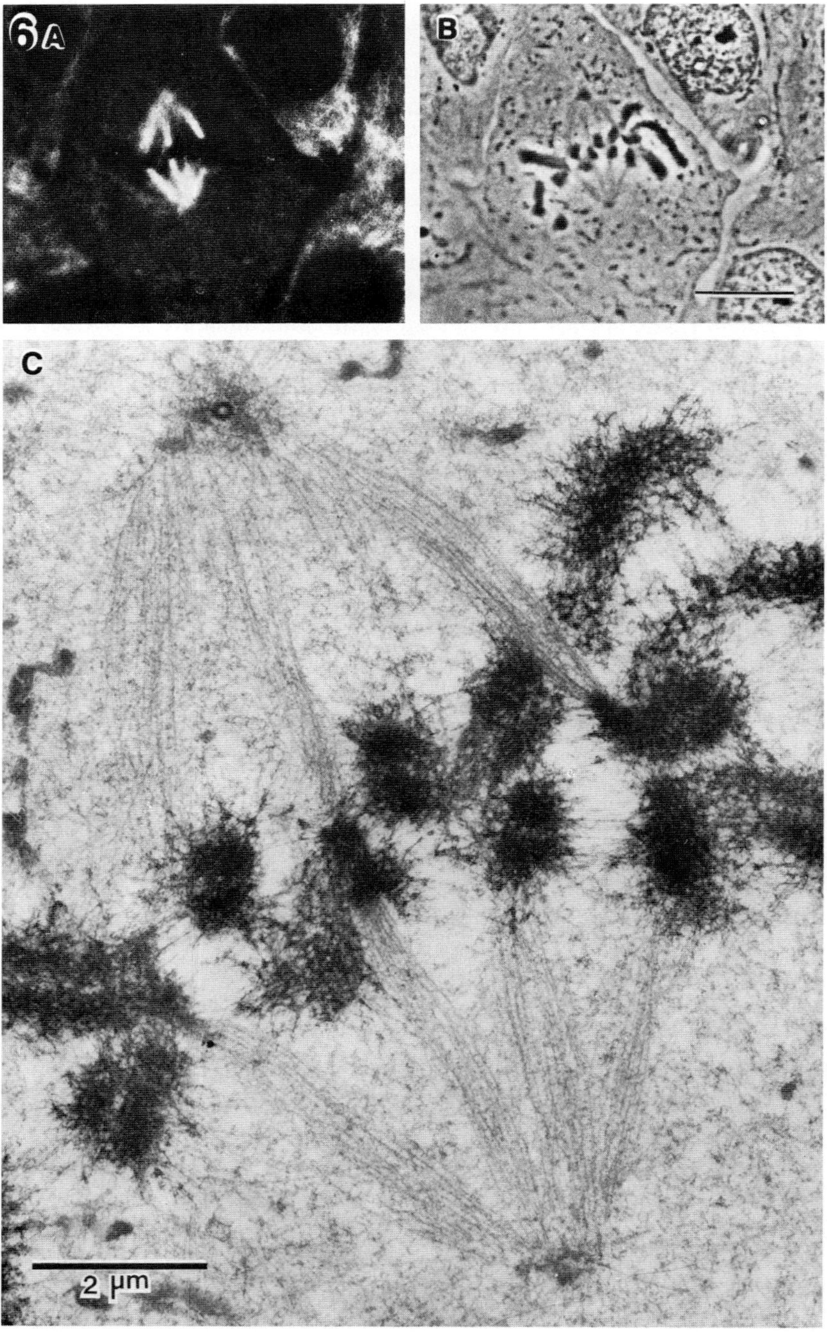
6A
B
C
2 µm

can be demonstrated by: (1) the fidelity of correlation between the fluorescent pattern of a known antigen and the corresponding ultrastructural distribution of the antigenic structure within the same section (e.g., microtubules, Fig. 6; keratin, Fig. 7), and (2) the use of fluorescent tertiary anti-immunogold antibodies to demonstrate directly the colocalization of the fluorescent and gold probes (Fig. 8; see also Alexander *et al.*, 1985).

By using the same section LM–EM approach outlined in the previous section, one can correlate the fluorescent image of a particular section with the ultrastructural data from the *same* section. This approach, therefore, makes it unnecessary to use the less precise method of correlating the underlying cellular ultrastructure, as determined from a 50- to 80-nm epoxy section, with the corresponding fluorescent image as determined from an adjacent 1- to 2-μm section (as in Peppe and Finck, 1961; Eneström and Hed, 1980, 1982; Rubin and Schwartz, 1984). In addition, the fluorescent localization of antigens in epoxy sections provides an alternative to the use of methacrylate (as in Franklin, 1982, 1983) in which specimens (especially cell monolayers) frequently become distorted during embedding (Pease, 1964).

Sections as thin as 50 nm, cut from specimens stained for the immunofluorescent localization of antigens prior to embedding, show high resolution patterns of fluorescence (Peppe and Finck, 1961; Rieder and Bowser, 1985). Unfortunately, these patterns rapidly photobleach and are therefore difficult to document by conventional 35-mm photography. However, the fluorescence within or on the surface of sections as thin as 25 nm can be easily visualized and recorded by newly developed video-enhanced LM techniques utilizing image intensification cameras and digital image processors (Coulter and Elde, 1986; C. L. Rieder and S. S. Bowser, unpublished).

The resolution of the same-section FLM–EM technique depends, in part, on how the antigen in question is distributed. Clearly, it is most useful when the antigen appears localized by FLM to a discrete structure within the section, visible by phase-contrast LM, which can then be positively identified by the EM (e.g., Figs. 6 and 7). Alternatively, if a particular antigenic site is restricted to a structure or domain which is not visible

Fig. 6. Epifluorescent (A), phase-contrast (B), and electron (C) micrographs of the same 0.25-μm section from a mitotic PtK_1 cell. This cell was processed for the indirect immunofluorescent (IMF) localization of tubulin, using a fluorescein-labeled secondary antibody, prior to embedding in Polybed 812 (see section on Methodology). A, unstained; B, C, stained with uranyl and lead. A comparison of these micrographs indicates that the IMF image correlates with the distribution of spindle microtubules; Bar in B, 10-μm. (From Rieder and Bowser, 1985.)

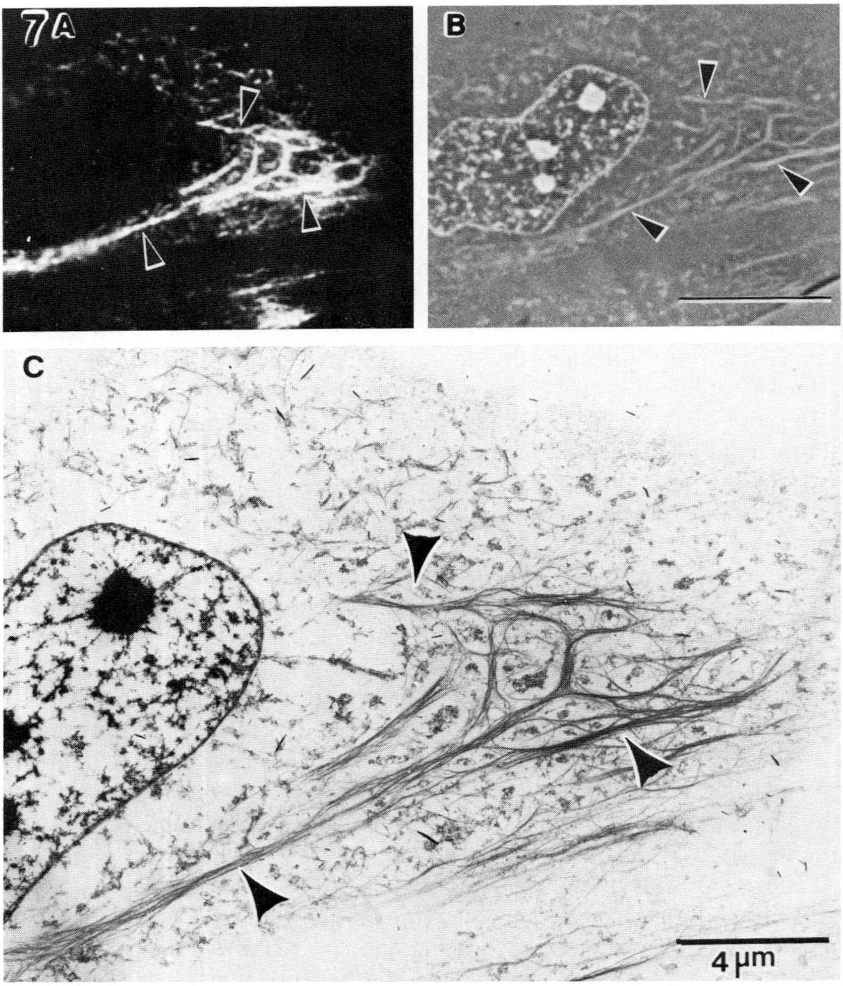

Fig. 7. Epifluorescent (A), negative phase-contrast (B), and electron (C) micrographs of the same 0.25-μm section from an interphase PtK_1 cell. This cell was processed for the indirect IMF localization of keratin, using a rhodamine-conjugated secondary antibody, prior to embedding in Epon-Araldite (see section on Methodology). A, unstained; B,C, stained with uranyl and lead. A comparison of these micrographs indicates that the IMF image correlates with the distribution of keratin filaments; bar in B, 10 μm. (Unpublished micrographs courtesy of E. Davison.)

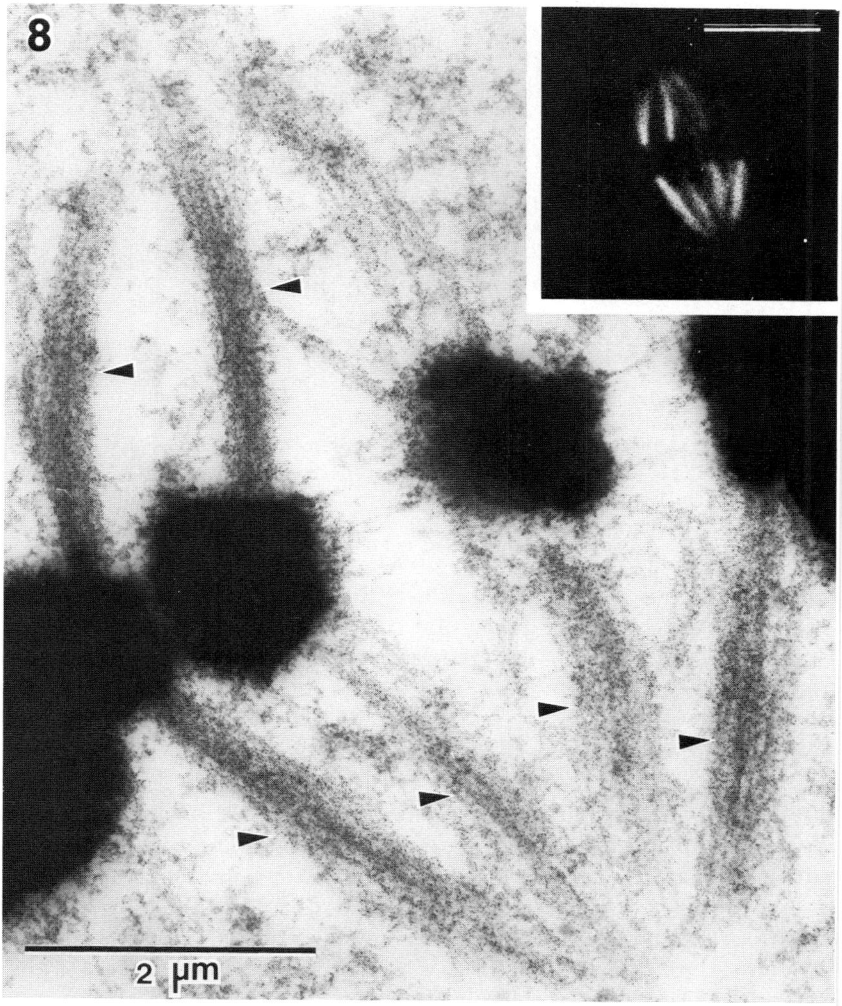

Fig. 8. Epifluorescent (inset) and electron micrographs of the same 0.50-μm section cut from a mitotic PtK_1 cell. This cell was processed, prior to embedding, for the indirect immunogold localization of microtubules. It was then incubated in a fluorescein-conjugated antibody directed against the immunogold label (see section on Methodology). The distribution of the fluorescent label within the section (inset) can be seen to correlate with the gold-labeled kinetochore fiber microtubules (arrows; cf. Fig. 6); bar in inset, 5 μm. (Unpublished micrographs courtesy of G. Rupp.)

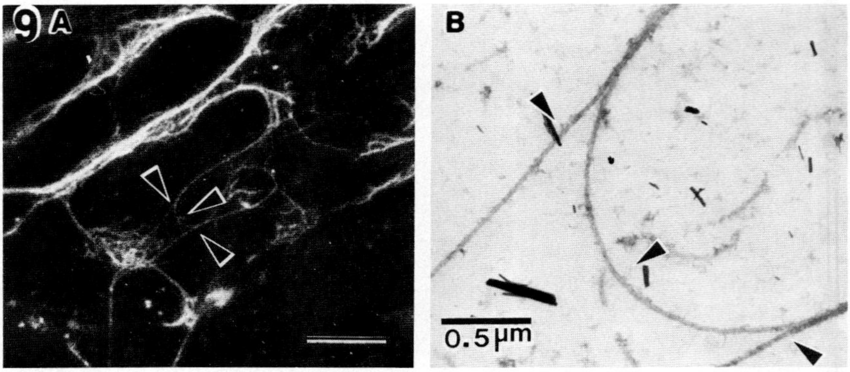

Fig. 9. Epifluorescent (A) and electron (B) micrographs of the same 0.25-μm section from an *Allogromia sp*. pseudopodial network. This cell was processed for the indirect IMF localization of tubulin prior to embedding (see section on Methodology). Many of the linear structures visible in the IMF image (e.g., arrowheads in A) correlate, at the EM level, with single microtubules (e.g., arrowheads in B); bar in A, 5 μm. (From Rieder and Bowser, 1985.)

within the section by phase-contrast LM, then same-section correlative FLM–EM can provide information concerning the precise region of the cell where the antigen is located. In addition, in many of these cases a fluorescent pattern is obtained from a section, which cannot be correlated to a structure visible by phase-contrast LM, but which correlates exactly with a structural element visible at the EM level. In these instances structures below the resolution limit of the LM can be detected by FLM within the section, and then positively identified in the EM. An example of such a case is illustrated in Fig. 9A which shows the fluorescent pattern obtained from a 0.25-μm section of an *Allogromia* pseudopod stained, prior to embedding, for the indirect immunofluorescent localization of tubulin. This fluorescent pattern can be clearly seen to correlate, at the EM level, with single microtubules (Fig. 9B) which could not be detected in phase-contrast micrographs of this section.

We recently used this same section FLM–EM technique to determine the ultrastructural localization of wound tumor virus (WTV) "assembly factories" within infected leafhopper cells (Rieder and Bowser, 1985). Epifluorescence LM of whole glycerol-mounted WTV-infected leafhopper cells, stained for the indirect immunofluorescent localization of viral core protein, reveals distinct fluorescent patches in the cytoplasm (e.g., see Fig. 1 of Nuss and Peterson, 1980). Similar fluorescent patches are also seen, after processing for EM, in serial 0.50-μm sections of these cells (Fig. 10). A comparison of FLM (Fig. 10A) and phase-contrast (Fig.

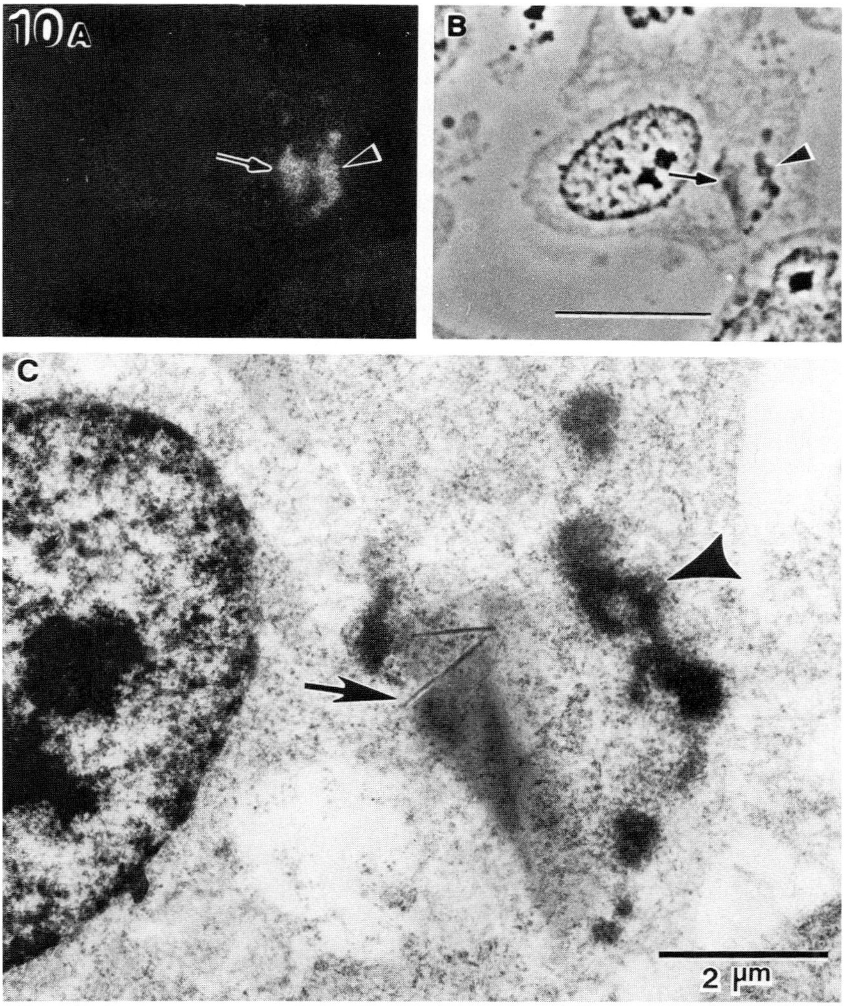

Fig. 10. Epifluorescent (A), phase-contrast (B), and electron (C) micrographs of the same 0.50-μm section from a WTV-infected AC-20 cell. This cell was stained for the indirect IMF localization of a viral core protein, using fluorescein-labeled protein A as a secondary label, prior to embedding. See text for details; bar in B, 10 μm. (From Rieder and Bowser, 1985.)

10B) images of these sections, with the high-voltage EM images of the same sections (Fig. 10C), clearly reveals that the WTV core protein is associated with two discrete cellular inclusions. The first of these appears at the ultrastructural level as large, moderately electron-translucent sheets of a homogenous material (arrow in Fig. 10C) while the second consists of electron-opaque aggregates of an amorphous material with which viral particles are frequently associated (arrowhead in Fig. 10C). These results could not have been obtained by LM of immunofluorescently stained whole mounts alone.

Rupp *et al.* (1986) used this same approach to investigate the composition of paracrystalline aggregates found in the pseudopodial networks of the marine foraminiferan *Allogromia* during network withdrawal. In 0.25-μm sections of this material, stained prior to embedding for the indirect immunofluorescent localization of tubulin, intense fluorescence occurs in regions that correlate, at the EM level, with paracrystalline aggregates (Fig. 11). Thus, the paracrystalline material in *Allogromia* reticulopods represents a naturally occurring tubulin-containing paracrystal.

Embedding in Epon, in contrast to paraffin, actually results in a long term (up to 10 years) preservation of the fluorophores induced by the fixation of biogenic amines with paraformaldehyde (Rubin and Coslo, 1978). Similarly, Eneström (1986) has shown that Epon embedding of immunofluorescently stained material stabilizes fluorescein and rhodamine fluorophores and that sections cut from such material can be stored for more than 3 years without demonstrable fluorescence decay. His preliminary results also suggest that the immunofluorescence within freshly cut Epon sections actually photobleaches *faster* than in older sections.

We find that the fluorescent image remains visible in freshly cut 0.25- to 0.50-μm sections, under 100× oil-immersion objectives, for up to 5 min. This is more than enough time for photography since our average exposure requires only 12–15 sec (see section on Methodology). Although *N*-propyl gallate does not appear to inhibit photobleaching in sections, 1% aqueous *p*-phenylenediamine appears to decrease this rate (Rieder and Bowser, 1985; see, however, Bock *et al.*, 1985). Finally, as noted above, sensitive and/or weak fluorescent patterns in sections can also be examined with image intensification video cameras, and image processed to provide additional resolution and detail which is not revealed by conventional 35-mm photography.

Correlative FLM–EM can also be conducted on sections, the surface of which has been stained with fluorescently tagged antibodies (i.e., postembedding staining; e.g., Kersken *et al.*, 1984). The advantages of this approach are that it can sometimes be used profitably on sections of

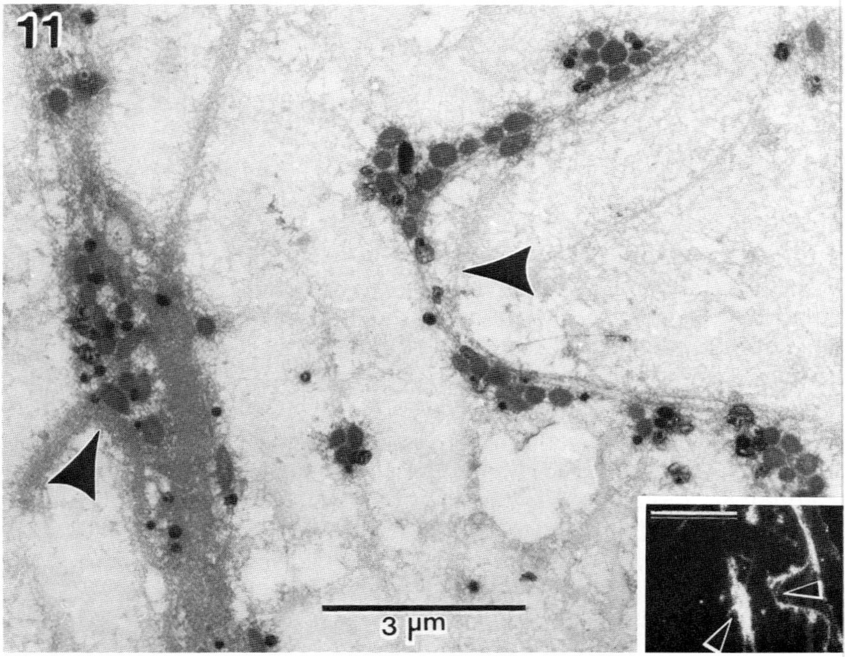

Fig. 11. Epifluorescent (inset) and electron micrographs of the same 0.25-μm section cut from a retracted pseudopodial network of *Allogromia sp*. This cell had been prepared for the indirect IMF localization of tubulin prior to embedding in Polybed 812. The fluorescent structures pictured within the inset (arrowheads) correspond, at the ultrastructural level, to pools of a paracrystalline material and intact microtubules. Bar in inset, 10 μm. (Unpublished micrographs courtesy of G. Rupp.)

osmicated material after sodium metaperiodate treatment (e.g., Hearn *et al.*, 1985), and that it provides access to target sites buried deep within a block of tissue (which may otherwise be inaccessible to the probe). However, a better correlation between the fluorescent and EM image of a section is expected with preembedding staining because the antigenic structure will be labeled throughout the section thickness. In this respect, same-section FLM–EM on material stained with fluorescent probes prior to embedding should be useful with thick or rounded specimens or cell components which show a confusing fluorescent image due to the overlapping of labeled constituents. In these cases a clearer picture of antigen distribution can be obtained from analyzing serial sections (e.g., see Figs. 1, 2 of Rupp *et al.*, 1984).

The fact that glutaraldehyde fixation does not quench fluorescein, rhodamine, or Texas red fluorescence (Rieder and Bowser, 1985) suggests that same-section FLM-EM will also prove useful for studying the distribution of fluorescently labeled proteins which have been microinjected into cells (i.e., fluorescent analogue cytochemistry; Taylor *et al.,* 1981; Salmon *et al.,* 1984).

There are, of course, some constraints which limit the usefulness of same-section FLM-EM. Not all fluorescent probes are stable to embedding in epoxy resins (seen section on Methodology). Furthermore, the technique provides information indirectly; methods utilizing colloidal gold or ferritin-conjugated antibodies are more direct and provide better resolution at the ultrastructural level. Finally, the depth to which even a fluorescent probe penetrates a tissue will be restricted by the conflicting demands of antigen accessibility and optimum structural preservation for EM (e.g., see Farnum and Wilsman, 1984). At present, attempts to overcome this restriction have focused on either post-embedment staining of the section surface or reversible embedding procedures (see section on Methodology, Same-Section Correlative FLM–EM).

We thank the following individuals for their contributions to this chapter: D. F. Parsons, A. Peterson, B. S. Eckert, G. Rupp, E. Davison, K. Roth, and S. Nowogrodzki. The projects described in this chapter were supported in part by NIH Biotechnological Resource Related grant RR 02157 (to CLR) and by Biotechnological Resource grant PHS 01219, awarded by the Division of Research Resources, PHS/DHHS, to support the Wadsworth Centers's high-voltage electron microscope as a National Biotechnological Resource Facility.

REFERENCES

Alexander, R. B., Isaacs, W. B., and Barrack, E. R. (1985). Immunogold probes for electron microscopy: Evaluation of staining by fluorescence microscopy. *J. Histochem. Cytochem.* **33,** 995.

Allen, R. D. (1985). New observations on cell architecture and dynamics by video-enhanced contrast optical microscopy. *Annu. Rev. Biophys. Biophys. Chem.* **14,** 265.

Aparicio, S. P., and Marsden, P. (1968). A rapid methylene blue-basic fuchsin stain for semithin sections of peripheral nerve and other tissues. *J. Microsc.* (*Oxford*) **89,** 139.

Bendayan, M. (1984). Protein A-gold electron microscopy immunocytochemistry: Methods, applications, and limitations. *J. Electron Microsc. Tech.* **1,** 243.

Bennett, H. S., Wyrick, A. D., Lee, S. W., and McNeil, J. H. (1976). Science and art in preparing tissues embedded in plastic for light microscopy, with special reference to glycol methacrylate, glass knives and simple stains. *Stain Technol.* **51,** 71.

Bergeron, M., and Thiéry, G. (1981). Three-dimensional characteristics of the endoplasmic reticulum of rat renal tubule cells, an electron microscopic study in thick sections. *Biol. Cell.* **42,** 43.

Bock, G., Hilchenbach, M., Schauenstein, K., and Wick, G. (1985). Photometric analysis of antifading reagents for immunofluorescence with laser and conventional illuminating sources. *J. Histochem. Cytochem.* **33,** 699.

Bowser, S. S., and Rieder, C. L. (1986). Section thickness terminology—a source of confusion in microscopy. *J. Microsc.* (*Oxford*) **143,** 319.

Bowser, S. S., McGee-Russell, S. M., and Rieder, C. L. (1985). Digestion of prey in foraminifera is not anomalous: A correlation of light microscopic, cytochemical, and HVEM techniques to study phagotrophy in two allogromiids. *Tissue Cell* **17,** 823.

Collins, B. A. (1979). A rapid bonding agent for Epon blocks. *Stain Technol.* **54,** 103.

Coulter, H. P., and Elde, R. P. (1986). Video-enhanced technique for detecting neurophysin, enkephlin, and somatostatin immunoreactivity in ultrathin sections of cat median eminence. *J. Histochem. Cytochem.* **34,** 1405.

DeBrabander, M., Geuens, G., DeMey, J., and Jonaiau, M. (1979). Light microscopic and ultrastructural distribution of immunoreactive tubulin in mitotic mammalian cells. *Biol. Cell.* **34,** 213.

DeBrabander, M., Bulinski, J. C., Geuens, G., DeMey, J., and Borisy, G. G. (1981). Immunoelectron microscopic localization of the 210,000 MW microtubule associated protein in cultured cells of primates. *J. Cell Biol.* **91,** 438.

DeMey, J. (1983). A critical review of light and electron microscopic immunocytochemical techniques used in neurobiology. *J. Neurosci.* **7,** 1.

DeMey, J., Moeremans, M., Geuens, G., Nuydens, R., and DeBrabander, M. (1981). High resolution light and electron microscopic localization of tubulin with the IGS (immuno gold staining) method. *Cell Biol. Int. Rep.* **5,** 889.

Eneström, S. (1986). Stabilization of fluorochromes after immunofluorescent staining of tissue and embedding in Epon. *Stain Technol.* **61,** 65.

Eneström, S., and Hed, T. (1980). Detection of immune deposits in glomeruli. *Pathol. Res. Pract.* **170,** 338.

Eneström, S., and Hed, T. (1982). Combined immunomorphological examination of Epon-embedded human kidney biopsies. *Am. J. Clin. Pathol.* **77,** 249.

Farnum, C. E., and Wilsman, N. J. (1984). Lectin-binding histochemistry of non-decalcified growth plate cartilage. *J. Histochem. Cytochem.* **32,** 593.

Franklin, R. M. (1982). Pre-embedding staining for immunohistochemistry at the light microscopic level. *In* "Techniques in Immunocytochemistry" (G. R. Bullock and P. Petrusz, eds.), Vol. 1, p. 251. Academic Press, London.

Franklin, R. M. (1983). Pre-embedment localization of antigens by immunofluorescent methods. *Ann. N. Y. Acad. Sci.* **40,** 165.

Glauert, A. M. (1965). Section staining, cytology, autoradiography and immunochemistry for biological specimens. *In* "Techniques for Electron Microscopy" (D. H. Kay, ed.), 2nd ed. pp. 254–310. Blackwell, Oxford.

Gorbsky, G., and Borisy, G. G. (1985). Microtubule distribution in cultured cells and intact tissues: Improved immunolabeling resolution through the use of reversible embedment cytochemistry. *Proc. Natl. Acad. Sci. U.S.A.* **82,** 6889.

Hearn, S. A., Silver, M. M., and Sholdice, J. A. (1985). Immunoelectron microscopic labeling of immunoglobulin in plasma cells after osmium fixation and plastic embedding. *J. Histochem. Cytochem.* **33,** 1212.

Hökfelt, T. (1965). A modification of the histochemical fluorescence method for the demonstration of catecholamines and 5-hydroxy-tryptamine, using Araldite as embedding medium. *J. Histochem. Cytochem.* **13,** 518.

Horisberger, M. (1979). Evaluation of colloidal gold as a cytochemical marker for transmission and scanning electron microscopy. *Biol. Cell.* **36,** 253.

Inoué, S. (1986). Videomicroscopy. Plenum, New York.

Kersken, H., Tiggemann, R., Westphal, C., and Plattner, H. (1984). The secretory contents of Paramecium tetraurelia trichocysts. *J. Histochem. Cytochem.* **32,** 179.

King, M. V., Parsons, D. F., Turner, J. N., Chang, B. B., and Ratkowski, A. J. (1980). Progress in applying the high voltage electron microscope to biomedical research. *Cell Biophys.* **2,** 1.

Larramendi, P. C. H. (1985). Method of retrieval of one to two micron sections from glass for ultrathin sectioning. *J. Electron Microsc. Tech.* **2,** 645.

Llewellyn-Smith, I. J., Costa, M., and Furness, J. B. (1985). Light and electron microscopic immunocytochemistry of the same nerves from whole mount preparations. *J. Histochem. Cytochem.* **33,** 857.

Locke, M., and Kirshnan, N. (1971). Hot alcoholic phosphotungstic acid and uranyl acetate as routine stains for thick and thin sections. *J. Cell Biol.* **50,** 550.

Lucocq, J. M., and Roth, J. (1984). Applications of immunocolloids in light microscopy. III. Demonstration of antigenic and lectin-binding sites in semithick resin sections. *J. Histochem. Cytochem.* **32,** 1075.

McEwen, B. R., Radermacher, M., Rieder, C. L., and Frank, J. (1986). Tomographic three-dimensional reconstruction of cilia ultrastructure from thick sections. *Proc. Natl. Acad. Sci. U.S.A.* **83,** 9040.

McGee-Russell, S. M. (1968). The method of combined observations with light and electron microscopes applied to the study of histochemical colourations in nerve cells and oocytes. *In* "Cell Structure and Its Interpretation" (S. M. McGee-Russell and K. F. A. Ross, eds.), p. 183. Arnold, London.

McNary, W. F., Rosan, R. C., and Kerrigan, J. A. (1964). Fluorescent microscopy of thin sections as an adjunct to electron microscopy. *J. Histochem. Cytochem.* **12,** 216.

Milroy, A. M. (1985). A method for eliminating unnecessary serial thin sectioning for electron microscopy. *J. Electron Microsc. Tech.* **2,** 399.

Moore, M. J. (1975). Removal of glass coverslips from cultures flat embedded in epoxy resins using hydrofluoric acid. *J. Microsc.* (*Oxford*) **104,** 205.

Nuss, D. C., and Peterson, A. J. (1980). Expression of wound tumor virus gene products in vitro and in vivo. *J. Virol.* **34,** 532.

Parsons, D. F., Marko, M., Radermacher, M., and Frank, J. (1985). Shape changes and polarization of cells migrating through tissue. A high voltage electron microscope and computer graphic study of serial thick sections. *Tissue Cell* **17,** 491.

Parysek, L. M., Wolosewick, J. J., and Olmsted, J. B. (1984). MAP 4: A microtubule-associated protein specific for a subset of tissue microtubules. *J. Cell Biol.* **99,** 2287.

Pease, D. C. (1964). "Histological Techniques for Electron Microscopy," 2nd ed. Academic Press, New York.

Peppe, F. A., and Finck, H. (1961). The use of specific antibody in electron microscopy. II. The visualization of mercury-labeled antibody in the electron microscope. *J. Biophys. Biochem. Cytol.* **11,** 521.

Rieder, C. L. (1981). Thick and thin serial sectioning for the three-dimensional reconstruction of biological ultrastructure. *Methods Cell Biol.* **22,** 215.

Rieder, C. L., and Bowser, S. S. (1983). Factors which influence light microscopic visualization of biological material in sections prepared for electron microscopy. *J. Microsc.* (*Oxford*) **132,** 71.

Rieder, C. L., and Bowser, S. S. (1984). A versatile system for illuminating flat-embedded specimens during block trimming. *J. Electron Microsc. Tech.* **1,** 207.

Rieder, C. L., and Bowser, S. S. (1985). Correlative immunofluorescence and electron

skeletal muscle sections which had been reembedded with Epon for TEM examination (Geissinger *et al.,* 1978), it does often result in artifacts, especially as far as TEM images of skeletal muscle are concerned (Vriend and Geissinger, 1980). Nevertheless, deparaffinized sections of biological tissues are still used to advantage by some workers (see Chapter 20, this volume).

Polyethylene glycol (PEG) is a water-miscible wax which has been used for the embedding of tissues examined with the LM for many years. It is currently in use for TEM sections in some laboratories (Wolosewick, 1984) because larger sections which have more contrast can be procured. Since PEG can be removed from the PEG-embedded section, these can be subjected to many histochemical and cytochemical techniques (Wolosewick, 1984). The PEG-embedded resinless section also should have great potential in obtaining SEM images with the SEM operating alternately in the secondary electron emission and transmission mode (Wolosewick, 1984).

As more refined methods for drying SEM specimens (e.g., critical-point drying; Cohen *et al.,* 1968) were used routinely in SEM preparation techniques, it became even more obvious that the hitherto unfamiliar SEM image often ought to be seen with microscopes (e.g., LM) which were more familiar to the histologist or histopathologist. Therefore, methods of LM→SEM correlation were evolved for obtaining conventional histological sections from the human cervix, after it had been critically point dried for SEM (Ayres *et al.,* 1971). Although the cumulative time for the post-SEM histology processing schedule was very long (158 hr), the morphological end results were sufficient to recognize and compare the different cell types in hematoxylin and eosin stained LM sections with the SEM image.

Shorter processing times for tissue which had been previously examined with the SEM improved the quality of LM sections which then could be profitably compared to SEM images. Thus, according to Fathalla *et al.* (1975), fixation of bovine endometrium in glutaraldehyde and subsequent washing in two changes of sucrose buffer made the sample ready for critical-point drying and subsequent SEM examination. Paraffin embedding and routine processing (12 hr) resulted in post-SEM histological sections which were very difficult to distinguish morphologically from routine histological sections, except for the fine metal coating which had been applied previously for SEM examination. As the microtomist for post-SEM material became more experienced, even better LM sections could be obtained; however, the reembedment of the paraffin section in Epon for eventual TEM evaluation did not result in good ultrastructural preservation (Fathalla *et al.,* 1978).

Quite adequate ultrastructural preservation (TEM and SEM) was achieved in another correlative study of endometrium a year later (Geissinger *et al.*, 1979), where the endometrial surface epithelium in the follicular and luteal phases of the estrous cycle in sows was examined with the SEM and TEM. Areas which were obviously devoid of cilia or microvilli turned out to be the bulging plasma membranes of secretory epithelial cells in the process of adding their cytoplasmic organelles to the secretion.

In this chapter the methodologies for specimen preparation which are currently used in our laboratory for the intermicroscopic correlation (LM, SEM, TEM) of skeletal muscle and endometrium are described and their efficacy discussed, giving examples where it is deemed appropriate. The instrumentation and general manipulations for correlative microscopy procedures have not materially changed in recent years, and have been fully described elsewhere (Geissinger, 1976).

MATERIALS AND METHODS FOR SKELETAL MUSCLE

OTOTO Method (LM→SEM→TEM)

Sample Manipulation and Fixation

Skeletal muscles are quickly excised from albino white mice which have been euthanized by cervical dislocation. The muscles are dissected into strips, approximately 2 mm wide, 5 mm long, and 1 mm high, rinsed at room temperature in 0.1 *M* phosphate buffer (pH 7.4), and pinned on corkboard in a relaxed or slightly stretched state. The specimens, which are still attached to the corkboard, are immersed at room temperature in Karnovsky's (1965) solution for a total of 4 hr. After the first hour in fixative, the muscle strips are removed from the corkboard and are placed in fresh Karnovsky's (1965) solution for the remaining 3 hr. The specimens are thoroughly washed in three 15-min rinses of 0.1 *M* phosphate buffer (pH 7.4) at room temperature.

Alternatively, small pieces of biopsy material from human myopathies, which have been fixed in Karnovsky's fluid (1965) for several hours and stored in 0.2 *M* phosphate buffer at 4°C for varying periods of time (up to 6 months), can be used (Geissinger *et al.*, 1980).

Preparation for LM and SEM

The muscles are removed from the buffer and are mounted on standard cryostat stubs with OCT compound (a cryoprotectant containing soluble glycols and resins; Lab Tek Products, Toronto, Ontario, Canada). The tissues on the stubs are frozen to −18°C, and sections are cut at 7–10 μm in a conventional cryostat. The sections are mounted on standard microscope glass slides which have been previously thinly coated with a "Sub Slide" solution which aids in the adherence of chemically fixed cryostat sections to the glass slides.

The solution is made up with 2 g granular gelatin powder, 200 ml distilled water, and 190 mg chrome alum. To the 200 ml of distilled water, add 2 g granular gelatin; do *not* stir. Place the solution in a 60°C oven for 1 hr. Remove the solution from the oven and add 190 mg of chrome alum (chromium potassium sulfate); stir well. Place clean, standard glass microscope slides into staining racks and dip them, two or three times, into the Sub Slide solution. Let the slides drain on paper towels while in the racks. Loosely cover the slides in the racks with more paper towels for added protection from dust particles. Air-dry overnight and store slides for later usage in dust-free slide boxes.

The slide mounted sections are exposed to room temperature for a maximum of 2–5 min to aid in their adherence and are placed in a minimum of three distilled water washes for a total of 15 min, to ensure complete removal of the OCT compound. Some of the preparations are stained with hematoxylin and eosin; the rest are left unstained. The slides are mounted in glycerol and examined with a LM equipped with polarizing optics (Pol) and photographed at various magnifications in order to find specific areas to be examined later under the SEM. The coverslips are floated off from the sections in three changes (20 min each) of 10% alcohol in order to remove all traces of glycerol. For subsequent SEM examination the slide preparations are treated as follows: They are either immersed in 1% aqueous OsO_4 for 10 min and washed for two 2-min periods in distilled water, followed by dehydration in graded ethanols, critical-point dried in CO_2 (de Bault, 1973) and sputter-coated with gold-palladium (Au-Pd) (Echlin, 1975), or are prepared using a modified thiocarbohydrazide technique (Malick and Wilson, 1975). In this case the slides are immersed in a freshly prepared saturated solution of 1% thiocarbohydrazide (TCH) for 5 min and followed by two 2-min washes in distilled water. The entire procedure is repeated and followed by placing the slides, once again, in the OsO_4 solution, resulting in the following treatment: Os-TCH-Os-TCH-Os (OTOTO). The slides are again placed in two 2-min washes of distilled water and are then dehydrated through an ascending series of ethanol concentrations, at room temperature, followed by a 1 : 1 Freon/100% ethanol solution for 15 min and two 10-min changes of Freon 113. Then the preparations are critical-point dried in CO_2 (de Bault, 1973).

SEM Examination

Slides are viewed in the SEM at 15 kV and scanned for areas of interest, which had been found previously under the LM, at low magnifications of 200–400×. Once an area is found, the magnification is increased to approximately 10,000–13,000× and photographs are taken using Polaroid P/N 55 film. The same microscopic field is photographed at succeedingly lower magnifications (6,000–100×) and the area of interest is marked on the Polaroid photographs for aid in correlation with the subsequent TEM images.

Preparation for TEM

The slides, with areas of interest, are placed in two 15-min changes of propylene oxide, followed by 30 min in 1 : 1 Epon/propylene oxide, and 30 min in 3 : 1 Epon/propylene oxide. Each of these solutions are drained off between changes by propping the slides up in a vertical position on filter paper. The slides are then repositioned on the flat bench top and fresh drops of Epon are placed on the sections for 1 hr. The slides are again placed vertically to allow the Epon to drain off. After this, the slides are once again positioned horizontally, and gelatin capsules, with small holes cut in the rounded tips, are inverted over the tissue and filled with fresh Epon, using a hypodermic syringe. The capsules on the slides are cured in a 60°C oven for 24–48 hr. Once the Epon has sufficiently cured, the slides with affixed capsules are removed from the oven and allowed to cool. The Epon surrounding the capsules is scored, and the undersides of the slides are slowly passed back and forth over the flame of a Bunsen burner several times. The Epon capsules with the attached tissues are grasped with a pair of forceps and are easily snapped off from the slides.

The Epon block is placed in a TEM block holder and viewed with a binocular microscope. The block is trimmed to remove any excess Epon from around the tissue edges, and is then positioned in the ultramicrotome and oriented so that the knife edge is parallel to the entire sectioning face. The area of interest, which has been photographed with the SEM, is located furthermost away from the block face. One-micrometer sections are cut and stained with toluidine blue and examined under LM and compared to the photographs taken with the SEM. Once a "landmark" has been found on the toluidine blue section and has been correlated with the SEM photographs, drawings are made of the block face and of the toluidine blue sections, and serial sectioning at 60–90 nm is performed using a diamond knife. The sections are placed on copper mesh grids and examined in the TEM at low magnifications, using the drawings made previously of the block face and the toluidine blue sections, in order to correlate certain structures with those in the SEM photographs. Subsequent higher magnifications are employed to observe structures of interest.

GTO Method (SEM→TEM)

Sample Manipulation and Fixation

Skeletal muscles are quickly excised from New Zealand rabbits or albino white mice which have been previously euthanized with sodium pentobarbital or by cervical dislocation. The muscles are dissected into strips and pinned onto corkboard as in the OTOTO method. The specimens, which are still attached to the corkboard, are immersed in a room temperature fixative made up of 3% glutaraldehyde, 8% tannic acid, with 0.1 *M* phosphate buffer (pH 7.4) solution for 2 hr (Bonilla, 1977; Vriend and Geissinger, 1980). After the first hour the muscle strips are removed from the corkboard and are placed in a fresh solution of the fixative as described above.

Following fixation, the specimens are washed in three 15-min washes of room temperature 0.1 *M* phosphate buffer (pH 7.4) and immersed in 1% OsO_4 in phosphate buffer for 2 hr. The entire fixation procedure is repeated and the muscle strips are mounted on cryostat stubs, sectioned at 7–10 μm, mounted on Sub Slides, washed in distilled water, dehydrated in ethanols and Freon, and critical-point dried, using the same procedure as outlined in the OTOTO method.

SEM Examination and Subsequent Procedures

The sections on the slides are viewed and photographed in the SEM at 15 and 35 kV. Areas of interest are photographed at magnifications of 10,000–13,000× maximum, using both secondary and backscattered electrons, followed by succeedingly lower magnifications of the areas which are photographed and marked on Polaroid prints. This procedure has been described more fully in the OTOTO method. The slides, with areas of interest are processed, cut for TEM, viewed, and photographed using the procedure described previously in the OTOTO method.

MATERIALS AND METHODS FOR ENDOMETRIUM

Sample Manipulation and Fixation

Porcine endometrial biopsies are surgically removed, immersed in 0.1 *M* phosphate buffer (pH 7.4), and pinned without stretching to corkboard. The samples are rinsed in fresh buffer and transferred to 6% glutaraldehyde in 0.1 *M* phosphate buffer (pH 7.2) for 2 hr at 4°C. The tissue is then cut into small (~5 mm × 5 mm × 1 mm high) pieces, trimming away the perimetrium and most of the myometrium in doing so. The remaining tissue is returned to fresh fixative for a further 2–3 hr at room temperature. Following primary fixation, tissue samples are given three 15-min washes in 0.1 *M* phosphate buffer (pH 7.4) and are postfixed in 1% OsO_4 for 2 hr. Following OsO_4 fixation, the samples are given three 15-min rinses in buffer and then stored in phosphate buffer at 4°C until further processing.

Preparation for SEM

Samples are removed from cold storage and brought to room temperature. After two 10-min rinses in fresh 0.1 *M* phosphate buffer (pH 7.4), the tissue is dehydrated in an ascending series of ethanol concentrations, critical-point dried in carbon dioxide (de Bault, 1973), and mounted on stubs using silver conducting paint. The samples are then sputter-coated under vacuum with Au-Pd alloy (Echlin, 1975) and viewed under the SEM at accelerating voltages of 15–25 kV.

Procedures for SEM→LM and TEM

Areas selected under the SEM are photographed at descending magnifications to aid in location of the desired area under the dissecting microscope (Wickham and Worthen, 1973). A final magnification of 60× is adequate. After SEM photography the tissue is removed from the SEM stub with forceps, and care is taken not to touch the epithelial surface of the endometrium. The silver paint, adhering to the underside of the tissue, is trimmed away with a sharp razor blade, and the working surface of the microscope stage is wiped clean of silver particles. The selected area of tissue is located under a dissecting microscope, with the aid of the SEM micrographs, and cut out as a slice of tissue approximately 1 × 1 × 1.5 mm. A rough drawing of this piece of tissue is made as an aid to orientation of the tissue during subsequent embedding for LM.

Embedding Media for LM and TEM

Three embedding media have been used with equal success in correlative procedures, i.e., a methacrylate based resin known as "JB-4 water miscible embedding medium" (Polysciences Inc. Warrington, PA), Spurr low-viscosity embedding medium, and Epon.

1. Methacrylate—Selected areas from the SEM stub may be placed directly into the infiltrating medium and stored at 4°C for periods of 8–16 hr. Specimens are oriented and embedded, according to manufacturers specifications, in JB-4 molds or sealed BEEM capsules, and are allowed to polymerize in a vacuum dessicator at room temperature for 24 hr. The blocks are then trimmed and serial sections are cut at 1–3 μm, stained, and examined under the LM until the area of interest is located and then photographed. Optimum results for SEM→LM correlation are obtained with tissue which has not been osmicated, because the resulting sections can be stained with any conventional histological stain. Ultrastructural preservation for eventual TEM examination is, of course, improved by osmication, which, however, limits the number of useful LM stains which may be employed. Toluidine blue is preferred for osmicated tissue which has to examined with the LM. For SEM→TEM correlation the block is further trimmed, using the last thick section to be cut as a guideline. Ninety-nanometer sections are cut and mounted on parlodion-coated grids and stained with uranyl acetate and lead citrate. The stability of the methacrylate under the electron beam of the TEM is increased by lightly coating the stained sections with carbon in a vacuum evaporator.

2. Spurr—Selected areas from the SEM stub may be placed directly into the infiltrating medium and stored at 4°C for 8–16 hr. Specimens are then oriented and embedded according to the manufacturers specifications. Blocks are cured at 70°C for 12 hr. Semithin sections are cut at 1–3 μm and stained with toluidine blue; ultrathin sections are cut at ~90 nm, mounted on uncoated copper grids, and contrasted as above.

3. Epon—Selected areas from the SEM specimen are placed in two 15-min changes of propylene oxide, followed by 30 min in 50/50 (v/v) propylene oxide/Epon and 60 min in 25/75 propylene oxide at room temperature and 24 hr in 100% Epon at 4°C. Tissue pieces are then embedded and orientated in fresh Epon and cured at 60°C for 24–48 hr. Semithin and ultrathin sections are prepared as described previously for Spurr.

Procedures for LM

The mounting of coverslips is necessary for good LM images at high, dry magnifications. We routinely measure the height of the surface epithelium (SE) and the glandular epithelium (GE) of endometrium using an ocular micrometer and 16× or 40× LM objective lenses. If toluidine blue sections are destined to be resectioned for TEM imaging, coverslips cannot be permanently mounted. Instead they are mounted with glycerol and the area of interest is photographed under the LM. Taking care not to move the LM stage, the slide is removed. To remove the coverslip, the

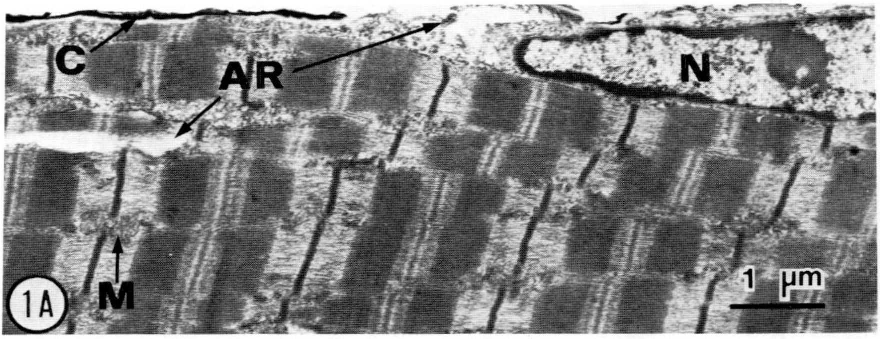

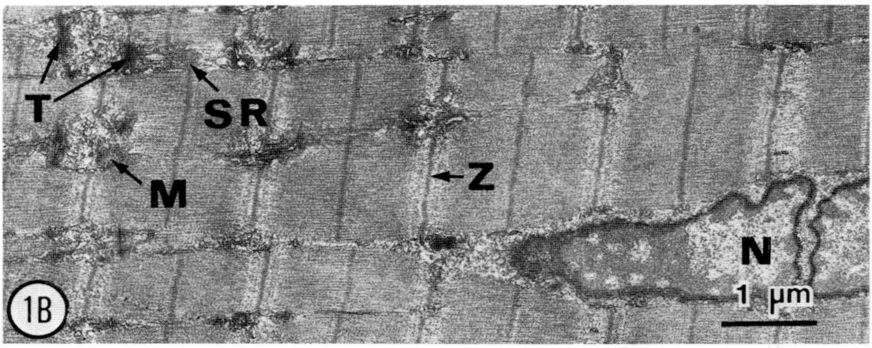

Fig. 1. TEM micrographs of skeletal muscle prepared in two manners. The micrographs are obtained from slide section specimens which had previously been examined with both LM and SEM. A, Au-Pd-coated cryostat section. Note the good preservation of the nucleus (N) and the myofibrils. Cryostat sectioning artifacts (AR) can be observed, as well as the metal coating (C). Membranous components, such as mitochondria (M), show adequate contrast. B, OTOTO prepared cryostat section. Note the excellent preservation of nucleus (N) and the myofibrils. The Z-lines (Z) appear lightly contrasted. However, membranous components such as the sarcoplasmic reticulum (SR), transverse tubules (T), and mitochondria (M) show very good contrast. (First published: Vriend and Geissinger, 1980, by permission.)

slide is soaked upright in 100% ethanol until the coverslip dislodges. After the slide has air-dried, it is returned to the LM stage and the area of interest is again located and marked by scoring a 1-mm diam. circle around the area with a diamond-tipped objective slide scorer.

If only high-power micrographs under oil immersion are required, coverslips are not necessary. A drop of immersion oil is placed directly over the section and the oil-immersion lens moved into place. This oil is later washed off with ethanol.

Preparation for TEM from LM Sections (LM→TEM)

Approximately 90-nm sections, of selected areas, may be obtained directly from toluidine blue sections by lifting either parts of, or entire sections from the glass slides. If more than one area of the toluidine blue section is required for TEM correlation, each selected area is scored with a diamond-tipped objective slide scorer. Dummy Epon blocks are trimmed and faced up with a diamond knife. The section is dusted lightly with cornstarch, a small drop of cyanoacrylate glue is applied to the face up surface of the blocks, and placed over the scored areas. After the glue has set, the dummy Epon blocks are snapped off the slide, including the chosen portion of the section. Warming the block/slide assembly in a 60°C oven for 5 min before attempting this, improves the success of this technique. The blocks are returned to the microtome chuck and ~90-nm sections are cut using a diamond knife.

If only one area on the toluidine blue section is selected for TEM examination, a simple method is to invert an Epon-filled capsule over the top of the whole section. The slide/capsule assembly is incubated at 60°C for 24 hr or until the Epon hardens. At this time, the underside of the slide is passed several times over the blue portion of a Bunsen burner flame and the block, with the attached section, is snapped off. It is then trimmed with a sharp razor blade, and carefully aligned with the cutting edge of a diamond knife. Sections are cut at ~90 nm thickness; all section fragments are retrieved from the knife boat and mounted on grids and are contrasted with uranyl acetate and lead citrate.

DISCUSSION OF METHODOLOGY AND SELECTED RESULTS

Correlative microscopy is technically very demanding, care and patience throughout each step cannot be overemphasized. A loss in technical quality at different levels of microscopy must be expected, since a compromise has to be reached when processing tissue for two or three levels of microscopy. Each step in the processing may detract from the final quality of preservation.

Although critical-point drying of tissue has limited detrimental effects on ultrastructural preservation in TEM images (Wickham and Worthen, 1973), the primary electron beam during SEM imaging may cause extensive damage. Therefore, to improve the TEM image it is advisable to limit the extent to which specimens are exposed to such damaging radiation; thus rapid correlation procedures are optimal (Geissinger and Kamler, 1972).

Skeletal Muscle

Fixation and Buffer

Fixation at resting length of carefully dissected out muscle bundles, which are longitudinally oriented to their long axis and fixed with Karnovsky's (1965) fixative, has yielded the best results in our laboratory, especially if details in the sarcoplasmic reticulum or mitochondria are to be correlated in the final TEM image. It has been our experience that provided tissue samples are small enough, room temperature fixation facilitates improved fixation. The incidence of buffer crystals on tissue surfaces to be examined under the SEM can effectively be reduced by allowing samples to warm up to room temperature and rinsing in fresh, room temperature buffer before beginning dehydration procedures.

The OTOTO method gives good contrast for TEM imaging without the need for contrasting with uranyl acetate and lead citrate. This method gives superior preservation of mitochondria, whereas the GTO method has proved to be effective for demonstrating T-tubules. The GTO method can be employed when banding patterns and T-tubules are of primary interest in the SEM image and T-tubules are of primary interest in the TEM image; however, mitochondrial details are not very well preserved. The method is very good for delineating the major bands (A and I) in myofibrils in the SEM image, so that a comparison to the LM (Pol)-image is not necessary, depending, of course, on what information is sought with the LM.

Preparation for LM

The cryosectioning procedure for skeletal muscle results in far superior preservation to paraffin sections at the level of TEM (Vriend and Geissinger, 1980). Therefore, it is routinely used in our laboratory when SEM and TEM images of muscle have to be correlated (Fig. 1). The cryoprotectant OCT can leave artifacts on SEM images unless samples are thoroughly rinsed in a number of changes of distilled water following cryostat sectioning. The use of Sub Slides has greatly reduced the loss of sections from the slide during the rinsing and dehydration processes. All other methods for affixing cryostat sections have proven inferior for the holding of the muscle fibers to the slides sufficiently well to withstand the subsequent washing process.

SEM Preparation and Examination

Coating the specimen with Au-Pd or the OTOTO procedure of impregnating the specimen are equally well suitable for SEM; for TEM better contrast is generally seen with the OTOTO method (Fig. 1). If possible, the specimen should not be too long under the electron beam of the SEM (Geissinger and Kamler, 1972) as damage to the specimen might result.

Tissues prepared using either the GTO or OTOTO methods can normally be viewed under the SEM without sputter coating with Au-Pd, unless charging of the specimen is a great problem. (Mounting specimens on Au-Pd-coated slides does not appear to reduce the incidence of charging during SEM examination. If charging during SEM examination should occur, the only practical solution to the problem is to sputter coat the section.)

Preparation for TEM

When embedding slide mounted muscle samples in Epon, the use of inverted gelatin capsules, with holes in the tips, allows traces of propylene oxide to escape and thus facilitates improved polymerization. Snapping off the capsules and tissues from glass slides requires practice. If the slide is passed over a Bunsen burner flame for too short a period, the block will not come off; too long exposure will result in burning of the tissue and the Epon.

The technically most difficult part of the procedure of SEM→TEM correlation is the ultramicrotomy required of a minute brittle specimen, which has already been examined with the SEM. Much expertise and patience are required there. Sometimes, parts of sections will shift on the slide during processing for TEM; therefore, it is most important that sufficient SEM micrographs of selected areas are taken and drawings of tissues on the slides are made before preparing the samples for TEM. They will be invaluable during later attempts to correlate SEM images with the TEM images.

Selected Results

The correlation method of muscle has been used on human biopsy material where a structure which had been seen with LM and SEM had been in doubt (Fig. 2). Only one TEM micrograph was needed to interpret the doubtful structure (Fig. 2B). More often, details which are seen with the SEM are difficult to interpret. A case in point was the question of whether the transverse ridges seen at approximate right angles to the longitudinal axis of myofibrils in SEM images represented T-tubules, sarcoplasmic tubules, contracted Z-band material, or mitochondria. Serial TEM sections were needed to settle this point (Fig. 3). Different areas in the same section taken at magnifications far in excess than those needed to answer the previous question showed some artifacts in the mitochondria (Figs. 4A,B), whereas details of the sarcolemma, T-tubules, myofilaments (Fig. 4A), and the fenestrated collar of the sarcoplasmic reticulum

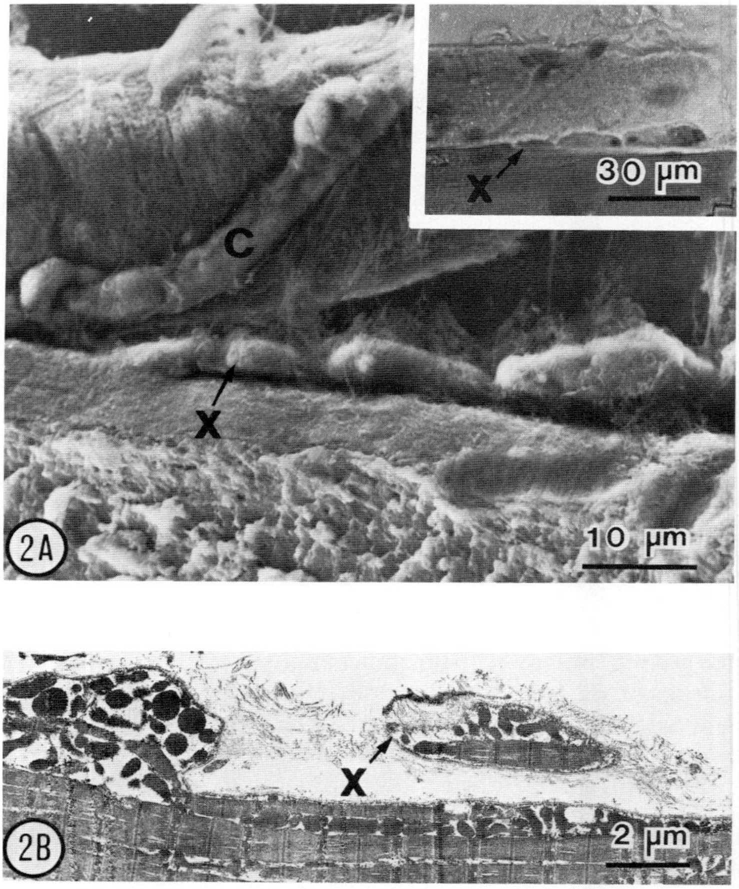

Fig. 2. Biopsy of normal appearing muscle from boy with limb-girdle muscular dystrophy demonstrating the use of correlative microscopy, i.e., the reinterpretation of an LM or SEM image. A, SEM of OTOTO slide preparation showing two myofibers, a capillary (C), and the questionable structure (X). LM (H&E stained) inset of same area. B, TEM of correlated area showing X to be a portion of a longitudinally cut myofiber. (First published: Geissinger *et al.*, 1980, by permission of Hemisphere Publishing.)

Fig. 3. SEM micrograph from a cryostat section of Karnovsky's-fixed mouse gastrocnemius muscle. This is an uncoated specimen which has been prepared for SEM by using the OTOTO procedure. Points a–f and point TB correspond to those same points in B–C. B–C, uncontrasted TEM micrographs which are excerpts from serial sections of the same area seem with SEM in A; a, procedural artifact; b, area where a portion of myofibril is broken off; c,e, transversely oriented mitochondria lying in the depression of the I-band; TB, T-tubule alongside mitochondria; d, tuft from a myofibril; f, a cleft which is an artifact of the cryosectioning procedure. Inset shows T-tubule at higher magnification. (First published: Vriend and Geissinger, 1982, by permission.)

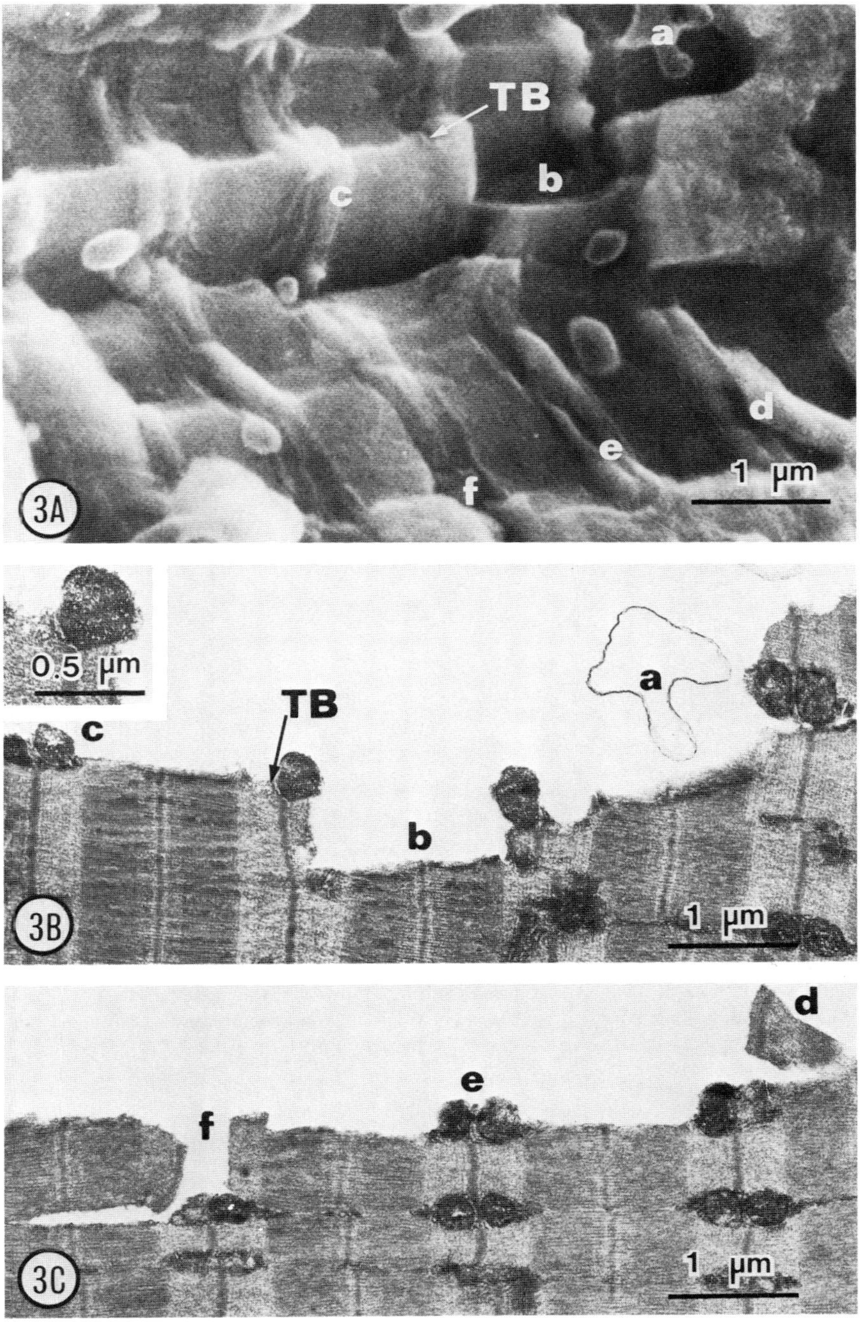
a
TB
b
c
d
e
f
1 μm
3A
0.5 μm
TB
c
a
b
1 μm
3B
d
e
f
1 μm
3C

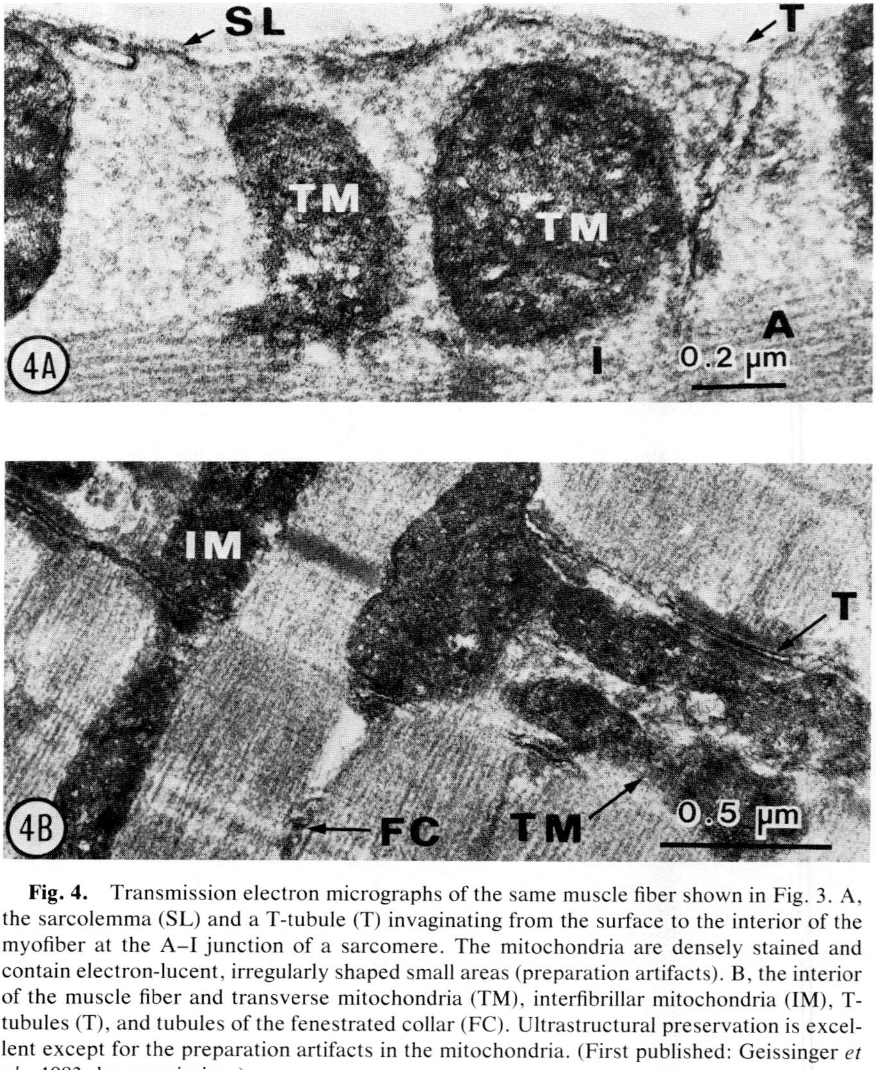

Fig. 4. Transmission electron micrographs of the same muscle fiber shown in Fig. 3. A, the sarcolemma (SL) and a T-tubule (T) invaginating from the surface to the interior of the myofiber at the A–I junction of a sarcomere. The mitochondria are densely stained and contain electron-lucent, irregularly shaped small areas (preparation artifacts). B, the interior of the muscle fiber and transverse mitochondria (TM), interfibrillar mitochondria (IM), T-tubules (T), and tubules of the fenestrated collar (FC). Ultrastructural preservation is excellent except for the preparation artifacts in the mitochondria. (First published: Geissinger *et al.*, 1983, by permission.)

at the H-band region (Fig. 4B) showed excellent preservation, considering the amount of processing and handling this piece of tissue had received.

Endometrium

Buffer Solutions

Phosphate buffer has proved to be adequate for all our processing needs when preparing tissue for EM. However, it is well known that phosphate buffer, if used to store tissue for periods longer than a few hours, can form crystals on the surface of tissue which may obscure important details and also detract from the aesthetic quality of the SEM image. Furthermore, complex salts may form within tissues and interfere with TEM image quality. Alternative hydrogen ion buffers such as *N*-2-hydroxyethylpiperazine-*N'*-2-ethanesulfonic acid, referred to as HEPES and piperazine-*N*,*N'*-bis(2-ethanesulfonic acid), referred to as PIPES (Good *et al.*, 1966), apparently eliminate these problems and are currently being investigated for use in correlative procedures. Indeed, these buffers play a vital role in the preservation of ultrastructural integrity (Schiff and Gennaro, 1979), particularly concerning the retention of lipid components, and may help to improve the overall quality derived from correlative techniques.

SEM Preparation

Silver conducting paint, to affix tissue to the SEM stub, ensures that the specimen is well grounded. This method is superior to tape or other methods of affixing SEM specimens in reducing the amount of charging. However, an excessive volume of paint underneath the specimens may result in a certain amount of shift under the electron beam which makes higher magnification focusing difficult.

It should be noted that critical-point dried tissue is very hygroscopic. If the SEM stubs have been exposed to the air for a long period of time, or if the dissection of selected areas from SEM processed tissue has involved excessive handling, the tissue pieces may absorb moisture. In this case it is advisable to drop the selected pieces into 100% ethanol before starting infiltration procedures. Methacrylate, being water miscible, does not require this step.

Embedding for LM and TEM

Three types of embedding media have been described; each has certain advantages which may be adapted to individual needs. Methacrylate, for example, has superior infiltrative qualities and allows for histological examination of tissue utilizing a variety of stains. Spurr offers the same infiltrative properties as methacrylate and is more stable under the electron beam. It is useful for stubborn specimens which are difficult to infiltrate. Epon is probably less toxic than the other two, and is a more familiar embedding medium for EM purposes. Although its infiltrative properties are not as efficient as methacrylate or Spurr, longer than usual infiltration times seem to compensate for this. In fact, it has been our experience that some specimens refuse to infiltrate efficiently regardless of the medium used.

The recipe of the chosen embedding medium should be adjusted to give a harder consistency than usual, to compensate for the brittle nature of critical-point dried tissue and thus make sectioning easier. The absolute proportions of resin ingredients is a matter of trial and error, and should be left to the preference of individual microtomists.

Preparation for TEM

Critical-point dried tissue is very brittle and great care must be taken when dissecting out selected areas. A certain amount of pressure must be applied, which may result in the propulsion of vital pieces of tissue into oblivion, unless precautions are taken to prevent extensive travel of these projectiles. It is advisable, since the tissue is black, to work on a light-colored surface (white shelf lining laid over the bench surface is adequate).

Selected Results

Since examination of sectioned material alone gives little or no topographical information, the SEM is an obvious choice for examining mucosal surfaces. However, very little information can be derived from the SEM image concerning the intracellular morphology of the specimen. Without the information, interpretation of the SEM image is often difficult.

For example, in a study of postpartum endometrium in sows, large patches of ciliated cells (Fig. 5A) were often seen to surround apparent invaginations in the SE. Examination of these "invaginations" under LM proved they were actually hypertrophied, tubular coiled glands (Fig. 5A,

---

Fig. 5. Porcine endometrium day 8 postpartum; × SE height = 12.2 ± 1.9 μm. A, a large gland opening (G) surrounded by a distinct patch of ciliated cells (arrow) and nonciliated cells. Inset shows a SEM→LM correlated micrograph of the same gland showing the surface epithelium (SE) overlying a dense *zona compacta* and loose (lighter stained) *zona spongiosa*. B, an SEM→TEM correlated micrograph of the gland shown in A illustrating a ciliated glandular epithelium (GE) with secretory vesicles (SV) in the apical cytoplasm; C = cilia. (First published: Rhodes *et al.*, 1983, by permission.)

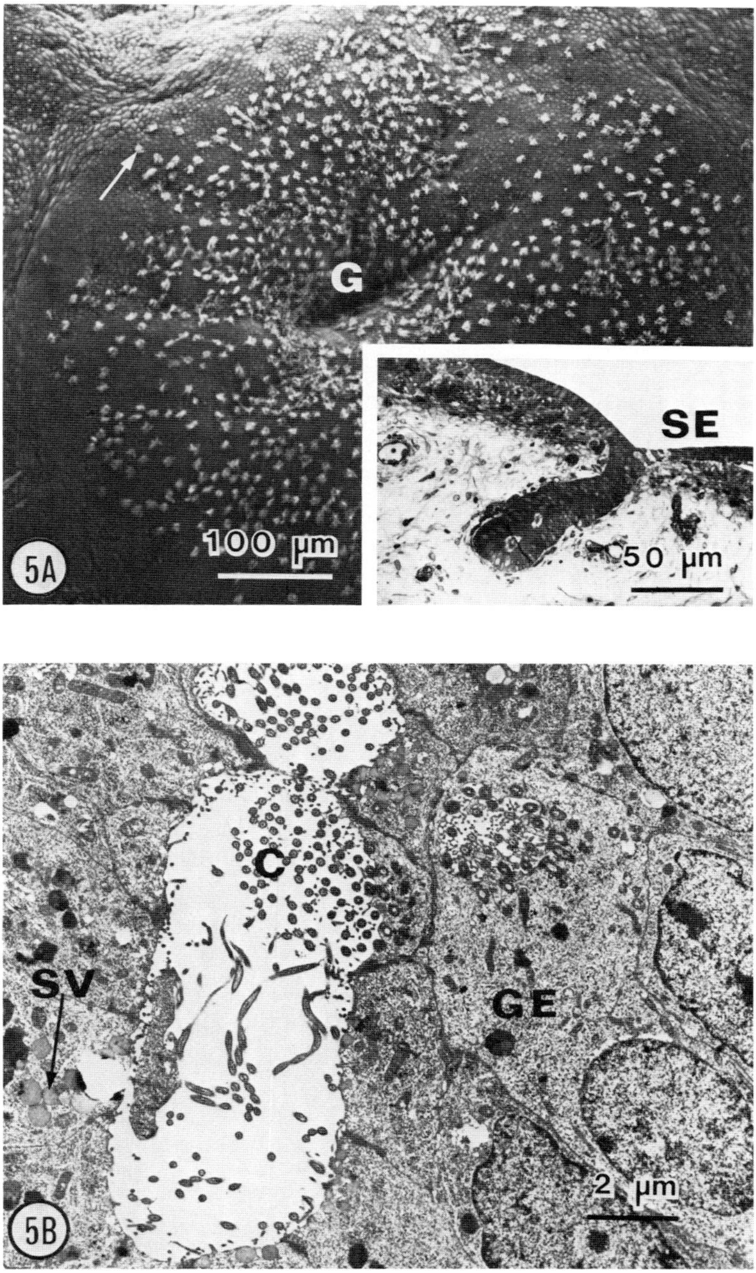
G
SE
100 μm
50 μm
5A
C
SV
GE
2 μm
5B

6A
20 µm

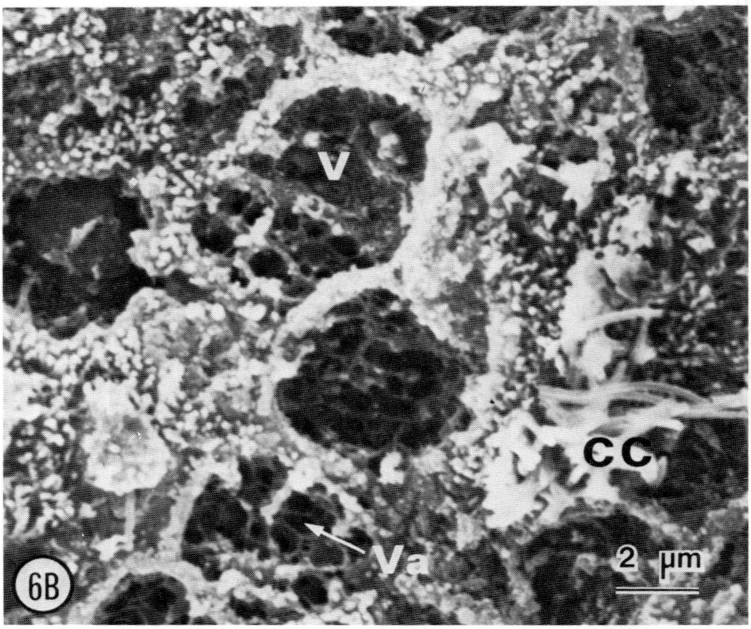
V
CC
Va
6B
2 µm

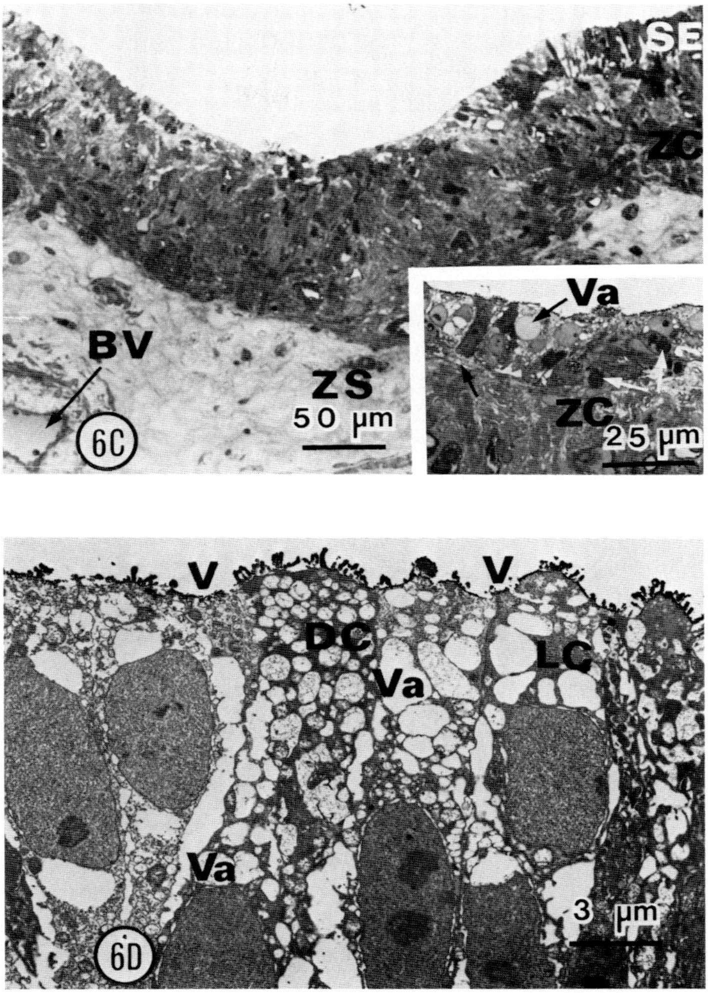

Fig. 6. Different region of same endometrium shown in Fig. 5. A, a region of extensive cellular breakdown. The boxed area is shown at higher power in B. B, large voids (V) in the surface epithelium featuring numerous vacuoles (Va) are evident; CC = ciliated cell. C, a SEM→LM correlated micrograph of the same region featured in A. It can be seen that breakdown of the surface epithelium (SE) is extensive in the "valley" region. ZC = *zona compacta;* ZS = *zona spongiosa;* BV = blood vessel. Inset shows numerous vacuoles (Va) within the cytoplasm of many cells. White arrow = lipid; black arrow = basement membrane; ZC = *zona compacta*. D, a SEM→TEM correlated micrograph of cells from the same region. The voids (V) appear to be a result of loss of apical portions of the SE cells. The vacuoles (Va) are probably grossly dilated endoplasmic reticulum. The nuclei remain relatively intact; DC = dark cell, LC = light cell. (First published: Rhodes *et al.*, 1983, by permission.)

inset). The same gland under TEM examination shows secretory vesicles in the ciliated GE (Fig. 5B). The ultrastructural preservation is good, despite SEM processing, although the preservation of organelles is slightly inferior to regular TEM processing.

In the same study, there were regions of extensive cell breakdown at the endometrial surface (Fig. 6A). Many "voids" in the SE surface were evident (Fig. 6B). It was not clear whether the "voids" represented desquamated cells or disruption in apical regions of individual SE cells. Therefore, this area of the SEM specimen was dissected out, embedded in Epon, and examined under the LM (Fig. 6C). Although extensive degradation of the SE occurred, it still appeared relatively intact (Fig. 6C, inset). TEM examination of the same area revealed extensive vacuolation in the SE cytoplasm. Despite the Au-Pd coating on the SE, it is evident that there are apical regions where the microvilli have been lost, and that these areas represent the "voids" seen under the SEM.

It should be noted in retrospect that the Au-Pd coating can present some problems during sectioning, and that sections should be cut away from the SE to prevent Au-Pd particles from being dragged across the surface of the section by the cutting edge of the knife. Furthermore, this coating may interfere with the efficiency in infiltration of the embedding medium, which is why extended infiltration times are recommended. OTOTO fixation of endometrial samples may eliminate the need for sputter coating, although some loss in SEM image quality can be expected.

This work was supported with financial assistance from the Natural Sciences and Engineering Research Council of Canada and from the Ontario Ministry of Agriculture and Food. The secretarial assistance of Dorothy Pharoah and the technical help of Kim McDonald-Taylor is gratefully acknowledged.

REFERENCES

Albrecht, R. M., and Wetzel, B. (1979). Ancillary methods for biological scanning electron microscopy. *Scanning Electron Microsc.* **3,** 203.

Ayres, A., Allen, J. M., and Williams, A. E. (1971). A method for obtaining conventional histological sections from specimens after examination by scanning electron microscopy. *J. Microsc.* (*Oxford*) **93,** 247.

Bonilla, E. (1977). Staining of transverse tubular system of skeletal muscle by tannic acid-glutaraldehyde fixation. *J. Ultrastruct. Res.* **58,** 162.

Cohen, A. L., Marlow, D. P., and Garner, G. E. (1968). A rapid critical point method using fluorocarbons ("Freons") as intermediate and transitional fluids. *J. Microsc.* (*Paris*) **7,** 331.

de Bault, S. E. (1973). A critical point drying technique for scanning electron microscopy of cells grown on plastic substratium. *Scanning Electron Microsc.* **3,** 317.

Echlin, P. (1975). Sputter coating techniques for scanning electron microscopy. *Scanning Electron Microsc.* **3,** 217.

Fathalla, M. A., Liptrap, R. M., and Geissinger, H. D. (1975). Combined scanning electron and light microscopy of biopsy samples of bovine uterus. *Can. J. Comp. Med.* **39,** 457.

Fathalla, M. A., Liptrap, R. M., Bast, T., and Geissinger, H. D. (1978). Steroid concentrations and correlative uterine microscopy in cows with cystic ovaries. *Scanning Electron Microsc.* **2,** 275.

Geissinger, H. D. (1971). Correlated light optical and scanning electron microscopy of Gram smears of bacteria and paraffin sections of cardiac muscle. *J. Microsc. (Oxford)* **93,** 109.

Geissinger, H. D. (1976). Intermicroscopic (LM, SEM, TEM) correlation. *In* "Principles and Techniques of Scanning Electron Microscopy: Biological Application" (M. A. Hayat, ed.), Vol. 5, p. 94. Van Nostrand-Reinhold, New York.

Geissinger, H. D., and Kamler, H. (1972). Precise and fast correlation of light microscopic and scanning electron microscopic images. *Can. Res. Dev.* **5,** 13.

Geissinger, H. D., Yamashiro, S., and Ackerley, C. A. (1978). Preparation of skeletal muscle for intermicroscopic (LM, SEM, TEM) correlation. *Scanning Electron Microsc.* **2,** 267.

Geissinger, H. D., Liptrap, R. M., and Ackerley, C. A. (1979). Correlative uterine microscopy and hormone levels in cyclic sows. *Scanning Electron Microsc.* **3,** 177.

Geissinger, H. D., Vriend, R. A., Ackerley, C. A., and Yamashiro, S. (1980). Correlative light optical, scanning electron, and transmission electron microscopy of skeletal muscle of muscular dystrophy and muscular atrophy: A pilot study. *Ultrastruct. Pathol.* **1,** 327.

Geissinger, H. D., Vriend, R. A., Meade, L. D., Ackerley, C. A., and Bhatnager, M. K. (1983). Osmium-thiocarbohydrazide-osmium versus tannic acid-osmium staining of skeletal muscle for scanning electron microscopy and correlative microscopy. *Trans. Am. Microsc. Soc.* **102,** 390.

Good, N. E., Winget, G. D., Winter, W., Connolly, T. N., Izawa, S., and Singh, R. M. M. (1966). Hydrogen ion buffers for biological research. *Biochemistry* **5,** 467.

Karnovsky, M. J. (1965). A formaldehyde-glutaraldehyde fixative of high osmolality for use in electron microscopy. *J. Cell Biol.* **27,** 137A.

Malick, L. E., and Wilson, R. B. (1975). Modified carbohydrazide procedure for scanning electron microscopy: Routine use for normal, pathological, or experimental tissue. *Stain Technol.* **50,** 265.

Poh, R. L., Altenhoff, J., Abraham, S., and Hayes, T. (1971). Scanning electron microscopy of myocardial sections originally prepared for light microscopy. *Exp. Mol. Pathol.* **14,** 404.

Rhodes, P. A., Liptrap, R. M., and Geissinger, H. D. (1983). A correlative study of porcine endometrium and hormone levels during early lactation and the late luteal phase. *Scanning Electron Microsc.* **4,** 1887.

Schiff, R. I., and Gennaro, J. F., Jr. (1979). The role of the buffer in the fixation of biological specimens for transmission and scanning electron microscopy. *Scanning* **2,** 135.

Vriend, R. A., and Geissinger, H. D. (1980). An improved direct intermicroscopic (LM SEM TEM) correlative procedure for the examination of mammalian skeletal muscle. *J. Microsc. (Oxford)* **120,** 53.

Vriend, R. A., and Geissinger, H. D. (1982). Mammalian skeletal muscle: an interpretation of the scanning electron microscopic image. *Trans. Am. Microsc. Soc.* **101,** 117.

Wetzel, B., Erickson, B. W., Jr., and Levis, R. (1973). The need for positive identification of leukocytes examined by scanning electron microscopy. *Scanning Electron Microsc.* **3,** 535.

Wickham, M. G., and Worthen, D. M. (1973). Correlation of scanning and transmission electron microscopy in a single specimen. *Stain Technol.* **48,** 63.

Wolosewick, J. J. (1984). Cell fine structure and protein antigenicity after polyethylene glycol processing. *In* "The Science of Biological Specimen Preparation" (J.-P. Revel, T. Barnard, and G. H. Haggis, eds.), p. 83. SEM Inc., AMF O'Hare, Illinois.

13

Correlative Microscopy of Isolated Glomeruli

S. B. OR AND LILY MA*

Electron Microscope Unit
**Department of Pathology*
University of Hong Kong
Hong Kong

INTRODUCTION

The scanning electron microscope (SEM), with its superior resolution and depth of field, is well suited for stereoscopic study of surface ultrastructure of biological samples. Buss and Krönert (1969) were the first to

CORRELATIVE MICROSCOPY IN BIOLOGY:
INSTRUMENTATION AND METHODS

use the SEM for studying kidney tissue, and 2 years later, Arakawa (1971) employed the SEM for the visualization of human glomeruli. Since then, scanning electron microscopy has been extensively used for the examination of fine structures of normal and diseased kidneys from both human and experimental animals (Arakawa and Tokunaga, 1972, 1974; Buss *et al.,* 1973; Wheeler and Herdson, 1973; Lehtonen *et al.,* 1973; Humphreys *et al.,* 1974; Andrews, 1975a,b, 1977; Lawler *et al.,* 1976; Arakawa *et al.,* 1977a,b; Jones, 1977, 1978, 1979a,b, 1983; Tarpey and Williams, 1980a,b; Ng *et al.,* 1982, 1983; Or *et al.,* 1983, 1984). It is obvious that the study of human glomeruli is increasing and the diagnostic application of scanning electron microscopy to renal biopsies is becoming important.

These percutaneous needle biopsies are usually small and each contains a few glomeruli or just one glomerulus alone. The conventional methods of razor blade sectioning (Buss *et al.,* 1973; Andrews, 1979), freeze drying, and cryofracturing (Wheeler and Herdson, 1973; Humphreys *et al.,* 1974; Tarpey and Williams, 1980a) and ultraplaning (Jones, 1977, 1981) expose the glomeruli by chance and are not suitable for the tiny biopsy specimens. The isolation of a single glomerulus is therefore essential for effective observation of the glomerular surface in the SEM (Tokunaga *et al.,* 1975; Arakawa *et al.,* 1977b; Jones, 1977). Moreover, as the mesangium and basement membrane are important reaction sites of glomerular lesions, it is desirable to examine also the cross section of a glomerulus. This is achieved by scanning electron microscopy of a single glomerulus, which is then embedded and sectioned for transmission electron microscopy and light microscopy (Or *et al.,* 1983, 1984). Apart from direct correlation of SEM observations with those of the transmission electron microscope (TEM) and light microscope (LM) of the same glomerulus, it also spares the limited amount of renal tissue for other morphological techniques such as immunofluorescent microscopy. This chapter presents a detailed description of the method for the preparation of an isolated glomerulus for comparative study with the SEM, TEM, and LM.

METHOD FOR CORRELATIVE MICROSCOPY

The preparation of a single glomerulus for correlative microscopy involves proper fixation of the specimen, isolation of a glomerulus, postfixation with osmium tetroxide, dehydration, and critical-point drying for SEM observation. This is followed by careful removal of the glomerulus from stud for epoxy resin infiltration and embedding. The embedded glomerulus is then sectioned for transmission electron and light microscopic evaluation.

Fixation and Isolation of Glomeruli

Fixation is an essential initial step in the preparation of soft tissues for microscopic examination. Careful and prompt fixation is necessary to arrest autolytic changes and to ensure good ultrastructural preservation. The best way to achieve this is by perfusion fixation followed by immersion fixation of the tissue in an appropriate fixative (Andrews, 1977, 1979; Hayat, 1981, 1986, 1987). Whenever possible, this fixation procedure should be employed, especially when working with experimental animals. However, as vascular perfusion is not applicable to human tissues, immersion fixation remains the method of choice for renal biopsy specimens. Currently we fix renal tissues in 2.5% glutaraldehyde in 0.1 *M* Na-cacodylate buffer (pH 7.2) at 4°C for 4 hr, or overnight if the specimens are received in the afternoon. This is followed by several changes of buffer wash and then glomerular isolation. Jones (1977) found that briefly fixed glomeruli isolated soon after biopsy were cleaner and less deformed than those with prolonged fixation.

Glomerular isolation is carried out by fine needle dissection (Ng *et al.*, 1982) under a dissecting microscope (Nikon SMZ-10) with oblique light illumination. The fixed kidney tissue is placed in a small amount of buffer on a clean surface, such as a piece of black Perspex or the glass cover of a coplin jar placed on a dark background. Glomeruli are identified under the dissecting microscope as small, round structures with a reddish tint in the renal tissue. The rough spherical surfaces of glomeruli in needle biopsies are often partly exposed so that they are easily recognized. In some cases, however, the entire tissue may appear whitish in color. Careful examination of specimen under the microscope is then necessary. Identification of glomeruli is facilitated by changing the direction of light incident on the specimen. This is accomplished by rotating the Perspex holding the specimen or by gently moving the tissue with a fine needle. Jones (1977) stained the fixed biopsy specimen in borax toluidine blue for 30 sec and then examined the tissue in buffer under a dissecting microscope for easy recognition of glomeruli. In rat kidney tissues, the glomeruli are smaller than those of human and they usually appear as faint red dots. Identification of glomeruli is much easier in thin slices of the tissue.

Once a glomerulus is located, it is carefully teased out from its Bowman's capsule by using a pair of fine needles. Fine-point tweezers or 27-gauge syringe needles are suitable for this purpose. Other glomeruli are isolated in the same manner with extreme care to avoid damage of glomeruli in this maneuver. Distorted glomeruli and tissue debris are then identified and discarded. The remaining ones are pipetted into a well in a hemagglutination plate (Baird & Tatlock) containing cacodylate buffer. If the specimen is not processed immediately, it should be rinsed several

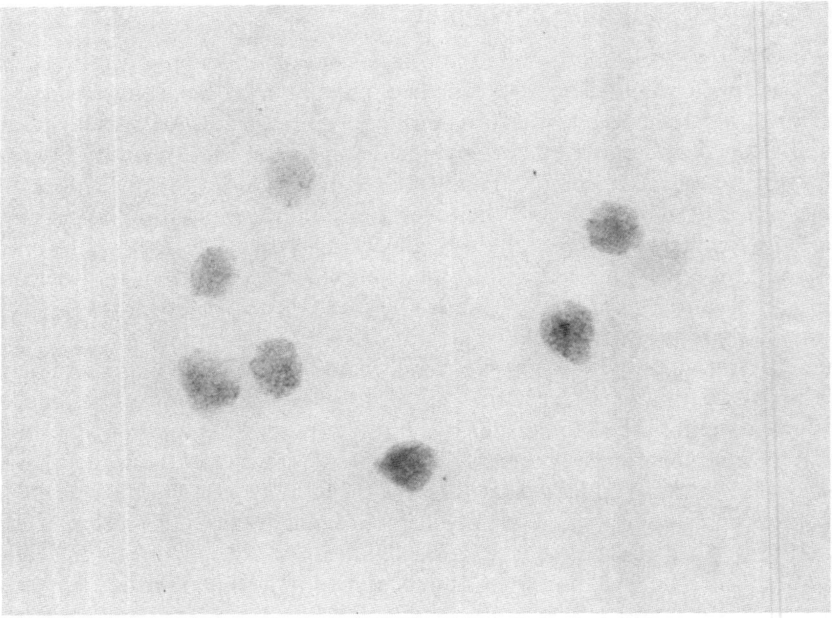

Fig. 1. The appearance, under a dissecting microscope, of several isolated human glomeruli in buffer (×38).

times in buffer and kept in the same buffer at 4°C until further processing. Several glomeruli in the well of a white porcelain plate are shown in Fig. 1.

It is important to have a good quality dissecting microscope with a bright light source so that identification and isolation of glomeruli are facilitated. A dark background helps in locating the glomeruli, which are usually whitish in color, once they are freed from the tissue. It should also be noted that the tissue and glomeruli are kept immersed in buffer at all times during the isolation process and should not be allowed to dry.

Preparation of Isolated Glomeruli for Scanning Electron Microscopy

Since the specimen is intended for correlative study by the SEM and TEM, postfixation of isolated glomeruli in OsO_4 is desirable. This is necessary to reduce volume change during preparation for scanning electron microscopy, prevent the leaching of plasmalemmal components (Boyde and Boyde, 1980; Boyde and Maconnachie, 1981), preserve lipids, and improve contrast of ultrathin sections.

The washed glomeruli in the well of a hemagglutination plate are post-fixed with 1% OsO_4 in 0.1 *M* Na-cacodylate buffer for 1 hr at room temperature. This blackens the glomeruli, making them readily visible with the naked eye. After osmication, they are rinsed in several changes of buffer and are pipetted onto a small piece of Chinese lantern paper (Fig. 2), which is then folded into an envelope. The wrapped package is loaded into a specimen container (Fig. 2), immersed in cacodylate buffer, and labeled. The specimen is then dehydrated through a graded series of ethanol as follows:

70% ethanol, 15 min
90% ethanol, 15 min
100% ethanol, three changes, 15 min each
20% Freon TF in 100% ethanol, 15 min
40% Freon TF in 100% ethanol, 15 min
60% Freon TF in 100% ethanol, 15 min
80% Freon TF in 100% ethanol, 15 min
Pure Freon TF, three changes, 15 min each

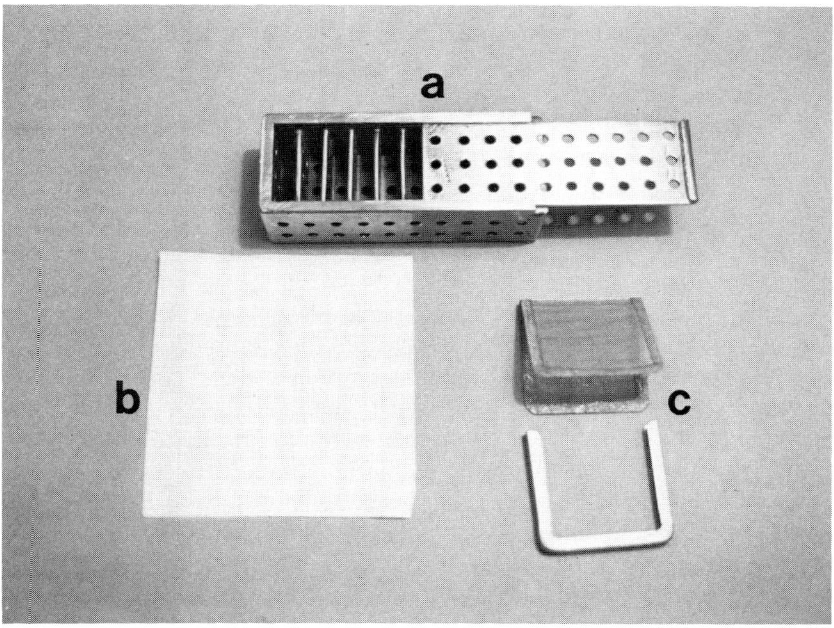

Fig. 2. Photograph of (a) a ten-compartment tissue cassette; (b) a piece of Chinese lantern paper; and (c) container for processing one package of glomeruli for SEM.

During dehydration, the specimen is agitated frequently to facilitate fluid exchange. Drying is completed in a critical-point dryer (Bomar SPC-1500) using Freon 13 as a transitional fluid. Alternatively, acetone can be used for dehydration. Following the final change of pure acetone, the specimen in acetone is transferred into the specimen chamber of a critical-point dryer and dried from liquid carbon dioxide. Cohen (1974, 1979) and Hayat (1978) have given a detailed account of the critical-point drying procedure.

Chinese lantern paper or any other paper such as filter paper can be used provided it is soft but durable when wet; resistant to immersion in water, ethanol, and Freon; not too porous to allow glomeruli to pass through; and allows fluid exchange during specimen processing. Tokunaga *et al.* (1975) and Arakawa *et al.* (1977a,b) used carbon-coated copper grids and grids hydrophilized by glow discharge for holding glomeruli during dehydration and critical-point drying. Grids of 300–400 and 150–180 mesh were found suitable for rat and human glomeruli, respectively. Jones (1977) employed small, fine mesh, nylon cloth bags for the same purpose.

To enable the processing of biopsy specimens in bulk, a stainless steel tissue cassette with 10 compartments (Fig. 2) was constructed (Ng *et al.*, 1982). It was designed to fit into a coplin staining jar and the specimen chamber of our critical-point dryer. A maximum of 10 samples are loaded each time and dehydration is done by hand in a coplin jar with the tissue cassette *in situ*. The latter is then transferred into the bomb of a critical-point dryer to complete the drying process. This saves time and trouble in handling individual samples. Similar containers may be constructed easily in other laboratories to suit their own purpose.

The critical-point dried package is then unfolded carefully under a dissecting microscope. Each glomerulus is transferred by means of a short hair and touched gently onto an adhesive silver tape, which is secured on an aluminium stud by a thin layer of quick-drying glue (Or *et al.*, 1983). The stud is then labeled on the undersurface for identification. Specimen studs with the adhesive tape on top may be prepared in advance and kept desiccated so that they are ready for use when needed. We prefer using silver tape to double-side tape as the latter attaches firmly to the stud and is difficult to remove later with the glomerulus intact. It should be noted that the glomerulus can drop easily and be lost during transfer. It is therefore advisable to hold the stud close to the glomerulus before picking it up with a hair. Also, a piece of clean filter paper on the baseplate of the dissecting microscope aids the location of the glomerulus if it is dropped.

The mounted glomerulus is ready for metal coating. This is undertaken to impart electrical conduction to the specimen surface to avoid charging artifact and thermal damage during electron bombardment. In addition, as most biological tissues are composed of light elements, a surface layer of heavy atoms also provides a high yield of secondary electrons, thus ensuring good signal. The coating is commonly done by depositing a thin layer of heavy metal onto the specimen surface by vacuum evaporation or by sputtering (Hayat, 1978; Echlin, 1978, 1981; Echlin *et al.*, 1982). The latter method is more popular because of its rapidity and simplicity. We routinely sputter a thin layer (15–20 nm) of gold-palladium onto the glomerular surface. This metallic film forms a continuous layer on specimen surface giving good electrical conduction. Metallic film of small grain size, which is critical for high-resolution microscopy, is obtainable by sputtering at low voltage in a good vacuum system with a low rate of deposition, low substrate temperature, and reduced film thickness (Echlin *et al.*, 1982). To minimize the thermal damage during coating, metal deposition may be done in a ''cool'' sputter coater having a cold stage and by coating specimens in several shots with cooling-off periods in between.

Observation of Glomeruli in the Scanning Electron Microscope

The coated glomerulus may now be introduced into the SEM (Cambridge Stereoscan 150) for stereoscopic examination and an accelerating voltage of 10–20 kV is normally used. It is recommended that freshly prepared specimens should be examined in the SEM as soon as possible after coating. The reason is that the volume change caused by dehydration and critical-point drying continues even after coating and during storage of specimens (Boyde and Boyde, 1980; Boyde and Maconnachie, 1981).

When examining renal tissues prepared by razor sectioning or freeze fracturing/drying, sometimes some glomeruli are totally obscured by their Bowman's capsules, only partially exposed for examination (Fig. 3), or they detach from the tissue. This imposes problems to renal biopsy specimens which are usually small and it is only by chance that a glomerulus is exposed for SEM observation. This problem can be circumvented by using the glomerular isolation technique. With the latter, a glomerulus may be continuously rotated and tilted so that the majority of the glomerular surface is available for visualization. Figure 4 shows an isolated rat glomerulus prepared by the present method and examined at 0° tilt. It is oval in shape and appears intact with a clean surface. The capillary loops with overlying podocytes are evident and the vascular pole is seen at bottom center. Figure 5 is the same glomerulus which has been rotated 180° and tilted 35°. Higher magnification (Fig. 6) shows clearly the inter-

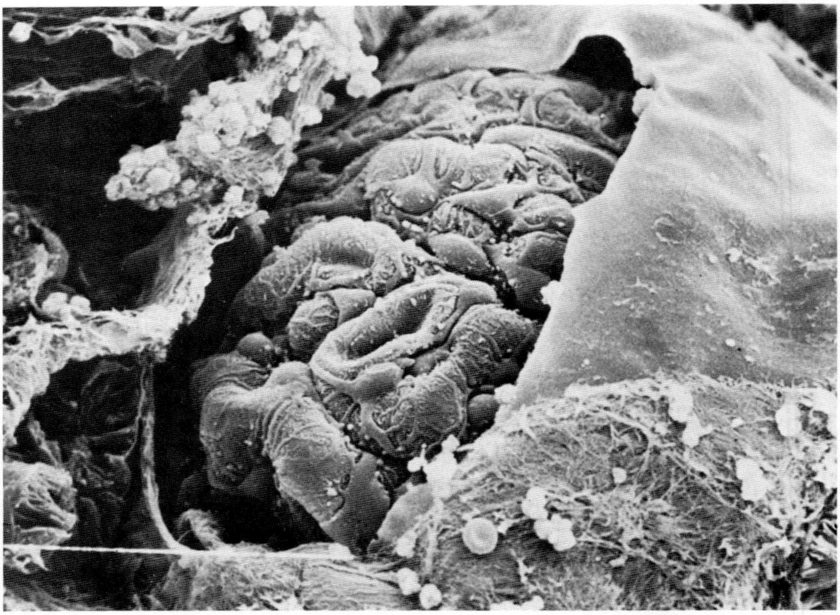

Fig. 3. Rat kidney tissue prepared by razor sectioning. A glomerulus is only partially exposed for SEM study (×1000).

digitating foot processes which are regularly arranged but vary in length and size. Short microvilli are seen arising from podocyte cell bodies and foot processes.

Preparation for Transmission Electron Microscopy and Light Microscopy

The scanned glomerulus is taken out from the microscope after SEM examination. With the aid of fine needles and a dissecting microscope, adhesive in the immediate surroundings of the glomerulus is carefully scraped away. The silver tape with the glomerulus *in situ* is then removed with care from the stud and immersed in 100% ethanol in a well of a white porcelain plate. The glomerulus is gently detached from the tape with fine needles and processed by hand for TEM and LM as follows:

100% ethanol, 5 min
Propylene oxide, 2 min
Epoxy resin/propylene oxide 1 : 1 mixture, 30 min

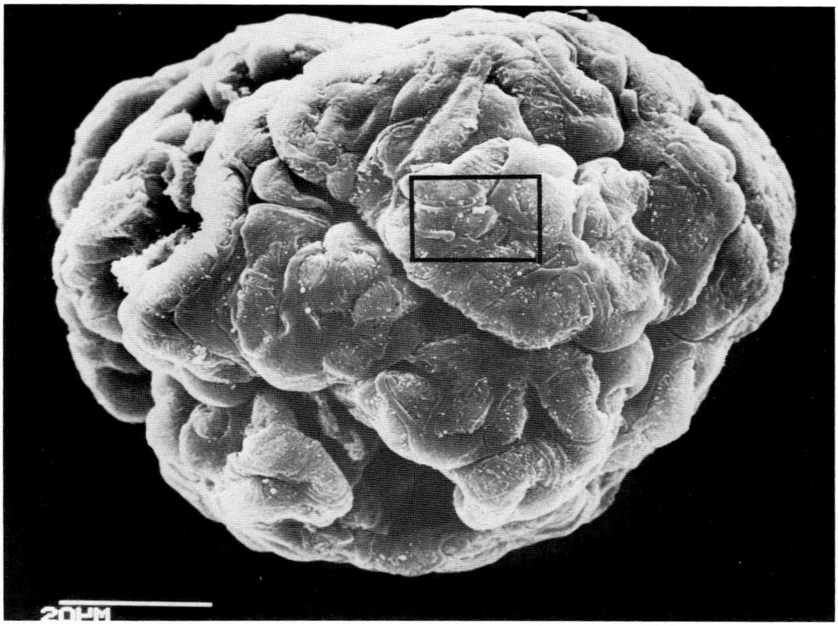

Fig. 4. Normal rat glomerulus, viewed at 0° tilt, showing the vascular pole at bottom center. Convoluted capillary loops with overlying podocytes are evident. The ruled area is shown at higher magnification in Fig. 6.

Epoxy resin/propylene oxide 3 : 1 mixture, 30 min
Pure epoxy resin, two changes, 30 min each

After adequate infiltration, the glomerulus is pipetted onto a disposable aluminium weighing dish or a plastic cap. A dried, labeled gelatin capsule filled with pure epoxy resin is then inverted over the glomerulus and the resin is polymerized at 60°C for 24 hr. The cured resin block is trimmed into a truncated pyramid and sectioned with a glass knife on an ultramicrotome to expose the glomerulus. Semithin sections are picked up on glass slides and stained with toluidine blue for general survey and to determine whether further cutting is necessary. When the required level is reached, a good section is obtained for toluidine blue staining. Ultrathin sections are then prepared with a diamond knife and mounted on 50-mesh Formvar-coated grids. They are stained with uranyl acetate and lead citrate as usual and viewed in a TEM (Philips EM 300) at 60–80 kV. Several semithin sections (0.5–1 μm) are cut following ultrathin sectioning and are mounted on albuminized glass slides for subsequent staining for light microscopy.

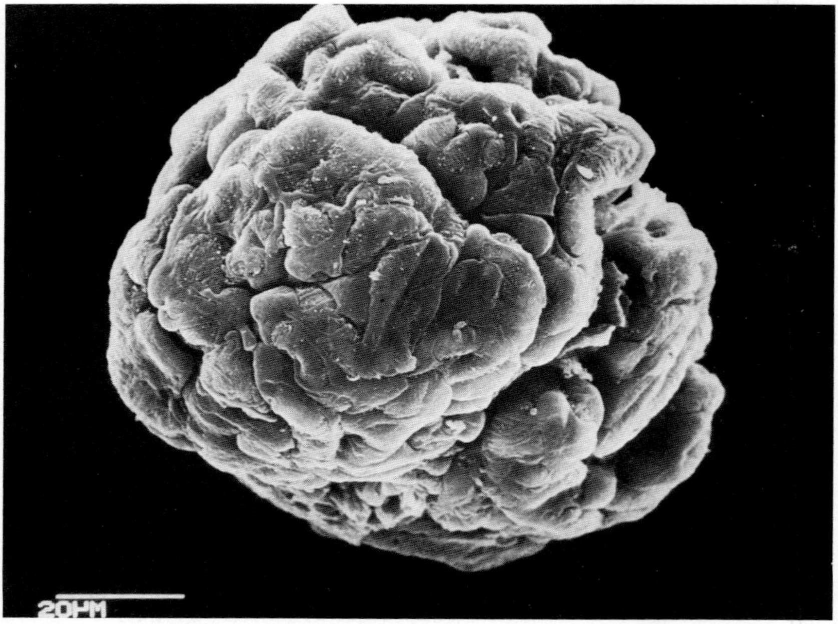

Fig. 5. The same glomerulus is rotated 180° and tilted 35°. A clean, intact surface allows thorough examination of the glomerulus by SEM.

Staining of Semithin Sections for Light Microscopy

The examination in the LM of semithin sections of plastic-embedded tissues, prior to ultrathin sectioning, has become a routine step in the preparation of specimens for electron microscopy. This is necessary for orientation and to assure that appropriate samples have been taken for electron microscopic study. The increased resolution, better preservation of tissues, and enhanced morphological details often provide additional information. As stated by Cardno and Steiner (1965), examination of semithin sections became increasingly useful as knowledge acquired by electron microscopy was applied to their interpretation. Light microscopic study of semithin sections was as informative as low-power electron micrographs and became an important adjunct in the diagnosis of glomerular pathology.

Although a great number of methods are available for staining semithin sections of osmicated, plastic-embedded tissues (Burns, 1978), not all of them are necessary for the study of renal tissues. In this connection, the most commonly used stains are toluidine blue, hematoxylin and eosin, methenamine silver, and periodic acid-Schiff (PAS).

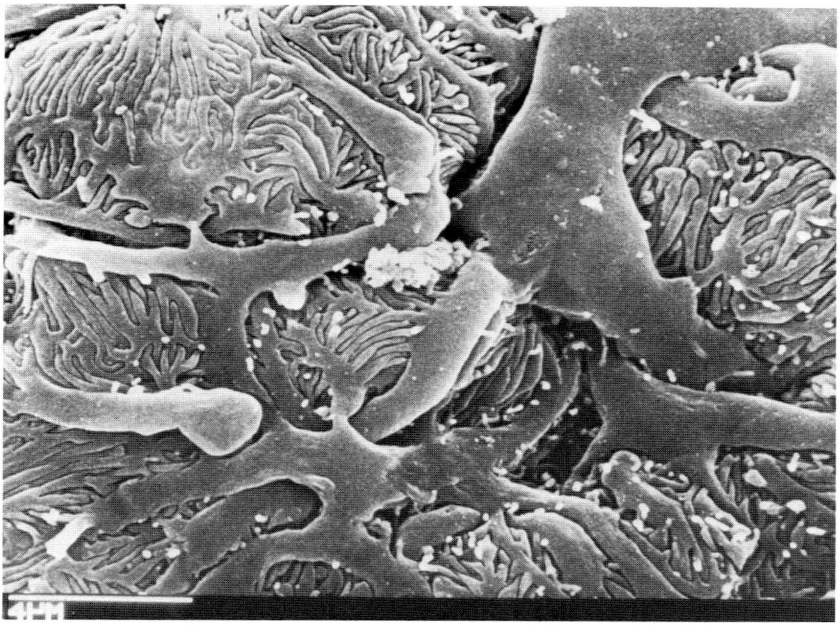

Fig. 6. Close-up view of Fig. 4 glomerulus showing the well-arranged interdigitating foot processes. Short microvilli are seen on podocyte surface.

Toluidine Blue

A double-filtered solution of 1% toluidine blue in 1% borax (Burns, 1978) is used. Semithin sections are mounted on uncoated glass slides and dried on a hot plate at about 60°C. They are stained for 10–20 sec on the hot plate, washed in water, dried on the hot plate, and mounted with cover glasses.

Hematoxylin and Eosin

Semithin sections are mounted on albumin-coated glass slides and dried on a hot plate. They are stained with hematoxylin and eosin according to Chang's method (1972).

Methenamine Silver

"Free-floating" sections are treated with periodic acid and methenamine silver according to Cardno and Steiner (1965). However, the handling of unmounted sections is tedious and may result in the loss of sections. An alternative method is to mount sections on albuminized glass slides. They are etched for 6 min in a freshly prepared solution of 2%

potassium hydroxide in equal volumes of propylene oxide and methanol (Chiu *et al.*, 1986), then rinsed in methanol and washed in running water. Afterward, the sections are treated as follows:

1% Periodic acid for 5 min, wash in water
5% Chromic acid for 15 min, wash in water
2% Sodium bisulfite for 2 min, wash well in distilled water
Stain with methenamine silver solution (Grocott, 1955) at 60°C for 20 min, then check microscopically at frequent intervals; basement membrane should be dark brown. Then wash well in distilled water
Tone in 0.2% gold chloride for 2 min, wash in water
Treat in 3% sodium thiosulfate for 3 min, wash in water
Lightly counterstain the sections in 0.2% light green, wash in water
Dehydrate, clear, and mount

Periodic Acid-Schiff

Mounted sections are etched in 2% potassium hydroxide in propylene oxide and methanol as described above. They are then treated with 1% periodic acid and stained with Schiff's reagent according to the method of McManus (1946). The nuclei are stained with Harris hematoxylin (Chang, 1972).

Alternative Method

As described in a previous section, a piece of adhesive silver tape glued on a stud is used for mounting the glomerulus. After SEM observation, the glomerulus is carefully detached from the tape, using fine needles under a dissecting microscope, for further processing. Such manipulation might cause damage to the glomeruli or specimen loss during fluid exchange in subsequent preparation for resin embedding. An alternative method is to use a small piece of resin support to hold the glomerulus so that after SEM study, the glomerulus is processed *in situ* on the resin for final embedding (Or *et al.*, 1984).

The dried glomerulus is mounted with a small amount of quick-drying glue on a piece of epoxy resin sheet, which is glued on a stud previously and electrically connected to the latter with colloidal silver or carbon. After drying of the glue, the glomerulus is sputter-coated with gold-palladium and viewed in the SEM. Afterward, the resin sheet with the attached glomerulus is removed from the stud. The glomerulus is processed *in situ* on the resin from 100% ethanol to pure epoxy resin. A gelatin capsule filled with fresh epoxy resin is then inverted over the infiltrated glomerulus, which is still attached to its resin support. Polymerization of the embedding medium is carried out at 60°C for 24 hr. Semithin and ultrathin sections are then prepared as described above.

The mounting of the glomerulus on a resin sheet should be done with caution and a minimum amount of quick-drying glue is used to fix the glomerulus. Excess glue tends to cover the glomerular surface and will ruin the preparation by obscuring surface details and impede infiltration of the glomerulus by epoxy resin in subsequent preparation. Provided that it is properly mounted, the glomerulus will remain attached to the resin sheet throughout the process. The handling of specimen during TEM preparation is thus facilitated. In addition, polymerization of epoxy resin may be accelerated by incubation in an oven at 100°C for $1\frac{1}{2}$ hr. This shortens the preparatory time for transmission electron microscopy.

ILLUSTRATED EXAMPLES

Five glomeruli from renal biopsy specimens and from a normal rat kidney are included here for illustration. One to two glomeruli were isolated from each biopsy and a few glomeruli were teased out from the rat kidney tissue. They were processed by the present method for correlative study.

Figure 7 is a semithin section of the rat glomerulus shown in Figs. 4–6. The capillary basement membrane is clearly outlined by the methenamine silver staining. A transmission electron micrograph of the same glomerulus is illustrated in Fig. 8. The internal structures are maintained with no discernible damage or distortion. Examination at a higher magnification reveals that the cytoplasmic organelles are well preserved.

The glomerulus from a case of diabetes is shown in Figs. 9–13. Scanning electron microscopy reveals flattening and marked effacement of foot processes (Figs. 9 and 10). Swelling of podocyte cell bodies, some of which have short microvilli, are noted in different areas. Semithin section of the same glomerulus stained with toluidine blue shows the characteristic feature of nodular glomerulosclerosis (Fig. 11). Several discrete round nodules with compressed mesangial cells on the outside are apparent. These nodules vary in size and are located mainly on the periphery of the glomerulus. Transmission electron microscopy shows clearly a similar feature (Fig. 12). Higher magnification reveals the accumulation of compacted mesangial matrix in the nodules (Fig. 13). Extensive fusion of foot processes and thickening of the tortuous basement membrane are also evident.

Figures 14–18 show the glomerulus from a case of minimal change glomerulonephritis. The chief complaint of the patient is proteinuria. Nu-

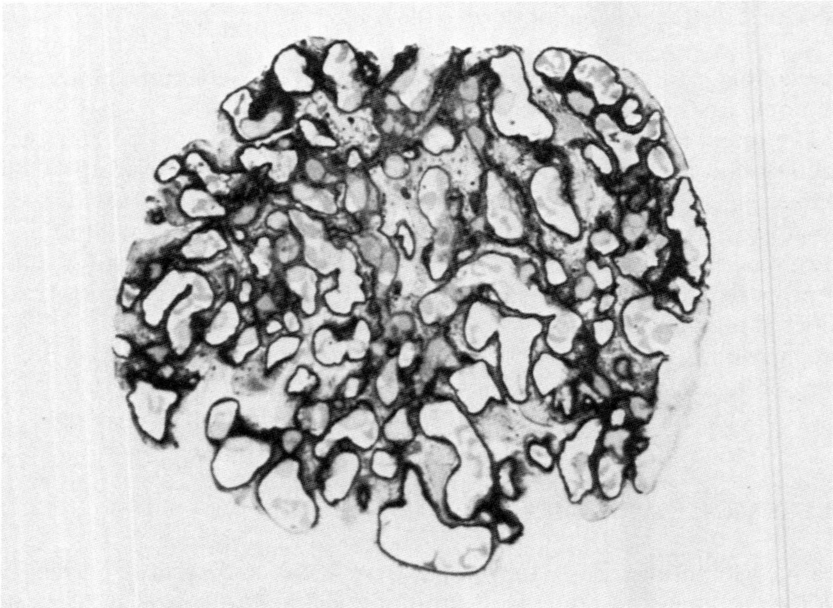

Fig. 7. Semithin section of the glomerulus shown in Fig. 4, stained with methenamine silver to demonstrate the capillary basement membrane (×760).

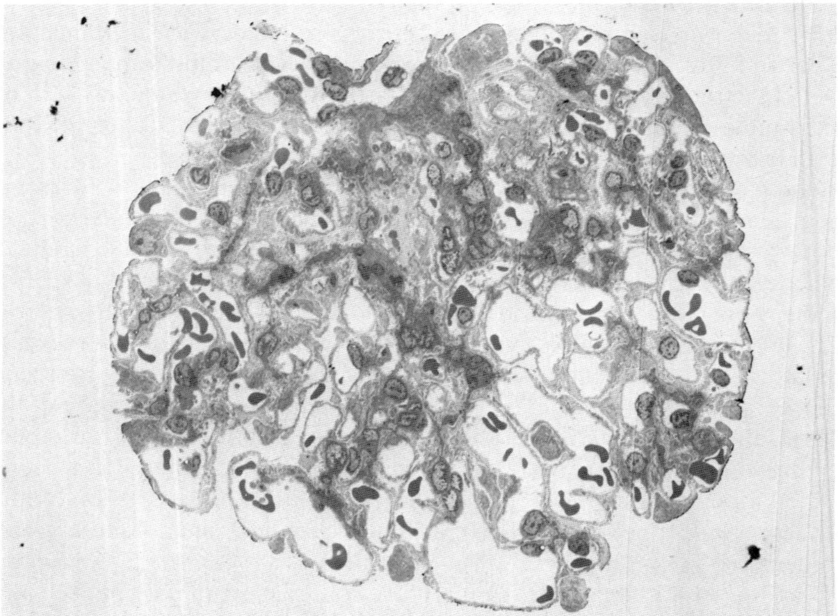

Fig. 8. TEM of Fig. 4 glomerulus showing the well-preserved glomerular features (uranyl acetate and lead citrate staining, ×830).

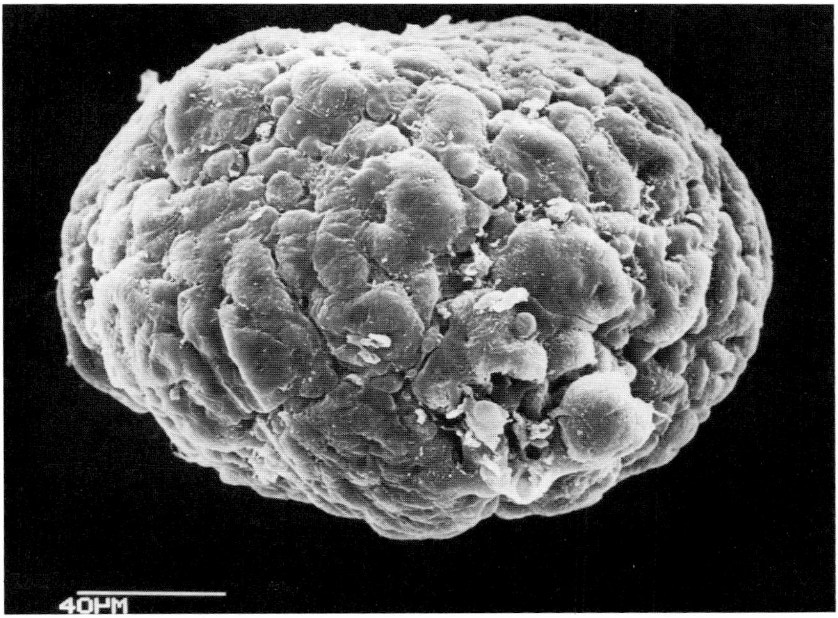

Fig. 9. Glomerulus from a case of diabetic glomerulosclerosis shows flattening of podocyte processes in most areas. A few podocytes with swollen cell body are also visualized.

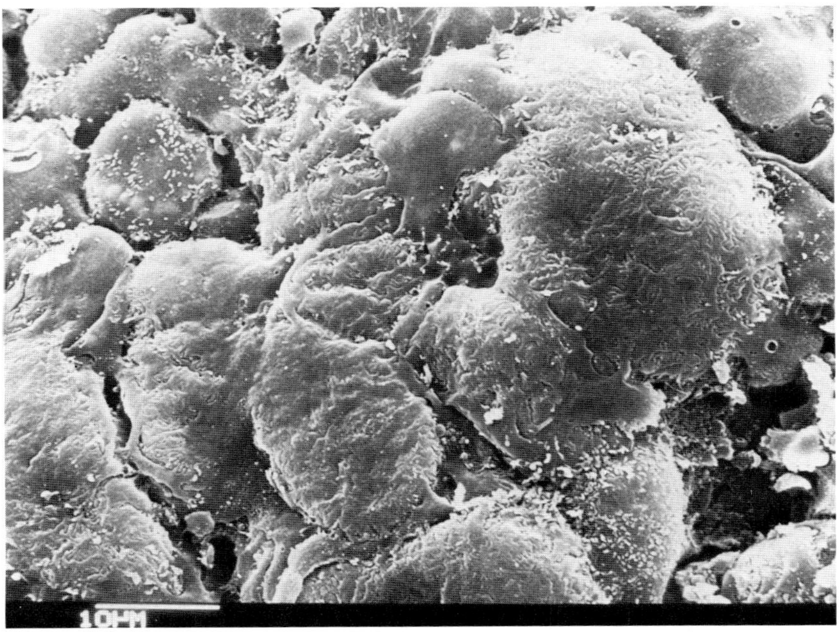

Fig. 10. Close-up view of Fig. 9 glomerulus showing marked fusion of foot processes. Short microvilli are present on the cell body of some podocytes.

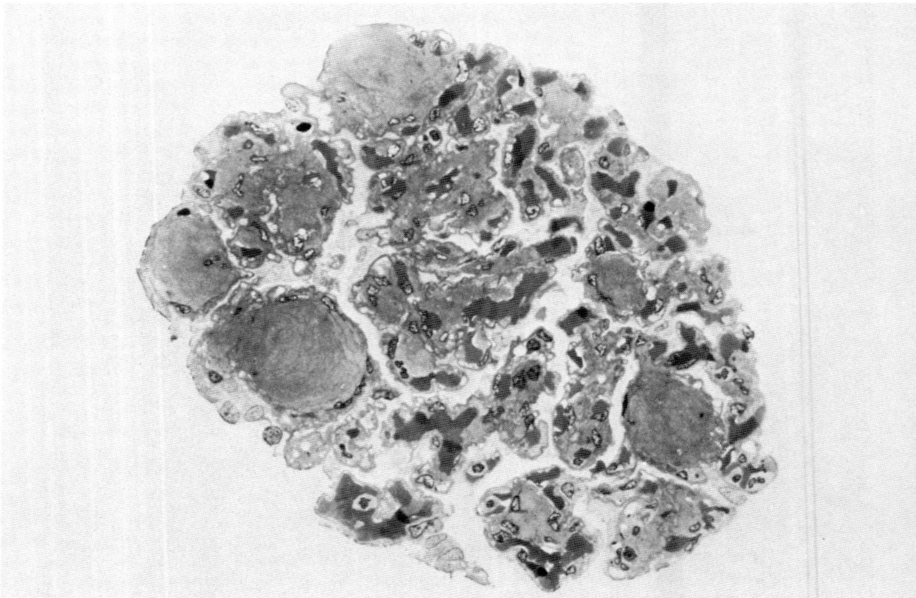

Fig. 11. Diabetic glomerulosclerosis. Toluidine blue staining on semithin section shows a few nodules on the periphery of the glomerulus (×440).

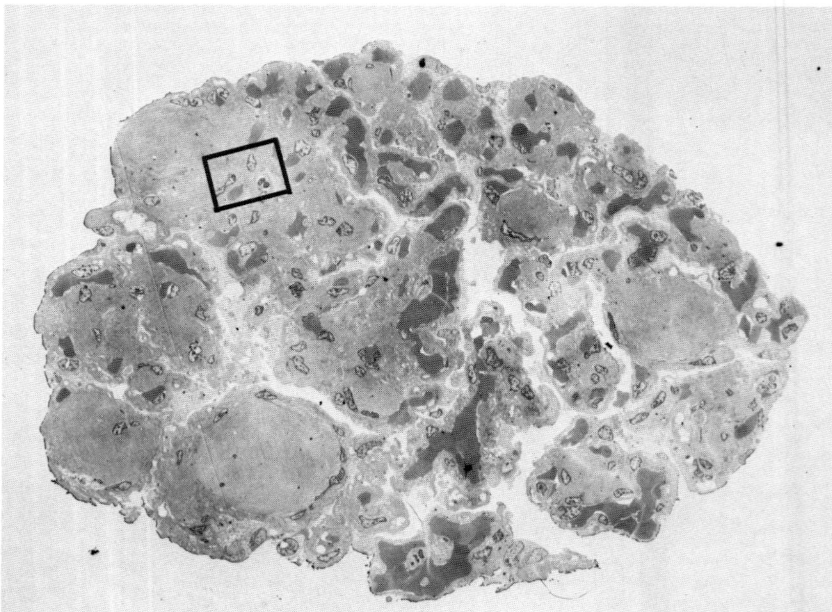

Fig. 12. TEM micrograph of Fig. 9 glomerulus showing prominent nodules of variable size. One nodule is surrounded neatly by mesangial cells. The enclosed area is shown at higher magnification in Fig. 13 (×500).

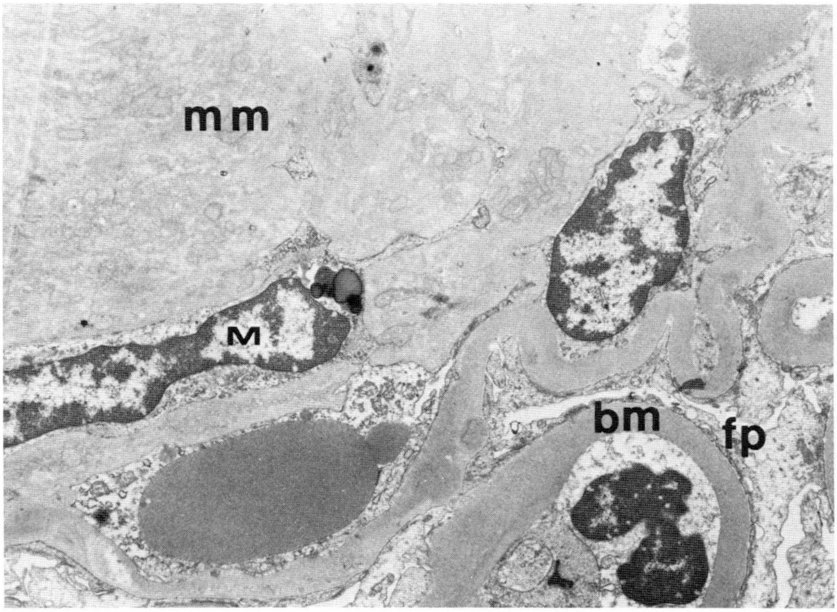

Fig. 13. Diabetic glomerulosclerosis showing conspicuous thickening of basement membrane (bm) and extensive effacement of foot processes (fp). The nodule (upper left) is composed of compacted mesangial matrix (mm). The elongated mesangial cell (M) is being compressed to the rim of the nodule (×6150).

merous slender microvilli on the surface of podocytes are seen in the SEM (Figs. 14 and 15). Effacement of foot processes is apparent although areas with intact podocytes are also noted. The glomerulus appears normal under the light microscope. The basement membrane, as demonstrated by the PAS staining, shows no apparent thickening (Fig. 16). Transmission electron microscopy reveals extensive fusion of foot processes with microvilli in the urinary space (Figs. 17 and 18). This is consistent with SEM findings.

Another case of minimal change disease is shown in Figs. 19–21. Small spherical bodies on podocyte surface and fusion of foot processes are visualized in the SEM. Microvilli not as numerous and slender as those seen in the previous case are also present. Light microscopy shows a relatively normal glomerulus (Fig. 20). Transmission electron microscopy reveals features similar to the previous case. Marked effacement of foot processes is evident but the basement membrane is not thickened.

Depicted in Figs. 22–26 is a glomerulus from a case of membranous

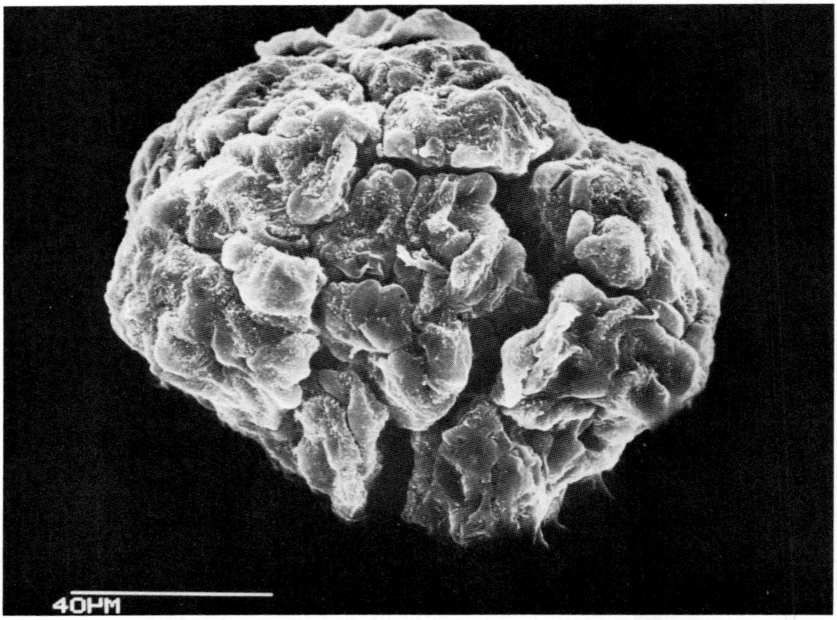

Fig. 14. Glomerulus from a case of minimal change glomerulonephritis showing podocytes with numerous microvilli overlying the convolutional capillary loops.

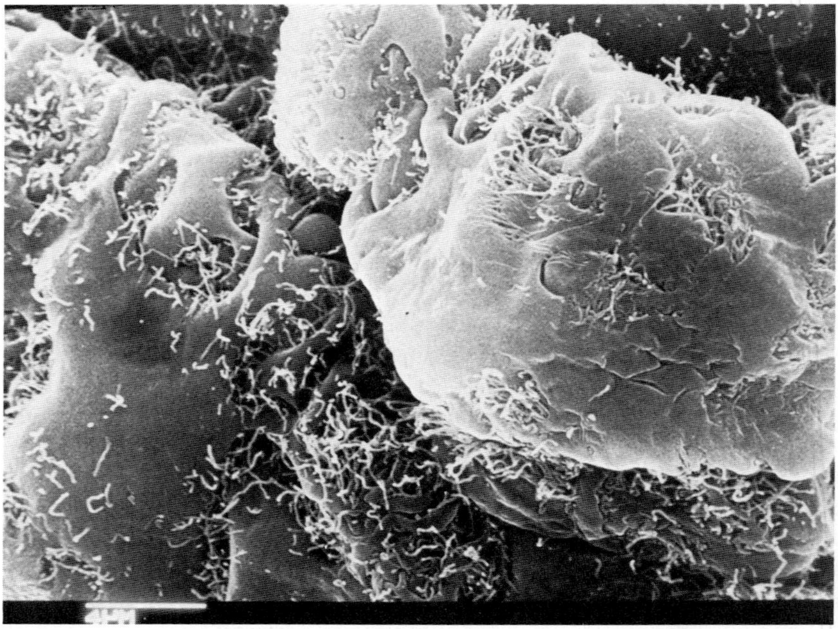

Fig. 15. Minimal change glomerulonephritis showing innumerable slender microvilli on the surface of podocytes. Effacement of foot processes is evident.

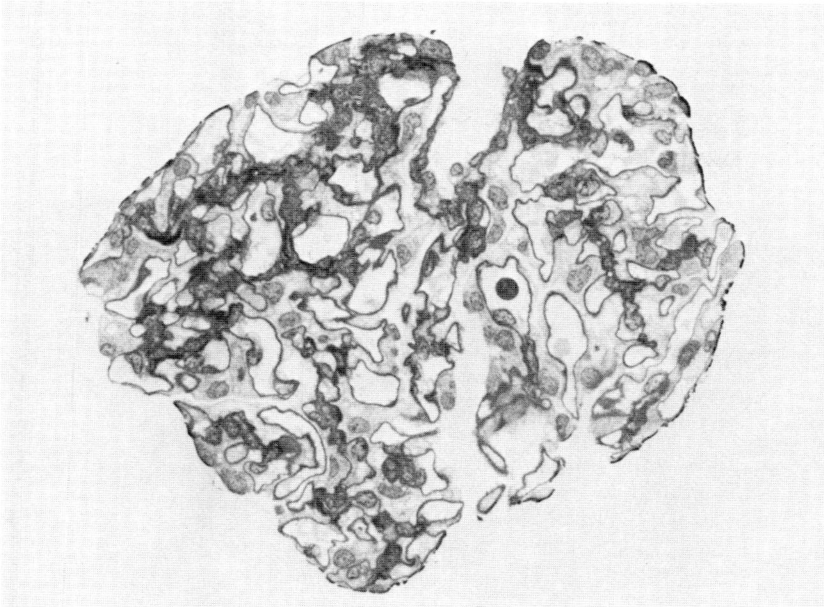

Fig. 16. Minimal change glomerulonephritis. Semithin section stained with PAS showing basement membrane of normal thickness. Nuclei stained with Harris hematoxylin (×550).

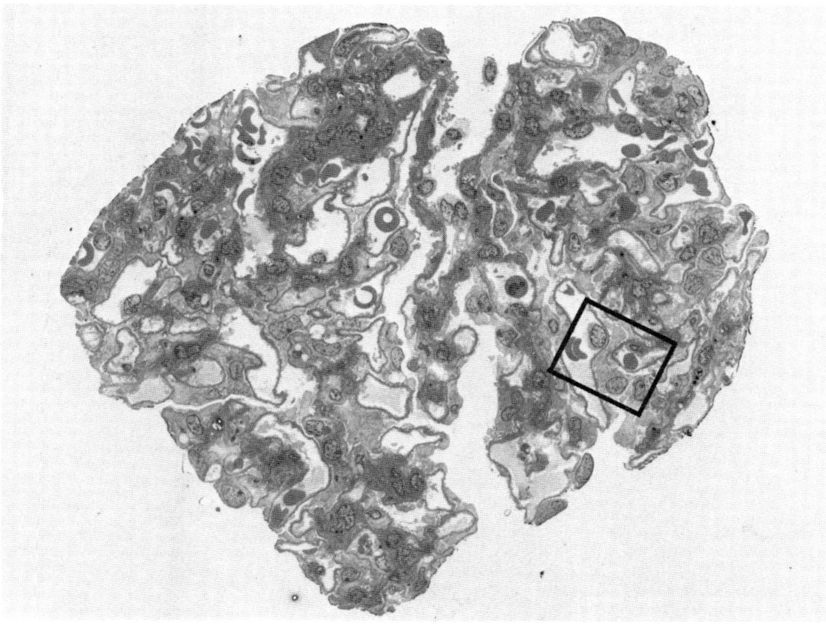

Fig. 17. TEM micrograph of Fig. 14 glomerulus showing a relatively normal glomerulus. The area in rectangle is shown in Fig. 18 (×680).

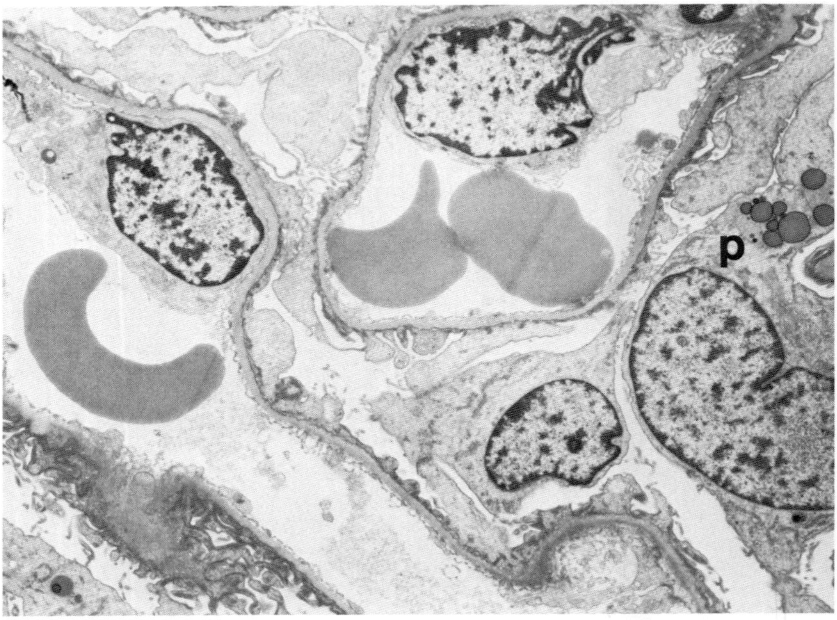

Fig. 18. Higher magnification of a consecutive ultrathin section reveals extensive effacement of foot processes. The podocyte (P) appears swollen and the basement membrane is normal (×4800).

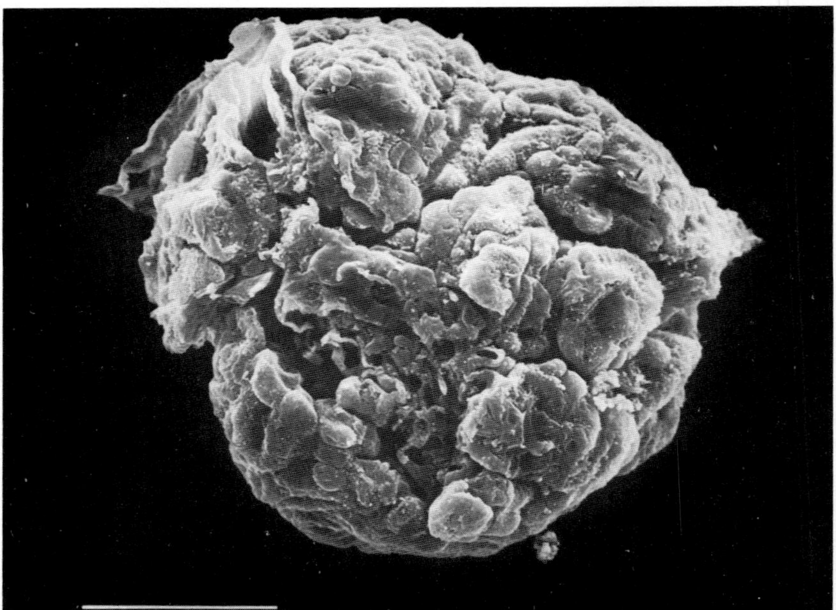

Fig. 19. Another case of minimal change glomerulonephritis. Small spherical bodies on podocyte surface and fusion of foot processes are visualized at higher magnification.

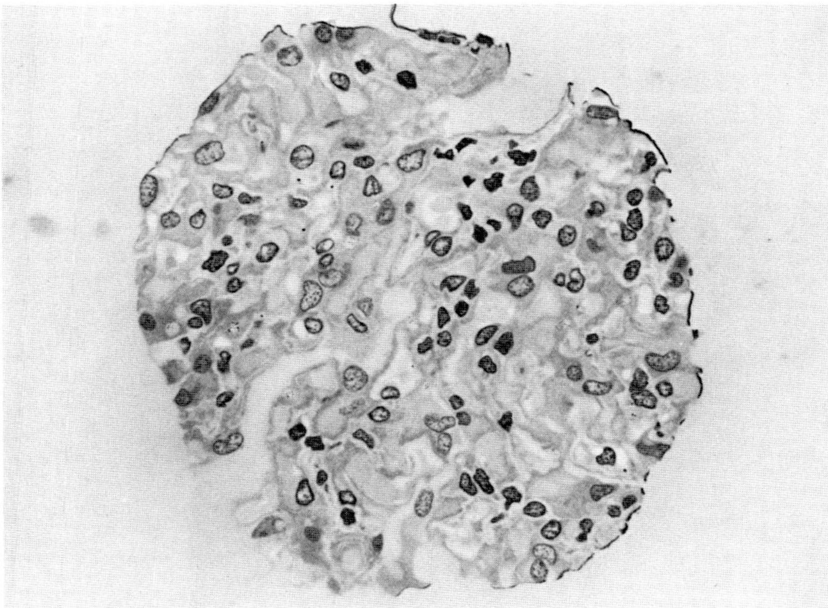

Fig. 20. Semithin section of Fig. 19 glomerulus stained with hematoxylin and eosin showing normal morphology. Gold-palladium coating is seen as a dark line on the surface of glomerulus (×680).

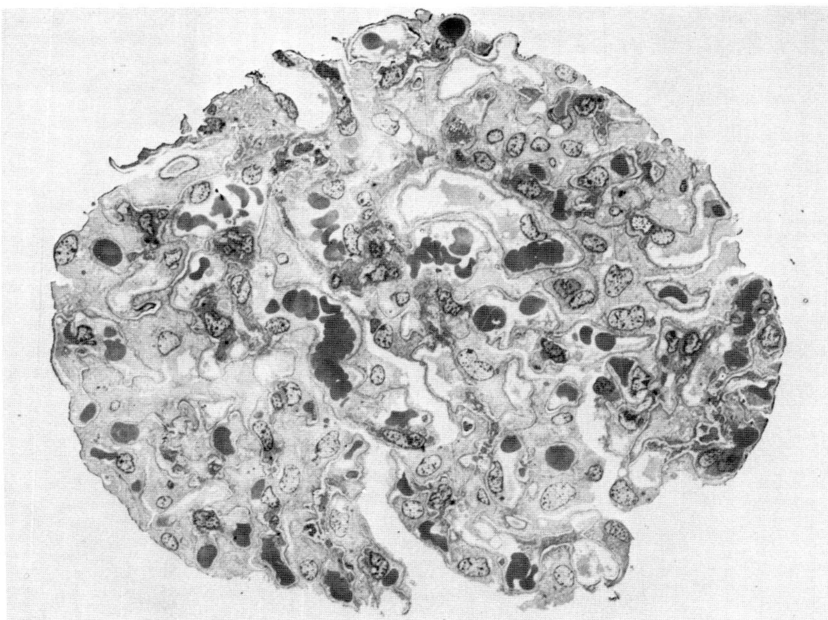

Fig. 21. Minimal change glomerulonephritis. TEM micrograph showing internal features of the glomerulus shown in Fig. 19. Examination at higher magnification reveals no apparent change apart from fusion of foot processes (×780).

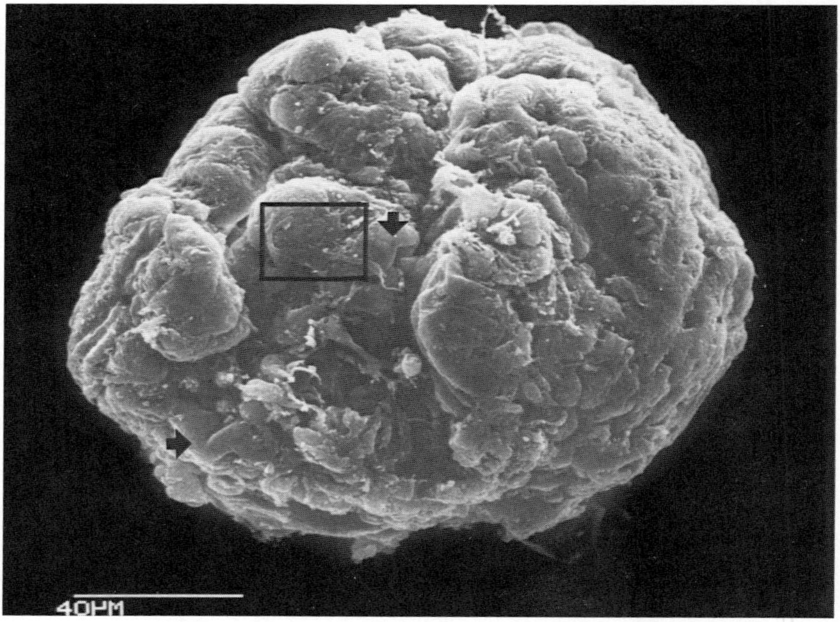

Fig. 22. Glomerulus from a case of membranous lupus nephritis showing focal severe injury. Necrotic podocytes and denuded basement membrane (arrows) are seen. The area in rectangle is shown in next micrograph.

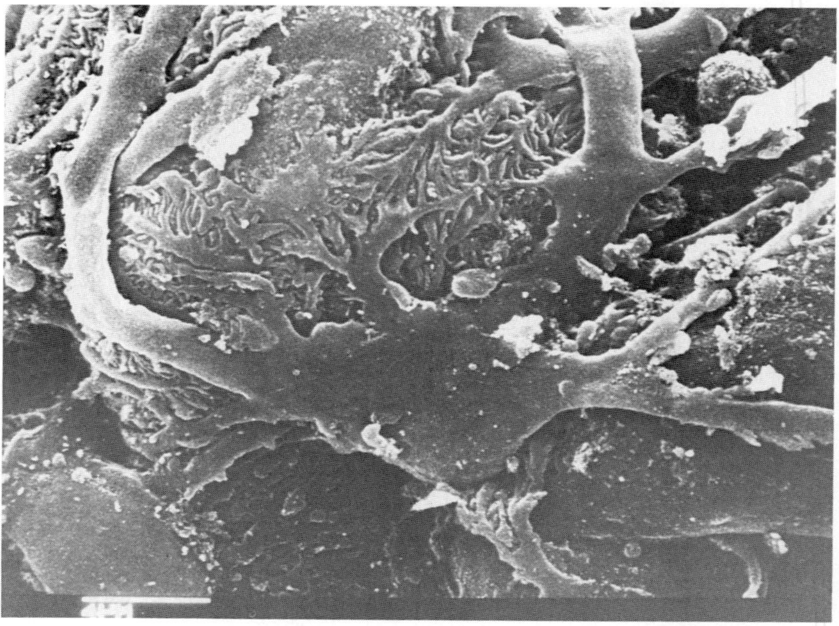

Fig. 23. Close-up view of Fig. 22 glomerulus showing desquamation of podocyte processes. The denuded basement membrane is coated with granular protein and tissue debris.

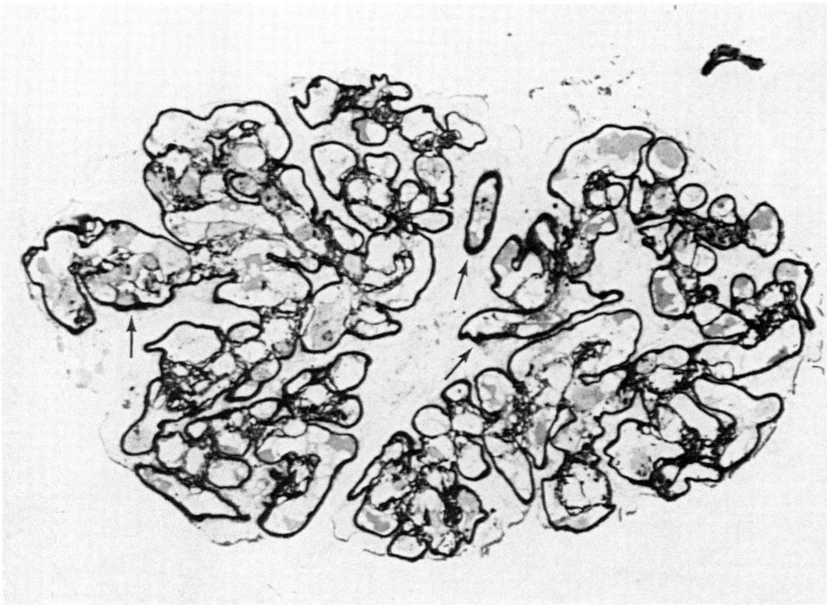

Fig. 24. Membranous lupus nephritis. Semithin section of the same glomerulus stained with methenamine silver. Focal thickening of basement membrane (arrows) is noted (×610).

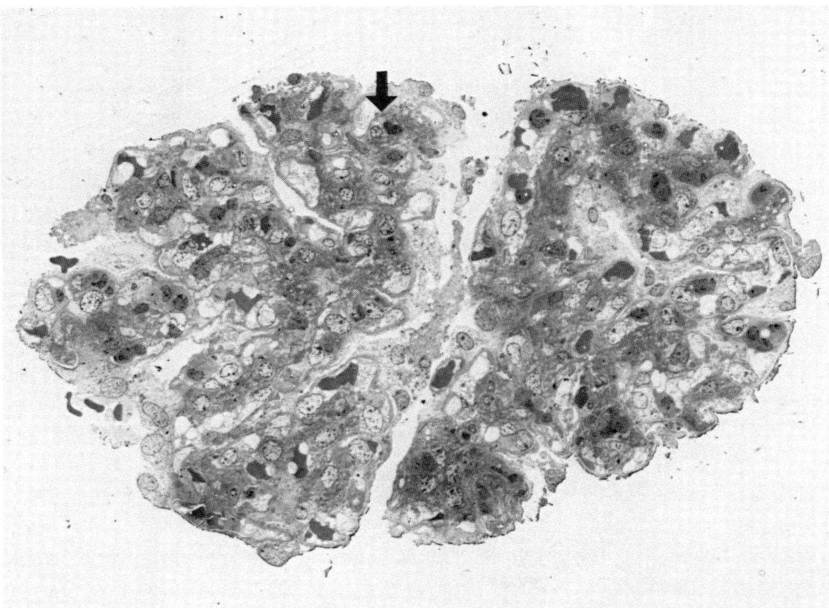

Fig. 25. TEM micrograph of Fig. 22 glomerulus showing segmental mesangial cell proliferation. The area marked by an arrow is shown in Fig. 26 (×600).

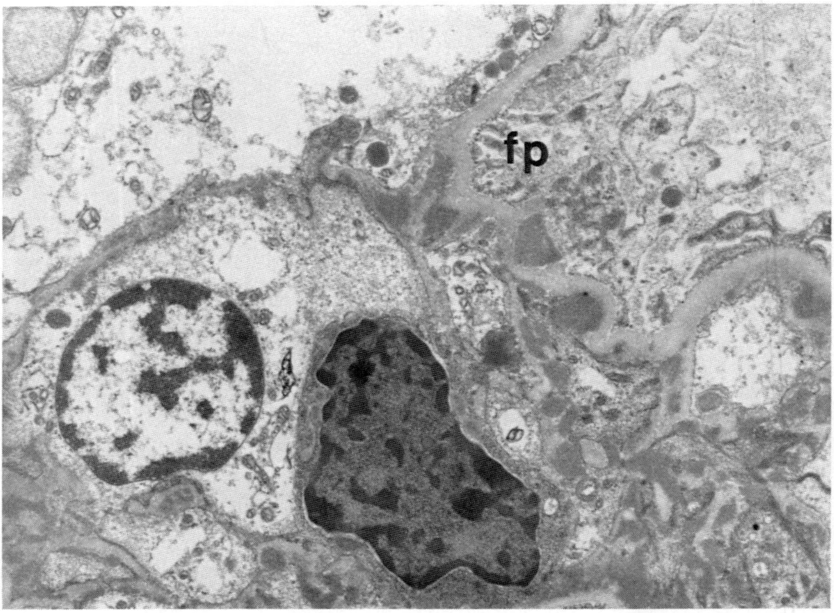

Fig. 26. Higher magnification shows subepithelial and mesangial deposits with fusion of podocyte foot processes (fp) (×9600).

lupus nephritis. The oval-shaped glomerulus shows areas of degeneration and flattening of podocytes with foot process effacement (Fig. 22). Desquamation of the necrotic podocytes results in loss of foot processes, thus exposing the underlying basement membrane (Fig. 23). Other areas with moderately well-preserved podocytes are also noted. Semithin section of the same glomerulus stained with methenamine silver shows focal thickening of basement membrane (Fig. 24). Transmission electron microscopy reveals the presence of subepithelial, subendothelial, and mesangial deposits (Figs. 25 and 26). There is focal thickening of basement membrane and fusion of foot processes. Segmental mesangial cell proliferation and an increase in mesangial matrix are also noted.

CONCLUDING REMARKS

The SEM is obviously a versatile and useful tool for visualizing surface morphology of tissues and cells. It gives excellent three-dimensional images of the cell surface provided that specimen preparation is carefully done. The perfusion fixation, which is necessary for good preservation of ultrastructure, is inapplicable to human tissues and the less ideal immersion fixation has to be used. Glomeruli fixed by the latter method were found to give acceptable ultrastructural details for morphologic evaluation with scanning and transmission electron microscopy.

In the study of renal tissues, scanning electron microscopy provides useful information for the understanding of glomerular diseases, especially the behavior of podocytes. However, as pointed out by Buss and Hollweg (1980) among others, scanning electron microscopy alone is of limited diagnostic value at the present state of knowledge. The changes occurring on podocyte surfaces are usually not characteristic of a particular renal disease. The formation of microvilli and effacement of foot processes, for example, are commonly seen in glomeruli in various glomerulonephritis. So far, the information provided by scanning electron microscopy is complementary to transmission electron microscopy and light microscopy. Many workers have advocated the need for correlative study (Andrews, 1975a; Arakawa *et al.*, 1977b; Jones, 1977; Hayat, 1978; Carr *et al.*, 1980; Carter, 1980; Ng *et al.*, 1983) in order to gain insight into the relationship between surface alteration and internal changes occurring in the tissue.

Comparative scanning and transmission electron microscopy has commonly been done using different fragments of tissue from the same specimen. However, true correlation of the same glomerulus as described herein is desirable because focal lesions are common and glomeruli from the same biopsy are often affected to a different extent. By sectioning through the embedded glomerulus at different levels, it is also possible to obtain a more comprehensive view of the pathological changes which are taking place in the glomerulus. The method is successful even when only one glomerulus is available. It is particularly suitable for renal biopsies which are usually too small to provide adequate portions of tissue for viewing with the SEM, TEM, LM, and immunofluorescent microscopy. By gaining more information and experience through extensive correlative studies, it would be possible in the future to establish SEM diagnostic criteria for glomerular diseases.

The authors would like to thank C. H. Ning for photographic assistance.

REFERENCES

Andrews, P. M. (1975a). Scanning electron microscopy of the nephrotic kidney. *Virchows Arch. B* **17,** 195.

Andrews, P. M. (1975b). Scanning electron microscopy of human and rhesus monkey kidneys. *Lab. Invest.* **32,** 610.

Andrews, P. M. (1977). A scanning and transmission electron microscopic comparison of puromycin aminonucleoside-induced nephrosis to hyperalbuminemia-induced proteinuria with emphasis on kidney podocyte pedicel loss. *Lab. Invest.* **36,** 183.

Andrews, P. M. (1979). The urinary system—kidney. *In* "Biomedical Research Applications of Scanning Electron Microscopy" (G. M. Hodges and R. C. Hallowes, eds.), Vol. 1, p. 273. Academic Press, New York.

Arakawa, M. (1971). A scanning electron microscope study of the glomerulus. *Am. J. Pathol.* **64,** 457.

Arakawa, M., and Tokunaga, J. (1972). A scanning electron microscope study of the glomerulus. Further consideration of the mechanism of the fusion of podocyte terminal processes in nephrotic rats. *Lab. Invest.* **27,** 366.

Arakawa, M., and Tokunaga, J. (1974). Further scanning electron microscope studies of the human glomerulus. *Lab. Invest.* **31,** 436.

Arakawa, M., Edanaga, M., and Tokunaga, J. (1977a). The application of scanning electron microscopy to the histological diagnosis of renal biopsy specimens. *Scanning Electron Microsc.* **2,** 171.

Arakawa, M., Edanaga, M., and Tokunaga, J. (1977b). Scanning electron microscopy of the isolated human glomerulus in normal, nephritic and nephrotic situations. *In* "Glomerulonephritis" (R. Kluthe, A. Vogt, and S. R. Batsford, eds.), p. 96. Wiley, New York.

Boyde, A., and Boyde, S. (1980). Further studies of specimen volume changes during processing for SEM: Including some plant tissue. *Scanning Electron Microsc.* **2,** 117.

Boyde, A., and Maconnachie, E. (1981). Morphological correlations with dimensional change during SEM specimen preparation. *Scanning Electron Microsc.* **4,** 27.

Burns, W. A. (1978). Thick sections: Techniques and applications. *In* "Diagnostic Electron Microscopy" (B. F. Trump and R. T. Jones, eds.), Vol. 1, p. 141. Wiley, New York.

Buss, H., and Hollweg, H. G. (1980). Application of scanning electron microscopy to diagnostic pathology. A critical review. *Scanning Electron Microsc.* **3,** 139.

Buss, H., and Krönert, W. (1969). Zur strukture des nierenglomerulum der ratte. Rasterelektronenmikorskopische untersuchungen. *Virchows Arch. B* **4,** 79.

Buss, H., Lamberts, B., and Brass, H. (1973). Orthology and pathology of the renal podocytes. *In* "Scanning Electron Microscopy," p. 573. IITRI, Chicago, Illinois.

Cardno, S. S., and Steiner, J. W. (1965). Improvement of staining technics for thin sections of epoxy-embedded tissue. *Am. J. Clin. Pathol.* **43,** 1.

Carr, K. E., McLay, A. L. C., Toner, P. G., Chung, P., and Wong, A. (1980). SEM in service pathology: A review of its potential role. *Scanning Electron Microsc.* **3,** 121.

Carter, H. W. (1980). Clinical application of scanning electron microscopy (SEM) in North America with emphasis on SEM's role in comparative microscopy. *Scanning Electron Microsc.* **3,** 115.

Chang, S. C. (1972). Hematoxylin-eosin staining of plastic-embedded tissue sections. *Arch. Pathol.* **93,** 344.

Chiu, K. Y., Or, S. B., Yau, W. L., and Ngai, H. K. (1986). Improved immunoperoxidase staining on semi-thin sections of epoxy-embedded tissues. *Med. Lab. Sci.* **43,** 93.

Cohen, A. L. (1974). Critical point drying. *In* "Principles and Techniques of Scanning Electron Microscopy" (M. A. Hayat, ed.), Vol. 1, p. 44. Van Nostrand-Reinhold, New York.

Cohen, A. L. (1979). Critical point drying—principles and procedures. *Scanning Electron Microsc.* **2,** 303.

Echlin, P. (1978). Coating techniques for scanning electron microscopy and x-ray microanalysis. *Scanning Electron Microsc.* **1,** 109.

Echlin, P. (1981). Recent advances in specimen coating techniques. *Scanning Electron Microsc.* **1,** 79.

Echlin, P., Chapman, B., Stoter, L., Gee, W., and Burgess, A. (1982). Low voltage sputter coating. *Scanning Electron Microsc.* **1,** 29.

Grocott, R. G. (1955). A stain for fungi in tissue sections and smears. *Am. J. Clin. Pathol.* **25,** 975.

Hayat, M. A. (1978). "Introduction to Biological Scanning Electron Microscopy," pp. 100, 131, 219, 255. University Park Press, Baltimore, Maryland.

Hayat, M. A. (1981). Fixation for Electron Microscopy. Academic Press, Orlando, Florida.

Hayat, M. A. (1986). "Basic Techniques for Transmission Electron Microscopy." Academic Press, Orlando, Florida.

Hayat, M. A. (1987). "Principles and Techniques of Electron Microscopy: Biological Applications." Macmillan Press, London.

Humphreys, S. W. J., Spurlock, B. V., and Johnson, J. S. (1974). Critical point drying of ethanol infiltrated, cryofractured biological specimens for scanning electron microscopy. *In* "Scanning Electron Microscopy," p. 275. IITRI, Chicago, Illinois.

Jones, D. B. (1977). Correlative scanning and transmission electron microscopy of glomeruli. *Lab. Invest.* **37,** 569.

Jones, D. B. (1978). Scanning electron microscopy of human hypertensive renal disease. *Scanning Electron Microsc.* **2,** 937.

Jones, D. B. (1979a). SEM of human and experimental renal disease. *Scanning Electron Microsc.* **3,** 679.

Jones, D. B. (1979b). The role of scanning electron microscopy in the study of normal and diseased glomeruli. *In* "Kidney Disease—Present Status" (J. Churg, B. H. Spargo, F. K. Mostofi, and M. R. Abell, eds.), p. 42, Williams & Wilkins, Baltimore, Maryland.

Jones, D. B. (1981). The ultraplaning technique for SEM specimen preparation. *Scanning Electron Microsc.* **2,** 77.

Jones, D. B. (1983). The complementary role of scanning electron microscopy in renal pathological diagnosis. *Scanning Electron Microsc.* **1,** 323.

Lawler, W., Tarpey, P., Williams, G., Acheson, E. J., and Mallick, N. P. (1976). Diseases and histological normality of the renal glomerulus: A clinicopathological study. *J. Clin. Pathol.* **29,** 380.

Lehtonen, E., Virtanen, I., and Wartiovaara, J. (1973). Visualization of human glomerular changes by scanning electron microscopy. *Virchows Arch. B* **13,** 259.

McManus, J. F. A. (1946). Histological demonstration of mucin after periodic acid. *Nature (London)* **158,** 202.

Ng, W. L., So, K. F., So, P. C., and Ngai, H. K. (1982). The preparation of glomeruli from renal biopsy specimens for scanning electron microscopy. *Pathology* **14,** 299.

Ng, W. L., Chan, K. W., and Ma, L. (1983). A scanning electron microscope study of isolated glomeruli in glomerulonephritis. *Pathology* **15,** 139.

Or, S. B., Ngai, H. K., and Yau, W. L. (1983). Correlative study of the same isolated glomerulus by electron and light microscopy. *Med. Lab. Sci.* **40,** 359.

Or, S. B., Yau, W. L., and Ngai, H. K. (1984). Improved method for sequential study of an isolated glomerulus by scanning electron, transmission electron, and light microscopy. *J. Clin. Pathol.* **37,** 1075.

Tarpey, P. A., and Williams, G. (1980a). The preparation of renal tissue for scanning electron microscopy. *Med. Lab. Sci.* **37,** 43.

Tarpey, P. A., and Williams, G. (1980b). Scanning electron microscope studies of various glomerulonephropathies. *Med. Lab. Sci.* **37,** 57.

Tokunaga, J., Edanaga, M., Masu, Y., and Fujita, T. (1975). Isolated renal glomeruli for scanning electron microscopy. *J. Electron Microsc.* **24,** 109.

Wheeler, E. E., and Herdson, P. B. (1973). Freeze fracturing and freeze drying of renal tissue for scanning electron microscopy. *Am. J. Clin. Pathol.* **60,** 229.

14

Correlative Microscopy of Cell Surface and Underlying Structures

JON B. REITAN,[1,2] KARI FEREN,[1,2] FRANZ THIEBAUT,[1,3,4] AND ALBRECHT REITH[1,3]

[1] *Laboratory for Electron Microscopy and Morphometry, The Norwegian Radium Hospital and Institute for Cancer Research, Oslo, Norway*
[2] *Medical Section, National Institute of Radiation Hygiene, Østerås, Norway*
[3] *Norwegian Cancer Society*
[4] *Laboratoire d'histologie du Cerchar, Verneuil-en-Halatte, France*

CORRELATIVE MICROSCOPY IN BIOLOGY:
INSTRUMENTATION AND METHODS

INTRODUCTION

The wide variety of microscopic techniques used in histopathological and experimental situations have proved to be more or less adequate for a proper description of different structures. The correlation between structure and function, which indicates the type of cell or tissue, cancerous or normal, etc., is mainly based on studies using light microscopy (LM) or transmission electron microscopy (TEM). With the evolution of scanning electron microscopy (SEM) operating both in secondary electron imaging (SEI), and backscattered electron imaging (BEI) mode, many surface structural features may be better described than by using transmission techniques alone. This includes, for example, the formation of pleomorphic microvilli in carcinogenic processes (Saxholm and Reith, 1979; Reith *et al.*, 1983) and the evaluation of the characteristic hexagonal surface plaques of the urinary bladder (Reitan and Feren, 1986).

The correlation between structural features seen with SEM and those seen with TEM or LM is a vital point. In most instances these modalities are used on neighboring parts or different parts of the same biopsy or experimental material. Consequently, the possibility that the cells and tissues viewed by the different techniques are not of the same type cannot be excluded. This uncertainty renders necessary techniques that can evaluate the surface and the underlying structures of the same material.

There are two main approaches to the problem of the correlation between images of the surface and the underlying structures:

1. Reembedding and sectioning of material originally processed for SEI, or removal of embedding medium of material processed for LM or TEM, followed by processing and coating for SEI (see Chapter 15, this volume). Usually these techniques result in two viewing directions of the surface: Light microscopic direction and transmission electron microscopic directions are at right angles to the SEI direction. This, however, does not apply to SEM of histological sections (see Chapter 20, this volume). The techniques may make use of micromanipulators, enabling a better definition of special areas of interest.
2. LM or BEI mode of SEM of the same material and with the same viewing direction as the SEI mode of SEM before or after the preparation for SEI.

This chapter describes methods belonging to category 2. The techniques are in use at the Laboratory for Electron Microscopy and Morphometry at the Norwegian Radium Hospital and Institute for Cancer Research.

TECHNIQUES USING LIGHT MICROSCOPY FOR CYTOLOGICAL FEATURES IN COMBINATION WITH SCANNING ELECTRON MICROSCOPY FOR SURFACE STRUCTURES

Correlative Light Microscopy and Scanning Electron Microscopy Using Feulgen Staining

During tissue culturing the varying appearance of the cells is obviously correlated to culture conditions but it is also correlated to the phase of the cell cycle. The most distinct feature is the rounding up of the cells during mitosis. However, the surface appearance as it appears in SEI also varies during the cell cycle. The correlation of the cell surface to chromosome structure can be elucidated by preparing the same cells for both SEI and LM using Feulgen staining.

Using Feulgen staining on cultured cells grown on glass coverslips, we have been able to identify mitotic cells in the light microscope, verify their mitotic stage, and find the same cells in the scanning electron microscope afterwards. This procedure (Feren and Reith, 1981; Feren, 1982) is a modification of Wetzel *et al.* (1978).

Procedure

1. Fix in 2% glutaraldehyde in 0.1 *M* cacodylate and 0.1 *M* sucrose (pH 7.2–7.4) at room temperature for 60 min (Collins *et al.*, 1977).
2. Rinse once in the same buffer as above.
3. Hydrolyze cells for 10 min at 60°C in 1 *N* HCl (use a waterbath).
4. Stain with filtered Schiff's* reagent for 30–45 min in the dark.
5. Rinse in metabisulfite† for 1 min.

* Schiff's reagent: A, dissolve 0.1 g pararosaniline in 15 ml 1 *N* HCl; B, dissolve 0.5 g $K_2S_2O_5$ in 85 ml distilled water. Mix A and B and let the solution stand for 24 hr. Add 350 mg medical charcoal and shake well. Wait for 2 min, then filter the solution through several layers of filter paper. The solution must be colorless. If not, more coal must be added and the solution refiltered. The solution may be stored in the dark at 4°C and used as long as it remains colorless.

† Metabisulfite must be made fresh every day: 5 ml 1 *N* HCL; 5 ml 10% $K_2S_2O_5$; 100 ml distilled water. (10% $K_2S_2O_5$ can be stored in a refrigerator for several months and is made by dissolving 10 g of $K_2S_2O_5$ in 90 ml distilled water.)

6. Rinse in metabisulfite for 2 min.
7. Rinse in metabisulfite for 2 min.
8. Rinse three times for $1\frac{1}{2}$ min in distilled water.
9. Place in cacodylate buffer. Mark the coverslips for easier identification of the cells in LM and SEM.**
10. Dehydrate in 70–90–95% ethanol, 2 min each, and three times for 5 min in 100% ethanol.
11. Critical-point dry (CPD) using liquid CO_2.
12. Store in the dark at 4°C.
13. Select cells of interest and take black-and-white photomicrographs using a 63× OD (ohne Deckglas, i.e., without coverslip) objective.
14. Mount on SEM specimen stub, and cover with 20 nm gold.
15. Examine the same cells by SEM using the light micrograph to identify the cells.

The Feulgen stain is stable in most reagents. The specimens can be stored for several days in the dark at 4°C before microscopy, but care should be taken that they do not rehydrate. NB: Feulgen stain is not *specific* for DNA and does not bind stoichiometrically when glutaraldehyde is used as a fixative.

This method involves a hydrolysis step which may be thought to interfere with the structural integrity of the cells and thus affect the surface structures. Our experience is, however, that this does not happen. When comparing samples hydrolysed according to this method with nonhydrolysed samples, no differences in surface structures could be seen.

Care should be taken during sputtering with gold after LM. The specimen stage of the sputter must be well cooled, otherwise the specimen might collapse or partially loosen from the substrate (Boyde *et al.*, 1972). We have not seen this happen to any great extent with Feulgen-stained specimens, compared to those that have just been dehydrated and critical-point dried.

** Marking may be done in several ways. Wetzel *et al.* (1978) described a method for identifying the cells in LM and SEM using wet LM. However, we have found it easier to mark the specimen just after staining and then do the LM after CPD. To do this we use a sharp wooden stick to scratch a grid system onto the coverslips, thus making separate cell fields. This is very easily done at this stage since the cells are colored red. After CPD an ink mark is placed on the upper part of the coverslip and the field nearest to this mark is numbered 1. By taking an LM survey photograph, all fields can be numbered and easily recognized in SEM. The mitotic cells are identified using first low magnification of the actual fields and then higher magnification to identify the separate cells and their mitotic stage (Fig. 1).

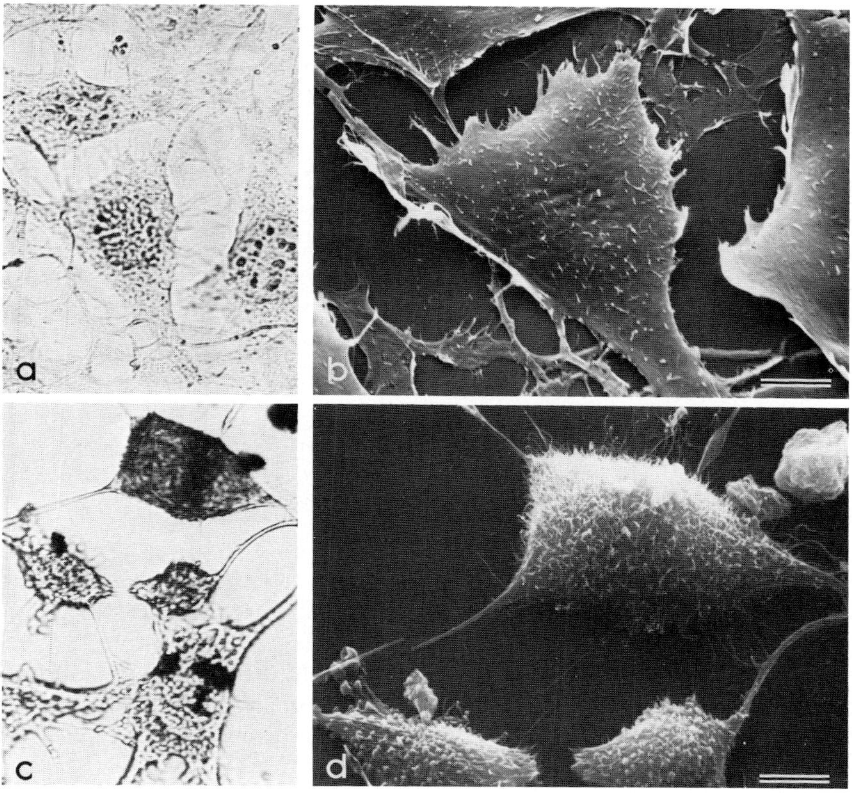

Fig. 1. C3H cells in prophase and metaphase as determined in Feulgen stain. a, LM of a prophase. The cell has not rounded up and is as flat as the interphase cells, allowing fair focusing during photography. b, SEI of the same cell as in a. Note relatively few microvilli on the surface; bar = 5 μm. c. LM of a metaphase cell. Due to the rounding up, sharp focusing of the cell and chromosomes during photography is impossible. A medium focusing plane is chosen to allow gross identification of the actual cell relative to a grid marking. Focusing by hand easily allows precise identification of the chromosomes and mitotic stage. d, SEI of the same cell as in c. Note the rounding up and the extensive expression of long microvilli, bar = 5 μm.

Correlative Light Microscopy and Scanning Electron Microscopy Using Papanicolaou Staining

Papanicolaou staining is a well-known method for LM of cytological smears. Since more interest is now being shown in surface structures of premalignant and malignant cells, in addition to cytological features, it is useful to know that this staining can be used for combined LM and SEM

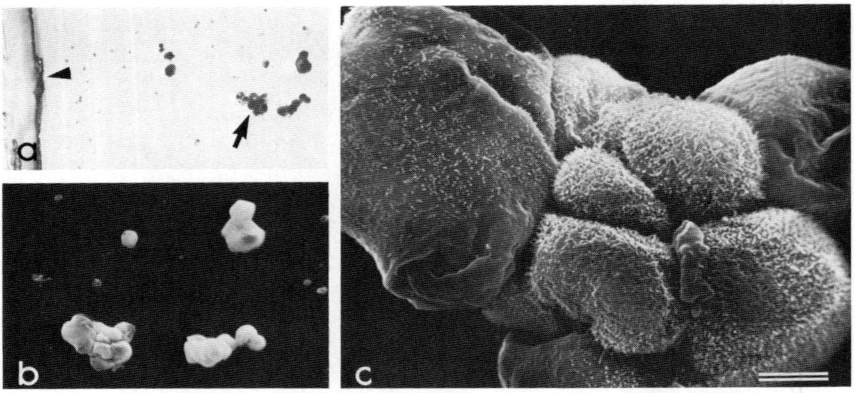

Fig. 2. Ascites fluid cells from a primary ovarian carcinoma. a, LM of Papanicolaou-stained cells in a wet chamber (low magnification). Note clump marked with arrow. Note grid line made with diamond marker (arrowhead) for identification purposes. b, SEI of the same cells as in a (low magnification). c, SEI of cell clump marked with arrow in a (high magnification); bar = 5 μm.

with only slight modifications. The method can be used for cytological specimens and *in vitro* cell cultures (Fig. 2). There are many modifications of the Papanicolaou stain for cytological smears, and the method described here is a modification used in our institution. The main modifications are: (1) glutaraldehyde is used as a fixative; and (2) xylol is omitted. However, the ordinary processing of cytological smears has been shown not to interfere seriously with the SEM result even when coverslips are removed with xylol (Kenemans *et al.*, 1981). In order to satisfy the cytological demands for good staining, and for optical reasons, the LM is done using a wet chamber or microcuvette (Adam and Czihak, 1964) (Fig. 3). The Papanicolaou stain has even been used in dry microscopy (Reith *et al.*, 1980) but most cytopathologists will prefer the wet chamber technique.

Procedure

1. Spin down the cytological specimen on polylysine-coated and marked plastic coverslips* or use tissue culture coverslips as given on

* For cell cultures, marking can be done as described for the preceding method. For cytological specimens, brush cytology specimens, ascites fluid, etc.; marked coverslips must be used. We usually make a grid system on square or round polystyrene coverslips (Thermanox, Miles Laboratories, Naperville, IL) with a diamond marker. After marking, the coverslips must be rinsed in alcohol to get rid of the plastic debris. We rinse them using an ultrasound cleaner (Bandelin Sonorex rapid, Bandelin Electronic KG, West Germany). Coat with poly-L-lysine (0.25 mg/ml, pH 6.9). Spin the cells down onto the coverslips using a

page 331. Fix for 60 min in 2% glutaraldehyde in 0.1 *M* cacodylate and 0.1 *M* sucrose (pH 7.2–7.4) at room temperature (Collins *et al.*, 1977).

2. Rinse once in the same buffer as above.
3. Stain for 2 min in Harris hematoxylin (filtered).
4. Rinse for 2 min in distilled water or until the water is clear.
5. Dip in 70% ethanol for 2 min.
6. Dip to a blue color in a solution of 97 ml 70% ethanol + 3 ml conc. ammonia.
7. Rinse twice in 70% ethanol.
8. Dip in 95% ethanol for 1 min.
9. Stain with OG 6 (orange G) for 1½ min.
10. Rinse twice in 95% ethanol.
11. Stain with EA 50 (filtered eosin-acridine) for 2 min.
12. Transfer the coverslip to a wet chamber† and do LM. For LM use an ordinary objective (up to 40×). Make photomicrographs which enables the identification in SEM afterwards.
13. Dehydrate for 2 min in 95% and two times for 5 min in 100% ethanol.
14. Critical-point dry using liquid CO_2.
15. Mount on SEM specimen stubs after light microscopy as described on page 332.

The main reason for using the wet chamber method instead of that given on page 332 is that the staining is not very stable after CPD. Moreover, the identification of cytological details is inferior in dry microscopy. Light microscopy has to be done within a short time (24 hr) because the color fades. A problem is the staining itself, since the same period in a color bath at times results in overstaining, at others in well-stained specimens, or even in weak staining, depending on the material. However, the staining is very rapid and more familiar to cytopathologists than Feulgen staining. If it is done by a well-trained person there should be no problem with overstaining, etc., and our modification need not replace the laboratory's routine staining methods, except for the glutaraldehyde fixation.

---

cytocentrifuge (Leiden centrifugation bucket, van-Driel-Kulker *et al.*, 1980). It is necessary to cut off a corner or otherwise mark the coverslip to ensure later identification of fields and cells.

† The wet chamber or microcuvette is a modification of that described by Adam and Czihak (1964). Make three tiny elongated pieces of plastic cover glass and fix to a slide glass with nail polish (Fig. 3). Fix another cover glass on top of this. The thickness of the nail polish will ensure sufficient room in the chamber for inserting the specimen coverslip. Fill chamber with buffer (e.g., phosphate-buffered saline) and insert the specimen coverslip into the chamber. Avoid any drying of the specimen.

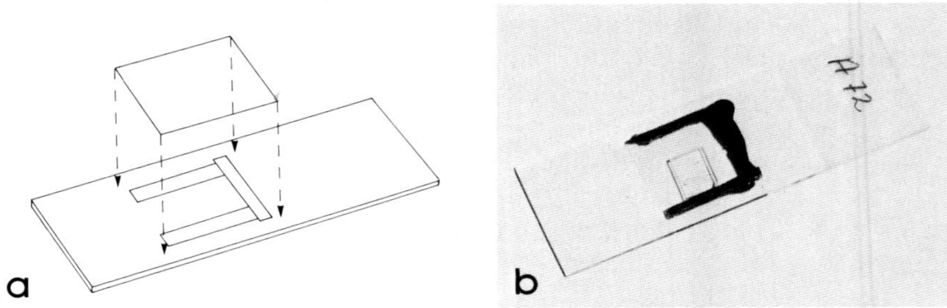

Fig. 3. Wet chamber or microcuvette. a, Line drawing. For details on preparation, see text. b, Photo of an actual chamber.

TECHNIQUES USING BACKSCATTERED AND SECONDARY ELECTRON IMAGING ON CELLS AND TISSUES PROCESSED FOR SCANNING ELECTRON MICROSCOPY

During the technical development of SEM, various modes have emerged. The most widely used mode is the SEI of material coated with, for example, gold/palladium. The attraction of SEM may be increased when information can be obtained both about the surface and the cytological features of the same specimens. The use of the BEI mode allows observation of specimens with thicknesses up to about 8 μm and with higher resolution than using LM (Becker and Sogard, 1979; Ogura and Laudate, 1980; Thiebaut *et al.,* 1985). The basic physical and technical principles have been described in more detail elsewhere (e.g., Black, 1974). BEI can identify heavy metal deposits such as phagocytized particles. However, by using silver impregnation stains, argyrophilic structures may be visualized. Application of an AgNOR stain (Bloom and Goodpasture, 1976) as modified by Thiebaut *et al.* (1984a) has proved successful in the demonstration of whole nuclei (Thiebaut *et al.,* 1986b), nucleolar organizer regions (Thiebaut *et al.,* 1984c), and junctional complexes (Thiebaut *et al.,* 1986a).

Detection of Heavy Metal Particles inside Cells Using Backscattered Electron Imaging

All kind of particles having a high density can be detected using BEI, i.e., heavy metals, crystals of insoluble compounds, asbestos, etc. We have used the method for identification of heavy metal particles in primary rat tracheal cells, cultured as described by Pai *et al.* (1983) (Fig. 4).

Procedure

1. Rinse the specimen in a salt solution, e.g., phosphate-buffered saline or Hanks salt solution, before fixation to remove particles from the cell surface.
2. Fix the cells for 60 min using 2% glutaraldehyde in 0.1 *M* cacodylate plus 0.1 *M* sucrose buffer, pH 7.4, at room temperature.
3. Dehydrate and critical-point dry as usual.
4. Mount the specimen on stubs and cover with a thin (20–30 nm) layer of carbon.
5. Examine in SEM. Use SEI to focus and then switch to BEI to detect particles inside the cells.

It may be necessary to use higher voltage and/or larger spot size when using BEI in order to get enough electrons with high energy.

Backscattered Electron Imaging of Silver-Impregnated Structures in Cell Cultures Combined with Secondary Electron Imaging

For this method the culture cells are grown on small glass coverslips or mica sheets. The method can be used for characterization of the cell's surface structures in SEI and quantitation of intracellular structures, i.e.,

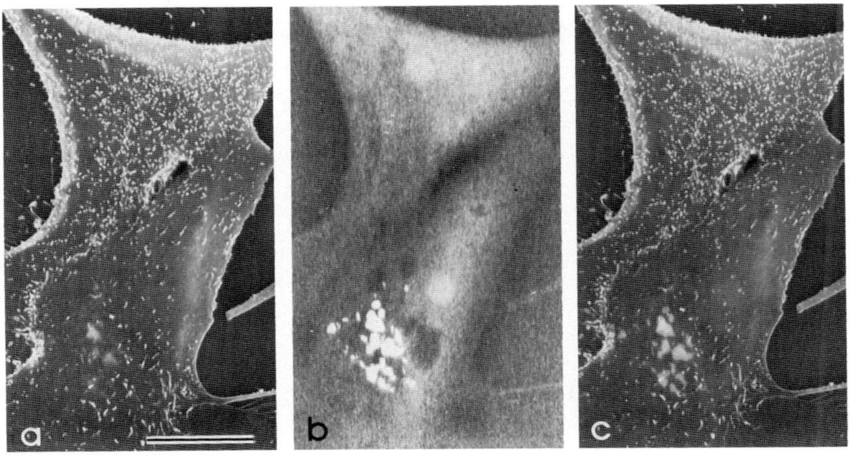

Fig. 4. SEM of primary rat tracheal epithelial cells in culture. a, SEI of two neighboring cells; bar = 10 μm. b, BEI of the same cells. Phagocytized $\alpha Ni_3 S_2$ particles as distinct white spots in the right cell. c, Double exposure combining images a and b, demonstrating surface structures and phagocytized particles in the same image.

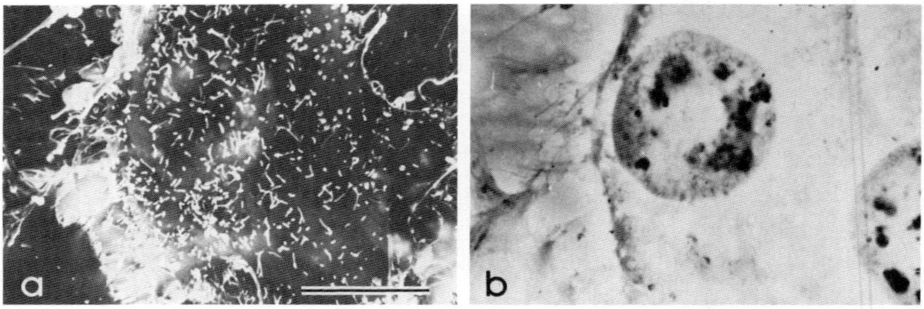

Fig. 5. C3H-cells AgNOR stained and coated with carbon. a, SEI of a cell and parts of neighboring cells; bar = 10 μm. b, Same cell and field as seen in BEI. Note identification of inhomogeneity of silver material in the nucleus; the dark areas represent nucleoli.

the size of nucleus and nucleoli and number of intranucleolar silver granules representing the nucleolar organizer regions with BEI and automatic image analysis (Thiebaut *et al.*, 1984b) as described on page 340 (Fig. 5).

Procedure

1. Fix the specimen in 2% glutaraldehyde in 0.1 *M* phosphate buffer plus 0.1 *M* sucrose, pH 7.4, for 1 hr at room temperature.
2. Postfix in a 1 : 3 solution of acetic acid and absolute ethanol for 1 hr at room temperature.
3. Rehydrate in graded alcohols, 96–90–70% for 5 min each.
4. To avoid undesirable spots, immerse in Schiff's reagent for 50 min in the dark.
5. Wash thoroughly in bidistilled water.
6. Prepare a silver nitrate solution ($AgNO_3$) by dissolving 0.5 g of $AgNO_3$ in 1 ml of bidistilled water. Filter through a 0.22–μm Millipore filter.
7. Incubate* for 15–19 hr at 62°C in a moist atmosphere.
8. Rinse the specimen thoroughly in bidistilled water, dehydrate in graded alcohols, and CPD by liquid CO_2.
9. Mount on SEM specimen stubs and cover with a thin (20–30 nm) carbon layer and examine as before.

The method can also be used for comparative LM and SEM (SEI and BEI). The specimens are then wet-mounted on object glasses after they

* Silver contamination of the specimen is a major drawback of this method and may be avoided in the following way: Several drops of the silver nitrate solution are placed in a Petri dish. A coverslip is then placed on each drop with the cells facing downwards and the Petri dish is incubated.

have been rinsed in bidistilled water and sealed with nail polish. After LM the further preparation is commenced.

Backscattered Electron Imaging of Silver-Impregnated Structures in Epithelial Tissues Combined with Secondary Electron Imaging.

This method has been used in our laboratory for staining liver, urinary bladder, intestine, and kidney and permits the detection of nuclear structures and junctional complexes (Thiebaut *et al.,* 1986a,b). The silver staining procedure is almost the same as that proposed by Bloom and Goodpasture (1976) and modified by Thiebaut *et al.* (1986a) (Fig. 6).

Procedure

1. Cut the tissue into small pieces, approximately 2 × 2 mm and 1 mm thick.
2. Fix in 2–2.5% glutaraldehyde in 0.1 *M* phosphate buffer, pH 7.2, at room temperature for 18–22 hr.
3. Rinse in bidistilled water for 4 hr.
4. The silver nitrate solution is prepared by dissolving 0.6 g $AgNO_3$ in 1 ml bidistilled water and filtering through a 0.22-μm Millipore filter.
5. Incubate the specimens in this solution for 30 hr at 52°C.
6. Wash thoroughly in bidistilled water, dehydrate, and CPD as usual for SEM.
7. Mount on specimen stubs and cover with carbon as described on page 338.

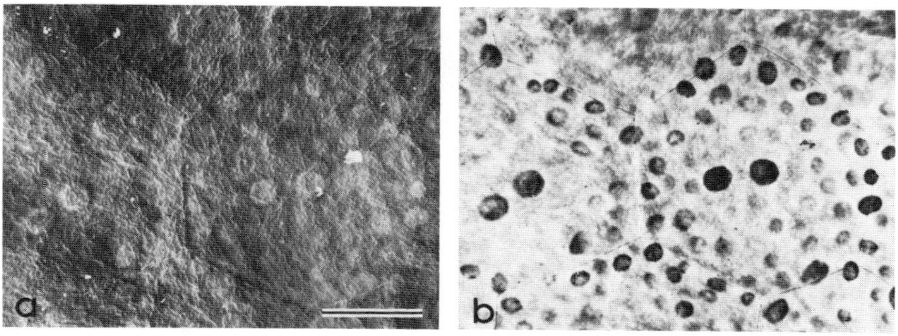

Fig. 6. Mouse urinary bladder epithelium, AgNOR stain and carbon coating. a, SEI surface view. Note intercellular elevated interdigitations; bar = 50 μm. b, Same field in BEI mode. The cell borders are visualized by silver deposits in the junctional complexes. The cell nuclei both of the superficial binuclear "umbrella" cells and deeper lying cells are visualized.

This silver staining technique gives satisfactory results over a wide range of temperatures (4–55°C). Higher temperatures increase the incidence of contamination at the surface as well as nonspecific staining. Using other temperatures provides a suitable modification of the incubation time. The formula which gives the relationship between time and temperature to obtain the same staining intensity is:

$$X^{(Q2-Q1)/10} = t1/t2$$

where $Q1$ and $Q2$ are the temperatures, $t1$ duration at $Q1$ and $t2$ duration at Q2. X is a factor dependent on the fixation and specimen, and must be experimentally determined in each material. For the tissues mentioned above, X=1.7. The final formula is then:

$$t = tr/1.7^{(Q-Qr)/10}$$

where tr is the reference time and Qr the reference temperature (Thiebaut *et al.*, 1986a).

Application of Digital Image Processing of Structures Identified with Backscattered Electron Imaging in Material Processed for Scanning Electron Microscopy

In recent years, digital image processing has evolved as a major research area concomitant with the technological development of computer hardware and software. In biomedical research such techniques may be applied both in optical (Gualtieri *et al.*, 1985) and electron microscopic (Rigaut, 1983, 1987) imaging. Digital analysis of SEM electronic signals is especially attractive and may be implemented on most microscopes, offering opportunities for contrast manipulations, morphometric and stereological analysis or, for example, subtraction of images obtained in different modes (Fig. 7). Images in SEI and BEI mode may be combined, but photographic combination by double exposure may often be easier (Fig. 4).

The construction of adequate image analysis systems from standard hardware components involves high technology. If the necessary knowledge and resources are available, such systems may be tailored to special needs. However, most laboratories will have to rely on commercially available systems. A multitude of such systems with various specifications and capacities are on the market. We have some experience with the Zeiss-Kontron IBAS system (Kontron, Munich, West Germany) which may be interfaced directly to the microscope. The use of image analysis and automatic registration of nuclear and nucleolar characteristics of silver-stained C3H cells has enabled us to register differences during the

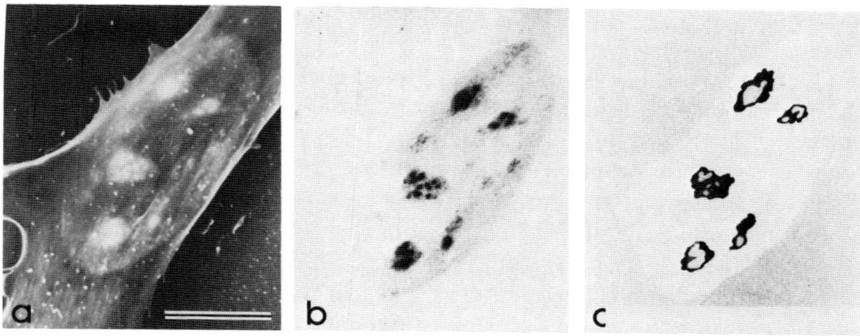

Fig. 7. C3H cell, AgNOR stain and carbon coating (by permission from Scanning Electron Microscopy Inc., AMF, O'Hare, Illinois). a, SEI surface aspect with few short microvilli; bar = 10 μm. b, BEI aspect with outlining of the nucleus and nucleoli. The dark intranucleolar spots represent AgNOR sites. c, image analysis by a Zeiss-Kontron IBAS demonstrates the nucleus in white and nucleoli in black. The light and dark grey shades correspond to the AgNOR sites.

transformation process undetectable by eye (Thiebaut *et al.*, 1984b). If suitable stereologic models are available, such morphometric analysis may be facilitated by automatic image analysis and permit studies which would otherwise have been prohibitive because of the enormous work load.

APPLICATIONS AND FUTURE PROSPECTS

The current applications have been briefly mentioned under the appropriate sections. For other applications some preliminary work has been done in our laboratory. It seems that application to materials other than those described may necessitate minor modifications if optimal results are to be obtained.

The methods using combined LM and SEM should probably be applicable to, for example, urinary cytology, an area in which surface structure may play an important role in diagnosis (Croft *et al.*, 1979). The choice between the two LM stains described must be based on the problems in question. By using Feulgen staining the mitotic stage may be defined. However, the use of dry microscopy after critical point drying gives poor representation of other cytological details despite adequate identification of the cells in both LM and SEM. Papanicolaou staining of cervical cytological smears, for example, identifies estrous stage of the cells and other cytological characteristics which may be correlated to specific diseases in the female genital organs (Kenemans *et al.*, 1981). Moreover, the use of wet microscopy gives a better optical resolution than that obtained in dry LM.

The techniques using combined SEI and BEI will probably be of interest in most surface epithelia, e.g., nasal epithelium and tracheal/bronchial epithelia. There may even be a place for such techniques in thick microtome sections or other preparations from deeper-lying tissues. In the liver, for example, spatial arrangement of cell nuclei around biliary canaliculi may be elucidated by BEI more easily than by serial sectioning.

Major prospects for the future are the wide number of applications made possible by the introduction of digital image analysis. Automatic three-dimensional analysis of stereotilt images may make possible the description of the spatial distribution of nuclei in a tissue, or spatial distribution of material within the nucleus. The possibilities may easily be appreciated by observing ordinary stereopairs of such preparations (Thiebaut *et al.,* 1985, 1986b). For correlative studies, combination with other techniques should be considered, such as immunocytochemistry on the electron microscopic level using gold marking.

We thank Ruth Puntervold for excellent technical assistance and Barbara Schüler for providing micrographs for Fig. 4.

REFERENCES

Adam, H., and Czihak, G. (1964). "Arbeitsmethoden der makroskopischen und mikroskopischen Anatomie." Fischer, Stuttgart.

Becker, R. P., and Sogard, M. (1979). Visualization of subsurface structures in cells and tissues by backscattered electron imaging. *Scanning Electron Microsc.* **2,** 835.

Black, J. T. (1974). The scanning electron microscope: Operating principles. *In* "Principles and Techniques of Scanning Electron Microscopy: Biological Applications" (M. A. Hayat, ed.), Vol. I, pp. 1–43. Van Nostrand-Reinhold, New York.

Bloom, S., and Goodpasture, C. (1976). An improved technique for selective staining of nucleolar organizer regions in human chromosomes. *Hum. Genet.* **34,** 199.

Boyde, A., Weiss, R. A., and Vesely, P. (1972). Scanning electron microscopy of cells in culture. *Exp. Cell Res.* **71,** 313.

Collins, V. P., Arborgh, B., and Brunk, U. (1977). A comparison of the effects of three widely used glutaraldehyde fixatives on cellular volume and structure. *Acta Pathol. Microbiol. Scand., Sect. A* **85A,** 157.

Croft, W. A., Nelson, C. E., and Nilsson, T. (1979). Scanning electron microscopy of exfoliated malignant and non-malignant human urothelial cells. *Scand. J. Urol. Nephrol.* **13,** 49.

Feren, K. (1982). SEM of mitotic C3H cells. Ph.D. Thesis, University of Oslo, Oslo (in Norwegian).

Feren, K., and Reith, A. (1981). Surface topography and other characteristics of non-transformed and carcinogen transformed C3H/10T 1/2 cells in mitosis, as revealed by quantitative scanning electron microscopy. *Scanning Electron Microsc.* **2,** 197.

Gualtieri, P., Barsanti, L., and Coltelli, P. (1985). Computer processing of optical microscope images. *Micron Microsc. Acta* **16,** 159.

Kenemans, P., Davina, J. H. M., de Haan, R. W., van der Zanden, P., Voovs, G. P., Stolk, J. G., and Stadhouders, A. M. (1981). Cell surface morphology in epithelial malignancy and its precursor lesions. *Scanning Electron Microsc.* **3,** 23.

Ogura, K., and Laudate, A. (1980). Comparative observation with a light microscope and a SEM in backscattered electron mode. *Scanning Electron Microsc.* **1,** 233.

Pai, S. B., Steele, V. E., and Nettesheim, P. (1983). Neoplastic transformation of primary tracheal epithelial cell culture. *Carcinogenesis* (*London*) **4,** 369.

Reitan, J. B., and Feren, K. (1986). Scanning electron microscopy of the irradiated mouse bladder urothelium. *Scanning Electron Microsc.* **2,** 773.

Reith, A., Puntervold, R., and Saxholm, H. J. K. (1980). A quantitative scanning electron microscope study of microvilli in interphase and mitosis of carcinogen-transformed fibroblasts. *Scanning Electron Microsc.* **3,** 23.

Reith, A., Rigaut, J. P., Boysen, M., and Marton, P. (1983). Pseudostratified, metaplastic, dysplastic and carcinomatous nasal mucosa in nickel workers: A study by scanning electron, transmission electron, and light microscopy. *In* "Nasal Tumors in Animals and Man" (G. Reznik and F. Sherman, eds.), Vol. 2. CRC Press, Boca Raton, Florida.

Rigaut, J. P. (1983). Image analysis in electron microscopy. *In* "Electron Microscopy in Human Medicine" (J. V. Johannessen, ed.), Vol. 11B, pp. 197–231. McGraw Hill, New York.

Rigaut, J. P. (1987). Analysing electron microscopic images by computer: A guided tour. *In* "Stereology and Morphometry in Electron Microscopy: An Illustration of Problems and Solutions" (A. Reith and T. M. Mayhew, eds.). Hemisphere, New York. (in press).

Saxholm, H. J. K., and Reith, A. (1979). The surface structure of 7, 12-dimethylbenz (a) anthracene transformed C3H/10T 1/2 cells. A quantitative scanning electron microscopical study. *Eur. J. Cancer* **15,** 843.

Thiebaut, F., Rigaut, J. P., and Reith, A. (1984a). Improvement in the specificity of the silver staining technique for AgNOR-associated acidic proteins on paraffin sections. *Stain Technol.* **59,** 181.

Thiebaut, F., Rigaut, J. P., Feren, K., and Reith, A. (1984b). Smooth-surfaced control and transformed C3H/10T-1/2 cells differ in cytology: A study by secondary electron, backscattered electron and image analysis. *Scanning Electron Microsc.* **3,** 1249.

Thiebaut, F., Rigaut, J. P., Feren, K., and Reith, A. (1984c). The application of the nucleolar organizer region silver staining (AgNOR) to backscattered electron imaging (BEI). *Biol. Cell.* **52,** 103.

Thiebaut, F., Rigaut, J. P., Feren, K., and Reith, A. (1985). Stereoscopic back-scattered electron imaging of silver-stained proteins in nucleoli. *Chromosoma* **91,** 372.

Thiebaut, F., Rigaut, J. P., Reitan, J., and Reith, A. (1986a). Spatial visualization of junctional complexes by backscattered electron imaging and silver staining. *Ultrastruct. Pathol.* **10,** 265.

Thiebaut, F., Reitan, J. B., Feren, K., Rigaut, J. P., and Reith, A. (1986b). An epidermal proliferative unit-like structure in mouse urinary bladder urothelium observed by backscattered electron imaging. *Cell Tissue Res.* **246,** 1.

van Driel-Kulker, A. M. J., Ploem-Zaaijer J. J., van der Zwan-van der Zwan, M., and Tanke, H. J. (1980). A preparation technique for exfoliated and aspirated cells allowing different staining procedures. *Anal. Quant. Cytol.* **2,** 243.

Wetzel, B., Kendig, E. M., Jones, G. M., and Sanford, K. K. (1978). A systematic scanning electron microscope (SEM) analysis of mitotic cell populations in monolayer culture. *Scanning Electron Microsc.* **2,** 1.

15

Correlative Light, Scanning, and Transmission Electron Microscopy of Deembedded Tissues

HISATAKE KONDO

Department of Anatomy
Kanazawa University
School of Medicine
Kanazawa, Japan

INTRODUCTION

The development of the scanning electron microscope (SEM) with its superior depth of field and the ability to scan large areas and reveal ultrastructural details has greatly expanded our understanding of the surface morphology of biological specimens. A logical extension is to observe the interior of biological specimens whose surface morphology has been revealed by this microscope. In order to accomplish a correlation of

SEM images with light (LM) or transmission electron microscopic (TEM) images of a given area of a tissue, it is crucial to remove the embedding medium from the tissue specimen, while at the same time retaining surface as well as internal morphological details. The removal of epoxy resins commonly employed as embedding media for transmission electron microscopy is usually accomplished with sodium or potassium hydroxide dissolved in methanol. There have been several reports using this type of removal for correlation of studies with the SEM, the LM, or the TEM (e.g., Erlandsen *et al.*, 1973). However, the length of exposure of biological specimens to Epon solvents must be carefully monitored with the LM since incomplete removal may result in the formation of artifactual structures such as "epoxy bridges," which link individual authentic biological components. On the other hand, excessive exposure to such caustic reagents results in surface contamination which almost completely obscures the surface details (Geissinger, 1976). Therefore, more reliable, simple, and less drastic methods for deembedment processes are needed to obtain high-quality correlative images.

Recently, a new methodology has been introduced to observe biological materials in embedment-free sections (Wolosewick, 1980). In this method, chemically fixed tissues are embedded for sectioning in polyethylene glycol (PEG), a highly water-soluble wax, which is subsequently removed from sections in water or ethanol. After critical-point drying with CO_2, these embedment-free sections present ultrastructural features quite similar to those in the conventional epoxy sections at low magnifications, although the appearance of the microtrabeculae is markedly enhanced in the cytoplasmic matrix at high magnifications. In addition, the complete removal of PEG from tissue blocks as a whole can be accomplished easily by soaking them in warm water or ethanol, and reembedding in Epon. The reembedded tissue blocks show little, if any, alteration in the cytoplasmic morphology when their ultrathin sections are observed in the TEM (Kondo *et al.*, 1982, 1983; Kondo, 1984a,b, 1985a,b). Therefore, high-quality correlation of images obtained with the SEM, LM, and TEM is expected when using PEG embedding and subsequent deembedding as a method for the tissue preparation (Kondo, 1985a). This chapter presents a detailed description of the PEG method for the correlation of such images as well as representative examples of this approach.

METHODOLOGY

Tissue or cell samples are fixed with 2.5% glutaraldehyde for 1–2 hr and then with 1% OsO_4 for 1–2 hr. The buffer used is 0.1 *M* Na-cacodylate (pH 7.4). In general, initial fixation is preferred by vascular perfusion to achieve superior quality of ultrastructural preservation. After fixation, the tissue blocks are dehydrated by increasing concentrations of ethanol (50–100%). Vials containing the tissue blocks in 100% ethanol are placed in an oven at about 60°C to warm the ethanol. The tissue blocks are immersed in a 50% solution of PEG-4000 (average mol. wt., 3000 to 3700) in 100% ethanol (v/v) prepared by mixing equal volume of the liquid PEG and ethanol. They are kept in the 50% PEG solution for 3 hr at 60°C. The tissue blocks are transferred to pure molten PEG contained in well-dried gelatin capsules. After the blocks have been sunk to the bottom of the capsules, each capsule is quickly immersed in swirling liquid nitrogen for 30 sec, resulting in solidification of the PEG.

The gelatin capsule is removed with a razor blade to expose the solidified PEG. Portions of the PEG including the tissue blocks are cut out and mounted with dental wax on supporting stubs fitting the collet of the microtome. The blocks are trimmed with a razor blade to obtain block faces of 1–4 mm, then sectioned with a well-dried glass or diamond knife. Knowledge and experience of sectioning of epoxy-embedded specimens is helpful to the PEG sectioning. With some practice, long ribbons of wrinkle-free sections can be easily obtained.

Several consecutive sections (about 200–300 nm thick) are cut; the first few sections are used with the LM and the remaining with the TEM. Sections are picked up with a platinum loop filled with 2.5% sucrose and mounted on a glass slide for light microscopy and on Formvar-coated grids for transmission electron microscopy. Both glass slides and grids have previously been treated with 0.1% poly-L-lysine in distilled water. For light microscopy, sections on a glass slide may be briefly stained with 1% toluidine blue. For transmission electron microscopy, sections on grids are transferred to a submerged grid holder in 90% ethanol, which is then quickly transferred to 100% ethanol. After several changes of 100% ethanol, the holder is transferred to a critical-point apparatus and dried with liquid CO_2. The dried grids are stored over silica gel until ready for observation with the TEM. For immunohistochemistry, tissues or cells fixed with 4% paraformaldehyde for 2 hr are processed for embedding and sectioning as described above. The sections on a glass slide are incubated for 20 hr at 4°C with a given antibody at an appropriate dilution. The sites of antigen–antibody reaction are visualized with the peroxidase–antiperoxidase (PAP) method (Sternberger, 1979).

After taking the final section from the block, the block with its mirror-smooth face is removed from the microtome and immersed in distilled water. The temperature of the water is then slowly raised to 40°C over a period of 15 min. The tissue is fixed with 2.5% glutaraldehyde containing

1% tannic acid in 0.1 *M* Na-cacodylate (pH 7.4) for 1 hr, followed by 1% OsO_4 in the same buffer for 5 hr. After the postfixation, the tissue blocks are dehydrated in a graded series of ethanol up to 100% and critical-point dried in CO_2. Dried tissue blocks are mounted on aluminum stubs with silver paste and coated with gold-palladium (Au-Pd) followed by carbon. The specimens are examined with the SEM.

APPLICATIONS

An example of the excellent preservation of surface morphology by means of the PEG method is shown in Fig. 1, in which the apical surface of ciliated cells and mucous cells are clearly visible from the luminal side of the intrapulmonary bronchus. The detailed arrangement of cilia and microvilli of each cell type is seen without any damage or distortion. This

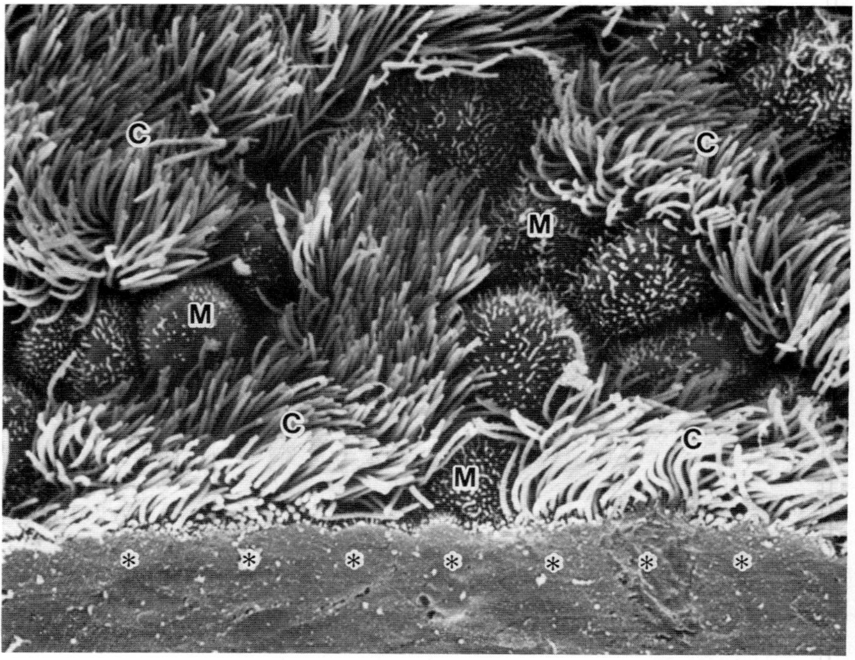

Fig. 1. A typical luminal surface view of the bronchial epithelium of the rat processed by the PEG method. The structure and arrangement of cilia (C) of the ciliated cells and microvilli (M) of the mucous cells are clearly demonstrated and well preserved. A smooth cutting face of the tissue block after cutting out thin sections is marked by asterisks; ×5000.

high-quality SEM image confirms that the PEG embedding and deembedding do not appreciably alter cell ultrastructure and that resulting images from the PEG method are regarded as quite acceptable (Kondo, 1985b).

Figures 2a,b are examples of the correlation between SEM and LM images from two consecutively sectioned tissue faces. Figure 2a is an immunolight micrograph of a section which has been incubated with the antiserum against the spot-35 protein and treated with the PAP procedure. The spot-35 is a soluble protein originally isolated from the cerebellum of adult rats and it has been found to be specifically localized in the Purkinje cells (Yamakuni *et al.,* 1984). Recently, cells showing the intense immunoreactivity for this protein have been detected in several loci outside the cerebellum such as epithelial cells of the distal tubules of the kidney, horizontal cells of the retina, and adrenal chromaffin cells (Iwanaga *et al.,* 1985; Kondo *et al.,* 1985). In this figure, the histological organization of the tubules of the rat kidney is well preserved. Intensely immunoreactive, uriniferous epithelial cells are clearly identified in the distal tubules and they are arranged in a mosaic pattern with immunonegative cells.

Figure 2b is a SEM image of the tissue block viewed from the sectioned face immediately after taking the section shown in Fig. 2a. The immunoreactive cells shown in the LM are identified easily and accurately in the SEM. The comparison in morphology of the luminal surface between the immunoreactive and immunonegative cells is quite possible.

The correlation between the SEM and the TEM is shown in Fig. 3. Areas enclosed by rectangles in the low-magnification SEM view of the intrapulmonary bronchial epithelium are correlated with the thin sections for the TEM. Since the thin sections are free from embedding media and unstained, the final contrast of the image is mainly due to the inherent electron-scattering property of individual osmicated tissue elements (Wolosewick, 1980). Although the indistinct appearance of the plasma membrane and the enhanced appearance of the microtrabeculae in the cytoplasmic matrix at a higher magnification are characteristic of the embedment-free section, the cytoplasmic organization of cellular components at such magnifications as in Figs. 3b,c is similar to the conventional TEM images of the same types of cells. It is clearly demonstrated that cells topped with microvilli in the SEM have ultrastructural features of the mucous cells in the TEM, while cells with cilia in the SEM have those of the ciliated cells in the TEM.

As demonstrated in this chapter, the exact correlation of morphological informations from the surface and the interior of a given portion of tissue is reliably done with the PEG method. This approach is particularly useful in correlation of immuno-LM with the SEM, and the three-dimensional understanding of immunoreactive cells is possible. With continued and

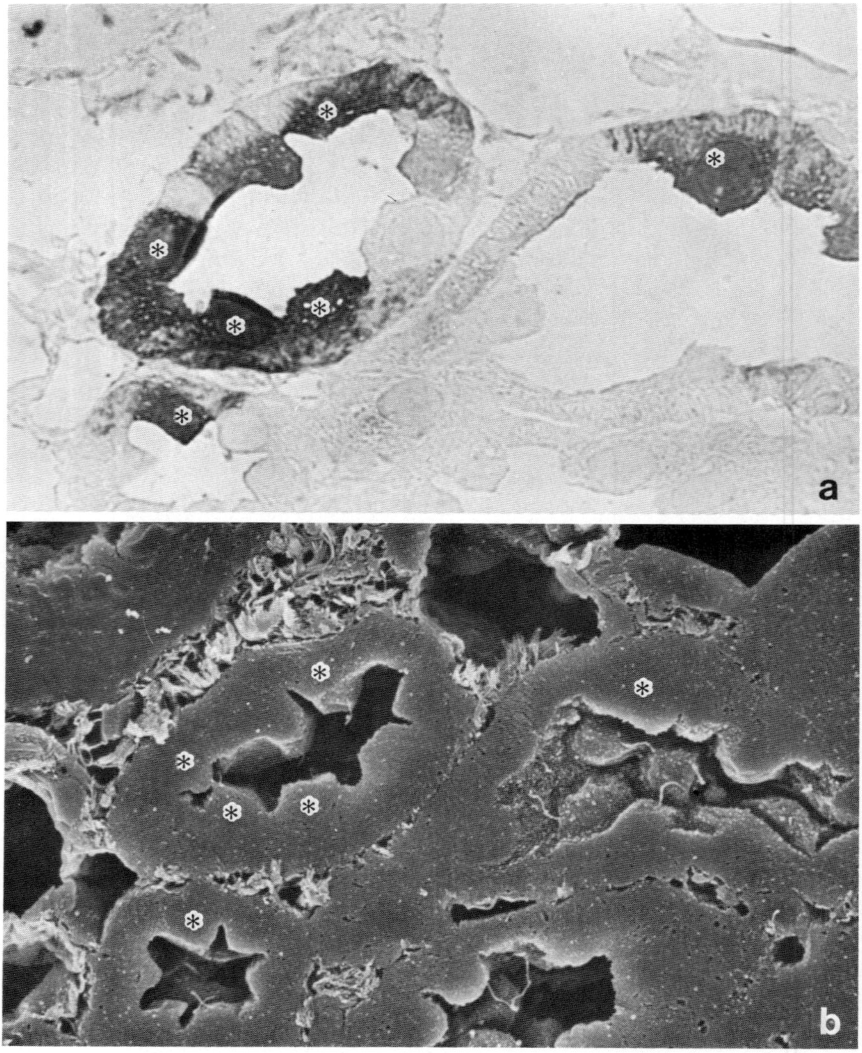

Fig. 2. a, A light micrograph of an embedment-free section of the rat kidney immunostained for the spot-35 protein. Several uriniferous epithelial cells of the distal convoluted tubules (*) are intensely immunostained and they show a mosaic arrangement with immunonegative cells; ×2100. b, A scanning electron micrograph of the same distal tubule after cutting the section shown in a. Good correlation of structure between a and b is evident and the comparison of surface structure between cells can be easily done; ×1900.

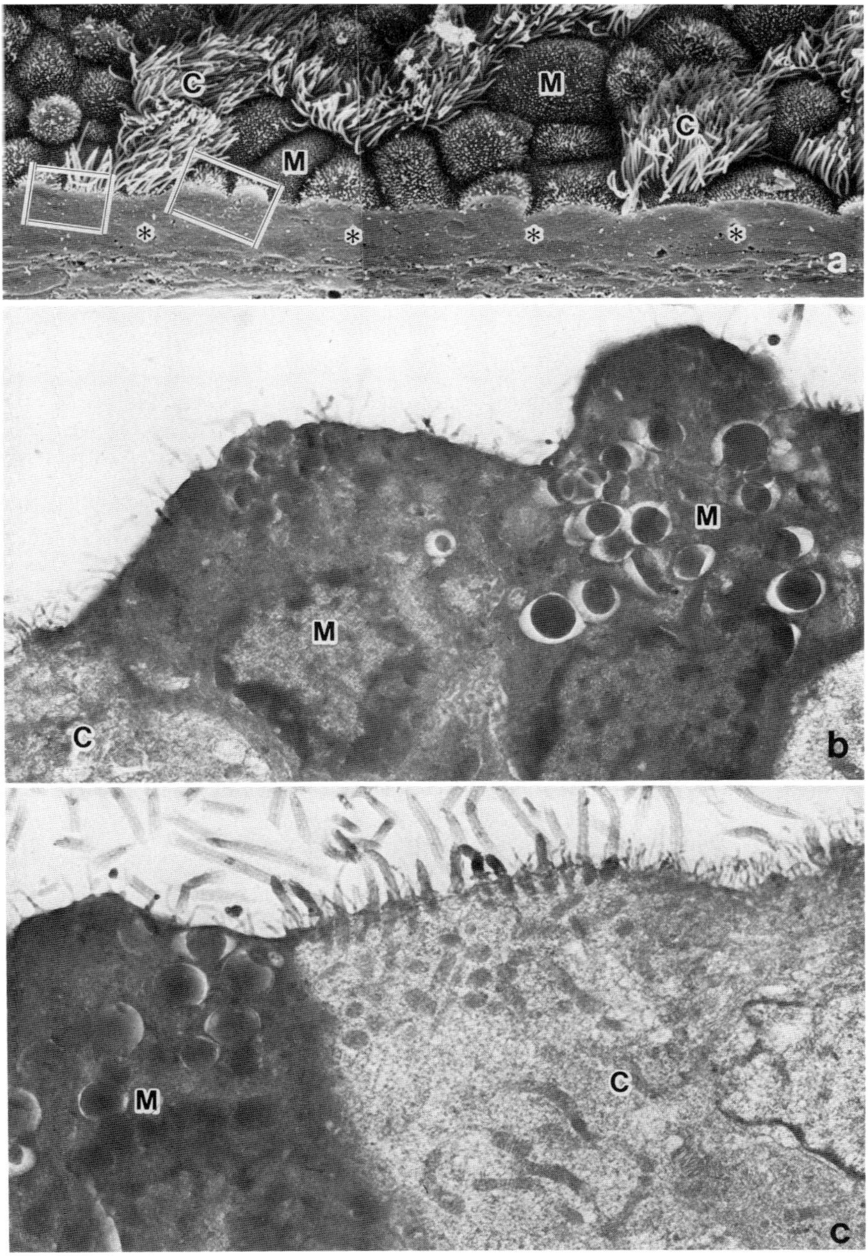

Fig. 3. A scanning electron micrograph of the rat bronchial epithelium (a) and transmission electron micrographs showing portions of the epithelium enclosed by rectangles in a (b,c). A cutting face after taking out the thin section shown in b and c is marked by asterisks. Fine structures of ciliated (C) and mucous (M) cells are clearly seen; a, ×1800; b,c, ×22,000.

wide application, the PEG method is expected to unravel a new level of histological details as a valuable adjunct to conventional microscopic techniques.

The author wishes to thank Dr. T. Ushiki for his collaboration in performing the SEM observation.

REFERENCES

Erlandsen, S. L., Thomas, A., and Wendelschaffer, F. (1973). A simple technique for correlating SEM with TEM on biological tissue originally embedded in epoxy resin for TEM. *Scanning Electron Microsc.* **3,** 349.

Geissinger, H. D. (1976). Intermicroscopic (LM, SEM, TEM) correlation. *In* "Principles and Techniques of Scanning Electron Microscopy" (M. A. Hayat, ed.), Vol. 5, pp. 94–121.

Iwanaga, T., Takahashi-Iwanaga, H., Fujita, T., Yamakuni, T., and Takahashi, Y. (1985). Immunohistochemical demonstration of a cerebellar protein (spot 35 protein) in some sensory cells of guinea pigs. *Biomed. Res.* **6,** 329.

Kondo, H. (1984a). Polyethylene glycol (PEG) embedding and subsequent deembedding as a method for the structural and immunocytochemical examination of biological specimens by electron microscopy. *J. Electron Microsc. Tech.* **1,** 227.

Kondo, H. (1884b). Reexamination of the reality or artifacts of the microtrabeculae. *J. Ultrastruct. Res.* **87,** 124.

Kondo, H. (1985a). Polyethylene glycol (PEG) embedding and subsequent de-embedding as a method for the correlation of light microscopy, scanning electron microscopy, and transmission electron microscopy. *J. Electron Microsc. Tech.* **2,** 457.

Kondo, H. (1985b). What is the microtrabecula? *J. Electron. Microsc.* **34,** 123.

Kondo, H., Wolosewick, J. J., and Pappas, G. D. (1982). The microtrabecular lattice of the adrenal medulla revealed by polyethylene glycol embedding and stereo electron microscopy. *J. Neurosci.* **2,** 57.

Kondo, H., Pappas, G. D., and Wolosewick, J. J. (1983). The cytoskeletal lattice of the neurohypophysial cells. *Biol. Cell.* **49,** 99.

Kondo, H., Kuramoto, H., Iwanaga, T., and Fujita, T. (1985) Cerebellar Purkinje cell-specific protein-like immunoreactivity in noradrenalin-chromaffin cells and ganglion cells but not in adrenaline-chromaffin cells in the rat adrenal medulla. *Arch. Histol. Jpn.* **48,** 421.

Sternberger, L. A. (1979). "Immunocytochemistry," 2nd ed., pp. 104–164. Wiley, New York.

Wolosewick, J. J. (1980). The application of polyethylene glycol (PEG) to electron microscopy. *J. Cell Biol.* **86,** 675.

Yamakuni, T., Usui, H., Iwanaga, T., Kondo, H., Odani, S., and Takahashi, Y. (1984). Isolation and immunohistochemical localization of a cerebellar protein. *Neurosci. Lett.* **45,** 235.

16

Immunoelectron Microscopic Mirror Method: A Correlative Study

HIROSHI TAKAGI

Second Department of Anatomy
School of Medicine
Kinki University
Osaka, Japan

INTRODUCTION

The immunoelectron microscopic mirror method was recently developed to identify the interaction between two neuronal elements that have been immunocytochemically defined (Kubota *et al.*, 1985, 1986a,b). This approach is useful for the analysis of neuronal circuitry at the level of the

CORRELATIVE MICROSCOPY IN BIOLOGY:
INSTRUMENTATION AND METHODS

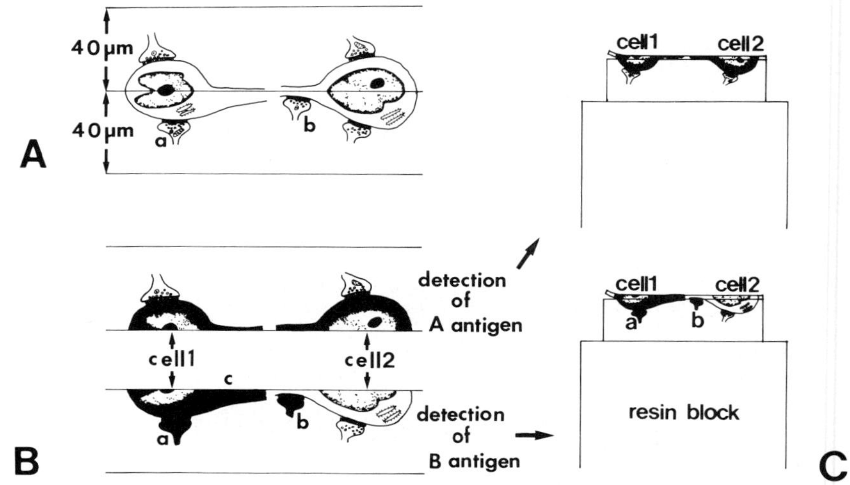

Fig. 1. Principle of the immunoelectron microscopic mirror method. A, tissue block is cut into 40-μm-thick serial sections. B, two consecutive sections are subjected to the PAP method using different antisera to detect A antigen and B antigen, respectively. C, after light microscopic identification of the cells split into consecutive sections, serial ultrathin sections are cut from the surface of paired sections. Cell 1 is immunoreactive both for A and B antigens whereas cell 2 is only immunoreactive for the A antigen. With this mirror method, synaptic contacts can be identified between cell 1 and axon terminal "a," and between cell 2 and axon terminal "b."

synapse where chemical neurotransmission occurs. The method is based on a combination of the correlated light and electron microscopic technique (Somogyi, 1978) applied to immunocytochemistry (Somogyi and Takagi, 1982) and the light microscopic adjacent section procedure (Kosaka *et al.*, 1985). The practical aspects of this method are described here.

Figure 1 shows the immunoelectron microscopic mirror method. Immunostaining is achieved in preembedding materials by the peroxidase–antiperoxidase (PAP) method (Sternberger, 1979). Although the principle of the immunoelectron microscopic mirror method is simple, it is important to obtain flat and smooth surfaced sections and cut ultrathin sections from both of the paired surfaces of two consecutive thick sections. Using

Fig. 2. A, immunocytochemical localization of TH in the caudal part of the dorsomedial medulla oblongata. AP, area postrema; C, central canal; dmnX, dorsal motor nucleus of the vagus; NTS, nucleus of the solitary tract; nXII, nucleus nervi hypoglossi. B, higher magnification micrograph of the framed area in A, taken from the lower surface of a Vibratome section. CA, capillary. C, light micrograph from the upper surface of the section adjacent to that in B, immunostained with CCK antiserum. Some of the CCK-immunoreactive cells (n1′, n2′, and n3′) look identical to the TH-immunoreactive cells (n1, n2, and n3) shown in

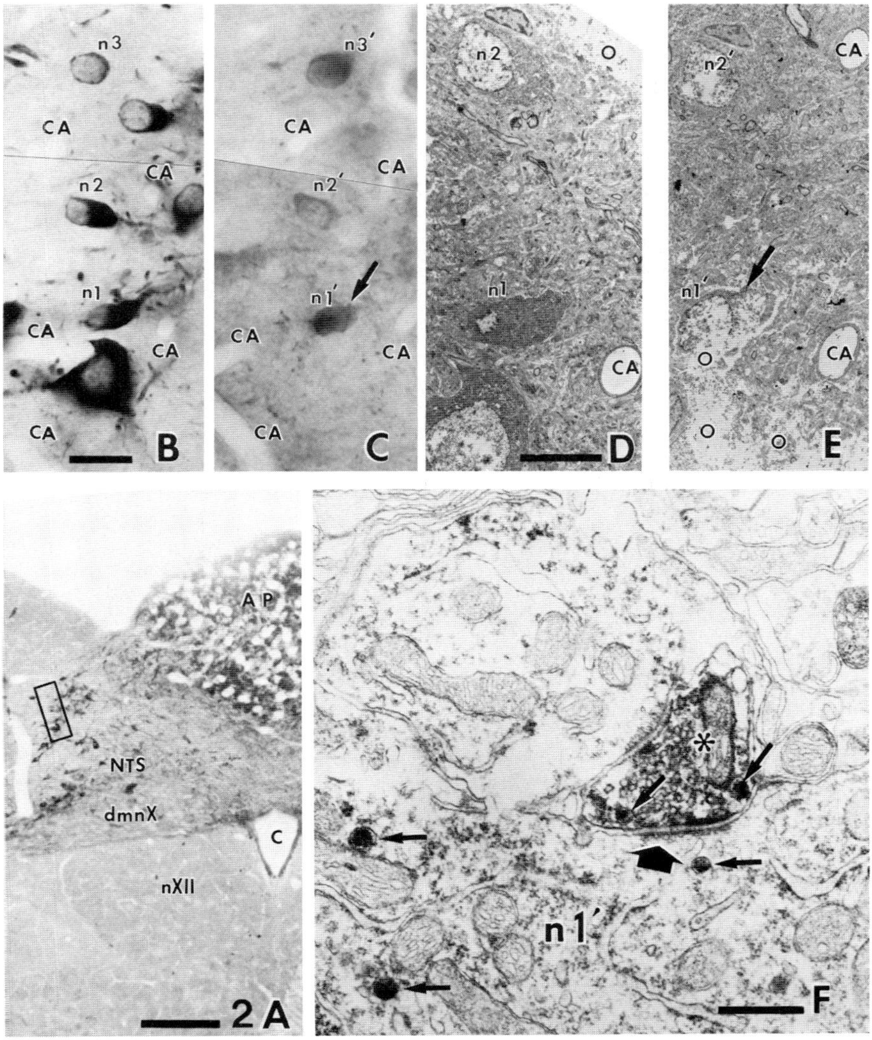

B. Capillaries (CA) are landmarks for correlating B and C. D,E, electron micrographs of parts of B and C, respectively. Vacant areas (circles) are a result of the micrographs being taken from the damaged surface of the sections. Cells n1 and n1′ are the same neuron, as are cells n2 and n2′. The arrow in E shows the same point indicated by the arrow in C. F, electron micrograph of a CCK-immunoreactive cell body (n1′ in C; the same cell, n1 in B, is also TH-immunoreactive) that is in synaptic contact (large solid arrow) with a CCK-immunoreactive axon terminal (asterisk) indicated by the arrow in E. F is several ultrathin sections away from E. Small arrows point at immunoreactive, large granulated vesicles. Scale bars: A, 200 μm; B,C, 20 μm; D,E, 10 μm; F, 0.5 μm. (From Kubota *et al.*, 1986b, with permission.)

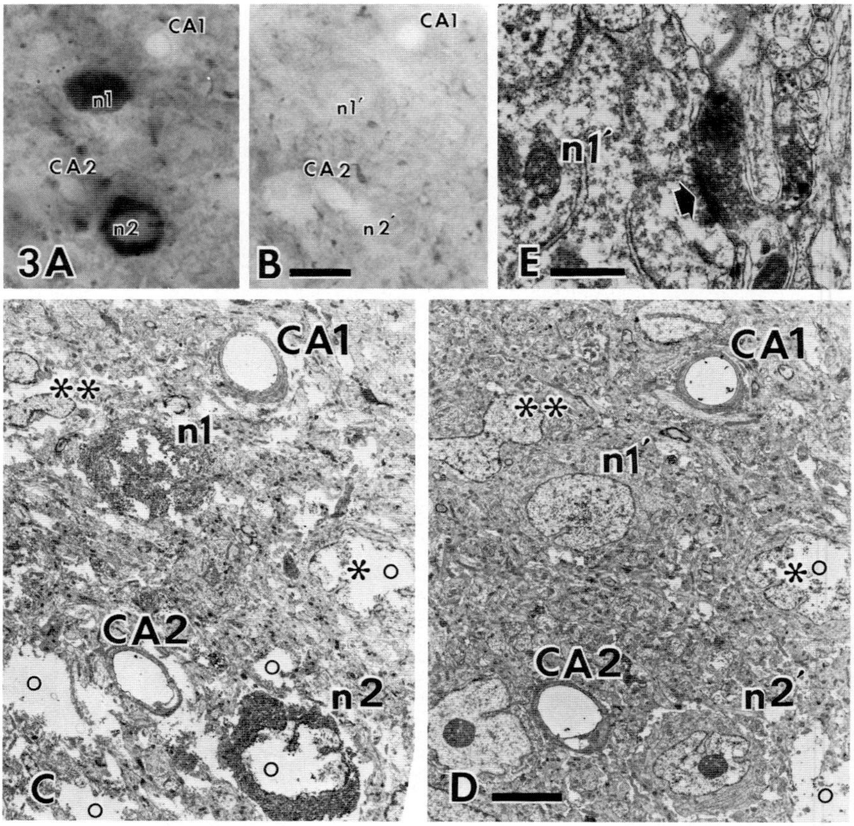

Fig. 3. A, light micrograph of TH-immunoreactive cells (n1 and n2) in the medial subnucleus of the NTS, taken from the upper surface of a Vibratome section. CA1 and CA2; capillaries. B, light micrograph from the lower surface of the section adjacent to that shown in A, which was immunostained with SP antiserum. Profiles of SP nonimmunoreactive cell bodies (n1′ and n2′) appear to be identical to TH-immunoreactive cells (n1 and n2) shown in A. C,D, electron micrographs of A and B, respectively. Capillaries (CA1 and CA2) and nonimmunoreactive cells (single and double asterisks) are landmarks for correlating C and D. Cells n1 and n1′ and n2 and n2′ were determined as the same cell, respectively. Vacant areas (circles) are a result of the micrographs being taken from the surface of the sections. E, electron micrograph of a synaptic contact (arrow) between SP-immunoreactive axon terminal and a nonimmunoreactive cell body (n1′ in B and D). E is 64 ultrathin sections away from D. Scale bars: A,B, 10 μm; C,D, 5 μm; E, 0.5 μm. (From Kubota *et al.*, 1985, with permission.)

this method, the synaptic interaction between neuronal elements containing different substances can be demonstrated. The cell immunoreactive for both tyrosine hydroxylase (TH; the enzyme that synthesizes catecholamine) and cholecystokinin (CCK) makes a synaptic contact with a CCK-immunoreactive axon terminal (Fig. 2), and the cell immunoreactive only for TH makes a synaptic contact with a substance P (SP)-immunoreactive axon terminal (Fig. 3).

METHODOLOGY

Tissue Fixation

Conditions of tissue fixation and subsequent processing including immunostaining must be worked out to obtain optimal localization of the two antigens to be examined. Tissue stabilization must be achieved by chemical fixation prior to subsequent tissue preparation to prevent antigen extraction or relocation and preservation of tissue structure (for review, see Larsson, 1983; Hayat, 1981, 1986). Various fixatives are recommended for immunoelectron microscopic studies and for each type of antigen. Glutaraldehyde is the most effective for tissue preservation of the nervous system. However, glutaraldehyde frequently destroys antigenicity and tends to reduce antibody permeation. Formaldehyde retains the immunoreactivity of many antigens but with poor preservation of fine structural details.

Currently, glutaraldehyde and paraformaldehyde in various ratios of combination are in common use. An example is the low-temperature, short time duration fixation procedure incorporating a formaldehyde (2%) and glutaraldehyde (3%) mixture in sodium phosphate buffer (pH 7.3). This particular formula is recommended for many neuropeptides (Larsson, 1981). If picric acid is included in the mixture with a lower percentage (0.05%) of glutaraldehyde, the fine structure is preserved well and the antigenicity for many neuropeptides (SP, CCK, somatostatin, and enkephalin, Somogyi and Takagi, 1982; Takagi *et al.*, 1984), enzyme proteins (TH, van den Pol, 1984; choline acetyltransferase, Wainer *et al.*, 1984; glutamate decarboxylase, Somogyi *et al.*, 1983), and histidine decarboxylase (Takagi *et al.*, 1986) is retained satisfactorily.

This fixative mixture with a slightly higher concentration (0.5%) of glutaraldehyde has been reported to be effective for (γ-aminobutyric acid (GABA) immunocytochemistry in the preembedding method (Somgyi *et al.*, 1985). Parabenzoquinone–formaldehyde–glutaraldehyde (PEG) (Larsson, 1977), glutaraldehyde–carbodiimide (Willingham and Yamada,

1979), periodate–lysine–paraformaldehyde (McLean and Nakane, 1974), and acrolein (King *et al.*, 1983) are also useful fixatives for immunoelectron microscopy. Sodium borohydrite treatment can partially restore the antigenicity in glutaraldehyde-fixed material (Eldred *et al.*, 1983). Osmium tetroxide destroys most antigenicities, thus being preferred as a postfixative after immunoreaction. The use of OsO_4 before immunoreaction has been reviewed by Hayat (1986).

Tissue Processing

Immunostaining is performed on relatively thick unembedded tissue sections. Vibratome and cryostat sections are utilized; the former results in better preservation of tissue structure. The penetration of immunoreagents is usually poor in well-fixed Vibratome sections. Although the immunostaining on the superficial layer of sections is important for the mirror method, good penetration of the reagent is desirable for light and electron microscopic observations of Golgi-stained nerve cells defined by neuroactive substances (Somogyi and Takagi, 1982; Takagi *et al.*, 1984. Rapid freeze-thawing facilities the penetration without substantial damage to the tissue fine structure caused by detergents such as Triton (Somogyi and Takagi, 1982; Vaughn *et al.*, 1981).

Immunostaining

Peroxidase is most commonly used as a marker in immunoelectron microscopy in the preembedding method. Of the detection techniques using this marker, the PAP method (Sternberger, 1979) and the avidin-biotin–peroxidase (ABC) method (Hsu *et al.*, 1981) are very sensitive. 3,3′-Diaminobenzidine is widely used as the chromogen for electron immunocytochemistry for its excellent properties. This reagent is insoluble in alcohol and is osmiophilic, but is thought to be carcinogenic and must be handled with care. Besides peroxidase, colloidal gold, which is later silver-intensified, has been introduced in light and electron microscopic immunocytochemistry for preembedded brain specimens (van den Pol, 1985; van den Pol *et al.*, 1985).

CORRELATIVE LIGHT AND ELECTRON MICROSCOPY

To confirm at the ultrastructural level that two parts cut from a single cell with a Vibratome are identical on paired section surfaces, comparison of the light and electron microscopic observations is required (Figs. 2 and 3). This also allows the electron microscopic observation of small structures, i.e., immunoreactive axon terminals seen under light microscopy and is, therefore, helpful in the identification of the synaptic relationship which is an aim of the immunoelectron microscopic mirror method (Fig. 2C,E,F).

General Method

Although methodological details including fixation depend on the type of antigen under study, the following protocol may be useful as a general guide for the application of the immunoelectron microscopic mirror method.

1. Animals are perfused through the heart or the ascending aorta with a mixture of picric acid (0.2%), paraformaldehyde (4%) and glutaraldehyde (0.5%) in 0.1 *M* phosphate buffer (pH 7.4) for 30 min following Tyrode's solution (gassed with a mixture of 95% O_2 and 5% CO_2) or saline for 1–2 min at room temperature. For a rat weighing 100–150 g, about 200 ml of the fixative is adequate.
2. After postfixation by immersion in the same fresh fixative for 1–3 hr at 4°C, areas of interest are cut into blocks (approximately 2 × 4 × 5 mm) and rinsed in several changes of 0.1 *M* phosphate buffer at 4°C.
3. The blocks are transferred sequentially to 10% and 20% sucrose in the same buffer at 4°C for cryoprotection until they sink. They are then frozen in liquid nitrogen and thawed in 20% sucrose in 0.1 *M* phosphate buffer at room temperature.
4. Serial sections (40 μm thick) are cut on a Vibratome (Oxford Instruments, Oxford) at 4°C. Each section is placed in a small vial and rinsed in several changes of 0.1 *M* phosphate buffer at 4°C.
5. Each section is individually subjected to the PAP* method with some modification. The section is incubated in the following order at 4°C unless otherwise stated: (a) 1 hr in 20% normal serum derived from the species producing the second layer antibody at room temperature to prevent nonspecific staining: (b) three 30-min rinsings in PBS; (c) alternate sections are incubated in the primary antiserum against "A antigen" (A antiserum)

* Phosphate-buffered saline (PBS) is used for all antibody dilutions. To further prevent nonspecific staining, the above normal serum (1%) used in (a) is included in the primary and secondary antisera and in each rinsing solution before PAP incubation. Furthermore, 0.25% carrageenan (Sofroniew and Glasmann, 1981) is also used in the second and third rinsings each time before PAP incubation. Sections are incubated and rinsed by gentle shaking to facilitate the penetration of immunoreagents or washing.

or against "B antigen" (B antiserum), respectively, at an optimal dilution; (d) three 30-min rinsings in PBS; (e) keep overnight in the antiserum against the IgG of a species producing the primary antiserum; (f) three 30-min rinsings in PBS; (g) keep for 3 hr in the PAP complex of a species producing the primary antiserum; (h) several rinsings; (i) antigens are localized by incubation with a 0.05% 3,3′-diaminobenzidine tetra HCl dissolved in 0.05 *M* Tris-HCl buffer at pH 7.6, plus 0.01% hydrogen peroxide for 3–20 min at room temperature following preincubation with the above solution (omitting hydrogen peroxide) for 1 hr in a dark room at 4°C.

6. After rinsing with 0.05 *M* Tris-HCl buffer followed by 0.1 *M* phosphate buffer, each section is postfixed for 1 hr in 1–2% OsO_4 in 0.1 *M* phosphate buffer.

7. When Durcupan (Fluka) recommended for flat embedding of the sections due to easy separation of slides from polymerized resin is used, the following dehydration time for each step is adequate: Each section is placed in 50% ethanol for 5 min and successively transferred through 70% and then 95% ethanol, two changes of absolute ethanol, two changes of propylene oxide for 15 min each except for the 70% ethanol. The section is stained with 1% uranyl acetate at the 70% alcohol dehydration state for 40 min to enhance contrast for electron microscopy. Then the section is stretched in 95% ethanol between the coverslip and microscope glass slide to prevent the section from bending at the last stage. The section is placed in a small glass vial during the dehydration stage except for the 95% ethanol. After the propylene oxide, the section is immersed into Durcupan overnight.

8. Alternate sections immunostained with A or B antiserum are placed in the same sequence on each of the slides, respectively. A small amount of Durcupan is used to fill the space between the coverslip and the slide without thickening the interlayer significantly. The sections are cured for 2 days at 56°C. For other epoxy resins, siliconized coverslips and slides should be used for easy removal from the polymerized resin.

9. The slide-mounted sections are first examined by light microscopy. The neurons and the cell body which have been immunostained with A- or B-antiserum on the section surface and having the same profile as the immunostained or unstained cells with antiserum different from the first on the adjacent surface of the next section are selected (Figs. 1–3).

10. To reembed a selected area, the coverslip is removed with a razor blade and stained structures are drawn using a camera lucida.

11. The selected area is cut from the section warmed on a hot plate and placed in a drop of fresh resin between a slide and a polymerized cylinderical block (5–7 mm high) with flat, smooth surfaces on both top and

bottom. The selected area should be faced on the top of the block, because the ultrathin sections are first cut from the top surface.

12. After curing for a 1 day in an oven, serial ultrathin sections are cut with a diamond knife from the surface of the pair-matched sections, one of which has been incubated with A antiserum and the other with B antiserum to give immunostaining in the superficial (5–10 μm) layer, and mounted on Formvar-coated, single-slot (2 × 1 mm) grids. The section surface should be adjusted carefully to keep it parallel to the knife edge, for comparison of the light and electron microscopic observations (Figs. 2 and 3). During the electron microscopic observation, high-power light micrographs and drawings help in identifying selected structures.

13. Sections are poststained with lead citrate (Reynolds, 1963) to enhance the contrast.

DISCUSSION

To demonstrate synaptic contacts between two chemically characterized neuronal elements, several methods, besides the immunoelectron microscopic mirror method, have been recently reported: (1) the combination of the PAP and uptake methods using autoradiography (Pickel *et al.*, 1984; Kiss *et al.*, 1984) or false transmitters (Yamano, *et al.*, 1985) in preembedding materials; (2) the combination of the PAP method and a direct marker method utilizing an internally labeled, radioactive, monoclonal antibody (Cuello *et al.*, 1982) in preembedding materials; (3) double-staining method using silver-intensified gold and peroxidase in preembedding materials (van den Pol, 1985; van den Pol *et al.*, 1985); and (4) combination of preembedding peroxidase method and postembedding method using colloidal gold (van den Pol, 1984) or peroxidase (Hisano *et al.*, 1984). They are all double-staining methods that are employed on a single section.

Each method has its advantages and disadvantages (van den Pol *et al.*, 1985). The uptake method may not always represent the localization of endogenous substances (Zucker *et al.*, 1984). Application of autoradiography to structures as small as axon terminals is technically demanding and time-consuming. In double-staining method, care must be taken to exclude the possibility of false-positive immunostaining due to artifactual crossreaction between immunoreagents used for the detection of different antigens. This is not necessary in the immunoelectron microscopic mirror method, since paired consecutive sections are immunostained separately. Furthermore, the preservation of the tissue structure except for that on the section surface is relatively good in the mirror method. However, by this method, the nonimmunoreactive counterpart of the cell which is immunostained on the adjacent section actually cannot be traced to the distal dendritic portion at the electron microscopic level. This may result in the limited survey on the cell bodies, and the proximal dendritic portions cut parallel to the section surface.

The possible application of the mirror method is that it can be combined with other marking methods. If it is used in combination with the double-staining method mentioned above, an additional level of information may be obtained; for example, about the interaction among three or more chemically defined neuronal elements. Furthermore, it may be applied to the immunocytochemistry combined with anterograde degeneration (Léránth and Frotscher, 1983) or retrograde peroxidase tracing (Ruda, 1982; Priestley and Cuello, 1983). These approaches appear very useful to understand the origin of inputs, or the projection area of the neurons characterized by the immunoelectron microscopic mirror method.

The author wishes to thank Yoshiyuki Kubota for his helpful advice and Michiko Nimura for typing the manuscript.

REFERENCES

Cuello, A. C., Priestley, J. V., and Milstein, C. (1982). Immunocytochemistry with internally labeled monoclonal antibodies. *Proc. Natl. Acad. Sci. U.S.A.* **79,** 665.

Eldred, W. D., Zucker, C., Karten, H. J., and Yazulla, S. (1983). Comparison of fixation and penetration enhancement techniques for use in ultrastructural immunocytochemistry. *J. Histochem. Cytochem.* **31,** 285.

Hayat, M. A. (1981). "Fixation for Electron Microscopy." Academic Press, New York.

Hayat, M. A. (1986). Glutaraldehyde: Role in electron microscopy. *Micron* **17,** 115.

Hisano, S., Okamura, Y., and Daikoku, S. (1984). Synaptic regulation of hypothalamic neurons containing ACTH by substance P. *Brain Res.* **308,** 162.

Hsu, S., Raine, L., and Fanger, H. (1981). Use of avidin-biotin-peroxidase complex (ABC) in immunoperoxidase techniques: A comparison between ABC and unlabeled antibody (PAP) procedures. *J. Histochem. Cytochem.* **29,** 577.

King, J. C., Lechan, R. M., Kugel, G., and Anthony, E. L. P. (1983). Acrolein: A fixative for immunocytochemical localization of peptides in the central nervous system. *J. Histochem. Cytochem.* **31,** 62.

Kiss, J., Léránth, Cs., and Halász, B. (1984). Serotoninergic endings on VIP-neurons in the suprachiasmatic nucleus and ACTH-neurons in the arcuate nucleus of the rat hypothalamus. A combination of high resolution autoradiography and electron microscopic immunocytochemistry. *Neurosci. Lett.* **44,** 119.

Kosaka, T., Hataguchi, Y., Hama, K., Nagatsu, I., and Wu, J.-Y. (1985). Coexistence of immunoreactivities for glutamate decarboxylase and tyrosine hydroxylase in some neurons in the periglomerular region of the rat main olfactory bulb: Possible coexistence of gamma-aminobutyric acid (GABA) and dopamine. *Brain Res.* **343,** 166.

Kubota, Y., Takagi, H., Morishima, Y., Powell, J. F., and Smith, A. D. (1985). Synaptic interaction between catecholaminergic neurons and substance P-immunoreactive axons in the caudal part of the nucleus of the solitary tract of rat: Demonstration by the electron microscopic mirror technique. *Brain Res.* **333,** 188.

Kubota, Y., Inagaki, S., Kito, S., Takagi, H., and Smith, A. D. (1986a). Ultrastructural evidence of dopaminergic input to enkephalinergic neurons in rat neostriatum. *Brain Res.* (in press).

Kubota, Y., Takagi, H., Morishima, Y., Kawai, Y., and Smith, A. D. (1986b). Relationships between catecholaminergic neurons and cholecystokinin-containing neurons in the caudal part of the dorsomedial medulla oblongata of the rat: Light and electron microscopic observations by the 'mirror technique.' *Brain Res.* (in press).

Larsson, L.-I. (1977). Ultrastructural localization of a new neuronal peptide (VIP). *Histochemistry* **54,** 173.

Larsson, L.-I. (1981). Peptide immunocytochemistry. *Prog. Histochem. Cytochem.* **13,** *1.*

Larsson, L.-I. (1983). Methods for immunocytochemistry of neurohormonal peptides. *In* "Handbook of Chemical Neuroanatomy" (A. Björklund and T. Hökfelt, eds.), Vol. 1, p. 147. Elsevier, Amsterdam.

Léránth, Cs., and Frotscher, M. (1983). Commissural afferents to the rat hippocampus terminate on VIP-like immunoreactive non-pyramidal neurons. An EM immunocytochemical degenerative study. *Brain Res.* **276,** 357.

McLean, I., and Nakane, P. K. (1974). Periodate-lysine-paraformaldehyde fixative—A new fixative for immuno-electron-microscopy. *J. Histochem. Cytochem.* **22,** 1077.

Pickel, V. M., Joh, T. H., Chan, J., and Beaudet, A. (1984). Serotoninergic terminals: Ultrastructure and synaptic interaction with catecholamine-containing neurons in the medial nuclei of the solitary tracts. *J. Comp. Neurol.* **225,** 291.

Priestley, J. V., and Cuello, A. C. (1983). Substance P immunoreactive terminals in the spinal trigeminal nucleus synapse with lamina I neurons projecting to the thalamus. *In* "Substance P" (P. Skrabanek and D. Powell, eds.), p. 251. Boole Press, Dublin.

Reynolds, E. S. (1963). The use of lead citrate at high pH as an electron opaque stain in electron microscopy. *J. Cell Biol.* **17,** 208.

Ruda, M. A. (1982). Opiates and pain pathways: Demonstration of enkephalin synapses on dorsal horn projection neurons. *Science* **215,** 1523.

Sofroniew, M. V., and Glasmann, W. (1981). Golgi-like immunoperoxidase staining of hypothalamic magnocellular neurons that contain vasopressin, oxytocin or neurophysin in the rat. *Neuroscience* **6,** 619.

Somogyi, P. (1978). The study of Golgi-stained cells and experimental degeneration under the electron-microscope: A direct method for the identification in the visual cortex of three successive links in a neuron chain. *Neuroscience* **3,** 167.

Somogyi, P., and Takagi, H. (1982). A note on the use of picric acid-paraformaldehyde-glutaraldehyde fixative for correlated light and electron microscopic immunocytochemistry. *Neuroscience* **7,** 1779.

Somogyi, P., Smith, A. D., Nunzi, M. G., Gorio, A., Takagi, H., and Wu, J.-Y. (1983). Glutamate decarboxylase immunoreactivity in the hippocampus of the cat. Distribution of immunoreactive synaptic terminals with special reference to the axon initial segment of pyramidal neurons. *J. Neurosci.* **3,** 1450.

Somogyi, P., Hodgson, A. J., Chubb, I. W., Penke, B., and Erdei, A. (1985). Antisera to γ-aminobutyric acid. II. Immunocytochemical application to the central nervous system. *J. Histochem. Cytochem.* **33,** 240.

Sternberger, L. A. (1979). "Immunocytochemistry," 2nd ed. Wiley, New York.

Takagi, H., Kubota, Y., Mori, S., Tateishi, K., Hamaoka, T., and Tohyama, M. (1984). Fine structural studies of cholecystokinin-8-like immunoreactive neurons and axon terminals in the nucleus of tractus solitarius of the rat. *J. Comp. Neurol.* **227,** 369.

Takagi, H., Morishima, Y., Matsuyama, T., Hayashi, H., Watanabe, T., and Wada, H. (1986). Histaminergic axons in the neostriatum and cerebral cortex of the rat: A correlated light and electron microscopic immunocytochemical study using histidine decarboxylase as a marker. *Brain Res.* **364,** 114.

van den Pol, A. N. (1984). Colloidal gold and biotin-avidin conjugates as ultrastructural markers for neural antigens. *J. Exp. Physiol. Cogn. Med. Sci.* **69,** 1.

van den Pol, A. N. (1985). Silver-intensified gold and peroxidase as dual ultrastructural immunolabels for pre- and postsynaptic neurotransmitters. *Science* **228,** 332.

van den Pol, A. N., Smith, A. D., and Powell, J. F. (1985). GABA axons in synaptic contact with dopamine neurons in the substantia nigra: Double immunocytochemistry with biotin-peroxidase and protein A-colloidal gold. *Brain Res.* **348,** 146.

Vaughn, J. E., Barber, R. P., Ribak, C. E., and Houser, C. R. (1981). Methods for the immunocytochemical localisation of proteins and peptides involved in neurotransmission. *In* "Current Trends in Morphological Techniques" (J. E. Johnson ed.), Vol. 3, p. 33. CRC Press, Boca Raton, Florida.

Wainer, B. H., Bolam, J. P., Freund, T. F., Henderson, Z., Totterdell, S., and Smith, A. D. (1984). Cholinergic synapses in the rat brain: A correlated light and electron microscopic immunohistochemical study employing a monoclonal antibody against choline acetyltransferase. *Brain Res.* **308,** 69.

Willingham, M. C., and Yamada, S. S. (1979). Development of a new primary fixative for electron-microscopic immunocytochemical localization of intracellular antigens in cultured cells. *J. Histochem. Cytochem.* **27,** 947.

Yamano, M., Bai, F.-L., Tohyama, M., and Shiotani, Y. (1985) Ultrastructural evidence of direct synaptic contact of catecholamine terminals with oxytocin-containing neurons in the parvocellular portion of the rat hypothalamic paraventricular nucleus. *Brain Res.* **336,** 176.

Zucker, C., Yazulla, S., and Wu, J.-Y. (1984). Non-correspondence of [^{3}H] GABA uptake and GAD localization in goldfish amacrine cells. *Brain Res.* **298,** 154.

17

The Detection of Amyloid Using Correlative Polarizing Light Microscopy and Electron Microscopy

DAVID C. COTTELL

Biological Sciences Electron Microscope Unit
University College
Dublin, Ireland

JOHN G. BANNIGAN

Department of Anatomy
University College
Dublin, Ireland

INTRODUCTION

The aim of correlative polarizing light microscopy and electron microscopy in the study of amyloid is twofold. First, in considering the problems faced by the pathologist in equivocal cases of amyloidosis it is important that he may have resource to electron microscopical analysis of the same material which he has examined in the polarizing microscope. Second, since the occurrence of the amyloid deposition may be focal in a given piece of tissue and an inherent flaw in the electron microscopical investigation is inadequate sampling, it would seem desirable to use a technique of broad screening potential, e.g., polarizing microscopy in combination with electron microscopy. The same objectives are also true for the morphological study of amyloidosis in experimental animal models.

The traditional histological procedure for the demonstration of amyloid in tissue, i.e., staining with congo red and detection of green birefringence in the polarizing microscope (Bennhold, 1922), has been modified and refined (Highman, 1946; Puchtler *et al.*, 1962). It should be noted that the sensitivity of the procedure was well demonstrated (Heller *et al.*, 1964) and its specificity endorsed (Cooper, 1969), and that as early as 1953, Missmahl and Hartwig predicted the fibrillar nature of amyloid based on their studies with the polarizing microscope. The ultrastructure of amyloid has been the subject of many investigations (Cohen and Calkins, 1959; Shirahama and Cohen, 1967; Cohen, 1967; Glenner and Page, 1976; Franklin and Zucker-Franklin, 1972).

The necessity for combining polarizing and electron microscopy studies was stressed by Shirahama and Cohen (1966) and Pearse (1968). However, there seems a paucity of such reports despite the fact that correlative microscopy of this kind has been deemed necessary by some workers (e.g., Ogg *et al.*, 1981) in the diagnosis of primary amyloidosis associated with the nephrotic syndrome.

Following staining with a cotton dye, e.g., congo red or sirius red, amyloid deposits are stained red when viewed with the conventional light

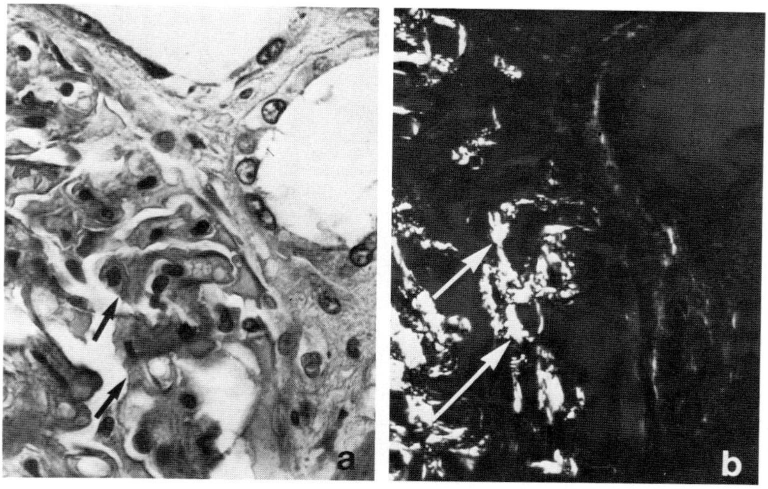

Fig. 1. Amyloidotic human kidney; 6-μm wax section stained with congo red (Puchtler *et al.*, 1962). a, Appearance with ordinary light microscopy. Note congophilic areas (arrows); ×380. b, Same section as a viewed between crossed polars. Note emerald green bifringence (arrows) in congophilic areas of a; ×380.

microscope (Fig. 1a). These deposits are seen to "light up" emitting emerald green birefringence when viewed with crossed polars in the polarizing microscope (Fig. 1b). Figure 1 is of a human kidney with massive amyloid deposits and represents the "traditional" diagnostic histological procedure for demonstrating amyloid. However, there have been many reports of false-positive results and the literature abounds with claims for specificity (e.g., Cooper, 1974) and for nonspecificity (Wolman, 1971; Klatskin, 1969). The ultrastructural appearance of the deposits is unequivocal, certain identification being based on the size, distribution, and extracellular location of the fibrils (Shirahama and Cohen, 1967; Eanes and Glenner, 1968; Glenner, 1980a,b). In Fig. 2, the typical appearance of the amyloid fibrils is seen at medium magnification in the electron microscope; this shows a human kidney fixed and processed for electron microscopy; note that the amyloid fibrils are confined to the basement membrane.

While the fine structure of amyloid as revealed in the electron microscope is, in the words of Pearse (1968), "The final arbiter of the presence or absence of amyloid," the occasional focal deposition of the substance and the low sampling potential of the electron microscope confers on this procedure such a high risk of false-negative results that Pirani (1977b) was prompted to say "Biopsies from nephrotic patients should be studied

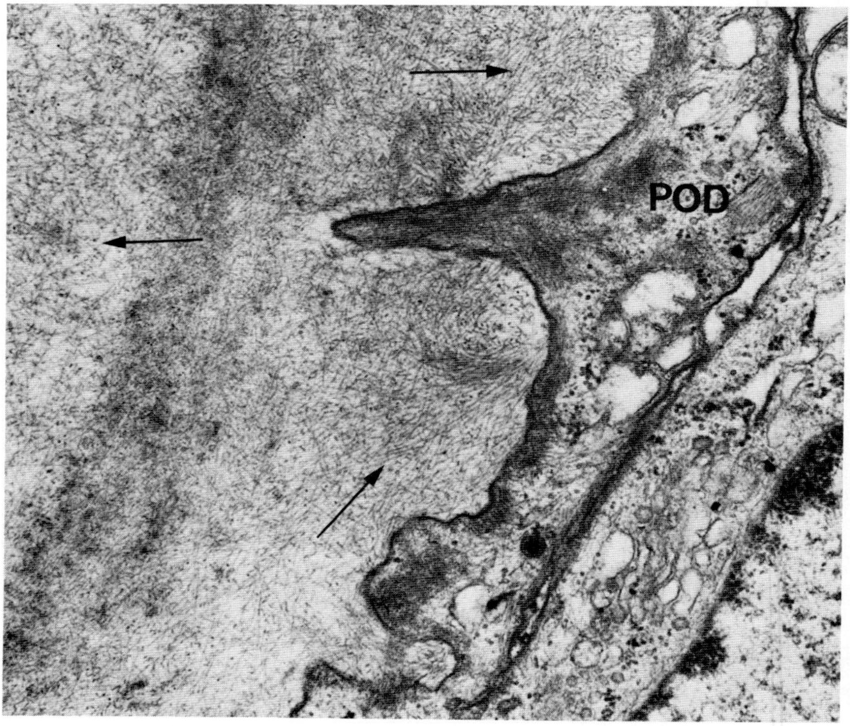

Fig. 2. Electron micrograph of human kidney biopsy showing typical appearance of amyloid fibers (arrows) deposited in the basement membrane. Foot process fusion of the podocyte (POD) is evident. Stained with uranyl acetate and lead citrate; ×3400.

with Congo-red staining and by electron microscopy. In my experience, the Congo-red birefringent stain has been more reliable than electron microscopy probably because, in these cases with very small amounts of amyloid, larger samples can be studied with Congo-red staining.''

It is clear that to fulfill the dual aim of correlative polarizing and electron microscopy is to reduce the chance on the one hand of encountering false-positive results by polarizing microscopy alone and on the other encountering false-negative results by using electron microscopy without polarizing microscopy.

AMYLOIDOSIS

Amyloid is an abnormal proteinaceous eosinophilic substance which is deposited between cells in certain organs. Amyloidosis is said to be primary when it occurs without any predisposing disease. In these cases the heart, tongue, skin, nerves, and gastrointestinal tract tend to be affected. Secondary amyloidosis occurs in association with prolonged or intermittent inflammatory diseases of infectious or autoimmune etiology. At present, rheumatoid arthritis is probably the commonest disease predisposing to amyloidosis. About 5–11% of people with rheumatoid arthritis develop amyloidosis (Arapakis and Tribe, 1963).

Secondary amyloidosis also occurs in association with tuberculosis, leprosy, syphilis, chronic osteomyelitis, ulcerative colitis, and Crohn's disease. In secondary amyloidosis the liver, spleen, kidney, and adrenal glands tend to be affected.

Amyloidosis may also occur in association with multiple myeloma, Hodgkin's disease, and hypernephroma. Familial forms of amyloidosis are also known. Thus, it may be seen in familial Mediterranean fever which affects predominantly, but not exclusively, people of Sephardic Jewish, Arab, and Armenian ancestry. There is also a familial entity affecting people of Portuguese origin in which amyloid deposits severely interfere with the function of peripheral nerve. For reviews on the tissue distribution and pathogenesis of amyloidosis see Pirani (1977a) and Glenner (1980a,b).

Electron Microscopy

On examination with the electron microscope, amyloid is seen as a skein of nonbranching fibrils sometimes occurring in parallel bundles of 10–15. Each fibril consists of two filaments 5–7.5 nm in diameter twisted around each other to form a double helix with a periodicity of 100 nm. The filaments have been shown by X-ray diffraction to consist of antiparallel β-pleated sheets (Eanes and Glenner, 1968; Glenner, 1980a,b). The affinity of amyloid for the cotton dye, congo red, and the interaction of amyloid with polarized light are dependant upon the β-pleated structure of the material. Thus, Harada *et al.* (1971) have shown that enzymatic destruction of the β-pleated sheet conformation results in the loss of congophilia.

Polarizing Microscopy

The behavior of light when it passes through a medium depends on an interaction between the electric vector of the light ray and the oscillating electrons of the chemical bonds of the substance of which the medium is made. These interactions may result in absorptions of certain wavelengths of the ray or in changes in the velocity of transmission. If the bond structure of the medium is completely random, incident light will be transmitted with equal velocity in all directions within the medium. As a corol-

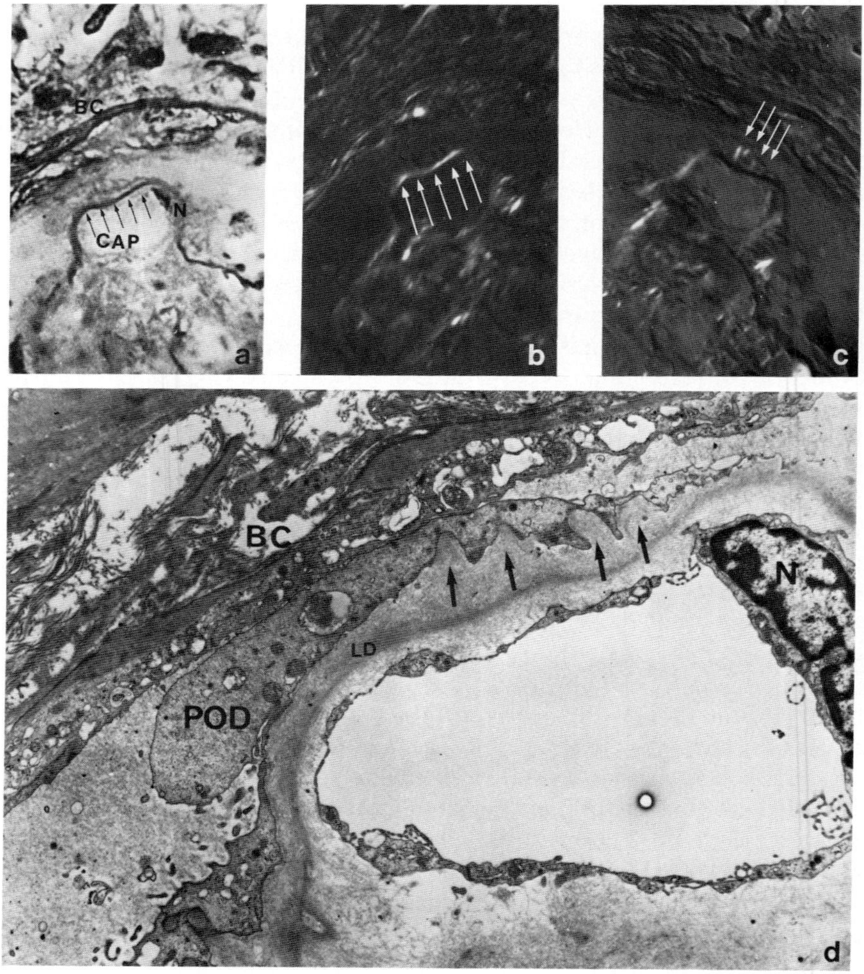

Fig. 3. A 6-μm Epon section of human kidney fixed in glutaraldehyde followed by osmium tetroxide; Epon removed and stained with congo red and hematoxylin (see text). a, Viewed with analyzer of polarizing microscope removed. A capillary (CAP), Bowman's capsule (BC), and an endothelial cell nucleus (N) are visible. The basement membrane (arrows) was stained red; ×1100. b, The same field as in (a) viewed between crossed polars. Note emerald green bifringence (arrows) in the basement membrane; ×1100. c, Same field as a and b but with specimen rotated 60° between crossed polars. Arrows indicate spikes of emerald green bifringence in the basement membrane; ×1100. d, Section serial to (a) viewed in the electron microscope. The spikes seen in c are now found to be projections of the lamina externa rara (arrows). LD, lamina densa; BC, Bowman's capsule; POD, podocyte; N, endothelial cell nucleus. Stained with uranyl acetate and lead citrate; ×8000. (From Cottell, D. C., 1983; reproduced by permission of the *Journal of Microscopy*, Oxford.)

lary, the plane of polarization of the incident light will not influence the velocity of transmission through the medium; a medium of this type is said to be isotropic. If the chemical bonds within a medium are arranged in a certain preferred direction, the interaction of the electric vector of the light ray with the bond electrons will result in the transmission of light with different velocities in different planes perpendicular to the axis of propagation. As a corollary, the plane of polarization of incident light will influence the velocity of propagation of the ray. Such a medium is said to be anisotropic, an example of which is amyloid. Since it transmits light with different velocities in different planes, an anisotropic body has more than one index of refraction and is said to be birefringent. In general, a birefringent body will have an axis of fast transmission and one of slow transmission perpendicular to each other.

When a ray of polarized light strikes a birefringent body at an angle of 45° to the two axes of transmission, it is resolved into two components each parallel to one of the two axes of transmission. When these rays emerge from the opposite side of the birefringent body, the component which was transmitted along the slow axis will be phase retarded with respect to the ray that was transmitted along the fast axis. If these two rays of light then pass through a piece of polarizing material, they will be recombined and will interfere with each other either constructively or destructively.

The nature of amyloid is such that crystals of congo red (themselves birefringent) are attached to the β-pleated sheet so that they are all more or less parallel (Cooper, 1974). When a ray of polarized light strikes a crystal of congo red it is resolved into a white component which is transmitted parallel to the crystal's fast axis of transmission and a red component transmitted more slowly. When the emerging rays pass through a polarizing filter arranged transverse to the one that produced the incident ray, the two are brought together in the same plane and, being out of phase, interfere with each other. The interference is destructive in the sense that the red moiety cancels the red component of the white element leaving the complimentary green color. This phenomenon is called dichroism and it helps to explain the emerald green color observed in congo red-stained amyloid placed between crossed polars. For a more comprehensive description, the texts of Wredden (1947), Bennet (1961), Schaeffer (1966), Shurcliff and Ballard (1964), Jenkins and White (1957), and Rochow and Rochow (1979) should be consulted.

The Correlative Concept

The fixation, dehydration, embedding, and staining of biological specimens is usually determined by the type of microscopical analysis contemplated in advance by the microscopist. Material which has been processed for light microscopy evaluation by wax sectioning and staining is usually less than satisfactory if it is reembedded in epoxy resin, sectioned, and examined in the electron microscope. This is associated with inferior fixation of the specimen, a situation which human pathologists are all too familiar with. On the other hand, specimens fixed and embedded for electron microscopical evaluation are extremely difficult to use for a large range of light microscopy techniques. These two seemingly irreconcilable requirements have been the subject of many ingenious and important reports.

Usually correlative light microscopy and electron microscopy entails taking a 1- to 2-μm resin section and staining with, e.g., toluidine blue, for examination in the light microscope. This simple but extremely useful procedure enables the microscopist to recognize the location of relevant features, (e.g., tumor cells or glomeruli) prior to ultrathin sectioning and electron microscopical analysis. A considerable number of staining procedures taken from the realms of light microscopy technique have been applied to resin-embedded specimens, as in the work of Cardno and Steiner (1965), Spurlock *et al.* (1966), Chang (1972), Bennet *et al.* (1976), Cottell and Livingston (1976), and Butler (1979). There remain, however, many staining procedures which so far have not been successfully applied to resin-embedded material.

Fortunately, the nature of the amyloid substance is such that it is not so sensitive to inferior fixation procedures usually used for light microscopy technique, nor does it autolyze quickly and is therefore suitable for ultrastructural analysis if reembedded for electron microscopy (Lehner *et al.,* 1966; Rossi *et al.,* 1970; Johannessen, 1977). Furthermore, it has been shown that material containing amyloid fixed and embedded for electron microscopy can be used with success and simplicity in the polarizing microscope (Shirahama and Cohen, 1966; Cottell, 1983). The following description of correlative microscopy of amyloid is largely based on these and other reports in the literature. Material, either biopsy, experimental, or taken at autopsy, can be screened to a very accurate extent using the polarizing microscope, and subsequently the ultrastructure of selected areas can be examined in the electron microscope. However, before describing some of the methods which may be employed, it is important to note that polarizing microscopy, although a well tried and trusted technique, requires the microscopist to be well versed in the theory and practice of this instrument. The caveat of Cooper (1969) concerning this point is just as valid today, and a dedicated polarizing microscope equipped with strain-free optical components should be used if possible.

METHODS

Fixation and Embedding for Electron Microscopy

Specimens may be fixed by standard methods (Hayat, 1986, 1987). A convenient and effective schedule is to place the specimens in 2.5% glutaraldehyde buffered with 0.1 *M* Sörensen's phosphate buffer (pH 7.4) for 1 hr at 4°C. The aldehyde is washed out of the tissue blocks by rinsing in buffer for 15 min with six changes. This is followed by postfixation in 2% OsO_4 in 0.1 *M* Sörensen's phosphate buffer (pH 7.4) for 1 hr at room temperature. The use of a rapid penetrating fixative (McDowell and Trump, 1976) is very useful if the specimen exceeds 3 mm in its shortest dimension. Large flat blocks of tissue (1 cm^2) may be used in this fixative provided they are sliced to 3 mm in depth. Specimens of this size are usually fixed for 24 hr at room temperature followed by washing in buffer and postfixed in 2% OsO_4 for 4 hr at room temperature. After double fixation, the blocks are washed in buffer for 15 min and dehydrated in ascending grades of ethanol. Embedding in Epon (Luft, 1973) is either achieved in capsules or flat-embedding molds (see Hayat, 1986; 1987).

Obtaining the Thick Presection

Since the detection of emerald green birefringence due to amyloid stained with cotton dyes is the aim of the procedure, the presection must be 5–6 μm thick (Cooper, 1969; Wolman and Bubis, 1965; Cottell, 1983). This thickness places severe stress on the glass knife edge and a knife of α-angle = 70° must be used. In some experimental or autopsy tissues the embedded sample may exceed 6 mm wide, which is the width of the conventional glass knife of the Latta and Hartmann type (1950), made on most knifemakers. In this case, the "Ralph" type knife should be used (Bennet *et al.*, 1976; Hayat, 1986).

Sections 5–6 μm are cut on a dry knife, picked up with fine forceps, and placed on a pool of distilled water on a clean glass slide. The slides should be acid washed and stored in alcohol; before use they are smeared with egg albumin. The section is dried fast to the slide at 60°C for 2–3 hr. The treatment before staining is severe and if the slide is not clean or the section has not adhered well, it will wash off during subsequent treatment.

Staining with Congo Red

Sections may be stained in congo red at this stage, after the method of Shirahama and Cohen (1966). However, the presence of the embedding medium seriously compromises the technique in that sometimes the Epon takes up the stain. Also, counterstaining with hematoxylin is not possible because this stain will not work with epoxy resins. To overcome these difficulties, the following schedule is found to be reliable:

1. Place the slides in a saturated solution of potassium hydroxide in 100% ethanol to remove or, at least, degrade the Epon (Imai *et al.*, 1968). this treatment will remove the Epon after 12 hr or sufficiently degrade it after 2 hr.
2. The slides are then rinsed in four changes of absolute ethanol for 5 min each and brought to water through 70% and 30% ethanol.
3. To stain the amyloid, a wide choice of procedures may now be used (Puchtler *et al.*, 1962; Shirahama and Cohen, 1966; Highman, 1946; Cottell, 1983; Llewellyn, 1970). All give very good results; the simplest method is as follows: Place the slides in 1% aqueous congo red (National Aniline Division C.1. 370) for 15 min at room temperature. Examine the wet section under a conventional light microscope and determine if there are definite dull red areas stained. Continue to stain until no further increase in staining intensity is seen. This is, perhaps, the most difficult part of the staining procedure since it relies on a visual judgment at a low magnification. To help overcome this, and as a matter of correct procedure, a control slide should always be processed with the one under investigation.
4. If differentiation is required, it can be achieved by using a rapid rinse in 0.2% potassium hydroxide in 80% ethanol (Highman, 1946). Differentiation, however, occurs rapidly and sometimes the section can be destained completely if it is allowed to proceed for too long. The section is then washed in gently running tap water for 5 min, stained for 2 min in Harris hematoxylin, washed again in running tap water for 5 min, followed by three rinses in distilled water. The slides are rapidly dehydrated starting with one rinse in 90% ethanol and two in absolute ethanol before immersion in the purest grade, clean xylene. The sections are mounted under thin coverslips (0.17 mm or less); a simple micrometer screw should be used to check the thickness of these. The dehydrating fluids, xylene, mounting medium, and coverslips must be perfectly clean, as almost any dirt particles will themselves "light up" between the crossed polars of the microscope.

Polarizing Microscopy

As mentioned earlier, a high-quality microscope with strain free optics is desirable. It should be noted that in some specimens only minute particles of amyloid will be present. These may have a very weak luminance which will make long photographic exposures necessary. Therefore, the microscope should be sited in a vibration-free area of the laboratory, preferably with blackout facility. Fine grain film, e.g., Kodak panatomic x, should be used even though the exposure time will be quite long. The use of faster film has the immediate term advantage of shorter exposure but the lifelong disadvantage of a grainy print or projection slide. It is, of course, of great importance that the film be processed in an appropriate developer and printed with precision.

The microscope should be set up to give Kohler illumination; all objectives should be centered relative to the center of rotation of the stage and all the optical components should be free from dust.

The section is now examined with the analyzer removed. Areas which appear red (Fig. 3a) are representative of amyloid congo red complexes. The analyzer is inserted and crossed with respect to the polarizer and the same areas examined for emerald green birefringence (Fig. 3b). Since amyloid sometimes exhibits a directional orientation, the stage should now be rotated while observing the image and further deposits of amyloid may be detected (Fig. 3c, rotated 60°). If the microscopist has not centered the objective to the center of the stage, this last maneuver will be very difficult, since the specimen will move out of the field of view when it is rotated.

In this case, a human kidney biopsy specimen (Fig. 3b) shows the deposition of the amyloid along the basement membrane of the capillary loop indicated, and (Fig. 3c) as spikes projecting from the lamina externa rara. Thus, small areas of emerald green birefringence are located in specimens embedded in plastic embedding media. The next part of the procedure is to examine the same area as Fig. 3 in the electron microscope.

Ultramicrotomy

The usually exacting task of cutting ultrathin sections is in this case made a little more difficult by two factors. First, it is the very first section of the block which is best correlated with the polarizing images. Second, it is very important when picking up this first section that no feature of the section is hidden behind a grid bar. Both of these restrictions can be overcome with practice and the use of large square or slot type grids, covered with Pioloform or Formvar film (Hayat, 1986, 1987). It should be stressed that the correlation depends critically on the resolution of these two problems.

The block face should be trimmed to contain the areas already identified in the polarizing microscope. The use of the well-known "mesa" technique of trimming is very useful in this instance, since it preserves the underlying tissue for subsequent investigation (Reid, 1976). The following procedure is now employed.

1. The block face is carefully aligned to the knife edge, both in the horizontal and vertical directions. A diamond knife is the most reliable for this task.
2. The specimen face is brought to within 1 μm of the knife edge, the trough filled with distilled water, and the advance mechanism of the microtome set to 40 nm. With such an increment, the first section will be cut within 25 cutting strokes.
3. The microtomist should now pick this section up onto a Pioloform-coated grid and nine subsequent sections, each on a separate grid.
4. These should be labeled 1 to 10. The tenth section should be stained with uranyl acetate and lead citrate and examined in the electron microscope. If the staining is clean, then the important first section should be stained without delay. If, however, the stains are precipitating, new stains should be made and tested on one of the less important grids, e.g., the ninth. In this manner one can usually ensure that the electron microscope investigation will not be hampered by contamination from these stains.

Electron Microscopy

Since the size and orientation of the amyloid fibrils are important distinguishing features, the microscope should be free of objective lens astigmatism and the magnification calibrated accurately. All the electron micrographs presented here were taken with a Philips 201C, accelerating voltage of 80 kV and objective aperture 30 or 50 μm.

The area of interest is very quickly located in the electron microscope. Figure 3d is a low-magnification picture taken from the first ultrathin section adjacent to the section examined in the polarizing microscope. It shows the same capillary loop as seen in Figs. 3a–c. At a higher magnification, the unmistakable amyloid fibrils can be seen (Fig. 4). Note that the fibrils are evident along the basement membrane, as demonstrated in the polarizing microscope (Fig. 3b). The "spikes" seen in Fig. 3c are here seen to be composed of large bundles of amyloid fibrils projecting from the lamina externa rara with an orientation completely consistent with the degree of rotation required to demonstrate them in the polarizing microscope (Fig. 3c).

Fixation and Embedding in Wax for Light Microscopy

Very often the microscopist, having examined a wax-embedded histological section, becomes aware of the need to look at the same structure in

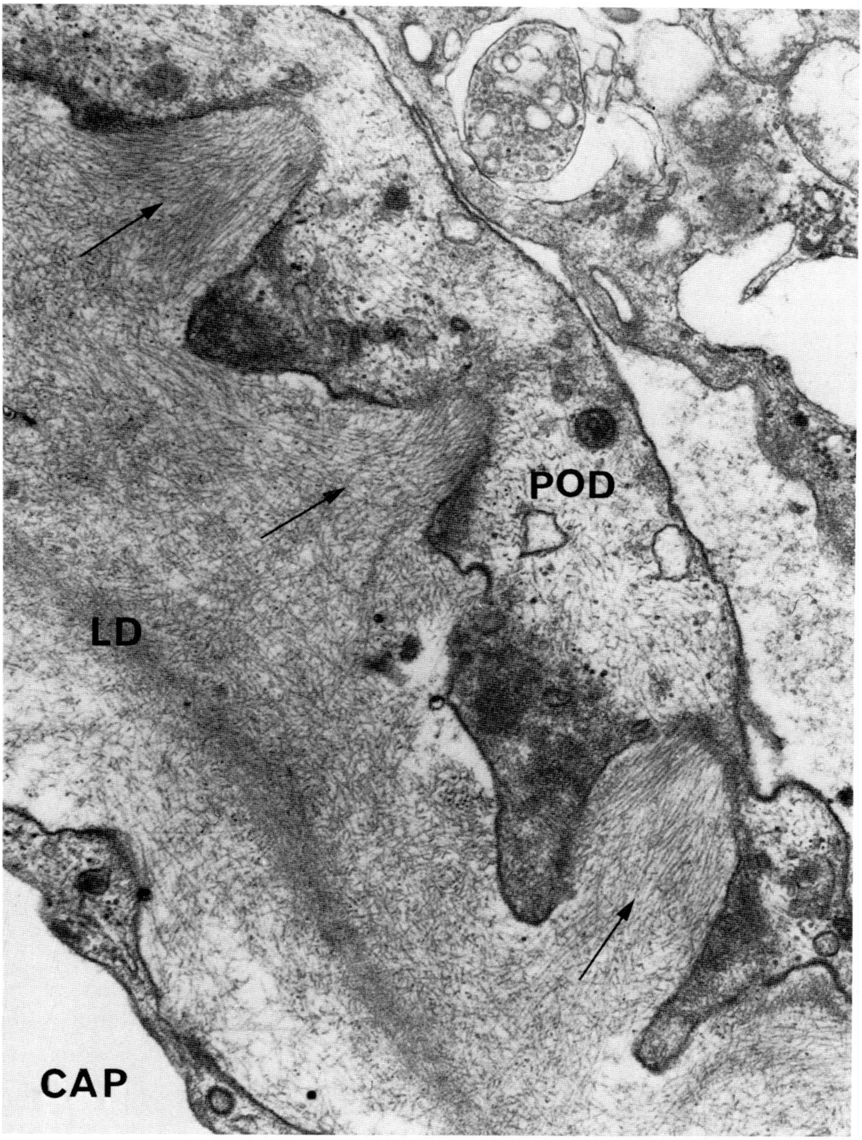

Fig. 4. In the electron microscope, the basement membrane is seen to be full of amyloid fibrils. Note that the spikes of the lamina externa rara of the basement membrane are occupied by fibrils that have a preferential orientation along the "axis" of the spike and clearly correspond to the birefringent spikes seen in Fig. 3c. Elsewhere in the basement membrane the amyloid fibrils have a more random distribution. Thus, the distribution of amyloid indicated by Figs. 3a–c correlate well with the distribution revealed by the electron microscope. POD, podocyte; LD, lamina densa; CAP, capillary. Stained with uranyl acetate and lead citrate; ×41,000. (From Cottell, D. C., 1983; reproduced by permission of the *Journal of Microscopy*, Oxford.)

the electron microscope. There is, in reality, every possibility of doing this provided it is accepted that the processing initially carried out for the wax embedding has probably damaged the ultrastructure to varying degrees. There are many ways that this approach may be made, e.g., dewaxing the block and reprocessing for electron microscopy (Lehner *et al.*, 1966), or taking an adjacent 5-μm section and reprocessing that. but perhaps the most useful and rewarding method is the use of the so-called "pop off" technique. This method, largely based on the work of Takeda (1969), Blank *et al.* (1970), Rossi *et al.* (1970), and Bretschneider *et al.* (1981), is given in the following section, but only in relation to the identification of amyloid deposits in the polarizing microscope and the subsequent ultrastructural study of those deposits. The procedure may be used on any histological section, e.g., human biopsy, autopsy, or experimental tissue.

Specimen Fixation and Staining

A piece of human peripheral nerve was taken as a biopsy and fixed in Bouin's fluid. The specimen was postfixed in 2% OsO_4 in order to demonstrate the node of Ranvier distribution, and was then processed by standard methods into wax. A 5-μm section was stained with hematoxylin and eosin (Fig. 5a). Some endoneurial deposits may be seen which appeared eosinophilic and waxy in the light microscope; these deposits were suspected of containing amyloid. A subsequent section was stained with congo red by the method of Puchtler *et al.* (1962) and examined in the polarizing microscope. The deposits were stained a diffuse red (Fig. 5b) with the analyzer removed and showed emerald green birefringence under crossed polars (Fig. 5c). In this case it was very important to have a more definite diagnosis of amyloid involvement. Also, the effect of picric acid molecules (Bouin's fluid) is known to alter polarization properties because they tend to bond to biological structures in ordered laminates; this cast some doubt on the appearance of emerald green birefringence in Fig. 5c. Since the deposits were rare in the tissue, it was decided to use the "pop off" technique and examine the ultrastructure of the actual deposits seen in Fig. 5b,c.

The "Pop Off" Technique

1. The coverslip is removed by soaking in xylene for 12 hr; the slide containing the exposed section is then placed in absolute ethanol. If the section is to be postfixed with osmium tetroxide, it is now brought to water and placed in 1% OsO_4 in 0.1 *M* Sörensen's phosphate buffer for 5 min, and then dehydrated to 100% ethanol.

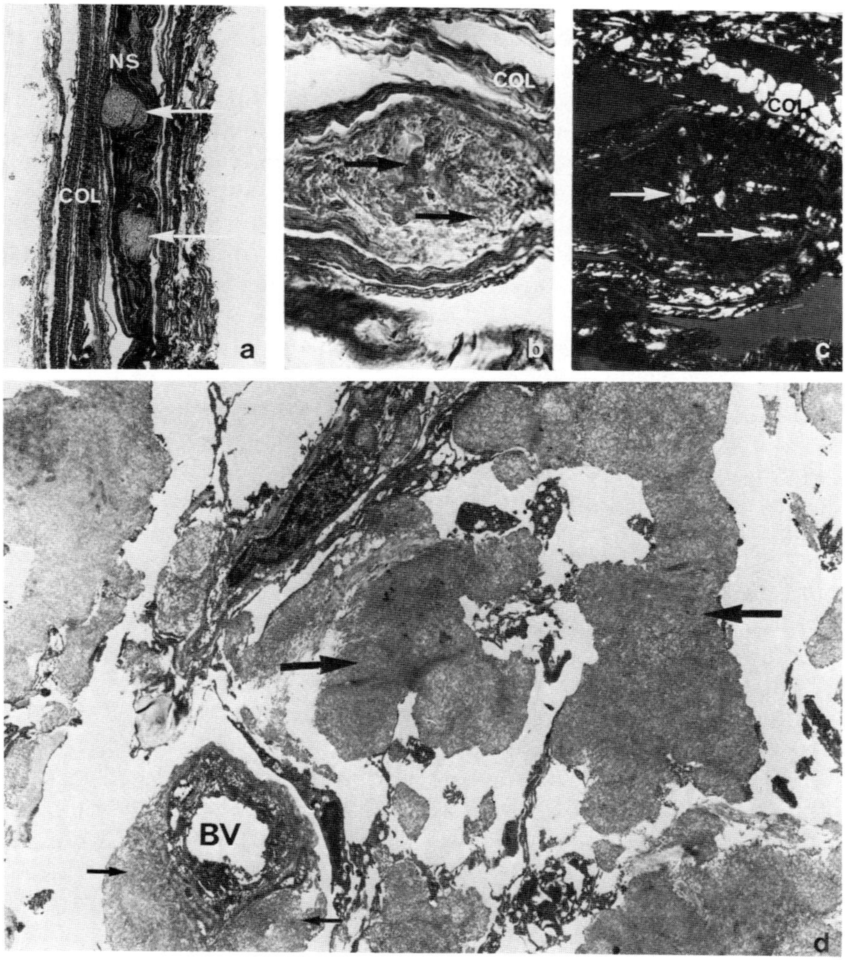

Fig. 5. a, A 5-μm wax section of human peripheral nerve (see text for details of processing) stained with hematoxylin and eosin. Note the collagen bundles (COL) in the perineurium, the nerve sheath (NS), and eosinophilic endoneurial deposits (arrows); ×48. b, A 5-μm sequential wax section from the same block as a stained with congo red (Puchtler *et al.*, 1962) viewed in the polarizing microscope with analyzer removed. Areas within the deposit (arrows) are stained red. Note the position of the collagen bundles (COL); ×240. c, Same section as b viewed between crossed polars. Areas previously stained red now exhibit intense emerald green birefringence (arrows). The collagen bundles (COL) exhibited yellow birefringence; ×240. d, Electron micrograph obtained by "pop off" method from the section illustrated in b–c. A small blood vessel (BV) is seen surrounded by fibrillar material (small arrows). Larger fibrillar deposits are also visible (large arrows). Note the rather poor structural preservation of the blood vessel. Stained with uranyl acetate and lead citrate; ×5600.

2. The slide is treated with propylene oxide for 5 min followed by a 1:1 mixture of propylene oxide and catalyzed Epon for 5 min, then catalyzed Epon at 37°C for 15 min.

3. A BEEM capsule is placed in a holder and filled to its brim with fresh catalyzed Epon. The slide is gently lowered onto the Epon of the capsule so that the area of interest in the section is in intimate contact with the Epon. This is left undisturbed to polymerize at 60°C for 24 hr.

4. The BEEM capsule–glass slide unit is removed from the BEEM capsule holder and the plastic of the capsule is cut away with a razor. The junction of the Epon mold and the glass slide is now lightly scored with a small diamond marker.

5. The unit is now heated to 60°C and while still hot the Epon mold is gently rocked back and forth. The glass slide part should be held flat on the bench while this is being done. The mold comes away, taking with it the section which is now forming the top of the mold. Sometimes the glass slide shatters during the rocking procedure, leaving small spicules of glass inconveniently adhered to the Epon. These are very easily removed by placing the mold in hydrofluoric acid for 5 min. Great caution is advised while using this acid.

6. The Epon mold now contains the section already examined in the light and polarizing microscopes. It can be further examined by placing the mold on the top lens of the substage condenser of a light microscope, section uppermost, and the location of the area of interest confirmed. For a more graphic description of the "pop off" technique, the paper of Bretschneider *et al.* (1981) should be consulted.

7. The surplus resin is now trimmed away from the block face and ultrathin sections taken for electron microscopy. There is no spare resin at the block face, so it should be borne in mind that the first ultrathin section taken contains the detail of interest. The same regime outlined on page 376 should now be used to secure good quality, clean, correlatable sections. This will leave about 4 μm of specimen which may be stored for further investigation.

Electron Microscopy

At a low magnification (Fig. 5d), a blood vessel surrounded by deposited material as well as larger deposits may be seen. Note the poor preservation usually associated with this procedure. At a higher magnification (Fig. 6), the deposits are seen to be composed of amyloid fibrils, interspersed with collagen. In some areas the fibrillar structure is interfered with by the presence of electron-lucent patches; these are the result of inappropriate fixation and wax embedding. Note also the wide variation in the width of the fibrils; this is probably artifactual and was discussed by

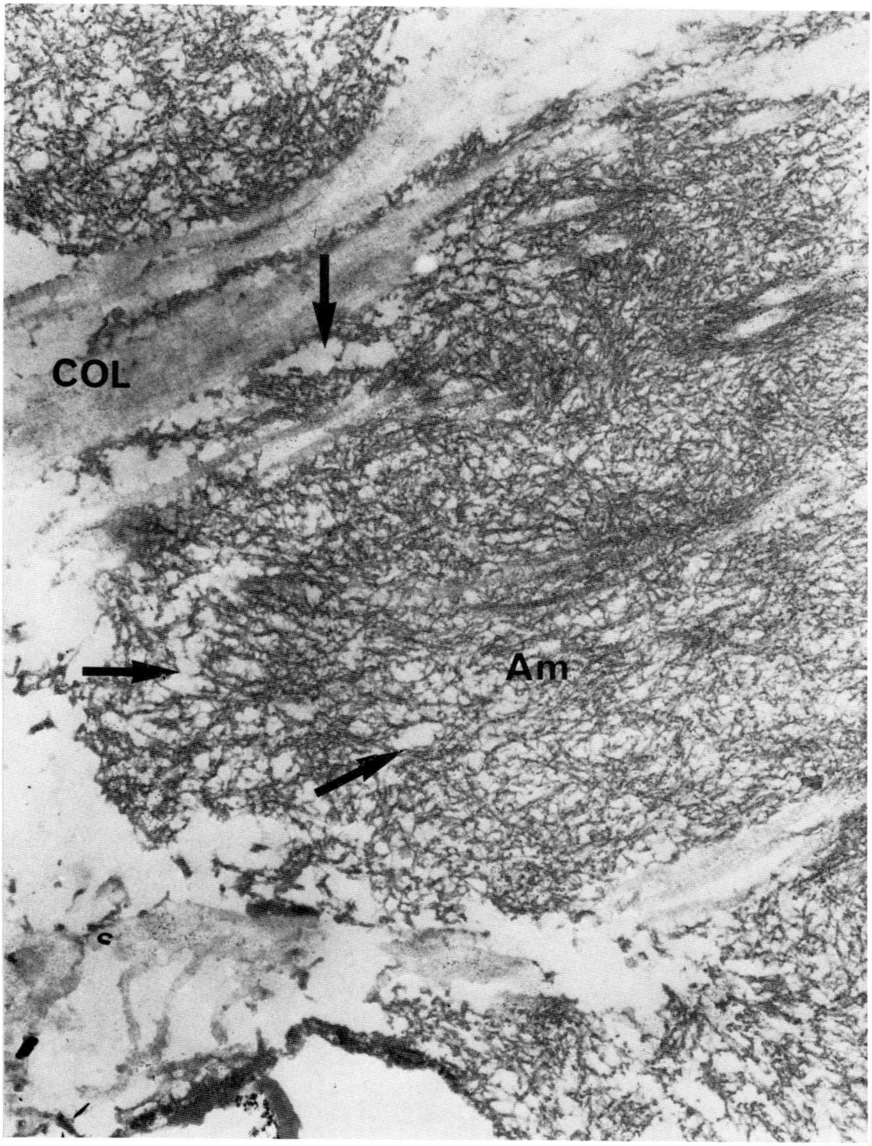

Fig. 6. Higher power electron micrograph of deposits seen in Fig. 5d. These are clearly amyloid fibrils (Am). They are distributed randomly and are interspersed with collagen fibers (COL). The electron-lucent areas (arrows) are the result of inappropriate processing, as is the variation in the widths of the amyloid fibrils. Stained with uranyl acetate and lead citrate; ×47,000.

Lehner *et al.* (1966). Despite the poor preservation, good correlation with the polarizing images (Fig. 5b,c) was achieved and a positive diagnosis of amloidosis was made.

CONCLUDING REMARKS

Correlative methods for detecting amyloid deposits using cotton dye staining, polarizing microscopy, and electron microscopy have been discussed in this chapter. With practice, these techniques used in combination can be made to yield information about the specimen which would go undetected if only a single method were used. A major drawback is the fact that specimens embedded in epoxy resins are not readily stained by light microscopy stains. It is to be hoped that the use of the recently developed resins, e.g., Lemix and L. R. White, will allow a broader range of staining procedures for amyloid such as van Gieson (Symmers, 1956), periodic acid Schiff (McAlpine, 1969), crystal violet (Highman, 1946), or the histochemical reactions of dimethylaminobenzaldehyde (Cohen, 1965; Adams, 1957) and sodium sulfate alcian blue (Lendrum *et al.,* 1972).

A major aim in correlative microscopy is to increase the sampling accuracy of electron microscopy. While this is achieved with large area, thin sections obtained from Epon embedded material, such material is somewhat difficult to cut with the Ralph-type knife. The new low-viscosity resins overcome this problem and they are also suitable for cutting thin sections for electron microscopy. This, added to their stainability, would seem to make them extremely useful in correlative work.

Amyloid needs to be treated by correlative procedures when comprehensive morphological work is to be undertaken. Although amyloid is resistant to autolytic change and tends not to be disrupted by light microscopic fixatives, its location as deposited material is often difficult to establish in the electron microscope if the specimen had originally been processed for light microscopy.

This problem can be minimized by using the universal fixative of McDowell and Trump (1976). To state a hope for future correlative studies, one would wish for a fixative and embedding medium common to both light and electron microscopy and with better applicability of light microscopic stains.

The authors express their gratitude to Peter Dervan of the Mater Misericordiae Hospital, Dublin, and Eoin Gaffney of St. James' Hospital, Dublin, for their helpful criticism and discussion. The authors also thank C. A. Cottell and M. Markey for typing the manuscript.

REFERENCES

Adams, C. W. M. (1957). A p-Dimethylaminobenzaldehyde-nitrite method for the histochemical demonstration of tryptophane and related compounds. *J. Clin. Pathol.* **10,** 56.

Arapakis, G., and Tribe, C. R. (1963). Amyloidis in rheumatoid arthritis investigated by means of rectal biopsy. *Ann. Rheum. Dis.* **22,** 256.

Bennet, H. S. (1961). The microscopical investigation of biological materials with polarised light. *In* "McClung's Handbook of Microscopical Techniques" (R. McClung-Jones, ed.), p. 591. Hafner, New York.

Bennet, H. S., Wynick, A. D., Lee, S. W., and McNeil, J. H. (1976). Science and art in preparing tissues embedded in plastic for light microscopy with special reference to glycol methacrylate, glass knives and simple stains. *Stain Technol.* **51,** 71.

Bennhold, H. (1922). Eine spezifische amloidfarbung mit Kongorot. *Muench. Med. Wochenschr.* **69,** 1537.

Blank, H., Davis, C., and Collins, C. (1970). Electron microscopy for the diagnosis of cutaneous viral infection. *Br. J. Dermatol.* **83,** 69.

Bretschneider, A., Burns, W., and Morrison, A. (1981). "Pop-off" technic. The ultrastructure of paraffin embedded sections. *Am. J. Clin. Pathol.* **76,** 450.

Butler, J. K. (1979). Methods for improved light microscope microtomy. *Stain Technol.* **54,** 53.

Cardno, S. S., and Steiner, J. W. (1965). Improvement of staining technics for thin sections of epoxy-embedded tissue. *Am. J. Clin. Pathol.* **43,** 1.

Chang, S. C. (1972). Haematoxylin-eosin staining of plastic embedded tissue sections. *Arch Pathol.* **93,** 345.

Cohen, A. S. (1965). The constitution and genesis of amyloid. *Int. Rev. Exp. Pathol.* **4,** 159.

Cohen, A. S. (1967). Studies on the amyloid fibril. *In* "The Connective Tissues" (B. M. Wagner and D. E. Smith, eds.), p. 81. Williams & Wilkins, Baltimore, Maryland.

Cohen, A. S., and Calkins, E. (1959). Electron microscopic observations on a fibrous component in amyloid of diverse origins. *Nature (London)* **183,** 1202.

Cooper, J. H. (1969). An evaluation of current methods for the diagnostic histochemistry of amyloid. *J. Clin. Pathol.* **22,** 410.

Cooper, J. H. (1974). Selective amyloid staining as a function of amyloid composition and structure. *Lab. Invest.* **31,** 232.

Cottell, D. C. (1983). A method for the detection of amyloid in human tissue embedded in epon using a combination of polarising light microscopy and electron microscopy. *J. Microsc. (Oxford)* **131,** 119.

Cottell, D. C., and Livingston, D. C. (1976). Flourescent reagent for the periodic acid-Schiff and Feulgen reactions for cytochemical studies. *J. Histochem. Cytochem.* **24,** 956.

Eanes, E. D., and Glenner, G. G. (1968). X-ray diffraction studies of amyloid filaments. *J. Histochem. Cytochem.* **16,** 673.

Franklin, E. C., and Zucker-Franklin, D. (1972). Current concepts of amyloid. *Adv. Immunol.* **15,** 249.

Glenner, D. G. (1980a). Amyloid deposits and amyloidosis. Part 1. *N. Engl. J. Med.* **302,** 1283.

Glenner, D. G. (1980b). Amyloid deposits and amyloidosis. Part 2. *N. Engl. J. Med.* **302,** 1333.

Glenner, D. G., and Page, D. (1976). Amyloid, amyloidosis and amyloidogenesis. *Int. Rev. Exp. Pathol.* **15,** 1.

Harada, M., Isersky, C., Cuatrecasas, P., Page, D., Bladen, H. A., Eanes, E. D., Keiser, H. R., and Glenner, G. G. (1971). Human amyloid protein: Chemical variability and homogeneity. *J. Histochem. Cytochem.* **19,** 1.

Hayat, M. A. (1986). "Basic Techniques for Transmission Electron Microscopy." Academic Press. Orlando, Florida.

Hayat, M. A., ed. (1987). "Principles and Techniques of Electron Microscopy." Macmillan Press, London.

Heller, H., Missmahl, H.-P., Sohar, E., and Gaffni, J. (1964). Amyloidosis: Its differentiation into peri-reticulin and peri-collagen types. *J. Pathol. Bacteriol.* **88,** 15.

Highman, B. (1946). Improved methods for demonstrating amyloid in paraffin sections. *Arch. Pathol.* **41,** 559.

Imai, Y., Sue, A., and Yamaguchi, A. (1968). A removing method of the resin from epoxy embedded sections for light microscopy. *J. Electron Microsc.* **17,** 84.

Jenkins, F. A., and White, H. E. (1957). "Fundamentals of Optics." McGraw-Hill, New York.

Johannessen, J. V. (1977). Use of paraffin material for electron microscopy. *In* "Pathology Annual" (S. S. Sommers and P. P. Rosen, eds.), p. 189. Appleton, New York.

Klatskin, G. (1969). Non-specific green birefringence in congo red stain tissues. *Am. J. Pathol.* **56,** 1.

Latta, H., and Hartmann, J. F. (1950). Use of a glass edge in thin sectioning for electron microscopy. *Proc. Soc. Exp. Biol. Med.* **74,** 436.

Lehner, T., Nunn, R. E., and Pearse, A. G. E. (1966). Electron Microscopy of paraffin embedded material in amyloidosis. *J. Pathol. Bacteriol.* **91,** 297.

Lendrum, A. C., Slidders, W., and Fraser, D. S. (1972). Renal hyalin. A study of amyloidosis and diabetic fibrinous vasculosis with new staining methods. *J. Clin. Pathol.* **25,** 373.

Luft, J. H. (1973). Embedding media old and new. *In* "Advanced Techniques in Biological Electron Microscopy" (J. K. Koehler, ed.), Vol. 1, p. 1. Springer-Verlag, Berlin and New York.

McAlpine, I. C. (1969). Electron histochemical study of 1:2 glycol groups in experimental animal amyloid. *Histochemie* **19,** 135.

McDowell, E. M., and Trump, E. F. (1976). Histologic fixatives suitable for diagnostic light and electron microscopy. *Arch. Pathol. Lab. Med.* **100,** 405.

Missmahl, H. P., and Hartwig, M. (1953). Polarisationsoptische Untersuchungen an der Amyloidsubstanz. *Virchows Arch. Pathol. Anat. Physiol.* **324,** 489.

Ogg, C. S., Cameron, J. J., Williams, D. G., and Turner, D. R. (1981). Presentation and course of primary amyloidosis of the kidney. *Clin. Nephrol.* **15,** 9.

Pearse, A. G. E. (1968). "Histochemistry: Theoretical and Applied" 3rd ed, Vol. 1. Churchill, London.

Pirani, C. L. (1977a). Tissue distribution of amyloid. *In* "Amyloidosis" (O. Wegelius and A. Pasternack, eds.), p. 390. Academic Press, New York.

Pirani, C. L. (1977b). Discussion on contribution by B. Kuhlback. *In* "Amyloidosis" (O. Wegelius and A. Pasternack, eds.). p. 33. Academic Press, New York.

Puchtler, H., Sweat, F., and Levine, M. (1962). On the binding of congo red by amyloid. *J. Histochem. Cytochem.* **10,** 355.

Reid, N. (1976). Ultramicrotomy. *In* "Practical Methods in Electron Microscopy" (A. M. Glauert, ed.). Elsevier. North-Holland, Amsterdam.

Rochow, T. G., and Rochow, E. G. (1979). "An Introduction to Microscopy by Means of Light, Electron, X-rays and Ultrasound." Plenum, New York.

Rossi, G. L., Luginbuhl, H., and Probst, D. (1970). A method for ultrastructural study of lesions found in conventional histological sections. *Virchows Arch. A: Pathol. Anat.* **350,** 216.

Schaeffer, H. F. (1966). "Microscopy for Chemists." Dover, New York.

Shirahama, T., and Cohen, A. S. (1966). A congo red staining method for epoxy embedded amyloid. *J. Histochem. Cytochem.* **14,** 725.

Shirahama, T., and Cohen, A. S. (1967). High resolution electron microscopic analysis of the amyloid fibril. *J. Cell Biol.* **33,** 679.

Shurcliff, W. A., and Ballard, S. S. (1964). "Polarized Light." Van Nostrand-Reinhold, New York.

Spurlock, B. O., Skinner, M. S., and Kattine, A. A. (1966). A simple rapid method for staining epoxy-embedded specimens for light microscopy with the polychromatic stain paragon-1301. *Am. J. Clin. Pathol.* **46,** 252.

Symmers, W. St. C. (1956). Primary amyloidosis: A review. *J. Clin. Pathol.* **9,** 187.

Takeda, M. (1969). Virus identification in cytologic and histologic material by electron microscopy. *Acta Cytol.* **13,** 206.

Wolman, M. (1971). Amyloid, its nature and molecular structure. *Lab. Invest.* **25,** 104.

Wolman, M., and Bubis, J. J. (1965). A cause of the green polarisation colour of amyloid stained with congo red. *Histochemie* **4,** 351.

Wredden, J. H. (1947). "The Microscope: Its Theory and Application." Churchill, London.

18

Correlative Microscopy of Human Vitreous Body

JUERGEN FAULBORN

Eye Department
Universitäts-Augenklink und Augenspital
Basel, Switzerland

INTRODUCTION

The vitreous body of a human is a highly hydrated tissue containing 99% water and 1% solid contents as collagen (Moses, 1975). Because of this high water content the vitreous is said to be one of the most difficult tissues for histological preparation. Trials with dehydration with acetone and embedding with celloidin have been made early in this century (Szent-Györgyi, 1917; Jokl, 1927). Although there have been surprisingly

CORRELATIVE MICROSCOPY IN BIOLOGY:
INSTRUMENTATION AND METHODS

good results, the method has its limitations because of shrinkage of the vitreous when using higher celloidin concentrations than 4%. Also, correlative microscopic examinations using electron microscopic methods were not available at that time. Because of these difficulties to prepare the whole vitreous body, in recent publications on this subject only special areas have been described, whereas modern histologic findings of the whole vitreous are not available. Only slitlamp biomicroscopic studies of the whole vitreous have been done (Eisner, 1971; Foos, 1972) without further correlative histological evaluation.

To overcome the shrinkage of the vitreous tissue which occurs when alcohol is used for dehydration, we developed a method using glycerol instead of alcohol for dehydration. By this method the tissues surrounding the vitreous body, i.e., sclera, choroid, and retina, are not removed during dehydration with glycerol, but such removal is necessary with alcohol. This is the most important refinement in the absence of which the delicate structures of this nearly liquid vitreous tissue are greatly disturbed or even lost when the surrounding tissues are removed before embedding is completed. Embedding is done in celloidin which mixes with small quantities of pure glycerol. The celloidin block thus obtained is further evaluated with stereomicroscopic, light microscopic, scanning electron microscopic, and transmission electron microscopic examination. This is possible because the block retains exact correlation of the vitreous structure and its spatial arrangement during different examination methods (Faulborn and Bowald, 1982a, 1985; Bowald and Faulborn, 1985).

METHOD

This method has been developed to study the vitreous body of human eyes, which are obtained by autopsy.

Fixation

Fixation of the whole autopsied eyes is carried out with formalin for 2 days or longer. Longer fixation times make the tissue harder which seems to be of benefit for the next steps. Also, eyes have been used after several weeks in formalin with no alteration of the result.

Dehydration

The glycerol used for dehydration is of reagent grade. After fixation, without rinsing in water, the cornea of the eye is cut off. The eye is placed in a 100-ml bottle half-filled with glycerol/water (1:1) without mixing glycerol with water. The heavier glycerol sinks to the lower half of the bottle and the eye stays suspended in the glycerol. The bottle is kept at room temperature. During the next days water and glycerol mix spontaneously and the eye sinks to the bottom as it becomes glycerinated, which enables a very slow dehydration of the eye. After the eye has sunk, the remaining half of the bottle is filled with glycerol; this step is repeated once more. The bottle is filled two more times with pure glycerol, each time waiting until the eye has again sunk to the bottom of the bottle. The glycerol dehydration takes at least 3 months.

Embedding

The fully glycerinated eye is transferred to a 100-ml bottle containing 8% celloidin in ether/alcohol, where it remains for 4 weeks. Subsequently, the eye is saturated with celloidin and is transferred in a small paper box to water-free chloroform for hardening. The hard celloidin block is stored in terpineol, where the celloidin becomes clear. The block can be stored in terpineol until it is used for further preparation.

Preparation for Stereomicroscopy

Because of the very low content of collagen material in the vitreous body, thick slices should be used; about 200- to 1000-μm-thick serial slices are cut. Staining is easily done with van Gieson's stain after removal of the terpineol with isopropyl alcohol which does not dissolve the celloidin. Isopropyl alcohol is also used for dehydration after staining. After staining, the specimens are again cleared with terpineol which is also used for mounting. The coverslip is fixed to the preparation slide with silicone sealant.

For stereomicroscopic examination, dark field illumination is preferred, which gives the best contrast and visibility of the fine anatomy of the vitreous body. Also, pathologic changes in the surrounding tissues, i.e., retina, optic papilla, ciliary body, and lens can be visualized using magnifications up to 50 times (Figs. 1 and 2).

Preparation for Light Microscopy

To obtain additional information from particularly interesting areas, small pieces are cut from the thick section under a stereomicroscope. These pieces are embedded in paraffin (Paraplast) as follows. The celloidin piece is taken through four changes of isopropyl alcohol during 24 hr to remove the terpineol. It is placed in chloroform for 48 hr and then in a

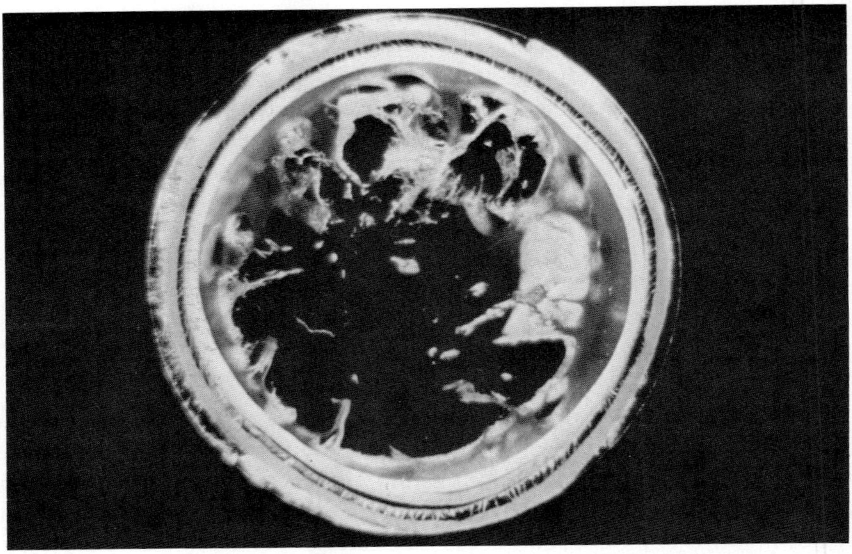

Fig. 1. Senile degeneration of the vitreous body; celloidin embedded coronary section (500 μm thick) stained with van Gieson; dark field image; ×3.6. (From Faulborn and Bowald, 1983b.)

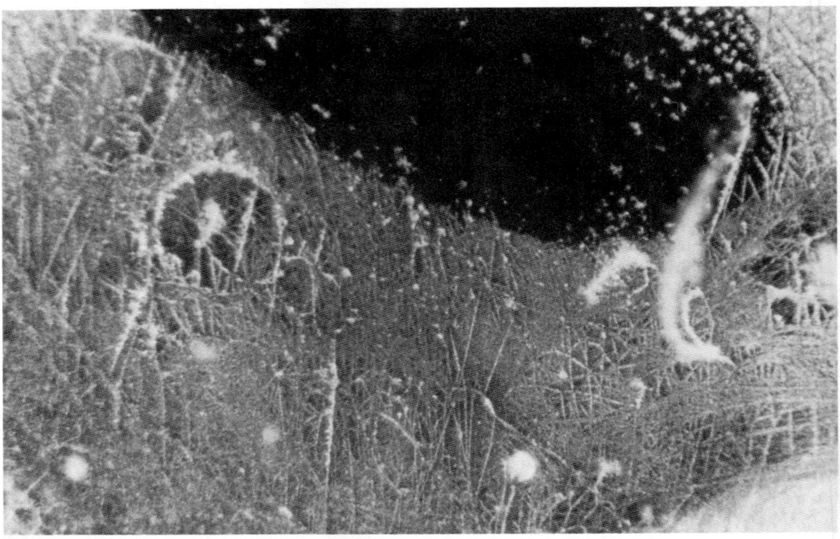

Fig. 2. Condensed vitreous structures at the wall of a degenerative cavity embedded in celloidin section (500 μm thick) stained with van Gieson; dark field image; ×57.5. (From Faulborn and Bowald, 1983b.)

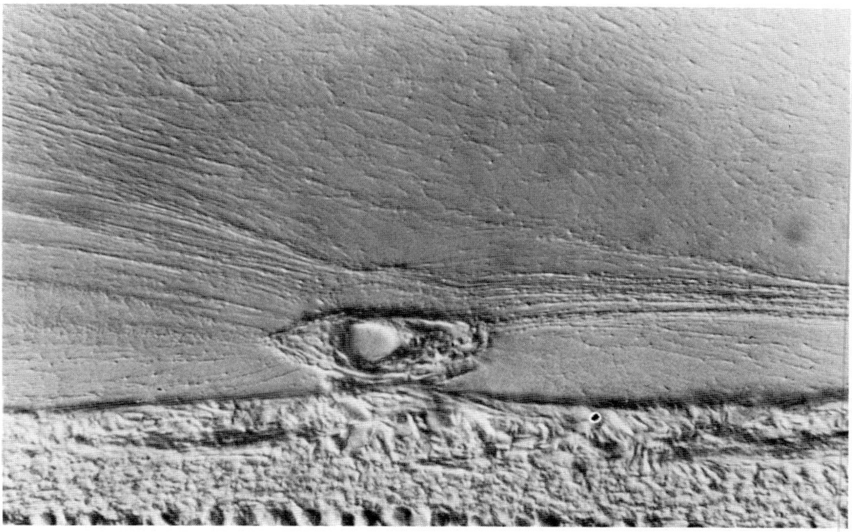

Fig. 3. Microproliferations growing from the retina into the vitreous body in an eye with proliferative diabetic retinopathy embedded in celloidin/paraffin stained with periodic acid Schiff; interference contrast image; ×800. (From Faulborn and Bowald, 1985.)

1:1 mixture of chloroform and paraffin for 48 hr in an oven at 58°C. Subsequently, the specimen is placed in pure paraffin for 72 hr with the glassware left open. Finally, the paraffin block, still containing the celloidin piece, is sectioned at 10 μm and stained as desired (Fig. 3). These sections can also be used for scanning electron microscopy.

Preparation for Scanning Electron Microscopy

For correlative scanning electron microscopy, the thick slices as well as the 10-μm-thick sections can be used. Thick celloidin sections are used for scanning electron microscopy without major difficulties. Celloidin is dissolved with acetone. Since the areas of particular interest are often not situated on the surface of the thick section, this problem can be solved by using thin, 10-μm celloidin/paraffin sections. The 10-μm-thick section is fixed on a glass slide and the celloidin removed by acetone, being very careful not to disturb the delicate vitreous structures. The specimen can be further evaluated by using critical-point drying with acetone/CO_2 and 20-nm-thick gold sputtering (Fig. 4).

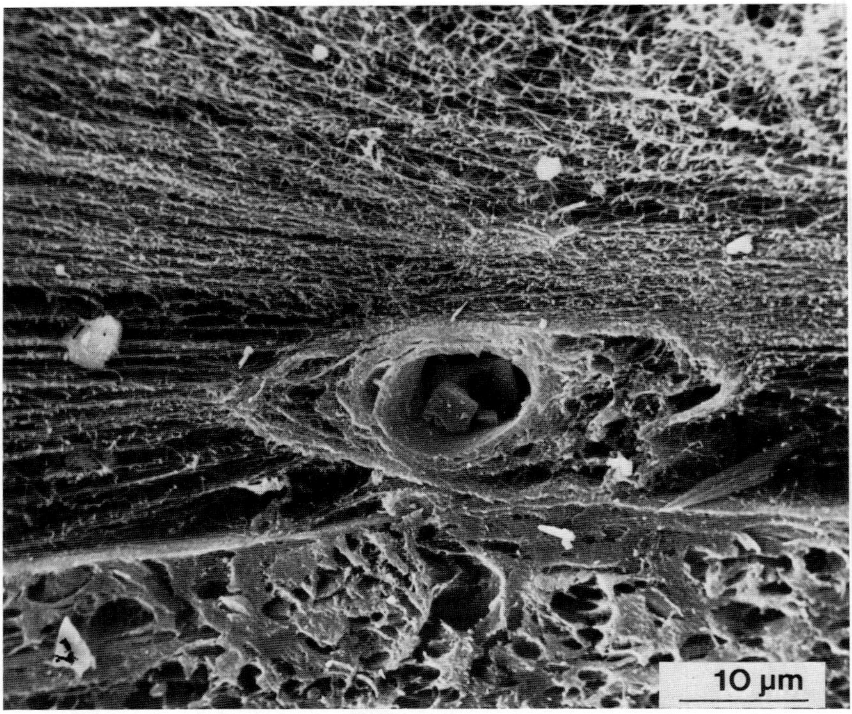

Fig. 4. Scanning electron micrograph of the celloidin/paraffin slide in Fig. 3; ×2700. (From Faulborn and Bowald, 1985.)

Preparation for Transmission Electron Microscopy

We have attempted to use the 10-μm-thick celloidin/paraffin sections for transmission electron microscopy. This attempt is rather difficult. Therefore, a neighboring thick section with similar changes should be used for this purpose. The area of interest is cut from the thick section under a stereomicroscope. The terpineol is removed with isopropyl alcohol (four changes in 2 days). The specimen is treated for 4 hr in 1% OsO_4 and dehydrated in isopropyl alcohol. The celloidin specimens are embedded in Epon for polymerization at 60°C for 3 days. The contrast of the ultrathin sections is enhanced with salts of uranium and lead (Fig. 5a,b).

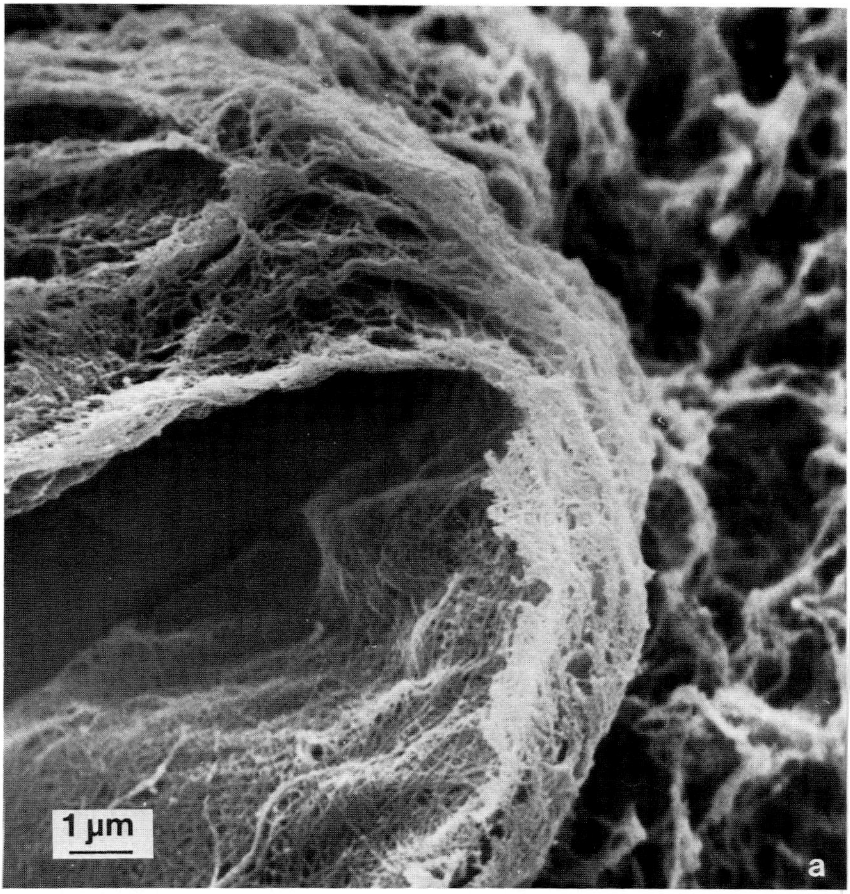

Fig. 5. Posterior vitreous border layer in an eye with posterior vitreous detachment. a, View with the scanning electron microscope; ×16,000. b, View with the transmission electron microscope; ×28,000. (From Faulborn and Bowald, 1983d.)

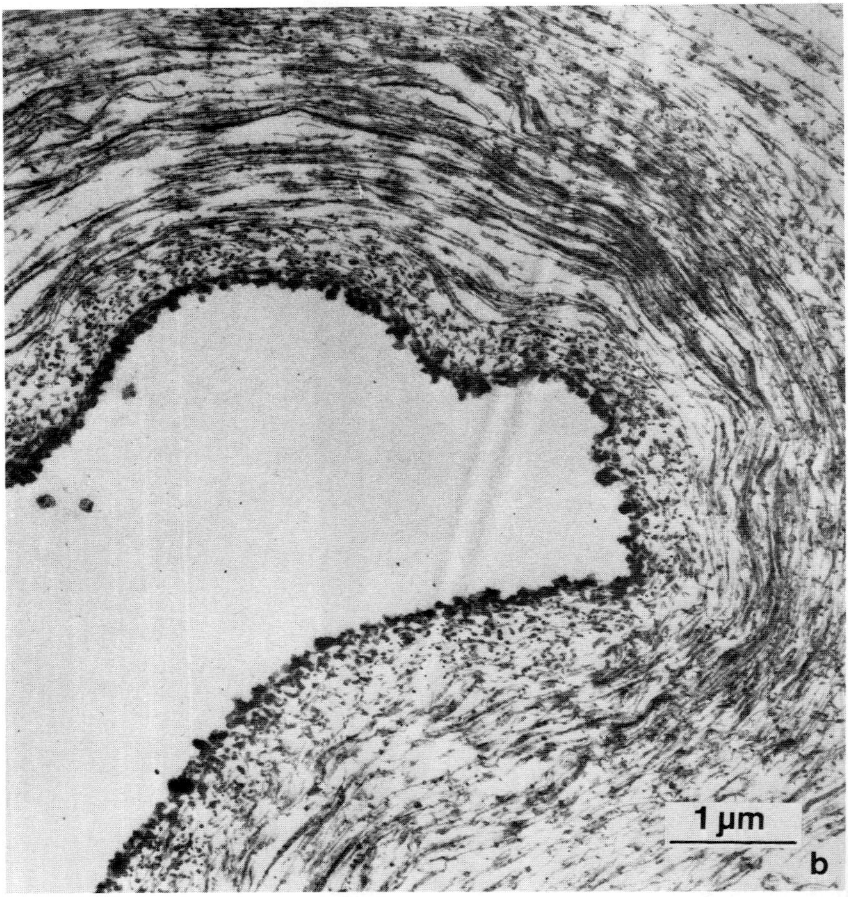

CONCLUSION

With the aforementioned method, exact correlative histological study of the vitreous boy becomes possible. Although this method is very time-consuming, it offers the great advantage of maintaining spatial arrangements of the vitreous structures and their relations to the neighboring tissues. By employing this method, several new findings concerning the anatomy, age-related changes, and pathologic alterations in diabetes of this organ have already been described (Faulborn, 1982, 1984; Faulborn and Bowald, 1982b, 1983a–d; Leuenberger *et al.*, 1985).

REFERENCES

Bowald, S., and Faulborn, J. (1985). Preparation of the vitreous body for comparative examinations with light microscopy (LM), scanning electron microscopy (SEM), and transmission electron microscopy (TEM). *Mikroskopie* **42,** 206.

Eisner, G. (1971). Slitlamp examination of the vitreous in autopsy eyes. I. Autopsy technique and examination method. *Albrecht von Graefes Arch. Klin. Exp. Ophtalmol.* **182,** 1.

Faulborn, J. (1982). Histology of the vitreous body. *Klin. Monatsbl. Augenheilkd.* **180,** 326.

Faulborn, J. (1984). Histology and treatment of diabetic tractional detachment. *Klin. Monatsbl. Augenheilkd.* **184,** 189.

Faulborn, J., and Bowald, S. (1982a). Combined macroscopic, light microscopic, scanning, and transmission electron microscopic investigation of the vitreous body. I. An improved celloidin embedding for histologic, SEM and TEM preparation of the vitreous body. *Ophthalmic Res.* **14,** 113.

Faulborn, J., and Bowald, S. (1982b). Combined macroscopic, light microscopic, scanning and transmission electron microscopic investigation of the vitreous body. II. The anterior vitreous cortex. *Ophthalmic Res.* **14,** 117.

Faulborn, J., and Bowald S. (1983a). On the structure of the anterior vitreous border layer. *Fortschr. Ophthalmol.* **79,** 422.

Faulborn, J., and Bowald, S. (1983b). Histological findings in senile vitreous degeneration. *Klin Monatsbl. Augenheilkd.* **182,** 486.

Faulborn, J., and Bowald, S. (1983c). Combined macroscopic, light microscopic, scanning and transmission electron microscopic investigation of the vitreous body. III. The structure of the anterior border layer of the vitreous. *Ophthalmic Res.* **15,** 11.

Faulborn, J., and Bowald, S. (1983d). Histological findings of the detached vitreous body. *Fortschr. Ophthalmol.* **80,** 329.

Faulborn, J., and Bowald, S. (1985). Microproliferations in proliferative diabetic retinopathy and their relationship to the vitreous: Corresponding light and electron microscopic studies. *Albrecht von Graefe's Arch. Klin. Exp. Ophthalmol.* **223,** 130.

Foos, R. Y. (1972). Posterior vitreous detachment. *Trans. Am. Acad. Ophthalmol. Otolaryngol.* **76,** 480.

Jokl, A. (1927). Vergleichende Untersuchungen über den Bau und die Entwicklung des Glaskörpers und seiner Inhaltsgebilde bei Wirbeltieren und beim Menschen. Dissertation, Uppsala.

Leuenberger, S., Faulborn, J., and Gülecek, O. (1985). Histologic studies of the effects of light coagulation of the retina on the vitreous body. *Klin. Monatsbl. Augenheilkd.* **186,** 272.

Moses, R. A., ed. (1975). "Adler's Physiology of the Eye: Clinical Application," 6th ed. St. Louis, Missouri.

Szent-Györgyi, A. (1917). Untersuchungen uber den Bau des Glaskörpers des Menschen. *Arch. Mikrosk. Anat. Entwicklungsmech.* **89,** 324.

19

A New Method for Bacterial Identification by Studying the Same Section with Light Microscopy and Scanning and Transmission Electron Microscopy

F. R. SAGLIE, J. C. SA FERREIRA, AND P. L. VALENTIN

Clinical Research Center for Periodontal Disease
School of Dentistry and Section of Electron Microscopy
School of Medicine
Center for the Health Sciences
University of California
Los Angeles, California

CORRELATIVE MICROSCOPY IN BIOLOGY:
INSTRUMENTATION AND METHODS

INTRODUCTION

Since the beginning of this century a controversy has existed about bacterial invasion as a pathogenic mechanism in periodontal disease. The controversy has continued mainly because of a lack of a methodology that could unify the capabilities of the three most common microscopes: light microscope (LM), scanning electron microscope (SEM), and transmission electron microscope (TEM). Since the early twenties, LM has been used to study bacterial penetration into the gingival tissues (Beckwith *et al.*, 1925; Cahn, 1929, 1935; Ray and Orban, 1948; Haberman, 1959). In the last decade, TEM and SEM have also been extensively used for detection of bacterial presence (reviewed by Saglie and Elbaz, 1983). The resolving power of these microscopes support the idea that bacterial invasion may be another pathogenic mechanism in periodontal disease. In spite of this fact, bacteria were not seen as often as should be expected because this microscope has several limitations for the purpose of detecting bacteria as will be analyzed later.

LIGHT MICROSCOPY, SCANNING AND TRANSMISSION ELECTRON MICROSCOPY AS SEPARATED SINGLE TOOL FOR DETECTING BACTERIA WITHIN GINGIVAL TISSUES

The LM has the advantage of quickly scanning large areas in numerous serial sections, but it lacks resolving power for a detailed study of bacteria. The SEM has a larger depth of focus and better resolving power than the LM and also allows for observation of rather large areas of tissue but shows only surface of the sample on a monochrome screen. The search and identification of bacteria with this type of microscope (SEM) has proven to be time-consuming, particularly in highly cellular and fibrous samples. The TEM provides ultrastructural details, which are useful for bacterial identification (Woo *et al.*, 1978, 1979; Lai *et al.*, 1981), but it is also time-consuming and allows the study of only small areas of tissue that do not provide an overview of the whole gingival biopsy, resulting in over- or understimulation of the degree of bacterial invasion. In order to obtain all the possible information provided by each microscope applied to the same histological sample, we have devised a simple and efficient method for bacterial identification in tissue sections with the LM, SEM, and TEM.

METHOD

Light Microscopy

After fixation with formaldehyde, the tissues are dehydrated through graded concentrations of ethanol, cleared in xylene, and embedded in paraffin. Sections are obtained with an AO Spencer 820 microtome at a thickness of 6 μm. A round microcoverslip (12 mm diameter) is attached to a 3×1 glass slide using Kerr sticky wax at two points. Egg albumin–glycerine mixture is applied to the coverslip to hold sections in place. Sections are then placed on a coverslip and allowed to dry overnight at room temperature. The next day the sections are deparaffinized in xylene, hydrated, and gram-stained. After staining and final clearing in xylene, stick wax is removed from the slide and coverslip. The coverslip is turned over and mounted on a slide with Permount. The tissues are examined using a Zeiss Ultraphoto II photomicroscope (see Fig. 1).

Prior to any photography, bacteria are localized under the highest magnification available ($\times 1000$). Once the bacteria are focused, pictures are taken at the lowest magnification (25x), where the whole tissue is seen, to the highest one, where, as said before, the entire bacterial colony is captured. Thus, the bacterial localization in the tissue is easily demon-

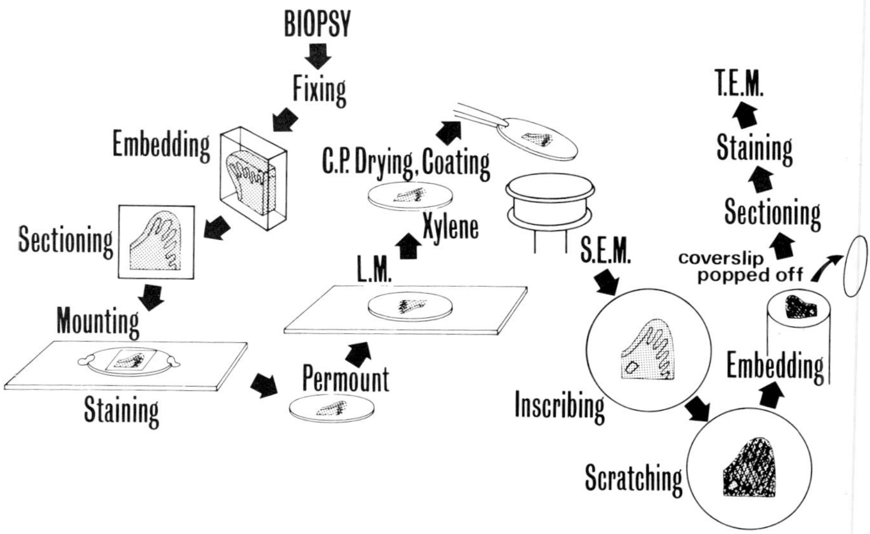

Fig. 1. Diagrammatic representation of the methodology used for studying the same paraffin section with the LM, SEM, and TEM.

strated by following the set of photomicrographs. Several colonies might be localized and photographed. The LM pictures are conditioned *sine qua non* when dealing with quick bacterial relocation in the SEM, as observed later.

Scanning Electron Microscopy after Light Microscopy

For observation in the SEM the microcoverslip is removed from the slide by soaking it in xylene. Following the removal of the coverslip with the tissues attached to it, it is rinsed in two changes of 50% xylene–50% ethanol solution and left in absolute alcohol until it is placed into a critical-point dryer. The tissue is not removed from the microcoverslip so that the original position is maintained. The microcoverslip is then mounted on an aluminum stub with silver conducting paint. The specimen is sputter coated with a coating of about 20 nm and is ready for observation in an ETC Autoscan SEM.

The section is oriented in the SEM in order to duplicate the orientation of the LM photomicrographs taken before by using acetate sheets over the LM prints and making sketches by means of lines, reproducing landmarks such as bundles of collagenous fibers, rete peges, holes, and also pointing out bacterial concentrations. For the drawings an All-Stabile pencil #8046 is used. The light photomicrographs taken at 25× magnification serve as a guide to tilt the tissue to the same orientation as that observed in the light photomicrograph.

With the acetate sheets presketched from each of the photomicrographs (100×, 160×, 400×, and 1000×) and by placing them on the viewing screen of the SEM (where the same magnification could be seen), it is easy to relocate the very same places photographed previously in the LM. Scanning electron micrographs are taken following the successive magnifications for comparative purposes. After taking the micrographs at 1000× magnification, the bacterial area are scanned and micrographs are taken at 4000× and 10,000× magnifications.

Transmission Electron Microscopy after Scanning Electron Microscopy

The scanning specimen mount is placed with the sample onto the Teflon specimen mount hold for processing. Several drops of buffer solution are added onto the entire coverslip, covering the specimen and placing it in the refrigerator for 15 min in a deep Petri dish with a lid. The buffer solution is drawn off with a filter paper by bringing it in contact with the side of the coverslip and solution. Several drops of 1% OsO_4 in buffer are placed onto the specimen, which is placed again into the Petri dish and transferred to the refrigerator at 4°C for 15 min. The osmium is rinsed off with the buffer using the same method as previously described. The rinse with buffer solution is continued until coloration disappears. Final drops of buffer are added and the dish is allowed to remain at room temperature for 5 min. The excess fluid is removed with filter paper and the mount is removed with forceps. The entire specimen mount is immersed into 50 ml of 50% ethanol with two changes for 5 min each. The specimen is placed in 50 ml of dehydration solution (acidified 2,2-demethoxypropane) for 10 min, with two changes. The specimen mount is then quickly removed from dehydration solution and immersed in 50 ml of 1:1 mixture of Epon and acetone. The mixture is infiltrated into the tissue for 15 min, and then the tissue is placed back on the mount holder. The excess of Epon mixture is removed by tilting the mount at an angle and allowing the mixture to run off onto the filter as described before. 100% Epon is added onto the microcoverslip making two changes, and in the final application Epon mixture is infiltrated for 20 min. Again the excess Epon is removed and the mount and holder are placed under a stereo LM to remove the microcoverslip disk. Using the sharp end of a prep blade between the coverslip and mount, the coverslip is pried off without fracture due to partially softened double-coated tape.

A BEEM capsule (size 00) is filled with pure Epon mixture slightly above the capsule rim and no visible air bubbles should be present. The capsule cap is removed and the filled capsule is placed in an oven at 70°C for 25 min to ensure the elimination of any trapped air bubbles.

The disk and specimen are inverted carefully onto the BEEM capsule while viewing the procedure under the stereomicroscope. The specimen is centered in place and a gentle pressure is applied on top of the microcoverslip until it comes fully in contact with the rim of the capsule, preventing any air leaking back under the coverslip disk. It is then polymerized overnight at 70°C. To pop-off the coverslip, the block is removed from the BEEM capsule mold and placed on a hot plate for 1 min at 110°C, removed, and quickly immersed in liquid nitrogen until the microcoverslip snaps off.

Prior to sectioning for the TEM, the area of interest is inscribed with a surgical blade #11, seen under a stereo LM with 30× magnification, locating it within 0.5 mm. With the aid of light and scanning photographs, one can easily use landmarks to pinpoint the areas using low magnifications (400× or 1100×) under the TEM. Now, the specimen is ready for trimming and sectioning. The block is sectioned with an ultramicrotome, and sections are stained with uranyl acetate and lead citrate. Micrographs are taken at magnifications ranging from 400× to 10,000×.

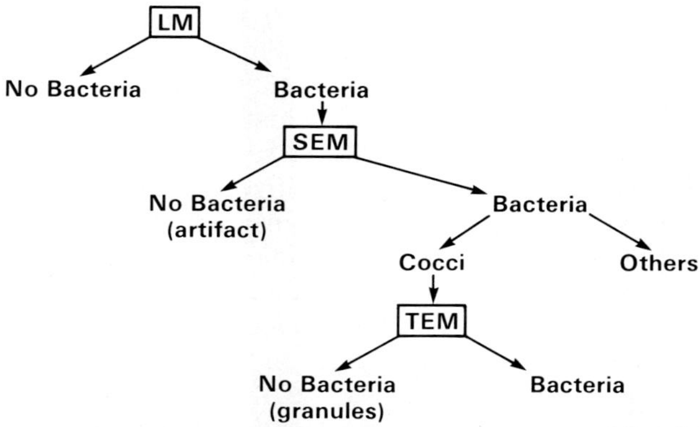

Fig. 2. Representation of the sequential methodology for bacterial identification in one paraffin section.

APPLICATION

The combination of LM, SEM, and TEM for studying the same histological section has proven to be a good tool for precisely identifying bacteria in disease gingival tissues (Fig. 2). Figure 3 is an example showing the usefulness of the methodology described, allowing in a simple way the study of the same paraffin section of a tissue sample or biopsy (same cells, connective tissue fibers, and bacteria) routinely stained for the LM, with the SEM and TEM. As shown in Fig. 3a, little information is provided by a gram-stained LM section, even the bacterial nature of the stained material seems doubtful. On the other hand, the LM section has the advantage in that several serial sections with a large surface can be quickly scanned one at a time for the presence of bacteria or bacteria-like material.

The SEM as applied in this method provides its large depth of focus and the possibility of achieving higher magnifications (Fig. 3b), allowing clear recognition of the shape, size, and number of those particles considered to be bacteria under the LM. Finding bacteria in the tissue with SEM is a long and tedious procedure. By using this method the gram-stained LM section can direct the observer rapidly to the areas of interest, saving considerable time and effort.

Although spirochetes, filamentous and rod-shaped bacteria, can be positively identified in the SEM (Saglie, 1977) requiring no transmission electron microscopy, several granular-shaped particles may resemble cocci forms (Fig. 3b), making necessary additional positive identification with

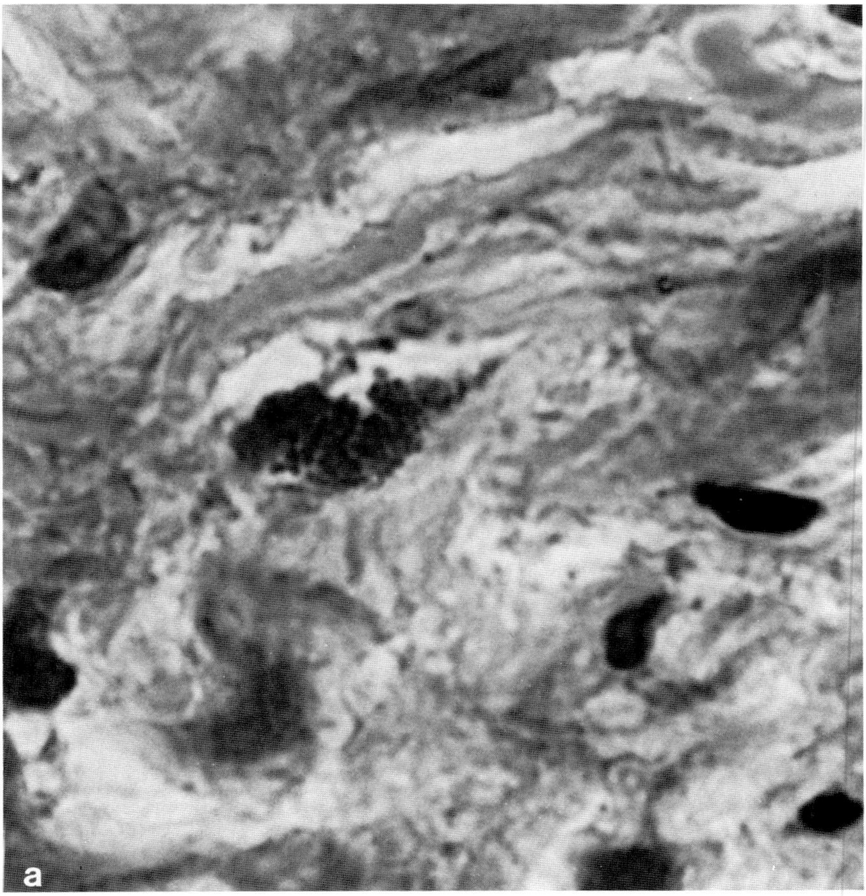

Fig. 3. Connective tissue of the lamina propria, adjacent to the oral epithelium in a patient suffering from periodontal disease. a, Light micrograph of a gram-stained section showing a "clump" of bacteria-like particles (cocci-forms) in between collagen fibers; ×900. b, Scanning electron micrograph of the same area of the same section shown in a; ×4000. c, Transmission electron micrograph of the bacteria-like clump area shown in b; bacterial morphology can be recognized; ×40,000.

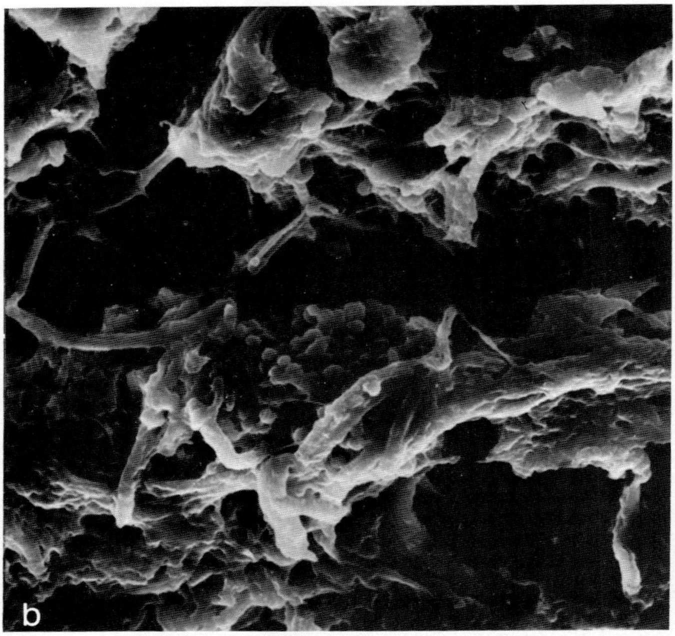

Fig. 3(b).

the TEM. The particles that resemble cocci forms include granules of mast cells, mitochondria, and other intracellular organelles and granules.

The final proof of the bacterial nature of these cocci particles is obtained from their characteristic fine structure seen after sectioning and staining the SEM sample for TEM (Fig. 3c). Despite the harsh procedures used for the LM and SEM preparations, ultrastructural details are generally adequately preserved for the purpose of bacterial identification.

REFERENCES

Beckwith, T. D., Simonton, F. V., and Williams, A. (1925). A histologic study of the gum in pyorrhea. *J. Am. Dent. Assoc.* **12,** 129.

Cahn, L. R. (1929). The penetration of the tissue by Vincent's organisms. A report of a case. *J. Dent. Res.* **9,** 695.

Cahn, L. R. (1935). The penetration of microorganisms into the gum as a cause of chronic and recurrent gingivitis. *Dent. Cosmos* **77,** 264.

Haberman, S. (1959). Inflammatory and non-inflammatory responses to gingival invasion by microorganisms. *J. Periodontol.* **30,** 190.

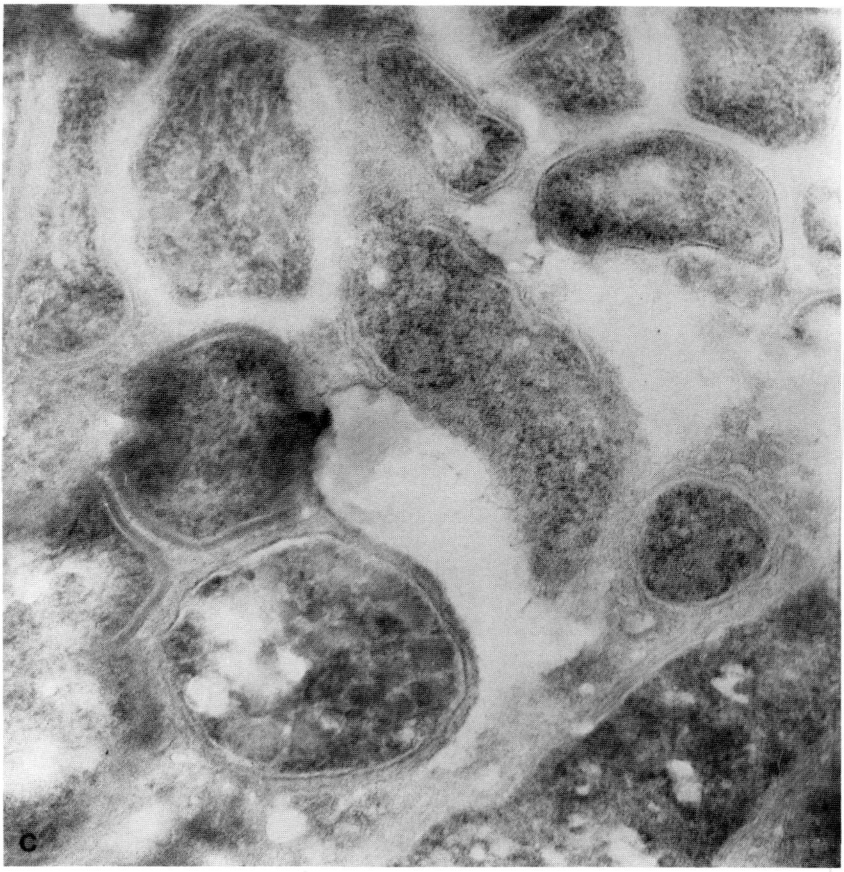

Fig. 3(c).

Lai, C. H., Listgarten, M. A., and Hammond, B. F. (1981). Comparative ultrastructure of leukotoxic and non-leukotoxic strains of Actinobacillus actinomycetemcomitans. *J. Periodontal Res.* **16,** 379.

Ray, H. G., and Orban, B. (1948). Deep necrotic foci in the gingiva. *J. Periodontol.* **19,** 91.

Saglie, R. (1977). A scanning electron microscopic study of the relationship between the most apically located subgingival plaque and epithelial attachment. *J. Periodontol.* **48,** 105.

Saglie, R., and Elbaz, J. J. (1983). Bacterial penetration into the gingival tissue in periodontal disease. *J. West. Soc. Periodontol.* **31,** 85.

Woo, D. D. L., Holt, S. C., and Leadbetter, E. R. (1978). Ultrastructural examination of the cell surface features of Bacteroides melaninogenicus subspecies. *J. Dent. Res.* **57,** 215.

Woo, D. D. L., Holt, S. C., and Leadbetter, E. R. (1979). Ultrastructure of Bacteroides species. *J. Infect. Dis.* **139,** 534.

20

Correlative Light and Scanning Electron Microscopy of Thick Sections and Isolated Cells

SUBHAS C. MOHAPATRA AND HAROLD E. PATTEE

Department of Biological and Agricultural Engineering
and U.S. Department of Agriculture
North Carolina State University
Raleigh, North Carolina

INTRODUCTION

Correlative microscopy of biological specimens through sequential application of light microscopy (LM),* transmission electron microscopy (TEM), and scanning electron microscopy (SEM), or any two of the three techniques offers advantages not available through the use of any one of these techniques alone. The indispensibility of LM for general cytological, anatomical, and histochemical investigations cannot be overemphasized. Although the highest available magnification for LM is >1000×, the operational magnification (particularly for photomicrography) is much lower than 1000×, thus necessitating the use of TEM for high magnifications. On the other hand, while TEM provides magnifications in excess of 100,000×, the lowest operational magnification is usually between 3000 and 5000×. Thus a gap exists between the highest magnification of LM and the lowest magnification of TEM. This gap is conveniently filled by SEM, which provides continuous magnifications between 10 and 100,000× (Buchanan, 1983). Availability of a wide range of continuous magnification and, more particularly, three-dimensional features has led to recent efforts for correlative application of SEM with LM, and to a limited extent with TEM. In order to take maximum advantage of correlative microscopy, it is imperative to have a working knowledge of the similarities and differences between LM, TEM, and SEM with regard to operating principles and specimen preparation procedures.

OPERATING PRINCIPLES

Details on the principles of LM, TEM, and SEM are beyond the scope of this chapter and can be found in several textbooks (e.g., Gray, 1973; Hayat, 1974a,b, 1975). However, the following brief descriptions, along with Fig. 1 and Table I, should provide a general understanding of the similarities and differences between the three techniques.

* The "M" in LM, TEM, and SEM denotes microscope, microscopy, microscopic, or micrograph depending on usage in a sentence, in the text, tables, and figures.

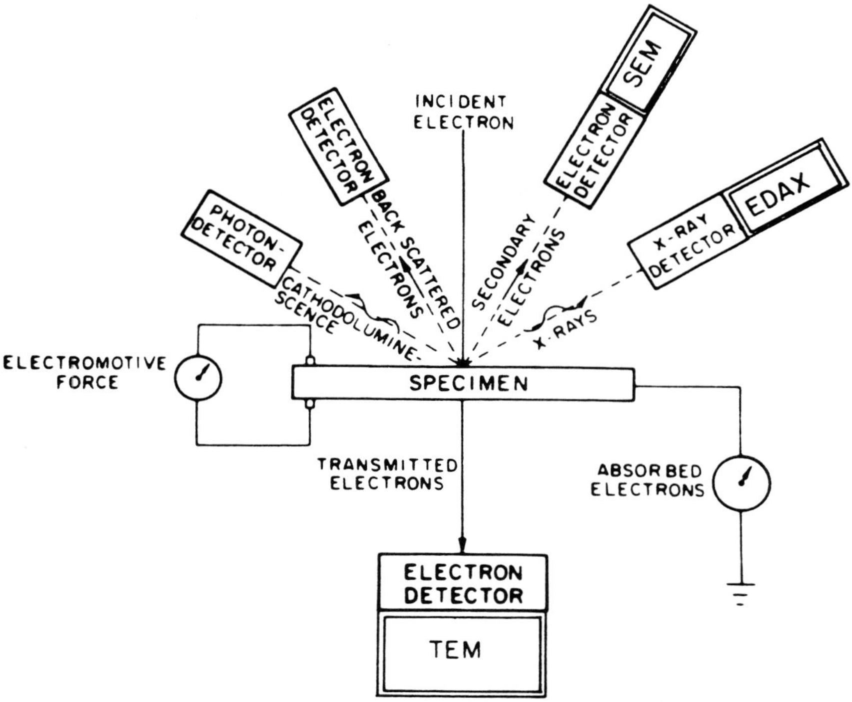

Fig. 1. Events following the contact of the incident electron beam with the specimen. Types of electron/energy used for TEM, SEM, and energy dispersive analysis of X rays (EDAX) are shown in double-lined boxes. The charging effect usually results from the accumulation of absorbed electrons and the electromotive force.

Light Microscopy

Specimens examined with the aid of LM can be fresh or fixed, and whole or sectioned, the latter being used for anatomical and cytological investigations. Sections are obtained through free-hand sectioning with a razor blade or through microtomy. Microtomy of delicate specimens require embedding in paraffin or any other suitable medium that can be sectioned. Since paraffin is insoluble in water, specimens must be fixed and dehydrated prior to embedding and sectioning. Section thickness for LM ranges between 2 and 20 μm, 10–14 μm being the most commonly used range. Prior to LM, sections are mounted on a glass slide, deparaffinized, stained, and permanently mounted. Light microscopy is conducted through transmitted light obtained through hypoillumination, and photomicrographs are obtained through a conventional camera body mounted onto the microscope.

TABLE I

Comparative Features of Light and Electron Microscopes

Feature	LM	Stereo LM	TEM	SEM
Wavelength	0.4–500.0 nm		0.00547 nm	
Numerical aperture	1.5		10^{-3}	
Limit of resolution	100–200 nm		0.5–1.0 nm or less	10.0 nm
Available magnification	20–3270×	20–280×	1,000–200,000×	10–100,000×
Photographic magnification	<500×	<200×	3,000–200,000×	10–100,000×
Image-forming radiation	Light	Light	Electron	Electron
Medium of travel	Air	Air	Vacuum (10^{-4} to 10^{-5} mmHg)	Vacuum
Nature of lenses	Glass	Glass	Magnetic field	Magnetic field
Object mounting	Glass slide	Intact or prepared	Thin film	Intact or prepared
Main source of contrast	Absorption and transmission	Absorption and reflection	Scattering and transmission	Scattering, reflection, and emission
Focusing	Mechanical	Mechanical	Electrical	Electrical
Adjusting magnification	Exchanging lenses	Exchanging lenses	Changing lense current	Changing lense current
Photograph	Transmitted light	Reflected light	Transmitted electrons	Secondary electrons and X-ray

Transmission Electron Microscopy

Fresh specimen TEM can not be performed on a routine basis because of the need for section thickness between 50 and 70 nm. To obtain these ultrathin sections, specimens are embedded in a plastic support medium (Spurr, 1969) and sectioned with a glass or diamond knife fitted to an ultramicrotome. Since plastic is insoluble in water, TEM specimens also need to be fixed and dehydrated like LM specimens prior to embedding. The dehydration procedure for both TEM and LM is identical; however, the former requires fixation with glutaraldehyde and postfixation with osmium tetroxide, whereas formalin-acetic acid-alcohol (FAA) is the common fixative for the latter. Further, optical staining used for LM bears no significance for TEM, where differential staining is obtained in the form of differential electron scattering through the use of lead citrate, uranyl acetate, silver, etc. in addition to OsO_4. Like LM, TEM photographs give flat images, thus requiring laborious serial reconstruction to obtain three-dimensional information (Elias, 1971). Based on the above considerations it is obvious that correlative application of LM and TEM is not straightforward, although it can be accomplished with appropriate specimen preparation techniques.

Scanning Electron Microscopy

Although fresh specimens have been used occasionally for SEM (Heslop-Harrison, 1970), routine application of SEM to biological specimens requires fixation, dehydration, and conductorization. Chemical fixation and dehydration for SEM is accomplished in the same way as for LM or TEM except that critical-point drying (CPD) constitutes the final step of dehydration followed by conductorization and SEM. Unlike TEM, image formation in SEM involves the use of secondary electrons, thereby providing three-dimensional information on the specimen being examined. For thin section SEM, the embedding medium must be removed from the sections to reveal their depth. Thus, deparaffinized sections prepared for LM can also be used for SEM. These features, in addition to continuous magnification between 10 and 100,000×, make SEM more ideal for correlative application with LM than with TEM. As shown in Fig. 1, X rays are also produced during SEM of biological specimens. Since X rays are specific to a given element in the periodic table, qualitative and quantitative histochemistry of inorganic constituents can be studied through energy dispersive analysis of X rays (EDAX) and X-ray photography. However, current technology limits the application of EDAX to elements having atomic numbers greater than 16.

APPLICATION OF CORRELATIVE SCANNING ELECTRON MICROSCOPY

In early studies SEM was used alone or in correlation with LM to study the external morphology of whole organisms, organs, isolated cells, or organelles. Its application to study internal structure was chiefly limited to the cut surface (Echlin, 1971; Hearle *et al.*, 1972). More recently, SEM has been used in correlation with LM and TEM to study thin and ultrathin sections obtained from embedded specimens (Hayat, 1978).

Figure 2 is an example of correlative LM and SEM of *Nicotiana tabacum* leaf anatomy (Mohapatra and Johnson, 1974). Photographs taken at ca. 80× of LM and SEM (Fig. 2A,B) were strikingly similar except that

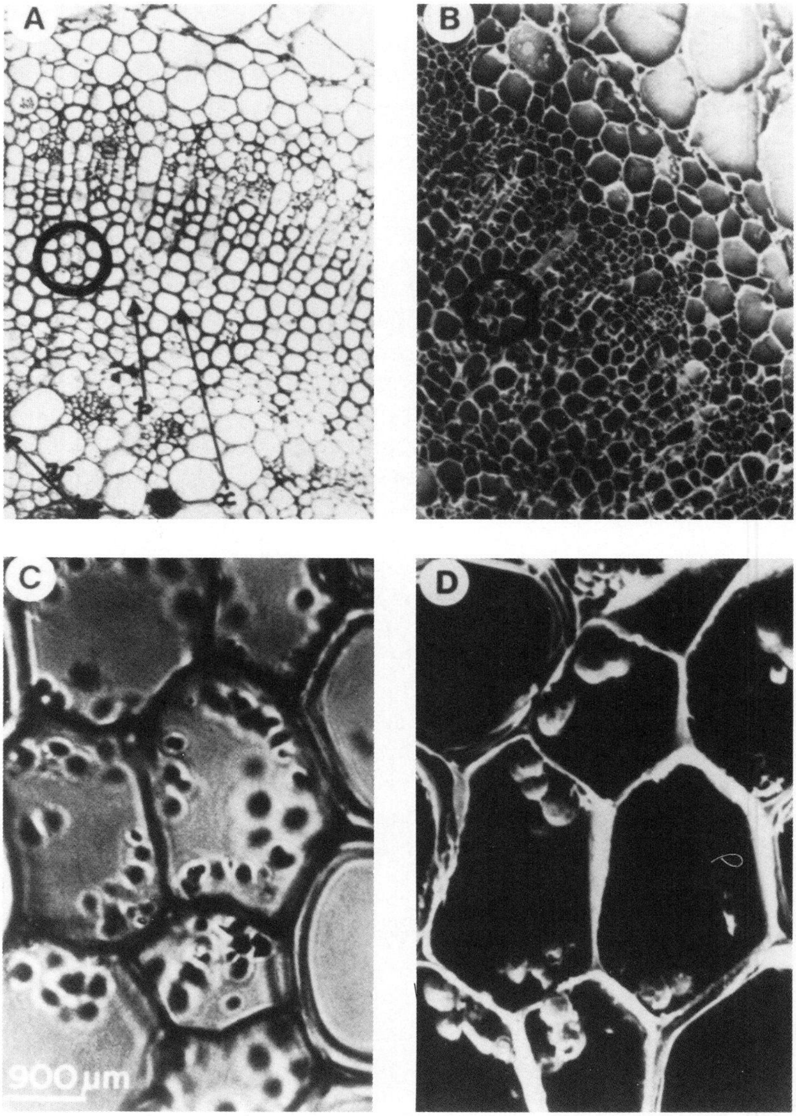

Fig. 2. Correlative LM and SEM of tobacco leaf cross sections. A, light micrograph of the midvein; ×90. B, scanning electron micrograph of the midvein; ×90. C, magnified view of the area encircled in A; ×900. D, magnified view of the area encircled in B; ×900. Note: Two adjacent sections were used: One was mounted on a glass slide for LM and the other was mounted on a metal slide for SEM. Differences between light and scanning electron micrographs are much more pronounced at higher magnification that at lower magnification. (From Mohapatra and Johnson, 1974.)

tissue delineation in the former was clearer because of staining, which had no effect on SEM. However, differences between LM and SEM became evident when the sections were examined at ca. 900× (Fig. 2C,D). Thus, LM facilitated tissue identification through differential staining while SEM revealed their three-dimensional features at the subcellular level. This feature of correlative SEM has been used by Mohapatra and Johnson (1974, 1975, 1976) to report polymorphic chloroplasts (Fig. 3), starch grain organization in the chloroplast (Fig. 4), and arrangement of calcium crystals in the cystolith (Fig. 5) in tobacco leaf cells. This information could not have been obtained with correlative or exclusive use of LM and TEM. Correlative SEM has been used further by Arcila (1978), Arcila and Mohapatra (1982, 1983a,b), Pattee *et al.* (1983, 1985) and Pattee and Mohapatra (1986, 1987) to study the anatomy and cytology of the tobacco seed

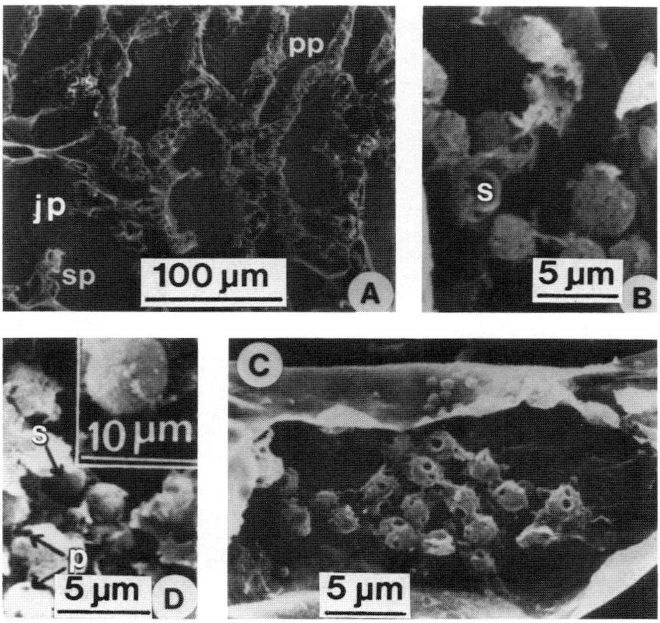

Fig. 3. Scanning electron micrographs of chloroplast polymorphism in the tobacco leaf cross section. A, general cross-sectional anatomy of the leaf showing arrangements of palisade parenchyma (PP), intermediary parenchyma (IP), and spongy parenchyma (SP); ×2700. B, Palisade parenchyma showing chloroplasts with starch grains (s); ×2700. C, spongy parenchyma showing chloroplasts with pores but without starch grain; ×2000. D, intermediary parenchyma showing both types of chloroplasts, some with starch grains (s) and others with pores (p); ×2300. (From Mohapatra and Johnson, 1976).

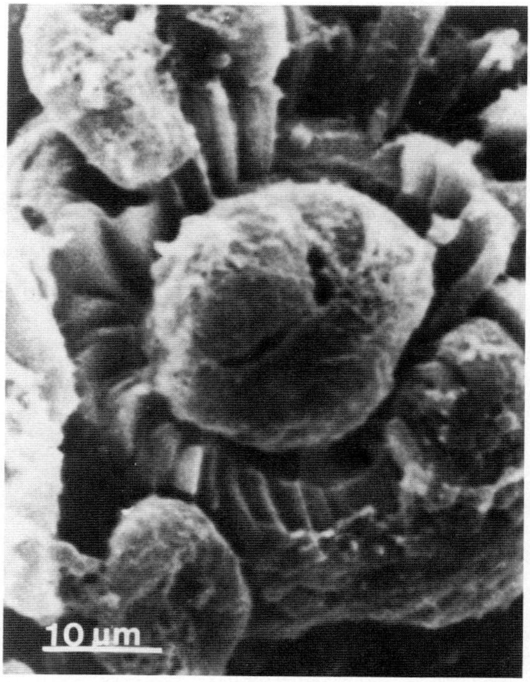

Fig. 4. Scanning electron micrograph of a starch grain-containing chloroplast. Note the concave arrangement of the chloroplast lamella where the starch grain is located. Also note the rod-shaped mitochondria-like bodies near the chloroplast; ×7000. (From Mohapatra and Johnson, 1976.)

and the peanut fruit. Scanning electron microscope histochemistry has also been used by Mohapatra and Johnson (1975) and Punja and Jenkins (1984) to study the structure and chemistry of previously unreported calcium crystals (Fig. 6).

While the foregoing are examples of the application of correlative SEM to thin sections of plant specimens, this technique has also been used to study thin sections (McDonald *et al.*, 1967; Geissinger, Poh *et al.*, 1971; Erlandsen *et al.*, 1973; Judge and Stubbs, 1975) and isolated cells (Kumon *et al.*, 1984; Finch *et al.*, 1985, and references therein) of animal tissues. Several factors need to be taken into consideration with respect to specimen preparation while applying correlative LM and SEM to biological specimens. These are discussed below for thin sections and isolated cells with separate emphasis on isolated cells where warranted.

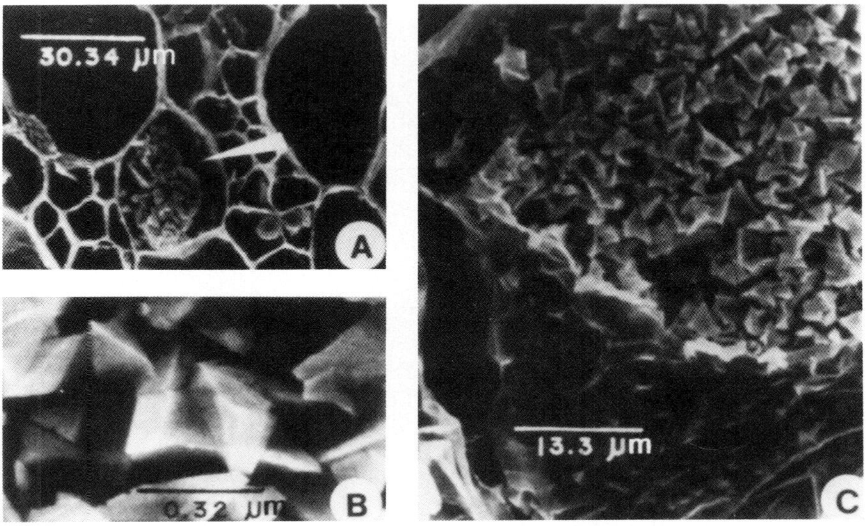

Fig. 5. Scanning electron micrographs of tobacco leaf crystals. A, a cystolith containing cell; ×660. B, crystalline structure in the cystolith shown in A; ×6320. C, another cell in the same thin section as A showing a membranous sheath enclosing the crystals; ×1510. (From Mohapatra and Johnson, 1975.)

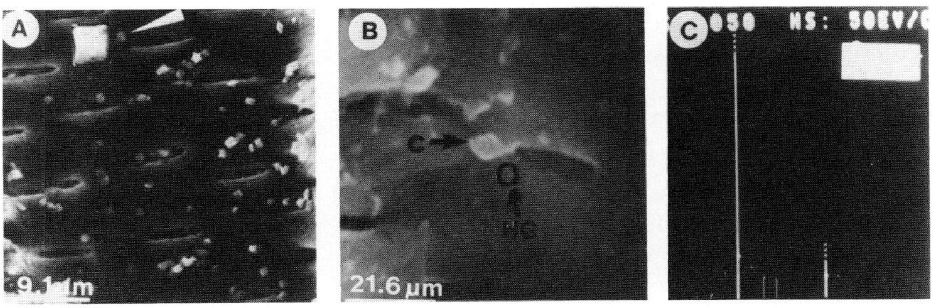

Fig. 6. Scanning electron microscopic histochemistry of tobacco leaf crystals. A, scanning electron micrograph of crystals and pits distributed in a longitudinal section of a xylem vessel; ×2210. The single large crystal is a magnification (6320×) of the crystal shown by the arrow. B, X-ray micrograph of the same crystal shown in A and the adjoining area without any crystal; ×5300. Compare crystal images in A and B and note the difference resulting from the use of secondary electrons and X rays, respectively. C, chemical identification of the crystal shown in A and B through energy dispersive analysis of X rays (EDAX). Note three types of signals; two bright bars scored at the top are standard calcium signals, several smaller paired bars are signals from the noncrystalline area encircled in B, and dots are signals from the single crystal shown in A and B. Signals from the calcium standard overlap the dot signals from the crystal but not the bar signals from the adjacent noncrystalline area, indicating that the crystal is calcium. (From Mohapatra and Johnson, 1975.)

Fixation and Postfixation

Isolated cells and organelles need to be sedimented prior to fixation through centrifugation, filtration, or by simply letting the solution stand for a prolonged period at a suitable temperature. Although several fixatives are available in the literature on electron microscopy (e.g., Hayat, 1981) and light microscopy (Gray, 1973; Berlyn and Miksche, 1976), fixation with glutaraldehyde and postfixation with OsO_4 are used routinely for TEM and SEM whereas fixation with FAA is used for LM. It is obvious that if LM and SEM are combined to study the same specimen, only one fixative (either a single chemical or a mixture of two or more chemicals) can be used. In as much as the role of the fixative is to arrest cell function and preserve cellular details, and clarity of cellular details is a function of magnification and associated resolution (Buchanan, 1983), these features need to be taken ito consideration while selecting a fixative.

Mohapatra and Johnson (1974, 1975, 1976), Arcila and Mohapatra (1982, 1983a,b), and Pattee *et al.* (1983, 1985) used FAA as the sole fixative (i.e., without postfixation with OsO_4) for correlative LM and SEM of thin sections of plant specimens and have used magnifications up to 6000× with satisfactory results. On the other hand, glutaraldehyde fixation and OsO_4 postfixation has been used rather routinely for correlative microscopy of thin sections (Geissinger, 1971; Poh *et al.*, 1971; Erlandsen *et al.*, 1973; Judge and Stubbs, 1975; Gogue *et al.*, 1976; Watari *et al.*, 1978; Pickett *et al.*, 1980) and isolated cells (e.g., Kumon *et al.*, 1984, and references therein) of animal specimens. The use of glutaraldehyde and OsO_4 for SEM appears to have been influenced by the use of these chemicals for TEM. Because of the potential risk associated with the use of OsO_4 and since SEM uses secondary electrons rather than transmitted electrons as in the case of TEM, the actual need for OsO_4 for SEM specimens should be given careful evaluation.

Following fixation, animal or plant specimens are chemically dehydrated and further processed for embedding in paraffin or another suitable medium for sectioning for correlative LM and SEM. Isolated cells/organelles are also chemically dehydrated following fixation and mounted on a suitable slide with the aid of poly-L-lysine solution (mol. wt. 85,000) (Kumon *et al.*, 1984). Alternatively, the cells/organelles can be mounted on the slide first followed by fixation and dehydration *in situ*.

Embedding and Deembedding

This step is not applicable to isolated cells/organelles or to sections obtained through free-hand sectioning, sliding microtomy, or cryomicrotomy. The vast majority of biological specimens, however, need to be embedded in a support medium for sectioning. Like fixatives, more than one formulation of embedding medium is available in the literature for both types of electron microscopy (Hayat, 1974a,b, 1975) and LM (Gray, 1973; Berlyn and Miksche, 1976). Usually paraffin is used for LM whereas a plastic embedding medium, such as Spurr's medium (1969), is used for TEM. When LM is used in correlation with TEM, plastic-embedded specimens are first sectioned at micrometer thickness. These sections are stained and used for LM. The remainder of the tissue block is then used for ultramicrotomy to obtain Ångstrom-thick sections.

Although plastic-embedded sections prepared for TEM have been used for SEM following the removal of the plastic to reveal the depth (Mayor *et al.*, 1961; Erlandsen *et al.*, 1973; Pachter *et al.*, 1973; Judge and Stubbs, 1975), they are not ideal for this purpose because removal of the plastic often results in loss of cell contents (Judge and Stubbs, 1975) and, even then, the depth available from Ångstrom-thick sections is not sufficient to reveal the depth for SEM. Erlandsen *et al.* (1973) used correlatively all three techniques (LM, SEM, and TEM) to study plastic-embedded sections of neonatal and adult rat tissue. They first cut 1-μm thick sections for LM and TEM, followed by removal of the plastic for SEM. Although the 1-μm thickness provided the depth needed for SEM, its usefulness for TEM is questionable, considering the fact that TEM requires sections having thickness around 50 nm. One possible approach for combining all three techniques, not reported to our knowledge, would be to fix specimens in glutaraldehyde followed by embedding in paraffin for sectioning at micrometer thickness for LM and SEM. Following SEM, the micrometer-thick deparaffinized sections can be postfixed in OsO_4 and further processed for embedding in plastic for ultramicrotomy and TEM. This way, a 14-μm-thick LM section should give between 20 and 25 sections (50 nm thick) for TEM.

Mounting Slide(s)

While glass slides are standard supports for the mounting of thick sections and isolated cells for LM, they are not ideal for electron microscopy because they, like the biological specimens mounted on them, are poor conductors of electricity and heat, and thus cause considerable charging (Buchanan, 1983). However, in absence of a better support material, glass slides have been used for the SEM of thick sections and isolated cells with conductorization with heavy metal vapor to minimize charging. Because of the difficulty associated with the identification of the same isolated cells for LM and SEM, several modifications have been introduced in mounting slides used for these specimens. Therefore, selection and use of mounting slides will be discussed separately for thick sections and isolated cells.

Thick Sections

Identity of paraffin-embedded sections is easily maintained during correlative LM and SEM because sections are serially arranged in the paraffin ribbon. However, if sections are obtained through free-hand sectioning, sliding microtomy, or cryomicrotomy, they need to be manually arranged serially on the mounting slide with appropriate record keeping. Charging during the SEM of thick sections is minimized through conductorization of the section(s) and the glass slide with a thin layer of carbon and heavy metal. The glass slide is broken into small pieces around the sections of interest and mounted on the specimen holder for conductorization and SEM (McDonald *et al.*, 1967; Poh *et al.*, 1971; Ayres *et al.*, 1971; Arnott and Webb, 1982).

Our experience has shown that it is often difficult to accomplish uniform conductorization of the glass slide because of its refractive surface. Further, even after careful etching with the diamond marker, the slide would not always break at the desired place, thus giving irregular cross-sectional thickness at the broken edge, which might break the continuity of the conductor layer between the slide and the SEM specimen holder (Fig. 7) during sputter-coating. Substitution of the thick glass slide by thin coverslips solved this "edge effect" but did not solve the problem of random fracture (S. C. Mohapatra, unpublished). Therefore, Mohapatra and Johnson (1974, 1975, 1976) used acid-washed thin brass film as the mounting slide for SEM of tobacco leaf sections. The metal slide permitted uniform coating with heavy metal and eliminated charging. This technique has been used successfully in subsequent studies on tobacco (Arcila, 1978; Arcila and Mohapatra, 1982, 1983a,b) and peanut (Pattee *et al.*, 1983).

The metal slide, however, did not permit the use of the same section for both SEM and LM, thus necessitating the use of adjacent sections. While this approach was satisfactory for general anatomical investigations, it was not ideal for cytological investigations because often subcellular components of interest, such as the peanut embryo at early stages of development (Fig. 8), would not be present in both the sections. Similar problems were also reported by Pickett *et al.* (1980) during correlative LM and SEM of inorganic crystals in human lung cells. To avoid this problem, Pickett *et al.* (1980) first conducted LM and then transferred selected sections from the glass slide directly to the SEM specimen holder with the aid of Krylon crystal clear acrylic spray. Although this method appears

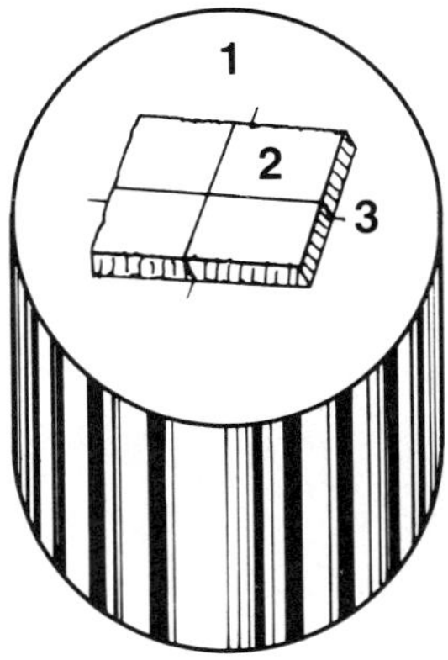

Fig. 7. Suspected edge effect during sputter coating of the standard glass slide. 1, SEM specimen holder; 2, a cut piece of a standard glass slide attached to the SEM specimen holder; 3, cross-sectional area at the broken edge which must be covered by the metal coating for complete conduction of the absorbed electrons and the electromotive force.

to be useful, the authors have reported loss of crystals during section transfer.

Isolated Cells

Because isolated cells/organelles are distributed randomly when smeared on the mounting slide, it is difficult to maintain the identity of individual cells/organelles for correlative LM and SEM. This has led to a large number of reports on methods to maintain cell identity. These fall under two general categories: use of ungridded and gridded slides. Takenaga *et al.* (1977) and Mikel and Johnson (1980) used ungridded glass coverslips but standard permanent mounting procedure to study cells from an ascitic fluid through LM. Following LM, the coverslip containing the cells was separated from the slide with xylene and the cells were dehydrated through critical-point drying. Then cells of interest were marked with a device attached to the nosepiece. The marked coverslip was cut to small pieces for conductorization and SEM. Individual cells were identified on the basis of the markings. Thierstein *et al.* (1971) also used ungridded slides in an earlier study on nannofossils. But for cell identification they solely relied on the scale and vernier of both LM and SEM using the right bottom corner of a rectangular coverslip as the reference point.

Among those who used gridded slides, Moshkovitz (1978) used a specially prepared coverslip with graticules. LM was conducted by dipping

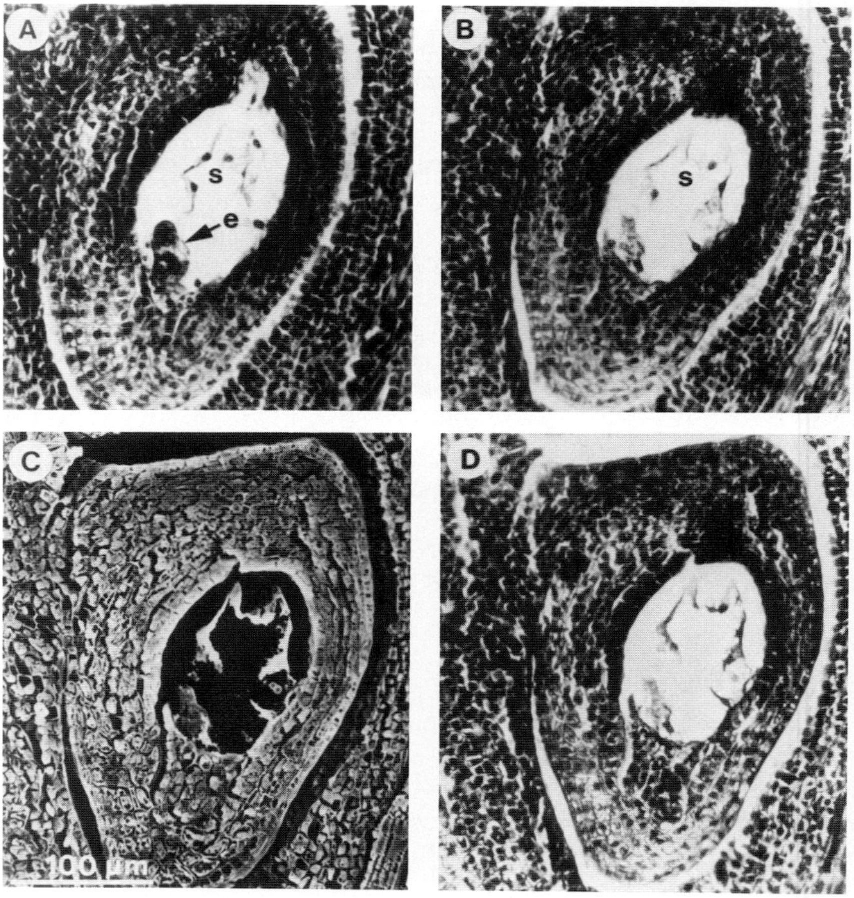

Fig. 8. Correlative LM and SEM of thick sections of developing peanut fruit. A, light micrograph of a longitudinal section mounted on a glass slide and showing the embryo (e) in the embryo sac (s); ×200. B, light micrograph of the section adjacent to that shown in A but mounted on a GelBond slide; ×200. Note that the embryo is missing in the embryo sac (s). No other difference is detectable between A and B, thus indicating that GelBond is a suitable replacement for the glass slide for LM. C, SEM following LM of the same section shown in B. Note that LM staining in B has no effect in C, which shows structural details of the embryo sac. Only the outlines of these structures can be seen in B. D, repeat LM of the same section shown in B following SEM in C; ×200. Note slight loss in stain intensity in D as compared to B. This effect of the thin metal coating used for SEM is, however, negligible. (From Pattee *et al.*, 1985.)

the coverslip in a suitable liquid medium (usually xylene) in a well-slide, also prepared specially for this purpose; a coverslip placed on top of the well protected the microscope optics from touching the liquid in which the graticule containing the cells were dipped. Following LM, the cells were critical-point dried while still attached to the graticule and processed for SEM. Individual cells were identified on the basis of their location on the graticule. Thornthwaite *et al.* (1976), on the other hand, used a computer program to develop a grid (with numbers in x-axis and letter in y-axis) which was transferred photographically to a standard glass slide. Bahr *et al.* (1976) also used photographic etching to prepare a gridded slide, but they etched the numbering system of a Maltwood finder (microscope field finder) into regular glass coverslips and smeared the cells to be investigated on these slides. Ewijk and Mulder (1976) prepared a special plastic film dish (PFD) to culture the cells directly on it, and then marked it with a cell marker fitted to an inverted microscope. During SEM the cells of interest were relocated on the basis of these markings.

Special chemicals and instrumentation needed for marking the PFD and for photographic etching of grids might not permit all laboratories to prepare gridded slides. Therefore, Domagala *et al.* (1978) and Nolte and Rudolph (1980) manually etched the grid on microcoverslips with the aid of a diamond marker. Because this technique did not permit etching of numbers on the coverslip, the authors made the top line heavier for easy identification and maintained a complete record of the grids. This is a simple procedure and can be used in any laboratory. However, because of the labor involved, it will be of limited application where large number of samples are to be examined. Saleh *et al.* (1981) recognized this limitation and improved the above method by attaching a ready-made copper grid to the coverslip instead of manually etching the grid. An alternative improvement of the Domogala method was introduced by Kumon *et al.* (1983, 1984) and consisted of forming a gelatin layer on a standard coverslip followed by etching lines on the gelatin layer with the help of a razor blade. The etched gelatin layer was fixed with OsO_4 to improve its conductivity. The authors further conductorized the specimens and the gelatin-coated slide with vaporized platinum and palladium.

It is obvious from the foregoing discussion on thick sections and isolated cells that the glass slide is the least desirable support material for correlative LM and SEM of the same specimen. At the same time, any alternative slide should have the same characteristics as the glass slide with respect to optical transmittance (i.e., refractive index), thermal resistance, and resistance against solvents, stains, and other chemicals used in microscopy. Further, it should be pliable enough to permit cutting into small pieces without random cracking, and thin enough to permit continuous metal coating between the slide and the SEM specimen holder without the edge effect shown in Fig. 7. This prompted Pattee *et al.* (1985) to conduct a systematic evaluation of various plastic films available commercially. They found that no brand of plastic except the GelBond brand available from the FMC Corp. (Rockland, Maine), and manufactured originally for slab-electrophoroesis, met all of the foregoing criteria.

Figure 8 compares photomicrographs of two adjacent cross sections from a peanut fruit, one mounted on a glass slide (Fig. 8A) and the other mounted on a GelBond plastic slide (Fig. 8B). Although the embryo is present only in one of the sections, the remaining anatomical features are similar in both the sections, thus suggesting that the GelBond plastic is a suitable replacement for the glass slide. The plastic can also be easily marked with a razor blade or needle to make grids needed for the study of isolated cells and organelles. Thus, a GelBond slide can be easily used for isolated cells without the need for custom preparation of plastic support materials. The fact that other commercial plastics tested did not have the same properties as glass slides might explain why plastic slides are not as common in microscopic laboratories as other plastic laboratory wares.

Staining

Anatomical and cytological investigations through LM depends on differential staining of cells and subcellular components with visible colors. Because the electron microscope cannot discern visible color, staining for TEM is obtained through differential electron scattering resulting from reaction between cell constituents and chemicals such as OsO_4, lead citrate, uranyl acetate, silver, etc. Like TEM, SEM cannot also discern visible colors. But it does not need special electron contrast for imaging because it gives true image through the use of secondary electrons. However, if LM is combined with SEM or TEM for correlative study of thick sections or isolated cells, an appropriate stain must be used, and the exact composition of the stain and timing of their application will depend on experimental objectives.

Sequence of Correlative Microscopy

Because LM represents low-magnification investigation and SEM represents high-magnification investigation, it is logical to conduct LM prior to SEM, although the reverse is true when whole specimens are studied through SEM and sections therefrom are studied through LM. However, when correlative LM and SEM is applied to thick sections or isolated cells either technique can precede the other. This is demonstrated in Fig. 8. The same section was first used for LM (Fig. 8B), then for SEM (Fig. 8C), and finally again for LM (Fig. 8D). Comparison of anatomical features obtained through LM before and after SEM shows little difference with respect to structural details. However, Fig. 8D shows slight loss in stain intensity, perhaps because of the presence of the conductive layer used for SEM. When LM is conducted first, removal of the coverslip and the mounting medium is necessary for SEM following LM. This will not be necessary if SEM is performed prior to LM; but interference from the conductive coating with LM needs to be carefully evaluated.

STEPWISE PROCEDURE

The following stepwise procedure is applicable to correlative LM and SEM of thick sections and isolated cells (and organelles) of both animal and plant specimens. The procedure described is based on the foregoing discussion on various factors and principles. An assumption is made that the reader has a working knowledge of the general principles and procedures associated with noncorrelative LM and SEM of biological specimens. For more details, the reader is referred to Hayat (1978) for SEM, and Gray (1973) and Berlyn and Miksche (1976) for LM.

Thick Sections

1. Fix tissue blocks in FAA. Preliminary trials may be necessary if any other fixative is considered for use. Infiltrate FAA into the tissues under partial vacuum, following which specimens are left overnight in the fixative at room temperature. Fixed specimens should be stored in 70% ethanol instead of the fixative until they are further processed for correlative microscopy.

2. Dehydrate the fixed specimen(s) by passing it sequentially through the following: ethanol (70%, 80%, 95%, 100%), ethanol:tertiary butanol (3:1, 1:1, 1:3, pure TBA). Keep the specimens in each step for at least 4 hr, and use three changes of pure tertiary butanol.

3. Place sample in 1:1 mixture TBA–paraffin oil for 4 hr with periodic application of partial vacuum to facilitate infiltration. Leave the samples overnight in fresh TBA–paraffin oil mixture. Pour melted paraffin (mp 56–58°C) to a vial and allow to cool just enough to solidify at the top. Place the sample on the solidified paraffin, add enough TBA–paraffin oil mixture to cover the sample, and leave overnight. Then transfer the sample to

pure melted paraffin; make eight changes of paraffin with at least 4-hr intervals between each change. These durations work well for plant specimens. For animal specimens, the duration for each step needs to be worked out through preliminary studies.

4. Cut paraffin-embedded specimens with the aid of a rotary microtome to give ribbons that contain sections of desired thickness (usually between 7 and 14 μm).

5. Cut the GelBond plastic film to the same size as the standard glass slide and attach a GelBond slide to a glass slide with a few drops of Haupt's adhesive. This is necessary to facilitate the handling of the flexible plastic slides. Cut the paraffin ribbon containing the sections to appropriate lengths and affix them with Haupt's adhesive to the GelBond slide. Allow to dry overnight. Detach the GelBond slide from the glass slide and clean both with 95% ethanol to remove Haupt's adhesive which held the two slides together. Use paperclips to hold the GelBond slide to the glass slide during the deparaffinization and staining process.

6. Deparaffinize the GelBond slide in xylene followed by several changes of xylene:ethanol (1 : 1) mixture. Rehydrate the slide in graded series (decreasing) of ethanol if aqueous stains are to be used. Stain the sections as necessary for LM. If nonaqueous stains are to be used, the slide should be passed through graded mixture of xylene and the stain solvent before transferring to the stain solution. Dehydrate the slide in graded series of ethanol up to 100% ethanol. At this time, a decision needs to be made on whether LM is to be performed first followed by SEM or vice versa. If LM is to be performed first, go directly to step No. 7. However, if SEM is to be conducted prior to LM, proceed as follows.

Cut the plastic slide to pieces of appropriate size and attach them to the SEM specimen holder with a conducting cement/glue. Sputter coat the specimens with heavy metal and carbon and conduct SEM. Following SEM, detach the pieces of plastic slides from the SEM specimen holder and clean thoroughly to remove the cement from its back. Put a layer of permanent mounting medium on a glass slide and arrange the pieces of plastic slide in the same sequence they were prior to cutting; take precautions to avoid trapping of air bubbles between the plastic pieces and the glass slide. Then place an additional amount of mounting medium to cover the GelBond pieces, and place a rectangular coverslip as is done during conventional LM. Let the slide dry at 35–40°C. Remove the excess mounting medium with xylene prior to LM and photomicrography.

7. Spread a layer of permanent mounting medium on the GelBond slide containing the deparaffinized sections, and mount a glass coverslip followed by drying at 30–40°C. Take care to remove air bubbles prior to placing the coverslip. Then remove the paperclips, detach the glass slide,

and clean the excess mounting medium with xylene. The plastic slide and the coverslip with the dried mounting medium in between should provide enough rigidity for handling the slide during LM and photomicrography.

8. Following LM, dip the GelBond slide containing the sections in xylene until the coverslip automatically separates from the slide. The excess mounting medium is washed off through several changes of xylene.

9. Cut the plastic slide to desired size, attach to the SEM specimen holder and sputter coat with carbon and heavy metal for SEM.

Comments

Because CPD is indispensible for the SEM of whole specimens, most researchers routinely pass the deparaffinized sections through this step. If it is determined that CPD is needed, this can be done following staining and prior to permanent mounting. However, our investigations showed that CPD does not contribute any detectable improvement over chemical dehydration for deparaffinized sections of plant specimens (compare Figs. 9A and B).

If LM is conducted prior to SEM, there may still be a need to preserve the cut pieces of the GelBond slide carrying the section(s) for further use at a later time following SEM. This can be done as described in step No. 6. It is, however, not advisable to store deparaffinized sections for a prolonged period without permanent mounting, because fluctuations in ambient temperature and relative humidity (even when the sections are stored under dessication) might lead to structural artifacts.

Isolated Cells

1. Prepare GelBond plastic slides of the same size as standard glass slides as described for thin sections. Mark grids on the GelBond slides with the help of a needle broken at the tip. Affix the GelBond slide to a glass slide with Haupt's adhesive as described for thick sections.

2. Concentrate isolated cells in suspension through centrifugation, and fix the cells overnight in FAA or 2% glutaraldehyde solution in phosphate-buffered saline (PBS; pH 7.0) at 4°C.

3. Spread the fixed cells on the GelBond slide and put a layer of 0.05% poly-L-lysine (mol. wt. 85,000) solution made with PBS; leave the slide at room temperature for 10 min to stabilize cell adhesion. Soak off excess poly-L-lysine with a blotter placed at one corner of the slide. Stain the slide with Giemesa stain or any other suitable stain.

4. Detach the glass slide support and clean the excess Haupt's adhesive

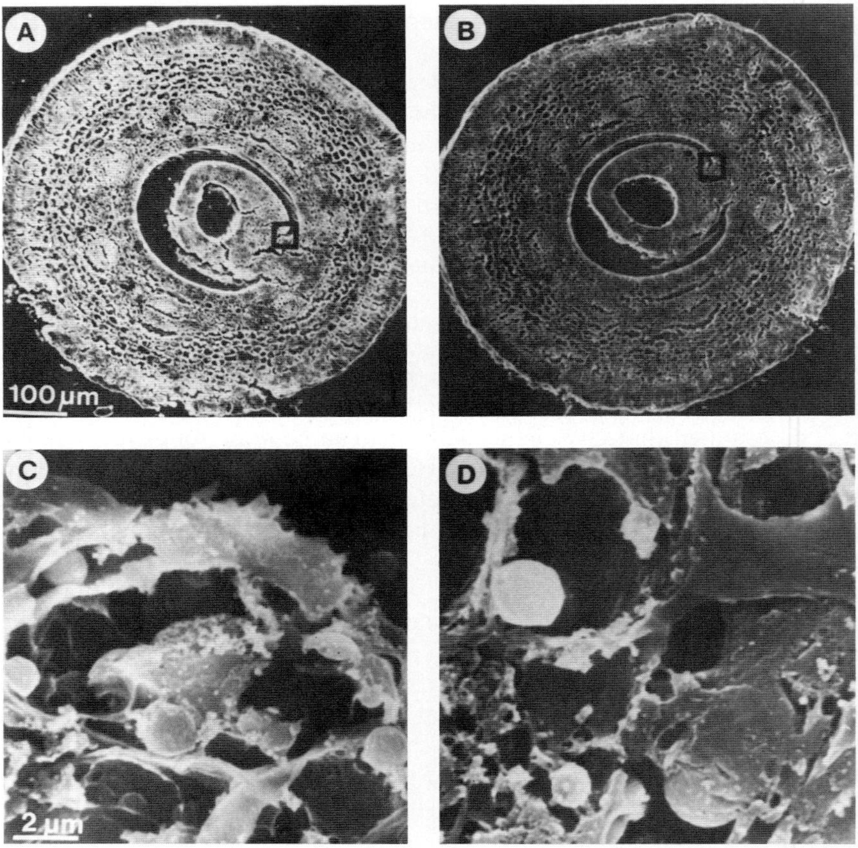

Fig. 9. SEM comparison of critical-point dried and air-dried microtomed sections of peanut peg. A, critical-point dried cross section; ×100. B, air-dried cross section of the thick section adjacent to that shown in A. C, higher magnification of the area marked in A; ×5000. D, higher magnification of the area marked in B; ×5000. Note that little difference can be detected between critical-point dried and air-dried specimens at the magnifications used. (From Pattee *et al.*, 1983.)

from the back of the plastic slide. Examine cells directly using a water immersion lens (Kaneshima *et al.*, 1977) without using a coverslip or, alternatively, following temporary mounting. Prepare temporary mounting of the plastic slide by placing a glass coverslip on PBS layered on the stained cells. Conduct LM and photography. Replenish the PBS to prevent drying of the cells. Maintain a record on the cells of interest on the basis of their location inside the grids.

5. Following LM, dip the mounted plastic slide in a dish containing PBS to separate the glass coverslip.

6. Place the slide in a covered Petri dish and postfix by adding a layer of 2% OsO_4 in PBS for 4 hr at 4°C. Do this only after the need for postfixation is determined through preliminary studies, and perform all steps under a fume hood.

7. Attach the postfixed GelBond slide to a glass slide with paper clips and dehydrate the cells by passing the slide sequentially through graded series of ethanol, isoamyl alcohol, and CPD with dry ice (solid CO_2) or Freon (Hayat, 1978).

8. Cut the gridded plastic slide with attached cells to small pieces, attach the slide pieces to SEM specimen holders, sputter coat for conductorization, and conduct SEM. Cell identification for LM and SEM can be maintained on the basis of cell location within given grids. It should be noted that even when LM and SEM are used correlatively to study isolated cells, both the techniques provide information on two different aspects: LM on internal cell morphology and SEM on external morphology. However, excellent information on the internal cell structure can be obtained through SEM if a fracturing technique (e.g., freeze-fracture) is introduced between LM and SEM.

9. Permanent slides from the cut pieces of plastic slide can be made for later use following SEM according to the method described for thick sections.

CONCLUSION

It must be recognized that no technique is applicable to all situations without appropriate modifications. The GelBond technique should be satisfactory for correlative LM and SEM of both paraffin-embedded specimens and free cells of animal and plant species. It must also be recognized that artifacts of one type or another frequently appear during the SEM of biological specimens (Clark and Glagov, 1976). These artifacts need to be identified carefully and completely in order to take maximum advantage of correlative LM and SEM.

REFERENCES

Arcila, J. (1978). Scanning electron microscope studies on tobacco seedling development. M. S. Thesis, Department of Botany, North Carolina State University, Raleigh.

Arcila, J., and Mohapatra, S. C. (1982). Specimen preparation techniques for scanning electron microscopy of tobacco seeds. *Tob. Sci.* **26,** 138.

Arcila, J., and Mohapatra, S. C. (1983a). Development of tobacco seedling. 2. Morphogenesis during radicle protrusion. *Tob. Sci.* **27,** 35.

Arcila, J., and Mohapatra, S. C. (1983b). Development of tobacco seedling. 3. Morphogenesis during plumule emergence and post-emergence development. *Tob. Sci.* **27,** 112.

Arnott, J. H., and Webb, M. A. (1982). Scanning electron microscopy and histochemistry of oilseeds. *In* "New Frontiers in Food Microstructure" (D. B. Bechtel, ed.), p. 182. Am. Assoc. Cereal Chem., St. Paul, Minnesota.

Ayres, A., Allen, J. M., and William, A. E. (1971). A method for obtaining conventional histological sections from specimens after examination by scanning electron microscopy. *J. Microsc.* (*Oxford*) **93,** 247.

Bahr, G. F., Bibbo, M., Mikel, U., Engler, W., Rao, C., and Weid, G. L. (1976). Correlation of light and scanning electron microscopy. A new method for exfoliative cytology. *Acta Cytol.* **20,** 239.

Berlyn, G. P., and Miksche, J. P. (1976). "Botanical Microtechnique and Cytochemistry." Iowa State Univ. Press, Ames.

Buchanan, R. (1983). SEM examination of nonconducting specimens. *Am. Lab.* **15,** 56.

Clark, J. M., and Glagov, S. (1976). Evaluation and publication of scanning electron micrographs. *Science* **192,** 1360.

Domagala, W., Kahan, A., and Koss, L. G. (1978). A simple method of preparation and identification of cells for scanning electron microscopy. *Acta Cytol.* **23,** 140.

Echlin, P. (1971). The application of scanning electron microscope to biological research. *Philos. Trans. R. Soc. London Ser.* **261,** 51.

Elias, H. (1971). Three dimensional structure identified from single sections. *Science* **174,** 997.

Erlandsen, S. L., Thomas, A., and Wendelschafter, G. (1973). A simple technique for correlating SEM with TEM on biological tissue originally embedded in epoxy resin for TEM. *In* "Proceedings of the Sixth Annual SEM Symposium" (O. Johari, ed.), p. 350. IITRI, Chicago, Illinois.

Ewijk, W. V., and Mulder, M. P. (1976). A new preparation method for scanning electron microscopic studies of single selected cells cultured on a plastic film. *Proc. Workshop Adv. Biomed. Appl. SEM,* Part V, p. 131.

Finch, G., McNeill, K., Democko, C., Lai, C., Bastacky, J., Haynes, T., and Fisher, G. (1985). A technique permitting correlative microscopy (LM, SEM, TEM, and HVEM) of cultured alveolar macrophage cells. *J. Electron Microsc. Tech.* **2,** 69.

Geissinger, H. D. (1971). Correlated light optical and scanning electron microscopy of gram smears of bacteria and paraffin sections of cardiac muscle. *J. Microsc.* (*Oxford*) **93,** 107.

Gogue, C. J., Rasmussen, H. P., and Hooper, G. R. (1976). Structure of a single tissue prepared for analysis by light, scanning and transmission electron microscopy. *J. Am. Soc. Hortic. Sci.* **101,** 224.

Gray, P. (1973). "Encyclopedia of Microscopy and Microtechnique." Van Nostrand-Reinhold, New York.

Hayat, M. A., ed. (1974a). "Principles and Techniques of Scanning Electron Microscopy: Biological Applications," Vol. 1, p. 171. Van Nostrand-Reinhold, New York.

Hayat, M. A., ed. (1974b). "Principles and Techniques of Scanning Electron Microscopy," Vol. 1, p. 273. Van Nostrand-Reinhold, New York.

Hayat, M. A., ed. (1975). "Principles and Techniques of Scanning Electron Microscopy," Vol. 3, Van Nostrand-Reinhold, New York.

Hayat, M. A. (1978). "Introduction to Biological Scanning Electron Microscopy." University Park Press, Baltimore, Maryland.

Hayat, M. A. (1981). "Fixation for Electron Microscopy." Academic Press, New York.

Hearle, W. S., Sparrow, J. T., and Cross, P. M. (1972) "The Use of the Scanning Electron Microscope." Pergamon, Oxford.

Heslop-Harrison, J. (1970). Scanning electron microscope studies of leaves of *Pinguicula*. *Science* **167,** 172.

Judge, F. J., and Stubbs, J. M. (1975). Scanning electron microscopy of tissue preparation after removal of epoxy resin. *Stain Technol.* **50,** 133.

Kaneshima, S., Koga, S., and Tanaka, K. (1977). A simple method for cellular identification of free cells by light microscopy and scanning electron microscopy. *J. Electron Microsc.* **26,** 335.

Kumon, H., Ohmori, H., and Tanaka, T. (1983). Human urinary cytology using a "direct mapping" technique: A combined light and scanning electron microscopic investigation. *Acta Med. Okayama* **37,** 233.

Kumon, H., Ohmori, H., and Tanaka, T. (1984). An improved and simplified method for combined light and scanning electron microscopy in human urinary cytology. *Acta Cytol.* **28,** 503.

McDonald, L. W., Pease, R. F. W., and Hayes, T. L. (1967). Scanning electron microscopy of sectioned tissue. *Lab. Invest.* **16,** 532.

Mayor, H. D., Hampton, J. C., and Rosario, B. (1961). A simple method for removing the resin from epoxy-embedded tissue. *J. Biophys. Biochem. Cytol.* **9,** 909.

Mikel, U. V., and Johnson, F. B. (1980). A simple method for study of the same cells by light and scanning electron microscopy. *Acta Cytol.* **24,** 252.

Mohapatra, S. C., and Johnson, W. H. (1974). Microscopic studies of bright leaf tobacco. III. Scanning electron microscopy of deparaffinized sections of tobacco leaf. *Tob Sci.* **18,** 80.

Mohapatra, S. C., and Johnson, W. H. (1975). Microscopic studies of bright leaf tobacco. IV. Scanning electron microscopy of crystals in leaf tissue. *Trans. Am. Microsc. Soc.* **93,** 413.

Mohapatra, S. C., and Johnson, W. H. (1976). Microscopic studies on bright leaf tobacco. V. Scanning electron microscopic observations on chloroplast structure and function. *Trans. Am. Microsc. Soc.* **95,** 228.

Moshkovitz, S. (1978). New types of cover-slip and mounting-slide with a graticule for examination of the same small object both by the light microscope and the scanning electron microscope. *Microsc. Acta* **80,** 161.

Nolte, I., and Rudolph, R. (1980). A method for morphological characterization of the same cell in light and scanning electron microscopy exemplified by blood leukocytes of sheep. *Blut* **40,** 407.

Pachter, B. R., Davidowitz, J., Zimmer, B., and Breinin, G. M. (1973). Scanning electron microscopy of etched epon extraocular muscle sections. *J. Microsc.* (*Oxford*) **99,** 85.

Pattee, H. E., Mohapatra, S. C., and Agnello, E. K. (1983). Specimen preparation techniques for scanning electron microscopy of developing peanut pegs. *Peanut Sci.* **10,** 93.

Pattee, H. E., Mohapatra, S. C., and Agnello, E. K. (1985). A technique for comparative light and scanning electron microscopy of the same deparaffinized microtomed section. *J. Microsc.* (*Oxford*) **137,** 111.

Pattee, H. E., and Mohapatra, S. C. (1985). Hypanthium and style senescence in relation to ovary development in *Arachis hypogaèa* L. *Bot. Gaz.* **147,** 302.

Pattee, H. E., and Mohapatra, S. C. (1987). Anatomical changes during ontogeny of the *Arachis hypogaea* L. fruit: mature gametophyte through heart-shaped embryo. *Bot. Gaz.* **148**(2), (in press).

Pickett, J. T., Ingram, P., and Shelburne, J. D. (1980). Identification of inorganic particulates in a single histologic section using both light microscopy and x-ray microprobe analysis. *J. Histotechnol.* **3,** 155.

Poh, T., Altenhoff, R. L. J., Abrama, S., and Hayes, T. (1971). Scanning electron microscopy of myocardial sections for the light microscopy. *Exp. Mol. Pathol.* **14,** 404.

Punja, Z. K., and Jenkins, S. F. (1984). Light and scanning electron microscopic observations of calcium oxalate crystals produced during growth of *Sclerotium rolfsii* in culture and in infected tissue. *Can. J. Bot.* **62,** 2028.

Saleh, K. M., Toner, P. G., and Carr, K. E. (1981). An improved method for sequential light and scanning electron microscopy of the same cell using localising microcoverslips. *J. Clin. Pathol.* **35,** 576.

Spurr, A. R. (1969). A low-viscosity epoxy resin embedding medium for electron microscopy. *J. Ultrastruct. Res.* **26,** 31.

Takenaga, A., Matsuda, M., Horai, T., Ikegami, H., and Hattori, S. (1977). Scanning electron microscopy in the study of lung cancer. New techniques of comparative studies on the same lung cancer cells by light microscopy and scanning electron microscopy. *Acta Cytol.* **21,** 90.

Thierstein, H. R., Franz, H. E., and Roth, P. H. (1971). Scanning electron and light microscopy of the same small object. *Micropaleontology* **17,** 501.

Thornthwaite, J. T., Cayer, M. L., Cameron, B. F., Leif, S. B., and Leif, R. C. (1976). A technique for combined light and scanning electron microscopy of cells. *Proc. Workshop Adv. Biomed. Appli. SEM,* Part V, p. 127.

Watari, N., Hotta, Y., and Mabuchi, Y. (1978). Two- or three-way observations of the identical cell elements within the same sections by LM, TEM and/or SEM. *Proc. Int. Congr. Electron Microsc. 9th, 1978,* **2,** 82.

Index